International Petroleum Encyclopedia 2002

by Pennwell Corporation
Box 1260
Tulsa, OK 74101
800-752-9764
sales@pennwell.com
www.pennwell-store.com
www.pennwell.com

Library of Congress CIP Pending
ISBN: 0878148388
Printed in the U.S.A.

INTERNATIONAL PETROLEUM

Table of Contents

ENCYCLOPEDIA 2002

Africa

Former Soviet Union

China

Asia-Pacific

IPE Key Stats & Tables

Key Stats

Statistical Tables

IPE Information Sources

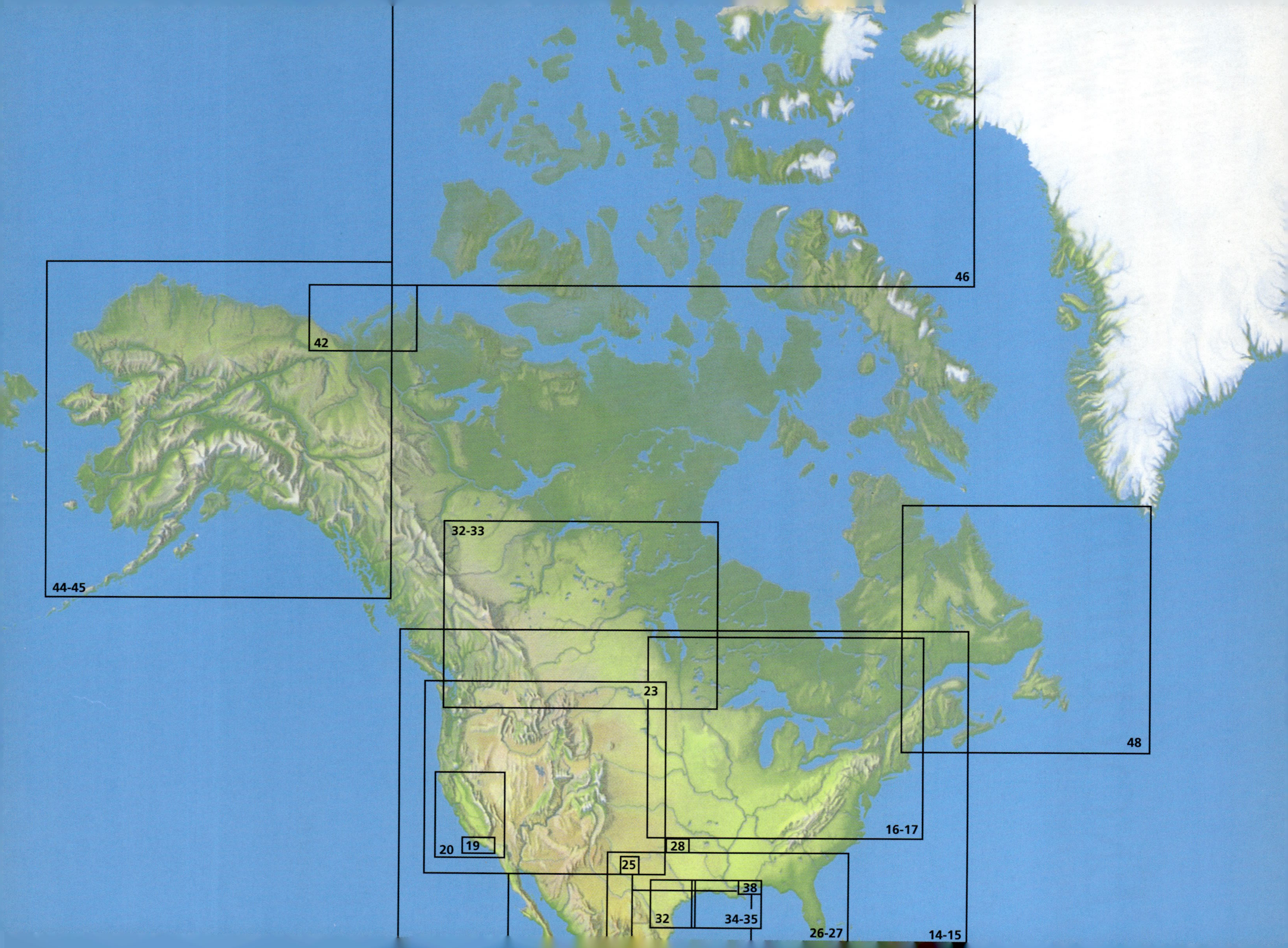
46
42
44-45
32-33
23
48
16-17
20
19
28
25
38
32
34-35
26-27
14-15

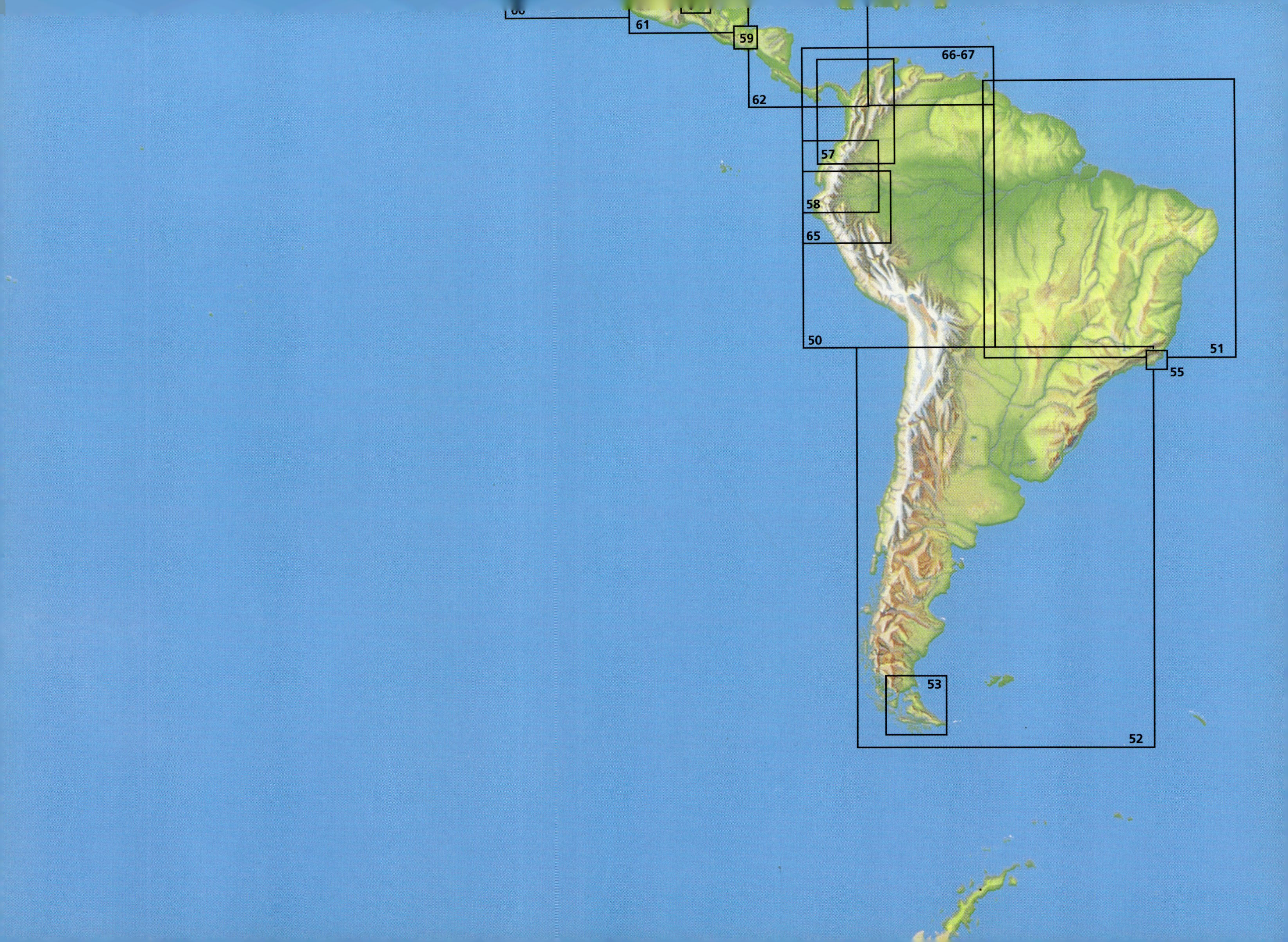
61
59
66-67
62
57
58
65
50
51
55
53
52

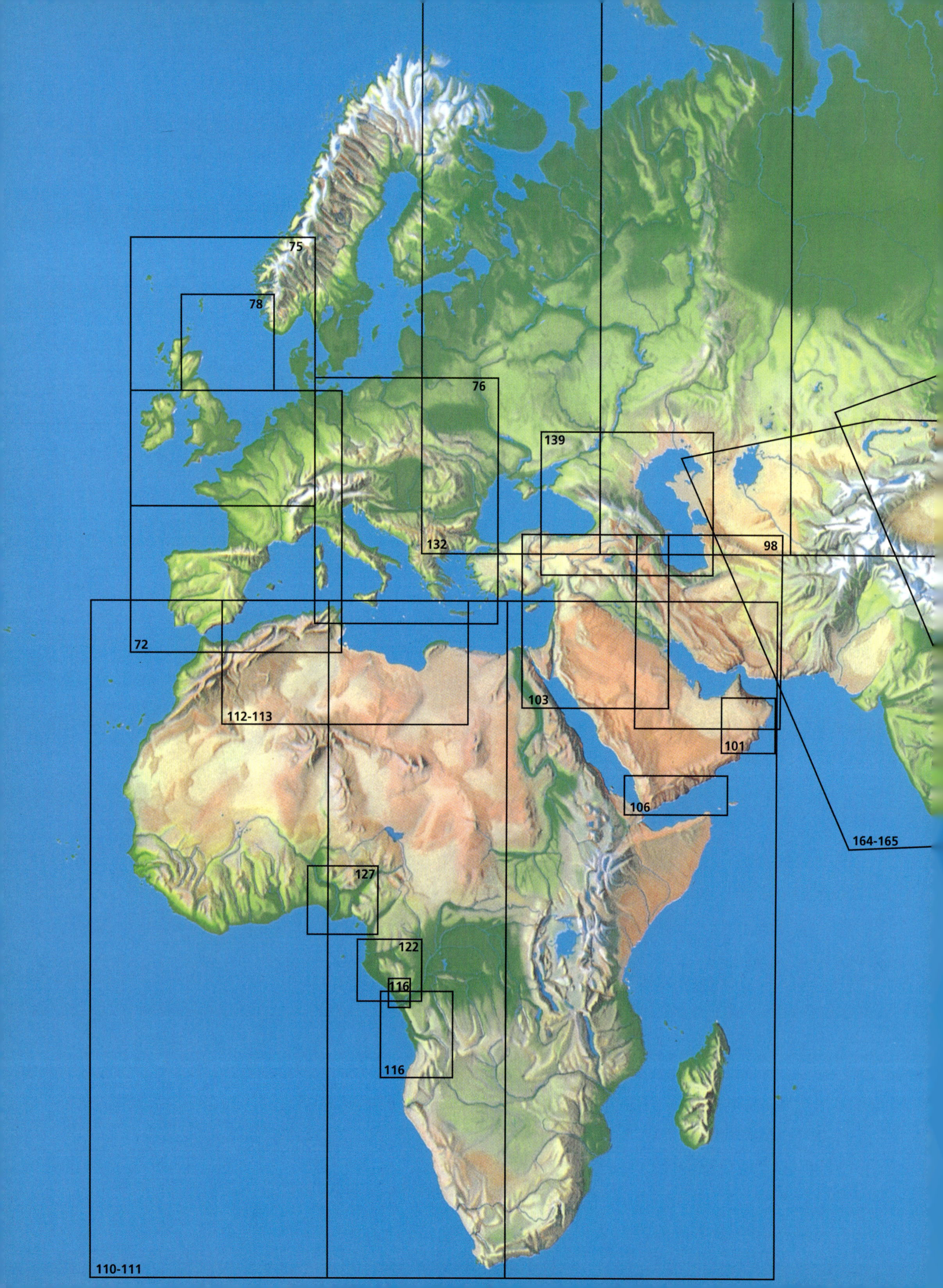
75
78
76
139
132
98
72
112-113
103
101
106
164-165
127
122
116
116
110-111

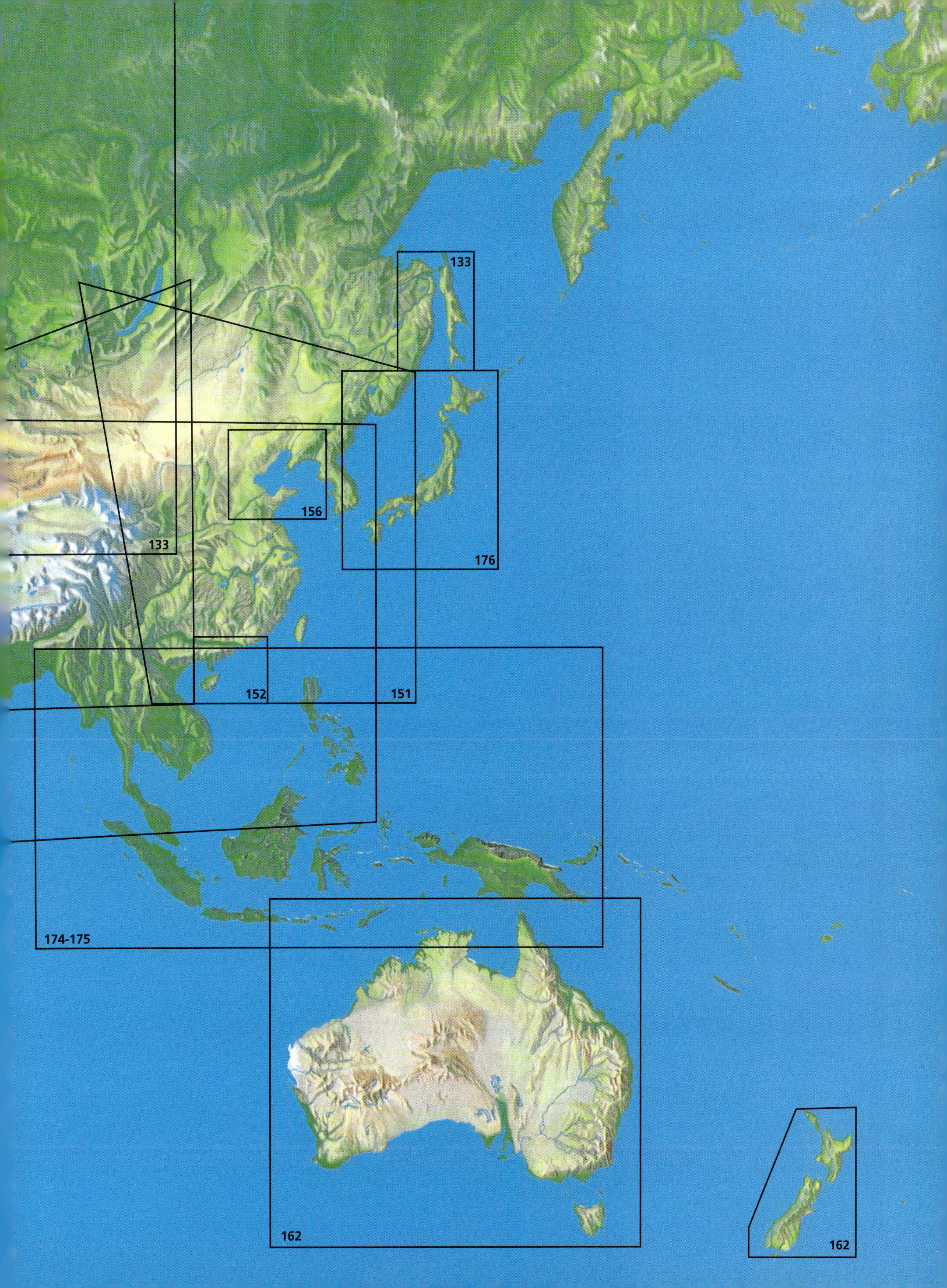
133
133
156
176
152
151
174-175
162
162

Editor
Rebecca L. Busby

Art Director
Clark Bell

Staff Editors
Kay Carstens
Mary Farrell
Patt Geith

Lead Illustrator
Kay Wayne

Staff Illustrators
Sandy Berenson
Mike Reeder
Wes Rowell
Andrea St. Clair

International Petroleum Encyclopedia
is published annually by
PennWell Corp., 1421 S. Sheridan Rd., Box 1260, Tulsa, Okla. 74101.
Phone: (918) 835-3161, Fax: (918) 831-9555.
Editorial offices: 1700 West Loop South, Suite 1000, Houston, Texas 77027.
Phone: (713) 621-9720, Fax: (713) 963-6285.

Chairman
Frank T. Lauinger

President and Chief Executive Officer
Robert F. Biolchini

President and Chief Operating Officer, Business & Industrial Division
James Pfister

Senior Vice-President, Finance and Administration, Chief Financial Officer
Thomas L. Stone

Publisher
Margaret Shake

Managing Editor
Marla Patterson

Advertising Production Coordinator
Mary Rhyne

PennWell®

IPE Essay
Coping With the Challenges of 2001 and Beyond – US Independent Producers' Perspective

Although the United States is the world's third-largest oil producer, its mature resource base makes production costs higher than in competing nations, which rely on petroleum sales for their national incomes and where production is controlled by government decisions and internal politics. Thus world crude oil prices are determined less by market factors than by global political decisions, resulting in significant risk to price-sensitive US producers and a shift in domestic production toward natural gas. US policy should recognize the need to increase both oil and gas production, especially from the perspective of national security since September 11th, and that a healthy oil production industry is essential to a healthy gas industry because the two are inherently intertwined.

by Lee O. Fuller
Vice President of Government Relations
Independent Petroleum Association of America

United States energy actions in 2001 built upon the supply and price issues of the winter of 2000, reacted to the global economic downturn, took on new meaning after the attacks of September 11th, and were ultimately confused by the Enron bankruptcy. Beneath these events, the US grappled with policy options that could change its use, supply, and mix of petroleum- and natural gas-derived energy. Many of these actions resulted in significant implications for the US independent oil and natural gas explorer and producer.

Natural Gas and the California Energy Crisis

Responding to the increased prices in 2000, domestic drilling activity had increased dramatically. By year-end 2000, domestic rig counts exceeded 1,100, but these were overwhelmingly directed toward natural gas development. Oil-directed drilling, which had been near 400 rigs in late 1997, never exceeded a monthly average of 250 after the oil price crisis of 1998-99. Yet despite this natural gas-directed drilling effort, the toll of low prices and reduced capital budgets was affecting natural gas supply. Natural gas prices – following cold November and December months – were exceeding $8.00 per thousand cubic feet (Mcf). These prices reflected supply constraints. Nowhere was the problem greater than in California.

The electricity crisis in California was a reflection of far broader problems. Characterized as a failure of electricity deregulation policies, it was more accurately the culmination of a series of energy-related decisions that failed to provide adequate resources to the state. In part, it reflected a policy structure that capped consumer prices of electricity while relying on the energy market place to keep supply prices in reasonable ranges. Consequently, combined with the regulatory permitting framework in California, there was little incentive for new generation capacity to be built over many years. What new construction occurred was largely natural gas-powered because of the positive environmental aspects of the fuel. However, within the state the natural gas pipeline network contained serious limitations on its ability to deliver natural gas volumes to certain key generation facilities. Thus when weather-related shortages of hydropower in the western US limited electricity supply, California was confronted with unusual constraints.

Taken together with the larger national supply issues, California electricity suppliers found themselves in a bidding war to acquire natural gas supplies, but unable to move gas efficiently within the state. This resulted in prohibitively expensive electricity. And, it raised questions regarding the use of natural gas as the principal source of new electric power generation. In 2001, natural gas provided about 16% of US electricity generation. Seven states obtain more than one-third of their generation from natural gas (Rhode Island, New York, Delaware, Louisiana, Texas, California, and Alaska).

More importantly, natural gas-fired electricity is projected to constitute about 90% of capacity additions between 1999 and 2020. The amount of natural gas used in electricity generation is projected

IPE Essay Guest Author

Lee Fuller,
Vice President of Government Relations

Independent Petroleum Association of America

Lee is responsible for coordinating the IPAA's legislative and regulatory activities, specifically those related to natural gas and environmental issues as well as federal tax and budget matters. He has 24 years of experience in US policy issues, including service with the Senate Committee on Environment and Public Works.

Fuller has also worked for several consulting firms on developing, implementing, and advocating policy strategies for clients involved with environmental legislation, including regulation of oil and gas production facilities through Superfund laws and the Clean Air Act.

Before coming to the US Capitol, Fuller worked for Exxon Co. USA at its Baytown, TX, refinery managing environmental compliance issues and providing technical support for process operations. He graduated from Lehigh University as a Chemical Engineer.

to triple by 2020. But, these assessments are subject to new scrutiny because of the California experience (and the subsequent implications of the economic downturn and the Enron collapse).

Equally significant, process uses of natural gas had to react to the supply and price issues. Probably the most significant impact fell upon the ammonia/fertilizer industry. Because natural gas is an essential feedstock to this international industry, the domestic price increase created significant economic problems. Much domestic production was shut down, creating a shift to overseas production. In general, this production remains curtailed.

By mid-year 2001, natural gas prices had largely returned to levels similar to the beginning of 2000. Collectively, changes to the market had brought it into balance. But, the events that took place have significant future implications. Reductions in natural gas demand had a greater influence than new supply. The US Energy Information Administration (EIA) estimated in early 2002 that consumption of natural gas declined in 2001, though this estimate is subject to revision. Some of the decline resulted from energy efficiency; some was from shifts in domestic production such as the ammonia/fertilizer effect; some was from the economic downturn. Significantly, despite record natural gas drilling rates, annual natural gas production increased a modest 400 billion cubic feet on a 19 trillion cubic feet (tcf) base. Much of the drilling was development to respond to the clear supply shortfall, but it also demonstrated that depletion rates continued to increase as more effective extraction took place.

National Energy Policy

The California electricity crisis following the gasoline supply problems the previous year, as well as high crude oil and natural gas prices, drew attention to national energy policy as the new Bush Administration took office. The administration convened a task force to review and propose a national energy policy. In May 2001, the policy was released addressing in part the need for more domestic energy supply. While it strongly emphasized allowing development of the Arctic National Wildlife Refuge, the policy also emphasized the need to improve access to the natural resource base of oil and natural gas, much of which underlies government- owned lands. The US is currently the third-largest oil-producing nation in the world, despite a 30-year decline in domestic production. Oil and natural gas are the dominant fuels in the US economy, providing 62% of the nation's energy and almost 100% of its transportation fuels. By 2020, EIA expects that the US will need about 50% more natural gas and one-third more oil to meet demand.

The federal government owns about 31% of the nation's land — it has a major role in increasing energy production in appropriate places. Government-controlled lands provide nearly 30% of annual national energy production and are estimated to contain a substantial majority of the nation's undiscovered domestic energy resources.

According to estimates from the US Geological Survey and the Minerals Management Service, oil resources underlying federal lands in the lower-48 states are estimated to be 4.1 billion barrels (bbl), and natural gas reserves are estimated to be 167 tcf.

Much of these resources are off-limits or subject to significant restrictions. For example, about 40% of the natural gas resource on federal land in the Rocky Mountain region has been placed off-limits. The US Department of the Interior has initiated a study to examine the energy potential and restrictions on development on federal lands in the lower-48 states.

The Bush Administration has every incentive to halt or reverse this lock-up of public lands. As Interior Secretary Gale Norton has pointed out, Interior takes in more money than any other government department bar Treasury. Energy development on federally managed lands, she pointed out, collected $11.3 billion in fiscal 2001 (almost $1 billion more than Interior's budget). About 64% of the cash came from offshore oil and natural gas leases, but there is no reason why onshore leases could not boost that figure and, with modern drilling techniques,

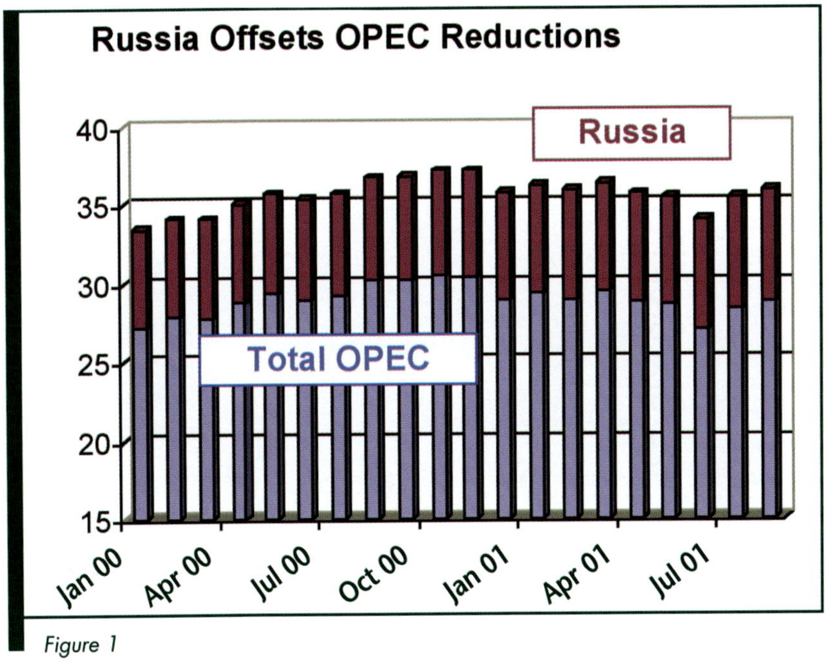

Figure 1

this would have little or no adverse effect on the environment.

Crude Oil, Economic Problems, and September 11th

When the Administration initiated its national energy policy effort, energy issues were clear and pressing, and the economy was strong. By May 2001, the winter energy problems had subsided, and economic conditions were changing. The world economy was sliding into a downturn. Among the effects were lower energy consumption and therefore weakening energy prices. But these effects were modest compared to the consequences of terrorist attacks on September 11th, attacks that led to worldwide military action against terrorism. The first effects involved dramatic limitations on domestic air travel, resulting in significant reductions in fuel demand. Effects began developing worldwide as the state of war exacerbated the economic downturn.

Crude oil supply, prices, and politics moved to the front of energy issues for domestic producers. Putting this in perspective, as a reaction to the low price crisis of 1998-99, the Organization of Petroleum Exporting Countries (OPEC) began to try to control production to keep its crude oil within a price band ranging from $22 per barrel to $28 per barrel. This range reflects a balance that OPEC believes provides their nations with the income necessary to meet their citizens' needs at the low end but maintains a price that does not hinder the world's consuming nations on the high end.

However, as 2001 began, the strain on worldwide production capacity limited OPEC's plan; prices exceeded $30 per barrel. Prices drove conservation, and the developing recession in the United States and elsewhere began to drive demand – and prices – down. OPEC initiated production cuts within its membership, but the commitment to meet quotas had waned, and targets were not met. Following September 11th, demand and prices began falling to levels not seen since 1999. OPEC sought to further reduce production to bring supply into balance with demand. However, it needed commitments from other key exporting nations – Mexico, Norway, and Russia – to reduce production as well.

Russia's role has become a linchpin to the course of future crude oil supply politics. It is now the second-largest oil producer in the world – trailing only Saudi Arabia. Its crude oil production had increased by year-end 2001 by more than 900,000 barrels per day since January 2000 (Figure 1). This increase essentially offset the reductions OPEC had made to stabilize crude oil supply. Without significant action by Russia to reduce its production, the world risked an oil price war that could result in adverse consequences to both global and national security.

In "Cheap Oil Comes at a Price," *Washington Post* reporter William Drozdiak observed that low oil prices can help bring nations out of a slump. "But lower oil prices also raise the danger of economic turmoil in parts of the world where the United States is seeking help in its war on terrorism. If the Organization of Petroleum Exporting Countries fails to persuade key non-OPEC oil producers to cut production, a further plunge in oil prices could destabilize much of the Persian Gulf region." It is a consequence that could impair future international efforts.

As 2001 ended, OPEC had gained the support from the non-OPEC countries that it needed to undertake new production limitations aimed at market stability.

The Role of US Independent Producers

The domestic upstream part of the oil and natural gas industry – exploration and production, or E&P – is characterized by thousands of companies operating in over 30 states and offshore. Overwhelmingly, these "independent" explorationists and producers receive revenues

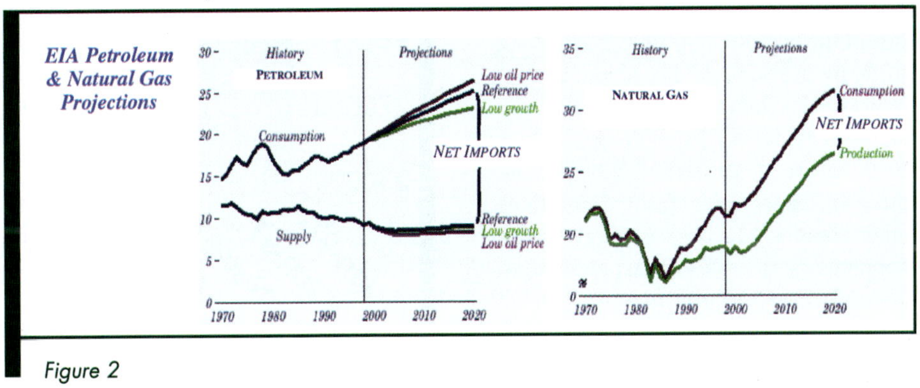

Figure 2

only from upstream activities. Most have fewer than 20 employees, but collectively, they are the link to future domestic supply.

US oil and gas production has changed over the years, particularly since the mid-1980s. Maturing production areas in the lower-48 states, along with the need to respond to shareholder expectations, have resulted in major integrated petroleum companies shifting their E&P focus toward the offshore in the US and other countries. More and more, these large companies must rely on large producing fields that are found only in frontier areas.

Consequently, domestic production in both the lower-48 states and in the near offshore are areas where the role of independents is increasing. For example, the independent share of lower-48 states oil production has increased from 45% in the mid-1980s to over 60% by 1995 – and these states, despite their mature fields, still accounted for 60% of domestic oil production as of 2001. Similarly, independent producers account for 65% of domestic natural gas production. These trends will continue. The US will need a strong independent exploration and production industry to meet it future production requirements.

Role of Independent Producers

Share of US Domestic Activities
Wells drilled: 85%
Natural gas produced: 65%
Oil produced: 40%
Lower-48 oil produced: 60%

In the larger energy picture, the US will remain dependent on oil and natural gas well into this century. EIA estimates that petroleum and natural gas will account for 66% of domestic energy consumption in 2015. A 1999 *Natural Gas* study by the National Petroleum Council projects that domestic demand for natural gas will increase by more than 30% by 2010 (Figure 2). At issue will be whether these increases come from domestic or other sources. To prevent more foreign dependency, the role of independent producers becomes more critical.

However, because they operate in the upstream, these independent producers are more susceptible to shifts in the commodity market prices of crude oil and natural gas. So when petroleum prices fell to historically low levels in 1998 and early 1999, the effect on these producers was markedly worse than the effect on large integrated petroleum companies. Capital expenditures fell by 30% or more, drilling rig counts dropped by 50%, and E&P employment fell by 65,000. As prices recovered, capital was reinvested in new exploration and production, but it was constrained by the recovery of skilled employees. And once again in 2001, the cycle of crude oil and natural gas prices has shifted to prices that threaten not just the development of new resources but the continuing production of existing ones.

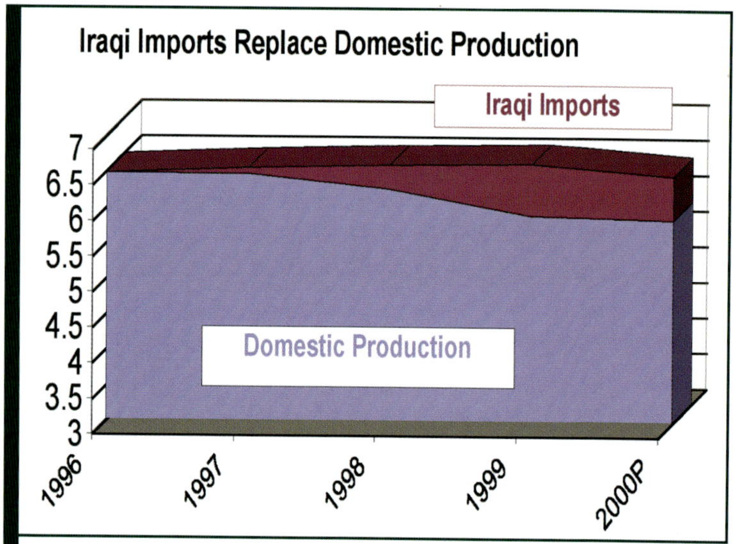

Figure 3

Thus as 2001 ended, the US needed to address key policy issues that will affect its domestic producer industry and, implicitly, the makeup of its future oil and natural gas supply.

Policy Questions

First, in making decisions regarding developing domestic petroleum resources, the nature of the world petroleum market must be recognized. Although the US remains the third-largest oil producer, it is operating from a mature resource base that makes the cost of production higher than in competitor nations. More importantly, most other significant petroleum-producing countries rely on their petroleum sales for their national incomes. For them, petroleum production is not driven by market

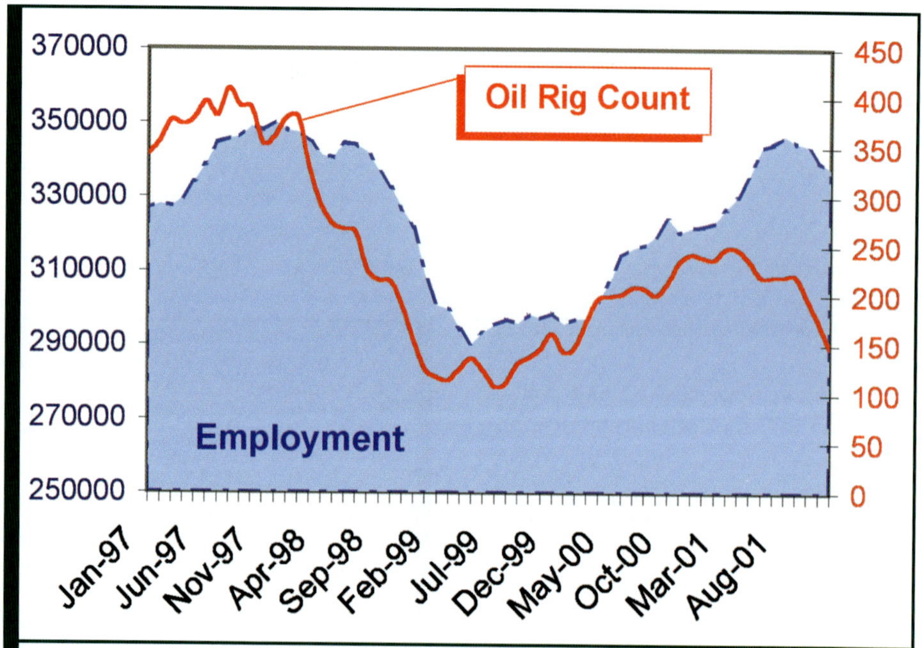

Figure 4

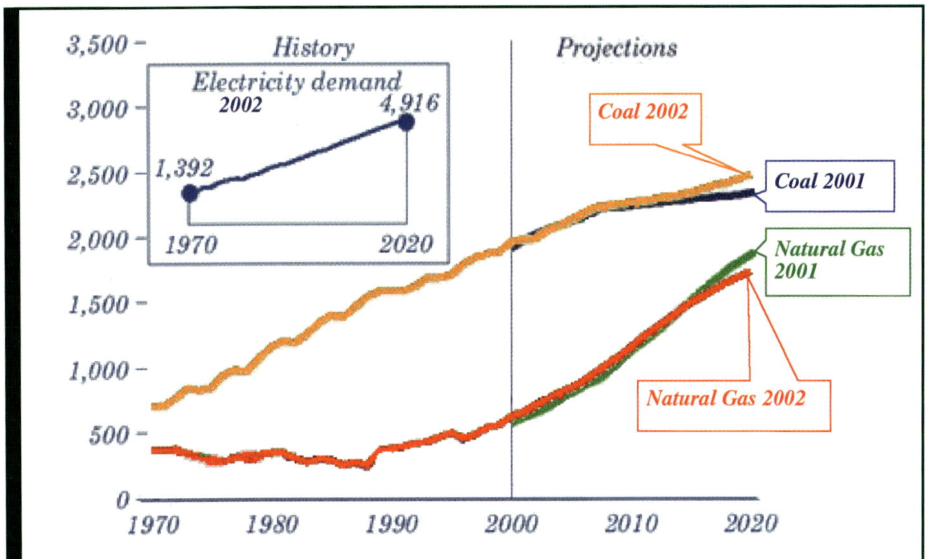

EIA Annual Energy Outlook Projections of Electricity Generation by Source (Billion Kilowatthours) -- Coal versus Natural Gas Projections ---- 2002 versus 2001 Projections

Figure 5

decisions; production is determined by government decisions.

Thus, the market price for petroleum will be largely framed by production decisions driven not by the market, but by the politics of these countries – both by internal issues and global objectives. US domestic policy decisions must reflect this reality – looking to this factor in taking actions that can affect domestic production and producers. And, it must recognize that a healthy domestic oil production industry is essential for a healthy domestic natural gas industry, because they are inherently intertwined.

One outcome of the 1998-99 oil price crisis appears to be a much lower interest in developing domestic crude oil – particularly onshore. Domestic oil has declined since 1970, but the most precipitous declines have come after the price crises of the mid-1980s. Prior to the oil price crisis in 1986, domestic oil production was about 8.5 million bbl/day (b/d). By 1997, domestic production had dropped to about 6.5 million b/d – a 2 million b/d loss. In 1998, the Clinton Administration's energy strategy called for a 500,000-b/d increase in domestic oil production by 2005 – moving to a 7 million b/d target. Now, as a result of the 1998-99 price crisis, domestic production has dropped below 6 million b/d.

The 1986 oil price crisis demonstrated that the US was the world's highest cost production area. In particular, the lower-48 states onshore is the highest cost area because it is such a mature area compared to the rest of the world. Combined with domestic policy changes, like the 1986 tax reform law that created the Alternative Minimum Tax, the desirability of domestic oil development in the lower-48 states onshore dropped dramatically. As a result, the major integrated oil companies revised their investment strategies. They shifted their investment plans to develop large "elephant" prospects. In the US, these are located offshore or in Alaska – frequently in areas where development has been prohibited. Thus, US policies led to a shift in capital deployment that encouraged foreign oil development over domestic.

About 20% of domestic production comes from marginal wells – wells that are the most vulnerable to low prices. Since the 1998-99 low price crisis, most of the domestic loss has come from closing marginal wells. It has largely been replaced by imports from Iraq (Figure 3). These implications on for the need to increase domestic production are significant at a minimum from the perspective of national security since September 11th.

Even as prices rebounded since 1999, the rig count never exceeded 250 for oil, from its high of 400 in 1997 (Figure 4). In part this is a reflection that certain oil-rich areas are off-limits, like the Arctic National Wildlife Reserve. In part, it reflects the nature of offshore development, where fewer rigs are used to develop production. However, it also reflects the sense of risk in the development of crude oil that its price will be determined by global political decisions and that domestic production will be the most sensitive to price swings. Compared to natural gas, which is essentially a North American resource, crude oil development is a far less secure option.

Natural Gas Prospects

Natural gas then has become the focal point of domestic producers. But here too, significant policy choices must be addressed – including the consequences of the California energy problems in early 2001. In December 2001, EIA released its *Annual Energy Outlook* 2002, which presents EIA's latest assessment of energy issues including projections through 2020. It includes estimates of electricity demand and projections of the fuels used to generate the power. Figure 5 compares the projections for electricity generation by coal and natural gas between the 2001 and the 2002 *Annual Energy Outlook*.

EIA projects that electricity demand will increase from its current level to 4,916 billion kilowatt-hours (kWh), representing an increase from the 2001 projection of 4,804 billion kWh. Natural gas continues to show a significant increase during the projection period, falling off somewhat in the 2020 time period compared to the 2001 projection. Total electricity demand is the large driver in determining natural gas demand growth. Based on EIA's 2002 *Annual Energy Outlook*, electricity demand and therefore natural gas demand is not predicted to drop in the near term, but it will not increase as quickly based on the

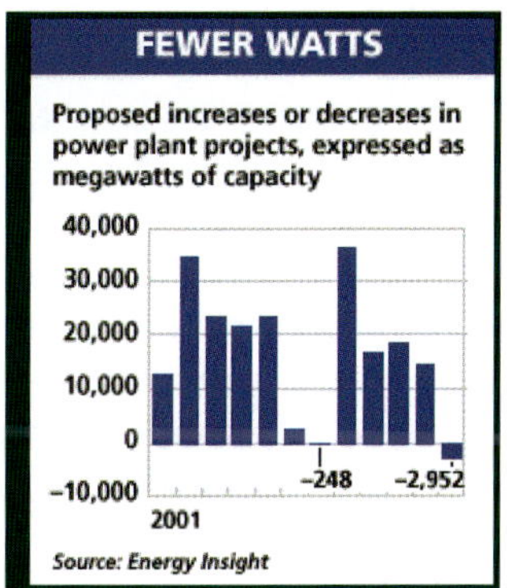

Figure 6

2002 *Annual Energy Outlook* compared to the 2001 projection.

However, the recent collapse of Enron has been creating a ripple of changes through the electric industry. Several companies have indicated that they intend to reduce capital spending on new generation capacity. A *Wall Street Journal* article, "Power Industry Cuts Plans for New Plants, Posing Risks for Post-Recessionary Period," January 4, 2002, makes the following observation:

> *New data show that some 18% of all announced projects already are effectively dead, nearly double the attrition rate a year ago. That could spell trouble when the U.S. emerges from recession and electricity demand again picks up. New plants not only support economic growth, but they replace older, dirtier plants — as required to satisfy increasingly stringent air-quality standards. . . . Data compiled by Energy Insight [Figure 6], a Colorado-based research unit of McGraw-Hill Cos., show that 91,139 megawatts of generating plants, out of a total announced portfolio of 503,780 megawatts, had been canceled or tabled by the end of 2001, amounting to 18% of proposed new additions. A year earlier, 30,909 megawatts of capacity had been jettisoned, amounting to about 10% of the announced total. Last month [December 2001] was the first month in recent memory in which plant cancellations significantly exceeded newly announced projects. A net deficit of 2,952 megawatts of capacity was equal to the loss of about six major power plants.*

In general, the implications of these capital spending cuts will not be understood for several years. Most are related to projects that do not come online until 2004 or later. However, these decisions could result in electricity generation shortfalls at least in certain areas in the future – bottlenecks similar to the problems seen in early 2001.

Over the longer term, natural gas remains a logical fuel choice for the United States. It is largely a North American resource, and there is ample reason to believe that the domestic elements of that resource base can be significant. But there are significant decisions that must be made. Unlike oil supply, natural gas supply is dependent on North American resources, with 80-85% coming from the US. However, much of this domestic supply is accessible only from government-controlled lands. The current restrictions affecting access to these lands differ depending on the area, but all must be altered to meet future demand.

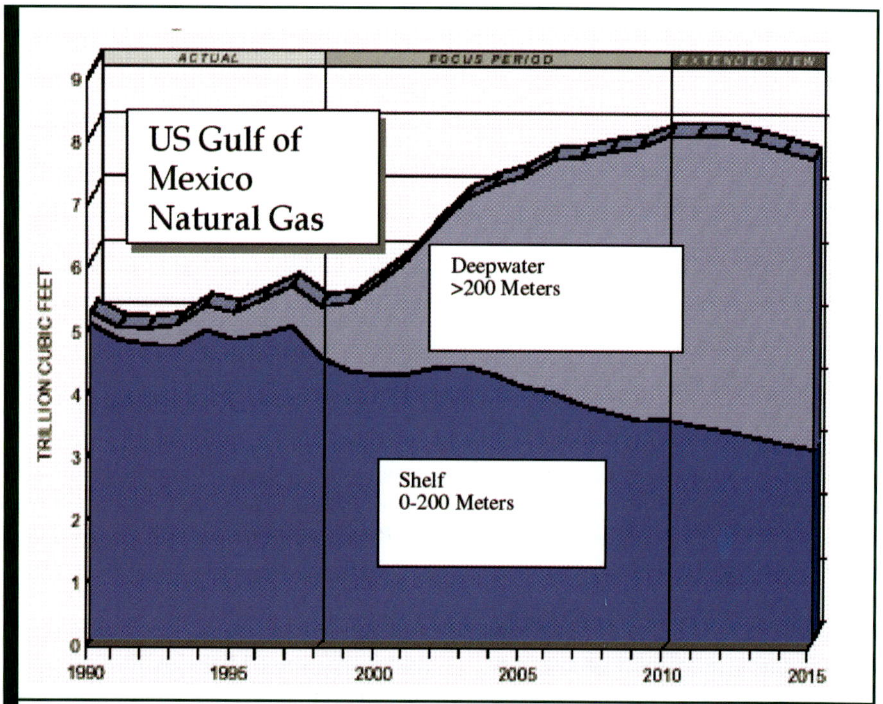

Figure 7

Offshore – Western and Central Gulf of Mexico

These portions of the Gulf of Mexico have proven to be a world-class area for natural gas as well as oil production, accounting for over 25% of domestic gas production. Production comes from the continental shelf, the deepwater, and the emerging ultra-deepwater (Figure 7). The National Petroleum Council's *Natural Gas* study projects that future production increases in these areas are essential to meet projected demand. However, these increases will hinge on federal offshore policies. The most significant of these relate to royalty policies.

First, federal royalty policies need to encourage the development of the resource. For example, the 1995 Deepwater Royalty Relief Act was extremely successful promoting activity in the deepwater Gulf. However, the 1995 program expired. The National Energy Policy recognized that offshore regulatory policies could inhibit the sound development of these resources. Its recommendations should be implemented, and incentives should be considered for deep drilling in all depths in the deepwater, deep drilling for natural gas on the shelf, and subsalt and highly deviated drilling. Second, offshore production is particularly suited for royalty-in-kind – paying the royalty with production instead of dollars. It is a more economical and fairer approach. Recent actions to fill the Strategic Petroleum Reserve could utilize 80% of this offshore royalty oil and should be encouraged.

Offshore – Eastern Gulf of Mexico, Atlantic Ocean, and California

Developing the substantial domestic natural gas reserves in these three areas is prohibited by moratoria. President Clinton extended these moratoria for another ten years in 1998, saying, "First, it is clear we must save these shores from oil drilling." This is a flawed argument ignoring the state of current technology; it results in these moratoria preventing natural gas development as well as oil. In fact, both the Eastern Gulf and the

Atlantic reserves are viewed as gas reserve areas, not oil – those coasts are not at risk. Too often, these policies are predicated on the events that occurred 30 years ago. For example, no Eastern Gulf of Mexico sale occurred from 1988 to 2001. The recent sale took place only under greatly reduced conditions, but it was an important first step. Federal moratoria policy needs to be reviewed. It needs to be based on a sound understanding of today's technology.

Onshore Restrictions – A Mosaic of Regulations and Prohibitions

Much of the onshore natural gas resource base is located in the Rocky Mountains where federal policy limits access to an estimated 138 tcf of natural gas (Figure 8). The constraints differ. Monument and wilderness designations prohibit access to some areas. Regulations like the Forest Service roadless policy and prohibitions in the Lewis and Clark National Forest are equally absolute. At the same time the permitting process to explore and develop resources can work to effectively prohibit access. These constraints include federal agencies delaying permits to revise environmental impact statements, habitat management plans overlaying one another to prohibit activity, and unreasonable permit requirements that prevent production. There is no single solution to these constraints.

What is required is a commitment to develop these access policies with a full recognition of the importance of developing the natural gas resource. The National Energy Policy recognized the magnitude of these limitations. Executive Orders to consider energy implications in federal decision making and to convene a task force to improve permitting are important first steps in developing a response. Similarly, the recommendation to expedite an inventory of resources under federal lands to determine where potential conflicts might exist will improve the debate over access.

Environmental Risk Is Not the Issue

Unfortunately, much of the natural gas production debate will rage over allegations of environmental risk. It should

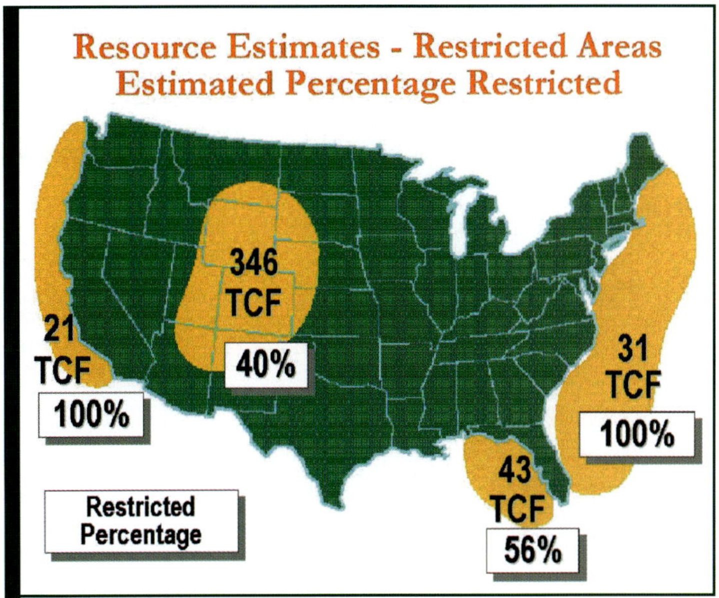

Figure 8

not be an issue. The Department of Energy's *Environmental Benefits of Advanced Oil and Gas Exploration and Production Technology* states the situation well:

> *Resources underlying arctic regions, coastal and deep offshore waters, sensitive wetlands and wildlife habitats, public lands, and even cities and airports can now be contacted and produced without disrupting surface features above them. Wildlife preserves and conservation easements are created and managed jointly by industry, environmental, and government stakeholders. In Alaska, such new approaches as ice pads and roads, multilateral completions, and annular injection of drilling wastes minimize environmental impacts while also reducing costs.*

During 2001, the US House of Representatives passed comprehensive energy legislation responding to the challenge of the national energy policy. As 2002 began, the US Senate was preparing for its consideration of energy policy. The actions taken in these efforts to address critical needs for access to the resource base and the capital needed to develop it could, in many respects, define the course available to US independent producers to meet domestic energy challenges. IPE

IPE Chronology
2001 In Brief

JANUARY

Markets
OPEC reduces member quotas by 1.5 million b/d
Natural gas prices fall 13% in 1 day ($10.10 to $6.293/MMBtu)
Discoveries
Petrobras claims Campos basin oil discovery in record water depth
Production
Sincor announces heavy syncrude plant startup in Venezuela
US deepwater Gulf rigs hit a record high of 40, MMS says
Burlington well sets Rockies depth record at +25,800 ft
Natural Gas
Pemex targets huge production increase through 2008
Pipelines
Startup of Thai line removes Myanmar gas bottleneck

Quote of the Month

"I would be pleased if tomorrow, industrialists from Monterrey could invest and extract from the abundance of gas in the Burgos basin, but the constitution, for the moment, does not let us." – Mexico Pres. Vicente Fox

Refining
Koch settles on emissions control for three US refineries
Premcor announces permanent closure of Blue Island plant
Government
George W. Bush becomes US President
Venezuela's new oil minister to continue defending OPEC price
France's ecotax on fuels is killed as unconstitutional
Companies
Halliburton to sell Dresser Equipment Group
AEP buys Enron's Texas intrastate gas system
Apache buys Repsol's Egyptian interests

FEBRUARY

Markets/Trading
China National Offshore Oil shares begin NYSE trading
Discoveries
BP, ExxonMobil find 1 billion bbl at Crazy Horse North in US Gulf
Inpex Masela finds gas, condensate in Indonesia's Arafura Sea
Production
Anadarko starts Tanzanite, Hickory production off Louisiana
ENI to operate Kazakhstan's giant offshore Kashagan oil field
Petroecuador declares force majeure on 1.8 billion bbl

Quote of the Month

"Our focus for Alaska is not to simply build a pipeline. Rather, our goal is to create a new gas exploration, development, and production industry that sits side-by-side with the existing oil industry in Alaska." – Robert Malone, BP Western Region Pres.

Pipelines
OCP consortium to build 450,000 b/d Ecuador export pipeline
Gulfstream gas system to serve Florida gets FERC approval
Natural Gas
Indonesia to sell Singapore up to 350 MMcfd over 20 years
Refining
Unocal keeps patent on RFG as Supreme Court denies appeal
EPA decides to implement tighter sulfur rules on diesel fuel
FTC finds no collusion on Midwest gasoline pricing
Pak-Arab's Mehmood Kot refinery starts up at 50% capacity
Government
VP Cheney to head US National Energy Policy Task Force
Companies
Phillips to acquire Tosco
FTC clears El Paso-Coastal combination

MARCH

Markets
OPEC cuts another 1 million b/d
More rolling blackouts hit California, gas enters spotlight
Discoveries
Caspian's Kashagan called world's largest oil find in 3 decades
Repsol finds oil off Spain near Casablanca complex
Production/Development
Deepwater Bonga development off Nigeria awards contracts
World's largest semi sinks off Brazil, killing 11 workers
Oxy Permian to recover 10 million bbl with CO_2 injection in Texas
Pipelines
Oil flows from Kazakhstan to Russia's Black Sea port
Natural Gas
Turkey to buy 233 bcf from Azerbaijan over 15 years
BP, ENI to build LNG plant on Egypt's coast
Qatar, UAE sign $4 billion Dolphin development deal

Quote of the Month

"Five years ago, money was the most important thing to us. Today, we understand the most important thing is knowledge and are looking to acquire this knowledge at the best price." – Mikhail Khodorkovsky, OAO Yukos CEO

Refining
Nigeria's first private refinery awards design contract
Venezuela awards Puerto la Cruz expansion contract
Government
US Pres. Bush won't regulate power plant CO_2
US EPA being sued over low-sulfur diesel rules
Companies
Royal Dutch/Shell battles for Barrett Resources
Pemex restructures, cutting investment and jobs

APRIL

Markets
IEA again lowers global oil demand forecast
Balkanization of US gasoline markets hampering flexibility
Discoveries
Kazakhstan's Tengiz reserves are doubled, to 368 million bbl
Mobil Cepu discovers Indonesia's Banyu Urip, >250 million bbl

Quote of the Month

"We won't offer a quick fix because there is none. Narrow, short-term thinking is what got the country into an energy crisis in the first place." – US VP Dick Cheney

Production
H&P orders 15 land rigs @ $9 million each for 2002-03 delivery
Gulf of Mexico sale exceeds $500 million for 547 tracts
World's biggest FPSO heads to Angola's Girassol field
Texas RR Commission to fulfill data requests through website
Pipelines
China to build oil pipeline, tank farm for Shanghai refineries
Natural Gas
Qatar's RasGas starts building 3rd train for largest-ever LNG deal
Refining
BP's Lavera refining-petchem complex being revamped
Phillips demonstrates S Zorb sulfur removal process in Texas plant
Government
World's largest free trade zone to be created in the Americas
Companies
PG&E, California's largest utility, files for bankruptcy
Germany's E.ON to take over UK's Powergen

MAY

Markets
Brent crude futures price peaks near $30
Natural gas futures plunge to 10-month low of $3.738/MMBtu
Discoveries
Trinidad 1-tcf gas discovery to fuel LNG exports
Alaska's Kuparuk oil field extended by Palm discovery
Production/Drilling
Ultra-deepwater drilling record, 9,687 ft of water, set in US Gulf
Floater market to grow to $42.5 billion by 2005, driven by FPSOs
Single hull tankers to be phased out by 2015, says IMO
Pipelines
Okeanos ultra-deepwater gas line planned for Gulf of Mexico
Bombs damage Indonesian gas pipeline, pump station
Natural Gas
Saudi Arabia picks 3 ventures for $25 billion gas initiative
Conoco to build GTL demo plant

Quote of the Month

"Anyone who has been in the business for some time knows he would be a fool to expect today's high prices to continue. As an industry, we've been swinging at curve balls for years. A company can't have long-term plans that are inflexible." –Michael Talbert, Transocean Sedco Forex Pres. and CEO

Government
US energy policy would cut regulation, promote production
Brazil fines Petrobras for oil spills
EU to bring France to court for failure to open gas market
California to sue FERC to force electricity price controls
Companies
Williams acquires Barrett Resources
Valero to buy UDS
Kerr-McGee to purchase HS Resources

JUNE

Markets/Trading
Crude, gasoline futures reach lowest levels in over a year
Int'l Petroleum Exchange to merge with Intercontinental Exchange
Discoveries
Spinnaker finds three fields plus deeper pool in Gulf of Mexico
Equatorial Guinea gas/condensate field reserves doubled by 2 wells
BP, Apache hit gas in Egypt

Quote of the Month

"Worrying about keeping prices in the $22-28/bbl band for the OPEC basket is a little like rearranging the deck chairs on the Titanic with the iceberg in sight." – Sarah Emerson, Managing Director, Energy Security Analysis Inc.

Production
Woodside begins oil production from Australia's Legendre
BP to double its 1-bcfd Trinidad gas output with new platform
Pipelines
Saudi Arabia seizes 1.6 million b/d Iraqi oil line to Red Sea
Natural Gas
North West Shelf to build 4th train, world's largest @ 4.2 million tpy
Qatar to build world's largest GTL plant using 640-720 MMcfd
Refining/Petrochemicals
China to revamp Nanjing plant to run more sour crude
Texas plant planned to produce 7 million lb/yr HDPE
Government
Brazil begins power rationing
UK re-elects Prime Minister Blair
EU to seek ratification of Kyoto protocol despite US rejection
California court rejects gasoline price-fixing allegations
Companies
Statoil takes part in IPO, 18% of company raises $2.9 billion
Conoco to buy Gulf Canada
Hunt Oil to acquire Chieftain

2001 In Brief–continued

JULY

Markets
Crude oil futures fall below $25/bbl, lowest in 14 months
OPEC cuts quotas by 4%, or 1 million b/d
IEA again slashes oil demand growth outlook
Discoveries
Sidetrack confirms world class oil find at N.Sea's Buzzard
BG's Rose finds more N.Sea gas
Statoil's Stine oil discovery drilled in record <10 days
Production
N.Sea Blake oil field on stream at 40-60,000 b/d, ahead of schedule
Phillips double-hull tanker loads first North Slope crude
Pipelines
Bomb attack again shuts down Colombia's 2nd largest oil line
Natural Gas
Shell, El Paso propose LNG terminal on Mexico east coast
Dynegy proposes Gulf of Mexico LNG terminal
Refining/Marketing
China bans service station construction by outsiders

Quote of the Month

"Everyone knew that Kyoto was dead, but, as at any funeral, it is necessary to be discreet in one's remarks about the deceased." – Thomas G. Burns, Principal, Ecoenergy

Government
Australia, East Timor agree on oil, gas production royalties split
FERC orders consolidation of US power grid into 4 RTOs
Indonesia gets new president
Companies
Amerada Hess to acquire Triton
BP buys 51% of Veba, German retail outlets
Anadarko to acquire Gulfstream

AUGUST

Discoveries
Mustang finds gas, condensate in Texas Gulf Stirrup prospect
Alpine satellites Nanuq, Fiord increase reserves on N.Slope

Quote of the Month

"ANWR is a moneymaker for the environmental community. They want to keep this [dispute] alive as a means of raising funds." – Ken Boyd, Arctic Power consultant

Production/Development
Algeria, BP agree to develop 7.2-tcf In Salah gas fields
Bohai Century FPSO to develop China's Qinhuangdao oil field
Queensland 130,000 bbl shale oil volume exceeds expectations
Deepwater US Gulf Typhoon field begins production
Unocal starts up Plamuk oil field off Thailand
N.Nemba off Angola begins oil production
Pipelines
Egypt, Libya form oil, gas pipeline company
Natural Gas
Shell proposes floating LNG plant for Australia development
Government
US extends Iran-Libya sanctions
Peru's new energy minister to accelerate privatization
FERC gets new chairman Wood
California sues EPA over gasoline additives requirement
Companies
Devon buys Mitchell Energy

SEPTEMBER

Markets/Trading
Energy, other trading briefly suspended after terrorist attack
World oil markets calm as Brent, US gasoline prices normalize
Crude, products futures fall to 2-year lows
Production/Development
Cerro Negro upgrading extra-heavy Venezuelan crude
N.Sea Franklin condensate/gas field on stream, adding to Elgin
Indonesia's Belanak gas field to get new FPSO
Azeri-Chirag-Guneshli oil field development to expand

Quote of the Month

"We expect these attacks to increase the desire for [US] energy independence, particularly for supply. This could mean more support for a gas pipeline from Alaska and increased drilling from restricted land." – Adam Siemenski, Global Energy Strategist, Deutsche Banc Alex. Brown

Natural Gas
Philippines' Malampaya gas-to-power project gets green light
World's largest methanol plant to use Trinidad gas
BP boosts India LNG project stake
Refining
US refiners scramble to store or blend unused jet fuel
12 oil pipelines explode at Mao-ming refinery, China's largest
TotalFinaElf revamping French, Belgian refineries
Government
US aviation halted by largest terrorist attack in world history
Companies
Chevron-Texaco merger approved
Global Marine, Santa Fe to merge
Devon Energy to acquire Anderson Exploration
Dominion Resources to buy Louis Dreyfus Natural Gas

OCTOBER

Markets
Crude oil futures continue slide
US retail gasoline prices fall, jet fuel demand plunges
Discoveries
Repsol's Trident prospect flows oil off Spain
Goliath find is first significant oil in Norwegian Barents Sea

Quote of the Month

"Whatever benefits globalization brings can and should be defended – there is more than just business interests at stake." – Peter Sutherland, BP PLC Chairman

Production/Development
Qinhuangdao in Bohai Bay comes on stream, to produce 65,000 b/d
Indonesia's 1st deepwater oil field development begins
N.Sea Mikkel, Sigyn gas-condensate fields under development
Petrobras to resume developing Roncador offshore field
ExxonMobil to develop Russian Far East oil, gas at Sakhalin
Pipelines
Caspian pipeline delivering oil from Kazakhstan to Russian port
China finishes northwestern gas line from Qinghai fields
Natural Gas
El Paso to buy LNG from Norway's Snohvit project
Qatar to expand Ras Laffan LNG plant
Government
FERC approves reopening mid-eastern US LNG facility
Companies
Burlington Resources to purchase Canadian Hunter Exploration
Duke to acquire Westcoast Energy
BG buys Enron India E&P assets

NOVEMBER

Markets
Crude oil futures hit 2-year low
OPEC plans reduction, pressures non-members to cut output
US retail gasoline prices lowest since mid-99
Discoveries
Sapphire confirms largest gas column discovered in Egypt
Production
Northstar field starts producing oil via 1st buried subsea Arctic line
N.Sea Leadon field comes on stream, to produce 45,000 b/d
Hamaca field begins producing extra-heavy crude for blending
Deepwater Gulf of Mexico Oregano oil field starts producing
Pipelines
Williams' deepwater Gulf of Mexico system near completion
Cleopatra gas lines to join Mardi Gras system in Gulf of Mexico
Natural Gas
Engineering begins on Nigerian GTL plant in Escravos

Quote of the Month

"We think the deepwater [Gulf of Mexico] is where the shelf was in the mid-1960s in terms of maturity. There is huge potential sitting right next to the largest energy consumer in the world." – Charlie Fischer, Nexen Inc. Pres.

Refining
Tyumen Oil opens cat cracker at Ryazan refinery
UDS' Quebec plant meets Can. low-sulfur rule in advance
Government
US Pres. Bush orders filling of SPR to 700 million bbl
New Venezuela law hikes royalties, reduces foreign stakes
Indonesia passes law to end Pertamina's monopoly
Brunei creates national oil firm
Companies
Phillips, Conoco to merge

DECEMBER

Markets
OPEC sets 1.5 million b/d cut after non-members promise reductions
Crude prices on NYMEX record 1-day jump of 8.4%
Discoveries
Shallow water off Trinidad yields significant oil resource
Shell finds Philippines' 1st commercial oil deposits

Quote of the Month

"We are in for a period of [economic] difficulty worldwide. As far as the energy sector is concerned, the impacts are going to be quite severe." – Fereidun Fesharaki, FACTS Inc. Pres.

Production/Development
Girassol begins oil production in deepwater off Angola
Work starts on platform, pipelines for Azerbaijan Caspian fields
Phillips to proceed with Bayu-Undan gas project in Timor Sea
Pipelines
Iran-Turkey gas line begins flowing at >100 bcf/yr
Qatar, Dolphin Energy to build 2-bcfd gas line to UAE
Bolivia hires group to build 777-MMcfd gas line to Brazil
Natural Gas
Engineering begins on China's Guangdong LNG project
Sempra to import Bolivian LNG to proposed terminal in NW Mexico
Refining
Bayernoil unit 1st to use BP gasoline desulfurization process
Brazil to invest record $8.9 billion in revamping 8 refineries
Government
Brazil opens fuels market to imports
Bangladesh court blocks gas exports to India
Companies
Enron files for bankruptcy
Veritas to merge with Petroleum Geo-Services

NORTH AMERICA

UNITED STATES

CAPITAL: WASHINGTON, DC
MONETARY UNIT: DOLLAR
REFINING CAPACITY: 16.6 MILLION B/D
OIL PRODUCTION: 5.8 MILLION B/D
OIL RESERVES: 22.045 BILLION BBL
GAS RESERVES: 177.4 TCF

After an optimistic start, the year 2001 ended with sharply lower oil and gas prices, which were rooted in an economic recession – America's first since 1991 – and then exacerbated by September's terrorist attacks, which worked to dampen oil demand worldwide. On top of that, Enron Corp.'s collapse threw a wild card into the energy trading marketplace, and OPEC's efforts to combat slumping demand threatened to ignite a price war.

Yet despite these unprecedented circumstances, the US oil and gas industry remained a bedrock of stability amidst the turmoil of world politics. Energy companies understand how to deal with today's price volatility and tomorrow's market uncertainties. Oil supply remained secure as buyers and sellers continued to work in a global market that is vastly more liquid and transparent than ever before.

Petroleum Market Movements

During 2001, production of crude oil held steady at around 5.83 million b/d, about the same as 2000 and down slightly from 1999's average of 5.88 b/d, reported the US Department of Energy (DOE) Energy Information Administration (EIA). Proved oil reserves increased to just over 22 billion bbl in 2001 from the previous year's 21.7 billion bbl. EIA said, "Domestic oil exploration and development spending by US major oil companies rebounded during 2001 from the deep cuts made during the oil price collapse of 1997/1998." The American Petroleum Institute (API) pegged 2001 US production slightly higher, at 5.85 million b/d, noting that the increase was the first in a decade and only the second since 1986.

Although 2001 began with healthy oil and gas prices, the cracks in America's economy widened, and energy demand began to soften as well, weakening prices. The terrorist attacks of September 11 sparked a brief oil price increase, due largely to uncertainty surrounding the impact on oil supplies. It soon became clear, however, that the US energy infrastructure was operating normally, and the Organization of Petroleum Exporting Countries (OPEC) reassured world markets that it would maintain plentiful supplies.

As a result, oil prices dropped to levels around $5/bbl lower than before the attacks. Between early September and the end of October, EIA reported, retail gasoline prices dropped 34¢/gal, or 22% – the steepest price drop over that period of time in at least a decade. In October 2001, gasoline cost 14% less than in October 2000, representing the biggest year-to-year plunge in almost 3 years. API reported that US consumer demand for petroleum products in 2001 declined for the first time in a decade, by nearly 1%, to 19.587 million b/d. With retail gasoline prices 20% lower at year-end, gasoline was the only product to log any increase in consumption, showing a slight 1.4% rise over 2000.

EIA stated that no oil supply shortages materialized in connection with terrorist attacks or subsequent military operations. To the contrary, world oil demand continued to fall, due to lower commercial jet fuel consumption, slower economic growth, and OPEC overproduction. This combination of factors led to the accumulation of oil inventories and, in general, reduced tightness in world oil markets, said EIA.

EIA also updated its US natural gas price and supply projections, saying prices would continue to decline from $4.09/Mcf in 2001 to $2.15/Mcf during

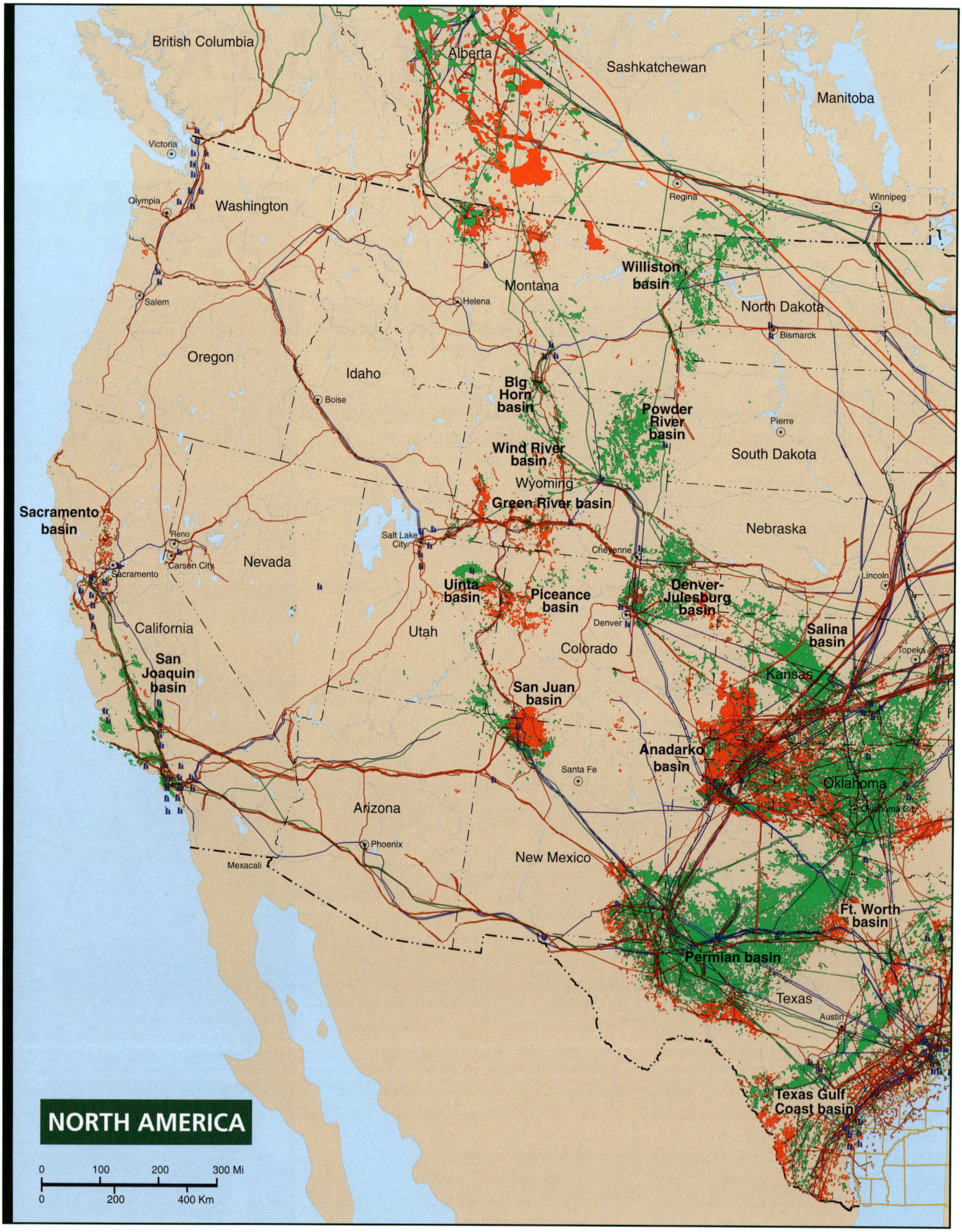
British Columbia
Alberta
Sashkatchewan
Manitoba
Victoria
Olympia
Washington
Regina
Winnipeg
Williston basin
Montana
Salem
Helena
North Dakota
Bismarck
Oregon
Idaho
Big Horn basin
Boise
Powder River basin
Pierre
South Dakota
Wind River basin
Wyoming
Green River basin
Sacramento basin
Nebraska
Reno
Salt Lake City
Cheyenne
Carson City
Nevada
Sacramento
Uinta basin
Piceance basin
Denver-Julesburg basin
Lincoln
California
Denver
Utah
Salina basin
Colorado
Topeka
San Joaquin basin
Kansas
San Juan basin
Anadarko basin
Santa Fe
Oklahoma
Oklahoma City
Arizona
Phoenix
Mexacali
New Mexico
Ft. Worth basin
Permian basin
Texas
Austin
Texas Gulf Coast basin
NORTH AMERICA
0 100 200 300 Mi
0 200 400 Km

Manitoba
Ontario
Quebec
New Brunswick
Nova Scotia
Winnipeg
Quebec
Fredericton
Augusta
Ottawa
Montpelier
Vermont
N.H.
Concord
Boston
Mass.
Providence
R.I.
Hartford
Conn.
Minnesota
Michigan
St. Paul
Michigan basin
Michigan
Toronto
New York
Albany
Wisconsin
Madison
Lansing
Iowa
Des Moines
Pennsylvania
Appalachian basin
Harrisburg
Trenton
New Jersey
Lincoln
Illinois
Indiana
Columbus
Maryland
Dover
Delaware
Springfield
Indianapolis
Ohio
West Virginia
Washington D.C.
Annapolis
Topeka
Charleston
Richmond
Frankfort
Virginia
Jefferson City
Illinois basin
Kentucky
Missouri
Raleigh
Nashville
Tennessee
North Carolina
Oklahoma
Arkansas
Oklahoma City
Little Rock
South Carolina
Arkoma basin
Black Warrior basin
Atlanta
Columbia
Mississippi
Alabama
Georgia
Jackson
Montgomery
East Texas basin
Louisiana
Louisiana-Mississippi-Alabama salt basins
Louisiana Gulf Coast basin
Baton Rouge
Austin
Tallahassee
Florida
Legend
Oil field
Oil sand
Gas field
Crude oil pipeline
Natural gas pipeline
Products pipeline
Pipeline planned or under construction
Refinery in operation
Refinery capacity in 1,000 b/d
600
Tanker terminal
Cities
Capital
International boundary
Water depth
0 to 200 m
200 m and deeper
NORTH AMERICA
0 100 200 300 Mi
0 200 400 Km

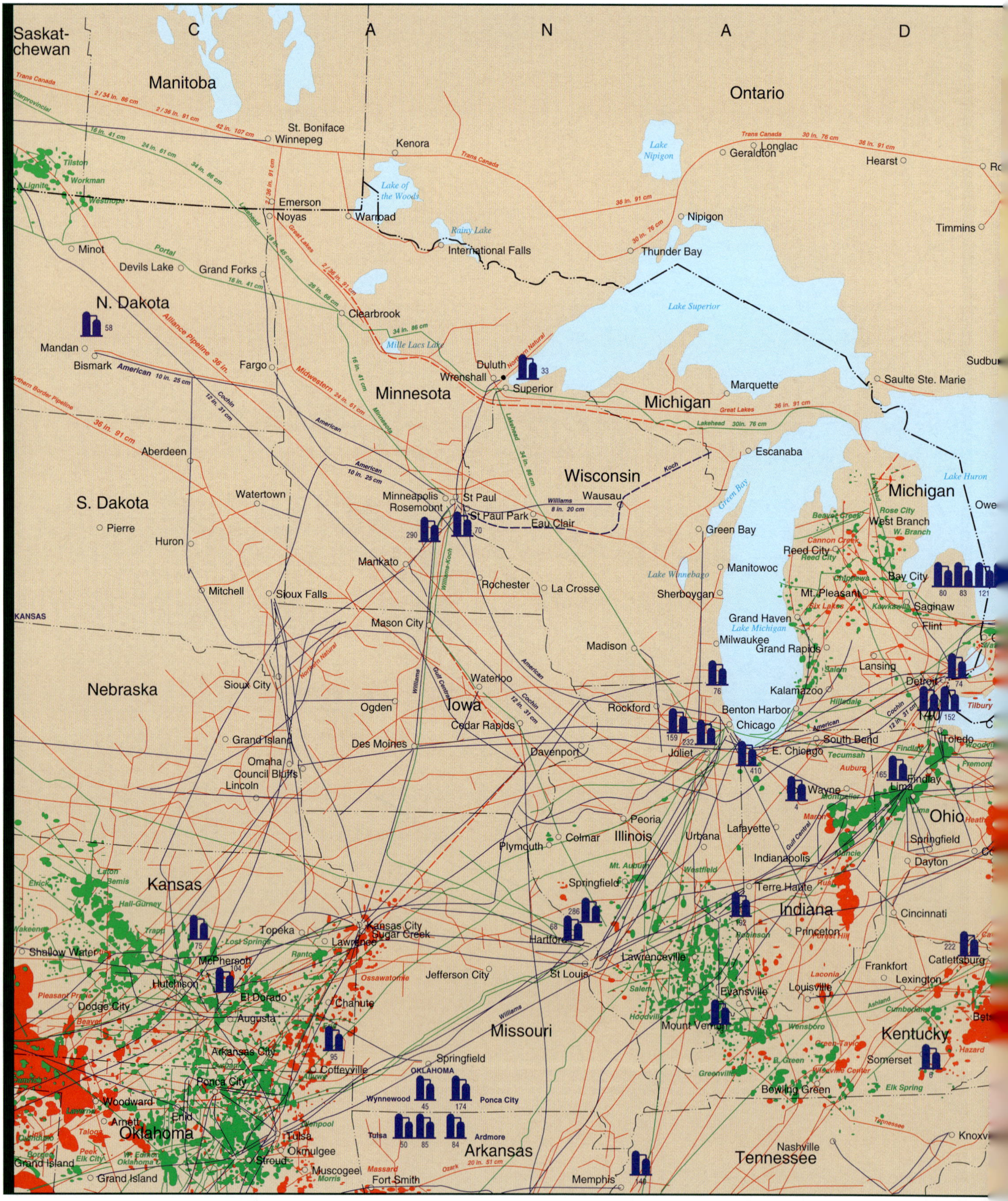
C A N A D A
Saskat-
chewan
Manitoba
Ontario
St. Boniface
Winnepeg
Kenora
Longlac
Geraldton
Hearst
Timmins
Lake Nipigon
Lake of the Woods
Emerson
Noyas
Warroad
Rainy Lake
International Falls
Nipigon
Thunder Bay
Lake Superior
Minot
Devils Lake
Grand Forks
N. Dakota
Clearbrook
Mille Lacs Lake
Mandan
Bismark
Fargo
Duluth
Wrenshall
Superior
Minnesota
Michigan
Marquette
Saulte Ste. Marie
Sudbury
Aberdeen
Escanaba
Wisconsin
Michigan
Lake Huron
S. Dakota
Watertown
Minneapolis
Rosemount
St Paul
St Paul Park
Eau Clair
Wausau
Green Bay
West Branch
Reed City
Pierre
Huron
Mankato
Rochester
La Crosse
Manitowoc
Lake Winnebago
Sherboygan
Bay City
Mt. Pleasant
Saginaw
Flint
Mitchell
Sioux Falls
Mason City
Grand Haven
Lake Michigan
Milwaukee
Grand Rapids
Madison
Lansing
Detroit
Nebraska
Sioux City
Waterloo
Iowa
Ogden
Cedar Rapids
Rockford
Kalamazoo
Benton Harbor
Chicago
South Bend
Toledo
Grand Island
Des Moines
Davenport
Joliet
E. Chicago
Omaha
Council Bluffs
Lincoln
Wayne
Findlay
Lima
Peoria
Ohio
Lafayette
Urbana
Illinois
Colmar
Plymouth
Springfield
Indianapolis
Dayton
Kansas
Springfield
Terre Haute
Indiana
Cincinnati
Topeka
Kansas City
Sugar Creek
Lawrence
Hartford
Princeton
Shallow Water
McPherson
Lawrenceville
Frankfort
Catlettsburg
Hutchison
Jefferson City
St Louis
Lexington
Louisville
Evansville
Dodge City
El Dorado
Chanute
Augusta
Mount Vernon
Missouri
Kentucky
Arkansas City
Springfield
Somerset
Coffeyville
OKLAHOMA
Ponca City
Bowling Green
Wynnewood
Ponca City
Woodward
Arnett
Enid
Oklahoma
Tulsa
Ardmore
Arkansas
Okmulgee
Nashville
Tennessee
Knoxville
Grand Island
Stroud
Muscogee
Fort Smith
Memphis
Grand Island

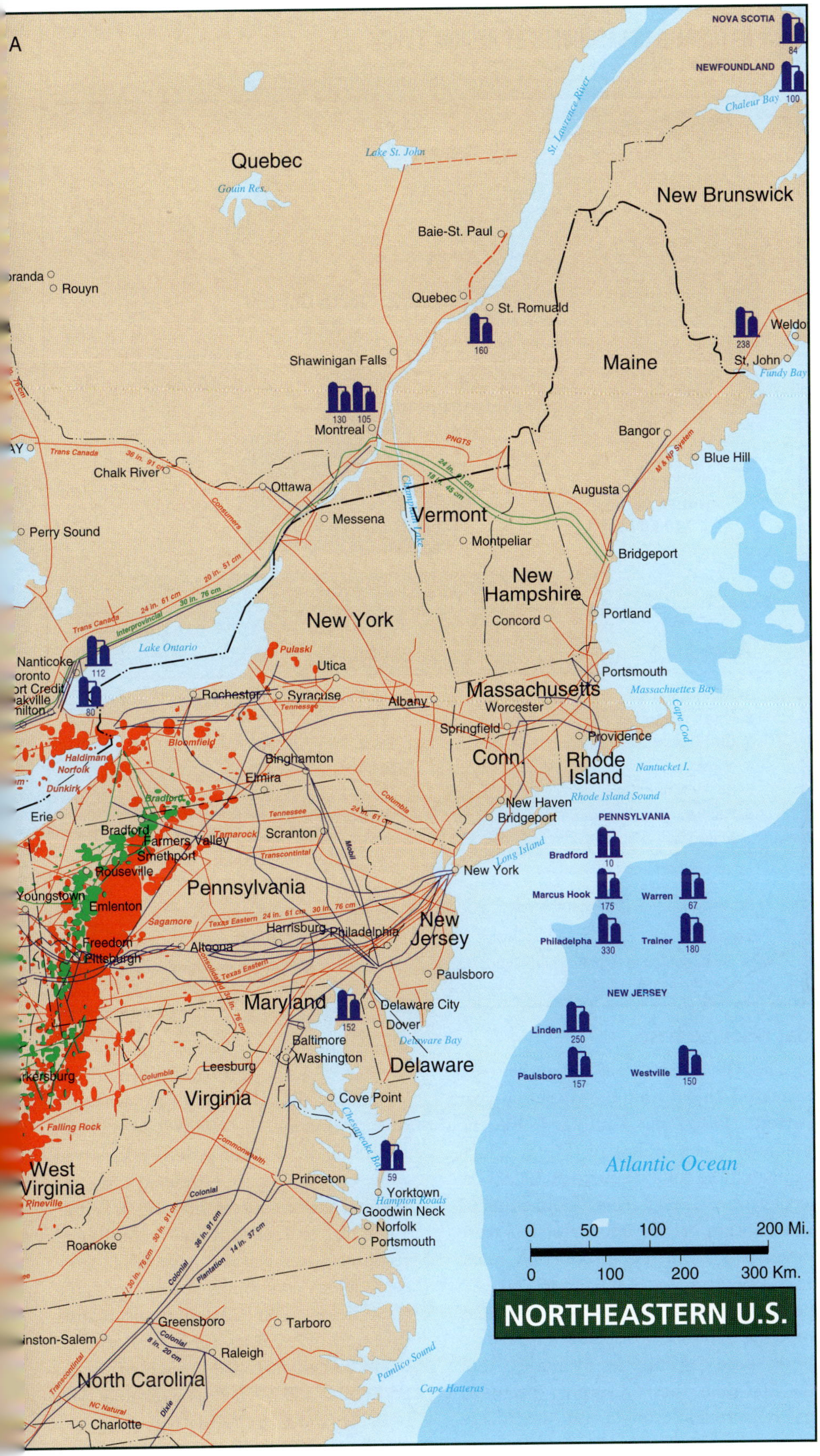

the 2001-02 winter and to an average just under \$2.00/Mcf for 2002. EIA expected supplies to increase from 22.45 tcf in 2001 to 23.53 tcf in 2002. EIA blamed the price drop on mild weather and the slowing economy, which have reduced gas consumption growth, and on additional drilling.

EIA re-estimated world oil demand growth during 2001 at only 200,000 b/d, which was less than one-quarter of its pre-September estimate, but projected growth of 800,000 b/d for 2002. US oil demand was expected to remain roughly flat during 2001, said EIA, then increase in 2002 by 0.9% or 180,000 b/d. According to Cambridge Energy Research Associates, world oil prices were expected to average \$6/bbl lower in 2002 than in 2001.

As efforts to stimulate America's economy stalled in Congress at year-end, the US oil industry prepared for another bout of low demand and prices but remained confident about longer term prospects. Although capital markets for exploration and production were tight, private financing remained abundant, having been freed from its devotion to the technology sector. In late 2001, OPEC won some production restraint from major non-OPEC oil exporters such as Russia, Mexico, and Norway, and the year ended on a positive note with rising oil futures prices as OPEC agreed to reduce its quotas by 1.5 million b/d through mid-2002.

The natural gas price outlook also took an upward turn, according to Raymond James & Associates Inc., which predicted a sharp jump in early 2002. And EIA projected a 32% increase in US energy demand over the longer term, by 2020, assuming no changes in federal laws and regulations.

The prospect of bigger global oil markets also received a boost from the swing in international political mood in evidence at the World Trade Organization's November 2001 meetings. With antiglobalization activists out of favor along with any other extremists, negotiations made more progress toward eliminating trade barriers and opening new markets, which would stimulate future oil and gas demand and industry growth.

Petroleum reserves and discovery rate

US proved reserves of natural gas increased a record 6% from 1999 to 2000, to 177.4 tcf, according to EIA, while proved reserves of crude oil and natural gas liquids also increased. The additions to dry gas reserves were the largest since EIA first began compiling its report in 1977. Total US crude oil proved reserves increased 1.3% during 1999-2000, to reach 22,045 million bbl.

However, the agency cautioned that the unusually large increase in gas reserves could be an anomaly and might not necessarily repeat itself. EIA cited two major reasons for this: 1) the number of frontier areas is diminishing, and 2) the unusual combination of a large increase in the number of exploratory wells and a large increase in total discoveries per exploratory well is unlikely to occur very often in the future, because an increase in the number of exploratory wells usually decreases discoveries per well.

"Anytime you're having a great run with a play, eventually it comes back down," noted John Wood, director of EIA's reserves and production division. Wood said that higher prices during 2000 helped encourage reserve numbers "but they weren't critical." He also attributed some effect to improved technology that allows producers to be much more efficient in choosing, drilling, and extending individual wells.

Gas reserves additions in 2000 replaced 152% of US dry natural gas production. Most of this gas came from Texas and New Mexico in the Southwest and from states with large coalbed methane reserves like Wyoming, Colorado, and Utah. "Coalbed methane is an important frontier area for US producers," Wood said. The agency expects natural gas demand and production to continue growing.

Crude oil reserves additions replaced 115% of oil production. The majority of this additional oil came from federal offshore leases in the Gulf of Mexico deep water. Large total discoveries of crude oil in 2000 also came mainly from the deepwater Gulf and from the Alaskan North Slope. Both provided very high discoveries per exploratory well, EIA said.

US NATURAL GAS RESOURCES BY REGION, TYPE

Lower 48 conventional

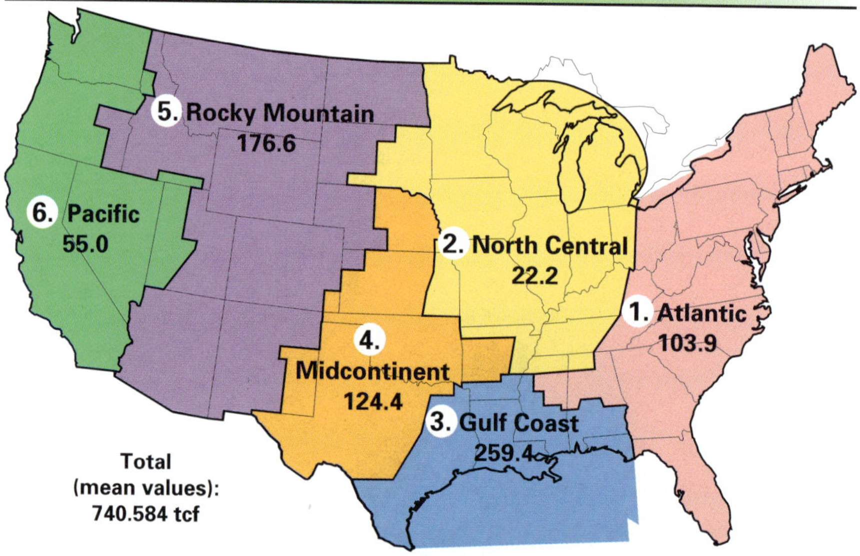

All regions, types

Area	Mean values, tcf 2000	Mean values, tcf 1998	Change, % 1998-2000
1. Atlantic	103.9	103.9	0
2. North Central	22.2	22.2	0
3. Gulf Coast	259.4	265.5	–2.3
4. Midcontinent	124.4	122.1	1.9
5. Rocky Mountain	176.6	150.0	17.7
6. Pacific	55.0	37.2	47.8
Alaska	193.8	193.8	0
Coalbed methane (all areas)	155.2	141.4	9.7
Proved reserves (DOE estimates)	167.4	167.0	
Total US	**1,258.4**	**1,204.5**	**4.5**

Note: Totals subject to rounding and slight differences due to statistical aggregation of distributions.
Source: Potential Gas Committee

Gas resource base

The US natural gas resource base increased 4.5% during the latest two-year period (1998-2000) analyzed by the Potential Gas Committee (PGC), which estimated the total resource at 1,258.4 tcf as of year-end 2000. This would equal 63 years of supply if the entire resource were developed at current production rates. The American Gas Association (A.G.A.) noted that expansion of America's gas pipeline network would be required to bring these gas supplies to market. For the first time, the PGC report discussed gas resources in Canada and Mexico, "in recognition that North America is becoming a single market for natural gas."

According to the PGC study, traditional natural gas resources in the US increased 4.4% and coalbed methane 9.7%, even though 38 tcf of gas was produced during the 1998-2000 period. Including 167.4 tcf of proved reserves in 2000, the PGC study estimated total traditional gas resources of 1,091 tcf, plus 155 tcf of coalbed methane. Traditional resources were classed as probable (current fields) containing 207 tcf, possible (new fields) with 332.2, and speculative (frontier) of 397.8 tcf, for a total of 935.8 tcf. Coalbed methane accounted for an additional 155.2 tcf (16.3 probable, 54.3 possible, and 84.6 speculative).

Changes in the comparison of resource estimates between 1998 and 2000 stemmed mainly from analysis of

new data for seven individual US regions. Regional data reflected the following increases and decreases, compared with year-end 1998:

- The Pacific region surged 47.8% from 37.2 to 55.0 tcf.
- The Rocky Mountain region climbed 17.7% from 150.0 to 176.6 tcf.
- The Midcontinent region increased 1.9% from 122.1 to 124.4 tcf.
- The Gulf Coast region dropped 2.3% from 265.5 to 259.4 tcf.
- Alaska was unchanged at 193.8 tcf.
- The Atlantic and North Central regions were unchanged at 103.9 and 22.2 tcf, respectively.

Energy policy and legislation

The year 2001 opened with high expectations as George W. Bush, a long-time supporter of oil industry interests, was inaugurated as President. Bush named several cabinet posts crucial to the oil industry, including Energy Secretary Spencer Abraham and Interior Secretary Gale Norton. Bush also quickly appointed Vice President Richard Cheney, who also has an oil industry background, to lead development of national energy policy by a cabinet-level task force.

These moves were accompanied by widespread talk of an "energy crisis," due to high prices for natural gas and oil during early 2001, and by debate over constructing a natural gas pipeline from Alaska's Prudhoe Bay field, which is estimated to contain 36 tcf, through Canada and into the continental US.

In May 2001, the president's energy task force recommended to allow leasing for exploratory drilling on the coastal plain of the Arctic National Wildlife Refuge (ANWR) east of Prudhoe Bay oil field in Alaska. Bush's energy plan also encouraged efforts on behalf of expeditious leasing of federal lands elsewhere and a review of strict regulations on refining operations. In August 2001, the US House of Representatives passed an energy bill containing many of the presidential energy plan's recommendations.

However, the US Senate failed to act on the legislation before the September 2001 terrorist attacks, and afterward the energy policy arena became dominated by the issue of domestic energy security. In late 2001, the Senate delayed debate on an energy bill until 2002. The Democratic bill to be discussed then would not include ANWR drilling but would allow federal incentives to build a pipeline to move Alaskan gas to the Lower 48 states. In late 2001, the state of Alaska began offering to sell North Slope natural gas directly to buyers for the first time.

Heightened concern over possible further terrorist attacks after September 2001 forced some adjustments in offshore, pipeline, and refining operations, as the US petroleum industry responded to new security measures and other developments. Oil and gas industry representatives met with government officials in Washington, DC, to coordinate ways to improve security. The Department of Transportation removed from its website detailed maps of the US pipeline infrastructure, and deliveries of LNG into Boston Harbor were suspended briefly during September and October.

President Bush created the Homeland Security Council to centralize plans for protecting the US. In November 2001, the Federal Bureau of Investigation (FBI) publicized terrorist threats to attack the US natural gas infrastructure and warned pipeline companies to be on alert. The US Federal Energy Regulatory Commission (FERC) disabled access on its website to pipeline maps and promised to act quickly on pipelines' requests for security investments, and a pipeline safety bill was introduced in Congress in late 2001. Offshore oil and gas operators agreed to meet in early 2002 with the US Coast Guard to establish a commission to set minimum security standards for Gulf of Mexico oil facilities.

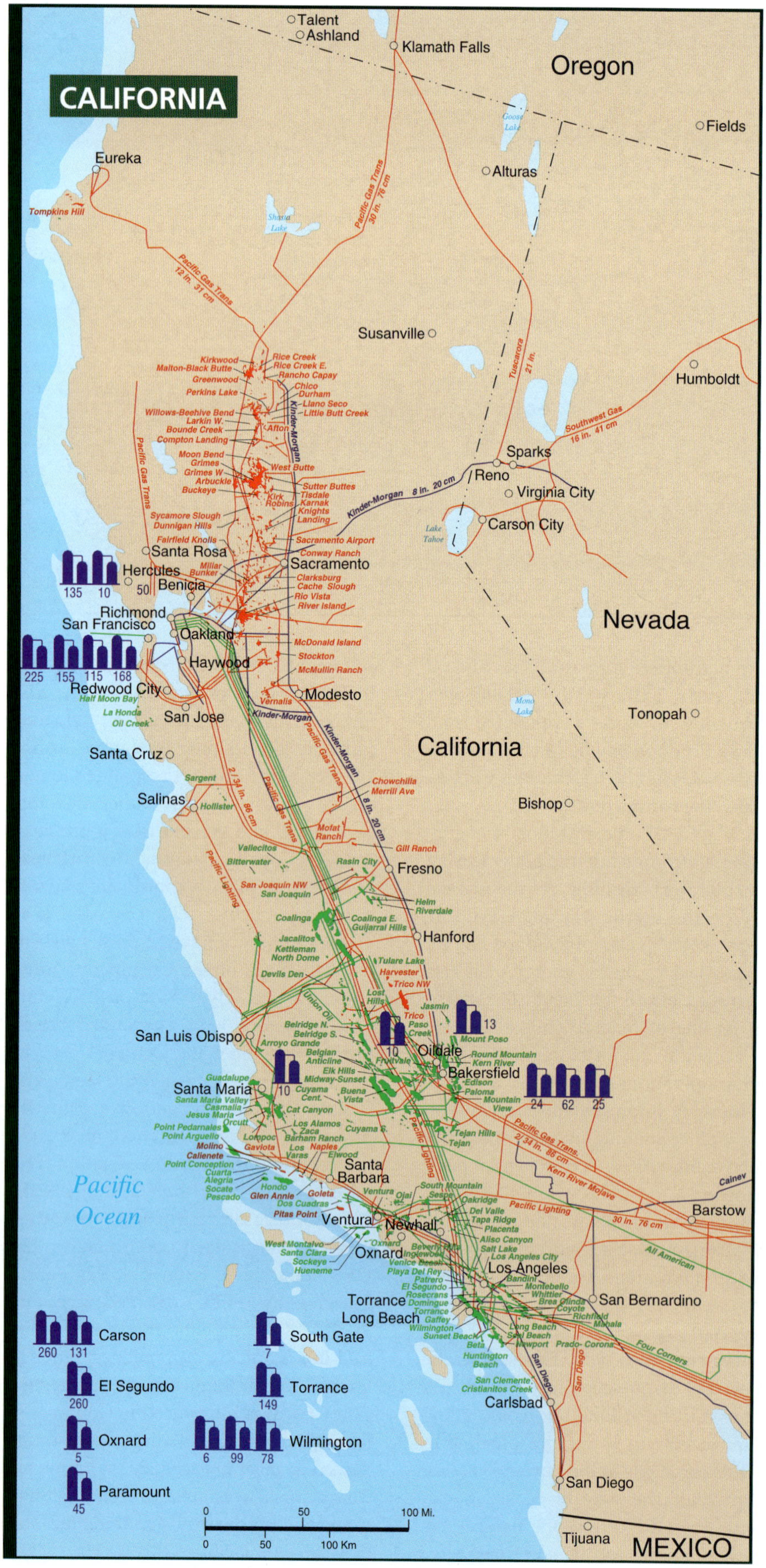

In late 2001, President Bush ordered that the Strategic Petroleum Reserve (SPR) be filled to its current physical capacity of 700 million bbl. Oil shipments into the SPR were expected to begin in April 2002 and be completed by 2005. As of late 2001, the SPR contained 547.2 million bbl of oil, equal to more than 50 days of oil imports, said EIA. The SPR has a maximum drawdown capability of 4.18 million b/d for 90 days, with oil beginning to arrive in the marketplace 15 days after a presidential decision to initiate a drawdown.

Economic sanctions affecting Iran and Libya were extended by President Bush in March 2001 for five years, although the President had sought only a two-year extension and noted that the sanctions should be reviewed frequently in the future. The Iran-Libya Sanctions Act (ILSA) allows penalties to be imposed on any company investing more than $20 million in either nation's energy sector. However, the legislation retained the president's power to waive sanctions in the national interest.

In early 2001, the Bush administration declared that the US has "no interest" in implementing or ratifying the Kyoto treaty, saying it would be too harmful to the US economy. The administration planned to pursue other ways to address the climate change issue.

Also in early 2001, President Bush decided not to try to reverse the low-sulfur diesel rule being implemented by the US Environmental Protection Agency (EPA). The plan requires refiners to slash highway diesel sulfur levels to 15 ppm by June 2006 from 500 ppm in 2001. While the EPA estimated that the rule would cost $4.2 billion/yr, the oil industry claimed a higher figure of $8 billion in up-front capital investment and a 12¢/gal increase at the pump to consumers.

In July 2001, EIA released an analysis of the rule's effect on the price and supply of highway diesel fuel. At the volumes needed to meet demand, costs were estimated at 5.4-6.8¢/gal plus 1.1¢/gal for distribution costs. The EIA pointed out that these costs could be sharply higher if supply falls short of demand. Over the longer term (post-2011), average marginal refining costs will narrow by about 0.5¢/gal as refineries make incremental improvements to produce ultra-low-sulfur diesel more efficiently, said EIA.

The Petroleum Industry Research Foundation Inc. (PIRINC) in a December 2001 report said that US supplies of motor vehicle fuel could be disrupted in the short term if the federal oxygenate requirement is maintained while quickly phasing out oxgenate methyl tertiary butyl ether (MTBE) at the state level. State initiatives banning MTBE were setting the stage for a massive increase in demand for ethanol as an oxygenate fuel additive, said the report. In late 2001, an unusual coalition of business and labor interests, including refiners, unions, and highway user groups, urged the US Senate to oppose legislation mandating ethanol use in gasoline, saying it would hurt consumers.

The Bush administration supported creation of the US Northeast Home Heating Oil Reserve, which was established in March 2001 by US Energy Secretary Abraham. The 2 million bbl reserve of government-owned heating oil for emergencies was to serve as an insurance policy against problems arising from extreme winter weather during 2001-02.

Power market deregulation

What began in early 2001 as a flashpoint of energy policy for President Bush – California's troubled electricity markets and power shortages – was gradually resolved by a combination of energy conservation by customers, new plants coming on line, temporary regional price controls enforced by FERC, and aggressive power purchases by a newly created state agency, as well as mild weather. As the year ended, California found itself with more power than it could use.

Also late in 2001, one-time financial and energy trading giant Enron Corp. filed for Chapter 11 bankruptcy protection – the largest-ever such filing –and sued former suitor Dynegy Inc. for terminating a last-minute merger. Until mid-2001, Enron was the top US natural gas and power trader with $100 billion in revenue in 2000. Despite Enron's former dominance of the energy trading business, US and worldwide trading proceeded virtually undisturbed as competitors moved in to handle increased volumes. In late 2001, the US Congress began investigating the causes of Enron's collapse.

The prospects for federal deregulation of the electricity industry dimmed during 2001, and several states postponed their deregulation plans. Although the California crisis and Enron's demise were blamed for stalling state and federal deregulation efforts, several other causes were also important, according to Scientech: the absence of national standards, the lack of transmission system upgrades, price restraints on the wholesale market, and dismal results for companies in the retail market.

Pipeline construction and planning

Building the Alaskan natural gas pipeline, with a price tag of about $10 billion, would be one of the largest single construction projects in the nation's history. The project could cost even more than originally expected, due in part to security concerns after the terrorist attacks. Senator Frank Murkowski (R-Alaska) said in late 2001 that the federal government might be needed to help build the pipeline.

The Interstate Natural Gas Association of America (INGAA) projected that the pipeline industry would need to invest more than $67 billion between 2002 and 2016, an average of $4.5 billion/yr. INGAA's study said that some pipeline projects could be delayed by financing constraints in the wake of Enron's bankruptcy filing.

While Alaska gas pipeline plans were being debated, the industry forged ahead with other projects that promised to meet growing demand for gas. In early 2001, FERC approved the first milestone in the Centennial gas pipeline conversion, giving Trunkline Gas, an affiliate of CMS Energy Corp., permission to abandon a gas pipeline before converting it to carrying petroleum products. Trunkline, Marathon Ashland Petroleum, and TEPPCO partners intended to convert the 720-mile, 26-inch line to carry gasoline, diesel, and jet fuel from Gulf Coast refineries to Midwestern markets. Centennial Pipeline will intersect TEPPCO's existing mainline in southern Illinois, where a petroleum products storage facility was under construction. The project was expected to be completed by year-end 2001.

Kinder Morgan Energy Partners LP bought Tejas Gas LLC, a 3,400-mile pipeline system inside Texas with a 3.5-bcfd capacity, and was purchasing Enron's 33.3% interest in Trailblazer Pipeline Co., increasing its ownership to 100%. The 436-mile Trailblazer pipeline, running from Rockport, CO, to Beatrice, NE, was recently increased in capacity to 846,000 dekatherms/day (Dth/d) and was due for an additional 324,000 Dth/d in 2002, all of which was fully subscribed.

Dynegy exercised its $23 million option to buy Enron's largest pipeline, Northern Natural Gas, which was excluded from the bankruptcy filing, in early 2002. The Northern Natural Gas pipeline system has a maximum capacity of 4.3 bcfd.

In May 2001, federal regulators issued El Paso Corp. affiliate El Paso Natural Gas a certificate to operate a pipeline that would provide 230 MMcfd of gas capacity to serve California and the southwestern US. This project would involve a $154 million conversion of 785 miles of an existing crude oil transmission pipeline to gas. The converted pipeline would begin near the California border at Ehrenberg, AZ, and extend to McCamey, TX. The project was scheduled to be operational in late 2001. El Paso was also holding open seasons for other pipeline projects under consideration in California.

In late 2001, El Paso began engineering studies for its proposed $1.6 billion Blue Atlantic pipeline, which would carry natural gas from fields off Nova Scotia to markets in eastern Canada and the Northeastern US. The 750-mile, 36-inch line would transport about 1 bcfd to landing points in New York and New Jersey.

Two other pipelines to move eastern Canada's Sable Island natural gas to the US Northeast received a positive final environmental impact statement in late 2001 and FERC approval early in 2002:

Phase three of the Maritimes & Northeast Pipeline, a 25-mile, 30-inch onshore pipe; and the Algonquin Gas Transmission HubLine, 30 miles of 24-inch offshore pipe and a 5-mile lateral. The project was designed to interconnect the Maritimes and Algonquin transmission systems.

During 2001, the Alliance Pipeline from British Columbia to south of Chicago and the Aux Sable natural gas processing plant completed their first full year of commercial operations. The 1.2-1.5 bcfd Alliance Pipeline operates at high pressure (1,740 psig) and with rich gas for dense-phase operation and more efficient hydraulics. Alliance can move twice the volume with much less fuel than a traditional 36-inch line. The Aux Sable plant, with a 2.1-bcfd design capacity, is the largest in the Lower 48 States. If future gas prices warrant more supply from western Canada, the Alliance pipeline could be expanded to reach a processing level of more than 2 bcfd, which Aux Sable would be able to handle easily.

In late 2001, Williams Cos. completed the first phase of its Transco pipeline system's expansion, which would boost gas transportation capacity into the US Northeast, and FERC approved the 424-mile US portion of the 551-mile Millennium pipeline project, which would bring 700 MMcfd of gas from Canada to the New York City suburbs. However, local activists vowed to fight the project.

FERC also approved in early 2002 the North Baja pipeline that will bring natural gas from Mexico to fuel power generation plants in southern California and northern Mexico. The 215-mile (346-km), $320 million pipeline would carry up to 500 MMcfd. FERC's approval applied to the 80-mile (129-km) US portion, while the 135-mile (217-km) Mexican segment had been approved earlier by Mexico's regulators.

Drilling trends

The average US well depth increased roughly 800 feet during 2001, or 14% over the previous year – the most significant single-year increase ever, according to Raymond James & Associates. Since 1980, average US well depth has increased nearly 40%. Contributing toward this trend was the focus on drilling for natural gas, which is typically found deeper than oil. Rigs drilling for gas accounted for more than 80% of the US rig fleet.

The US government collected a record $8.61 billion during the 2000-2001 fiscal year from royalties paid by energy companies drilling for oil and gas on federal leases. Also, producing US states received a record $656 million in mineral revenues during the first half of 2001, opening the possibility of setting a new high for the full year.

As US petroleum basins continued to mature, the economics of new wells to be drilled became more marginal, and drilling contractors responded by delivering efficient newbuild rigs and renovating existing rigs to enhance their capacity and efficiency. In 2001, Helmerich & Payne IDC (H&P) and Pioneer Drilling Co. Ltd. stood out as examples of contractors working to expand their fleets' capabilities and efficiency.

Pioneer mobilized the first of two new IRI 1700E series rigs to spud its first well. Believed to be the first application of AC-drive drawworks on a US land rig, the unit was designed for 18,000 feet and included a specially designed mud system, enclosed driller's cabin, and programmable logic control systems.

H&P upgraded its modular newbuild rig design, introduced in 1998, to include climate-controlled driller's cabins, larger mud pumps, and a V-ICIS electronic operating system on touch screen. The company began delivering these second-generation "flexrigs" in 2001 and planned substantial improvements for 2002-03, including a 25-foot substructure to accommodate larger blowout preventer stacks used for deeper wells. H&P said that its newbuild rigs have experienced 94-95% utilization.

Helmerich & Payne Inc. said that its newbuild rigs have experienced 94-95% utilization. The company began delivering its second-generation "flexrigs" in 2001 and planned substantial improvements for 2002-03. Photo courtesy of Helmerich & Payne.

Rocky Mountain renaissance

While much of America's exploration and production activity during 2001 focused on the Gulf of Mexico and Alaska, onshore discoveries in the Lower 48 also contributed their share to petroleum resources. The oil and gas potential of the Rocky Mountain region and several areas of Arizona has yet to be exploited fully, and coalbed methane basins contin-

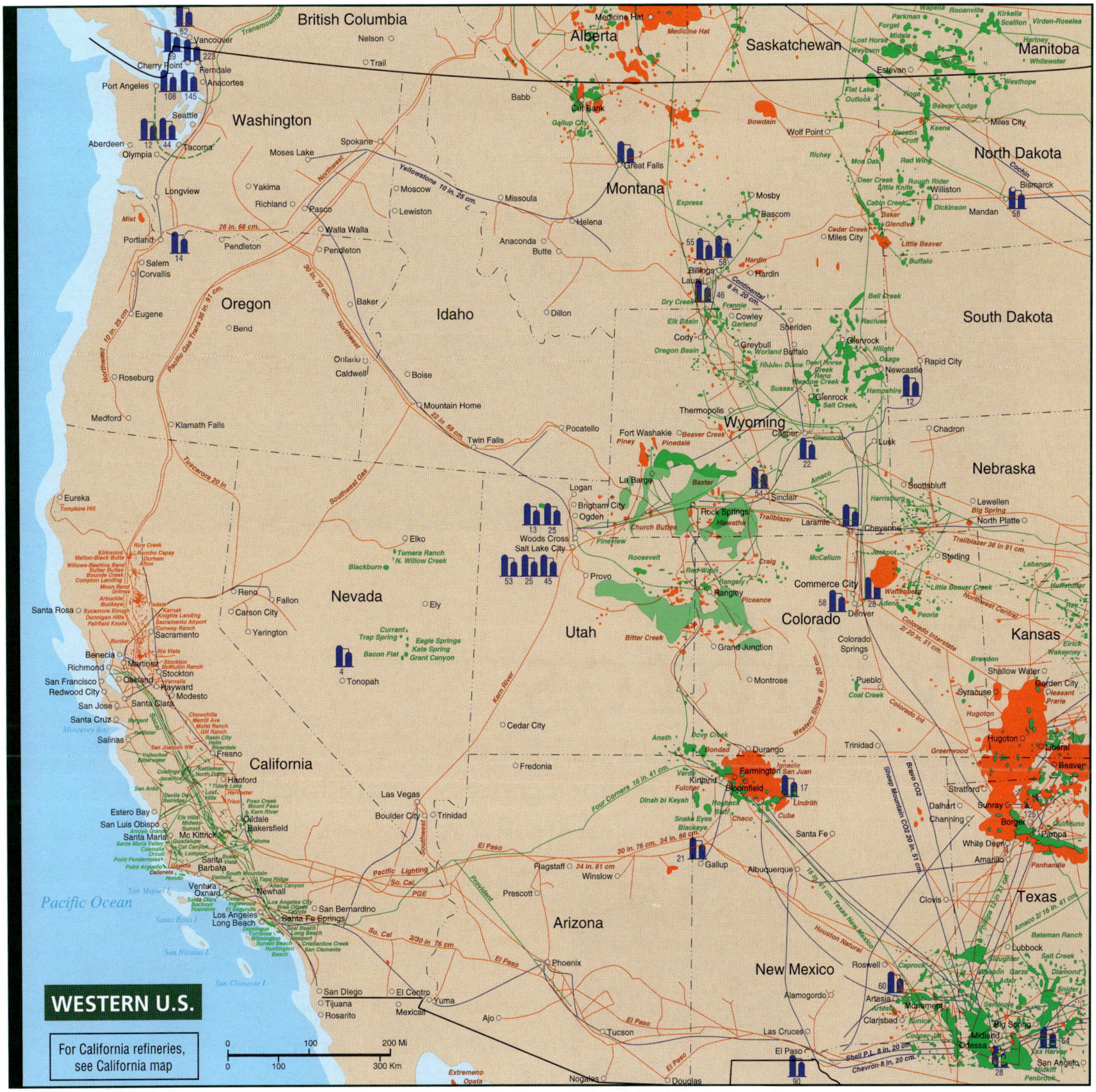

ue to hold the potential for significant gas reserves. Reducing federal restrictions on drilling in the Rocky Mountain states could expand available US gas supplies by 87 tcf, or 7%, said DOE.

In the case of the Rockies, eight fields could each contain more than 100 million bbl of oil or 1 tcf of gas remaining to be developed, qualifying them as "giants." A renaissance of development in the Greater Rocky Mountain Region was being fueled largely by the application of new technology to previous discoveries and by the merger of isolated coalbed methane deposits into giant fields, as well as the development of central basin gas deposits and their eventual merging into giant fields.

The largest oil field found in the Greater Rocky Mountain Region in the last 10 years is Cedar Hills, in the Ordovician Red River "B" Zone in the Williston basin, and its companion, East Lookout Butte. This field has been developed with horizontal drilling of an extremely thin reservoir (less than 10 feet thick). Some 669 horizontal wells have been drilled, resulting in 626 productive completions ranging in true vertical depth from 8,800 to 9,500 feet. Ultimate recovery was estimated to exceed 130 million bbl of oil.

Also in the Rocky region is Jonah, an overpressured gas field in the Green River basin with at least 2.5 tcf recoverable, and the Greater Cave Gulch gas and

condensate field in the Wind River basin. The region also contains coalbed methane fields in the Powder River and western Uinta basins and low-permeability gas sands in the Green River and Piceance basins.

A high level of successful drilling in the Jonah field and on the Pinedale anticline was reported by Ultra Petroleum Corp. The company participated in 16 Pinedale and 8 Jonah wells, all successful. Wells drilled during 2001 appeared to be capable of 10 MMcfd/well and contained an estimated 8 bcfe/well of reserves.

Permian tight gas play, other drilling

A handful of major and independent producers in West Texas began tapping tight natural gas deposits in 2001 using horizontal drilling technology. Activity in the Montoya and Devonian formations of Pecos and Reeves Counties could shape up as a potential boom in the Permian basin. Drilling also extended north into Ward County, northeast into Upton County, and into Terrell County to the southeast. Only a few wells were drilled or even producing in 2001, but the deep, tight formations were challenging current drilling technology. Some wells exceeded 14,000 feet, with horizontal laterals of as much as 5,000 feet that required stimulation.

South of the Rockies, the Arizona Geological Survey developed data on another eight areas with fair to excellent potential for oil, natural gas, coalbed methane, and other resources. Drilling density in Arizona has averaged about one well per 100 square miles and, in some basins, as little as one well per 250-395 square miles. All of the state's current production was from Navajo Indian land in the Paradox basin of northeastern Arizona, where the potential for discovery was considered good to very good.

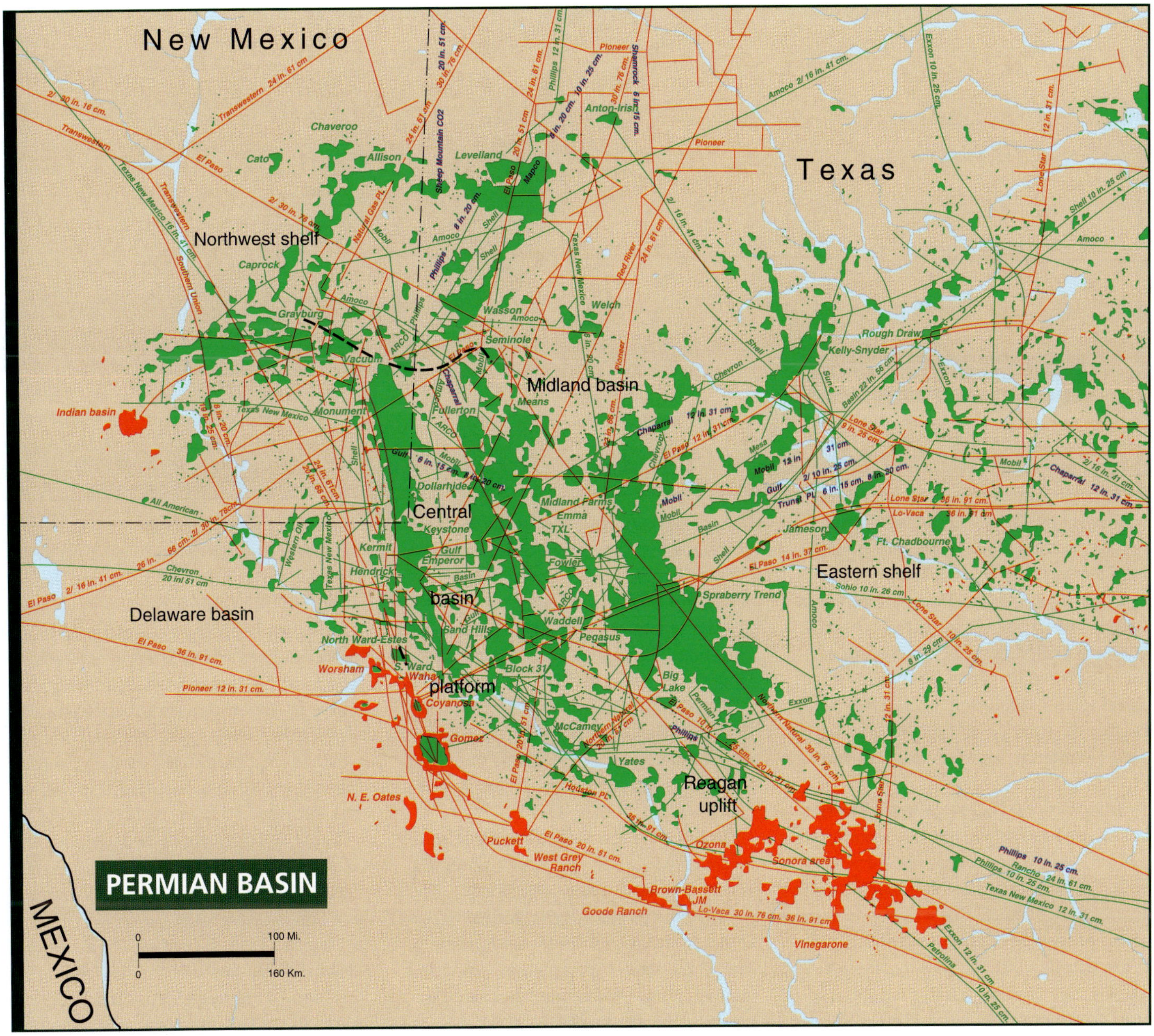

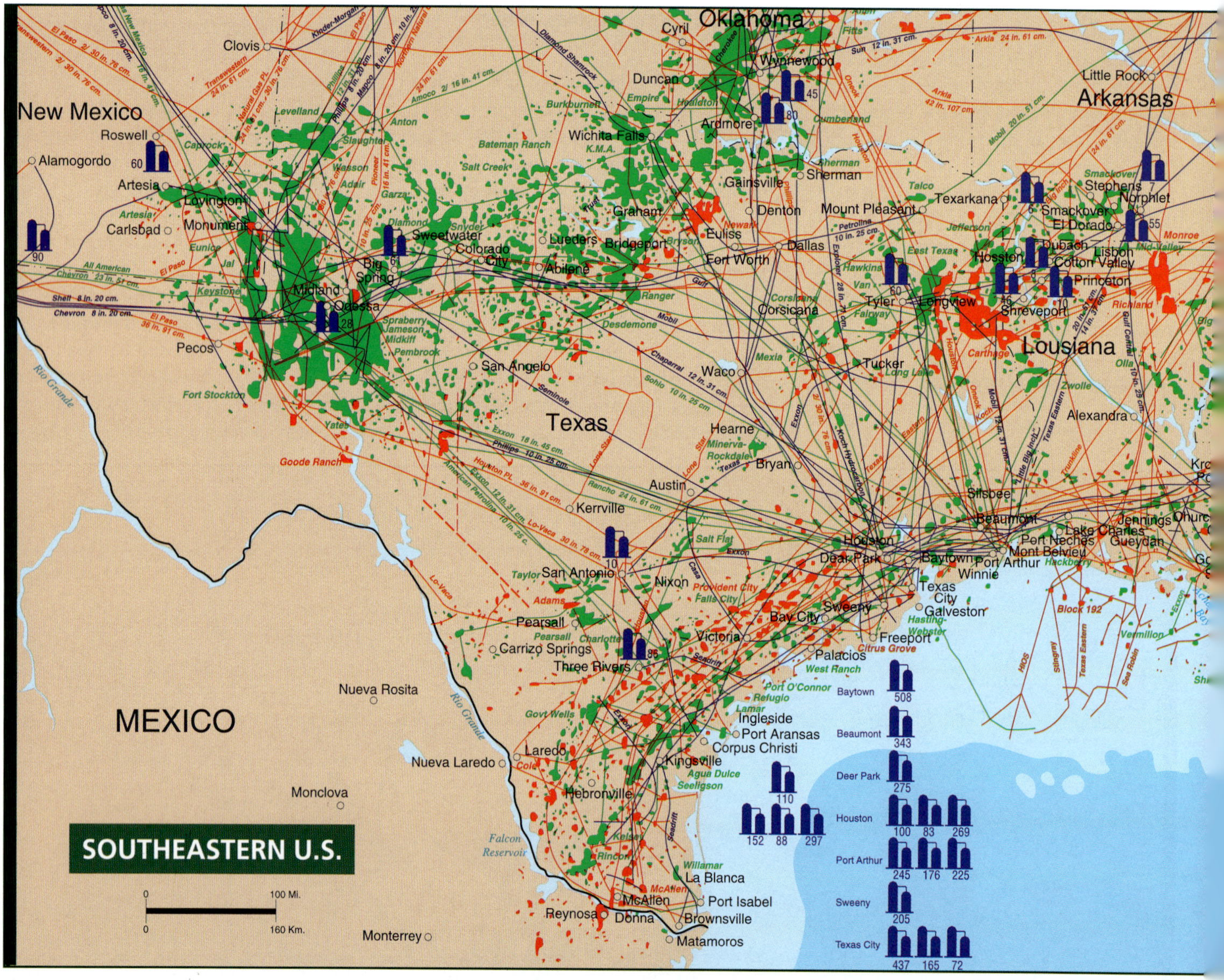

Nearby is the Black Mesa basin containing Cretaceous strata with very good to excellent potential for coalbed methane.

New Mexico's late-Paleozoic elevator basins, which were undrilled or only sparsely drilled, were also receiving new attention from geologists who have developed exploration tools that help predict the location and quality of associated petroleum source rocks and reservoirs.

Mississippi's venerable Black Warrior basin also has the potential for significant gas reserves in Cambro-Ordovician rocks. At Maben field, TotalFinaElf's unit Fina Oil & Chemical Co. was involved in an intensive development drilling program during 2001, expanding the field's areal extent and increasing gas production dramatically from the "Knox-Ordovician" (also called Stones River dolostone or "Snow Zone"). At year-end 2000 with production from three new wells, the field was yielding more than 500,000 Mcf/month, up from about 3,000 Mcf/month in early 1998. In March 2001, Fina drilled a Knox-Ordovician test well with a total depth of 16,640 feet. Additionally, Fina permitted two new development drilling locations to extend the gas production seen at the 1 Love Heirs well.

In an Oklahoma well, a stackable splitter allowed access to three individual zones, reported Baker Oil Tools. The field produces gas from a vertically oriented, multilayered subthrust formation. The three-lateral completion provided greater initial production rates than single borehole producers.

Gas production began in February 2001 from a well in the East Lost Hills area of California's San Joaquin basin, reported Berkley Petroleum Corp. During startup, flow was restricted to 9.6 MMcfd of gas and 475 b/d of condensate at 13,300 psi flowing wellhead pressure. Flow of 20 MMcfd was expected, and three other wells were planned for the area. Aera Energy and partner Ivanhoe Energy spudded a deep-gas test well on the Northwest Lost Hills prospect in August 2001. Nabors was drilling the well, with a total depth of about 20,000 feet, in an orchard in California's southern San Joaquin basin.

On the Texas coast in Goliad County, Esenjay Exploration began developing its 2 Pereira Children's Trust discovery in early 2001. The discovery was drilled on

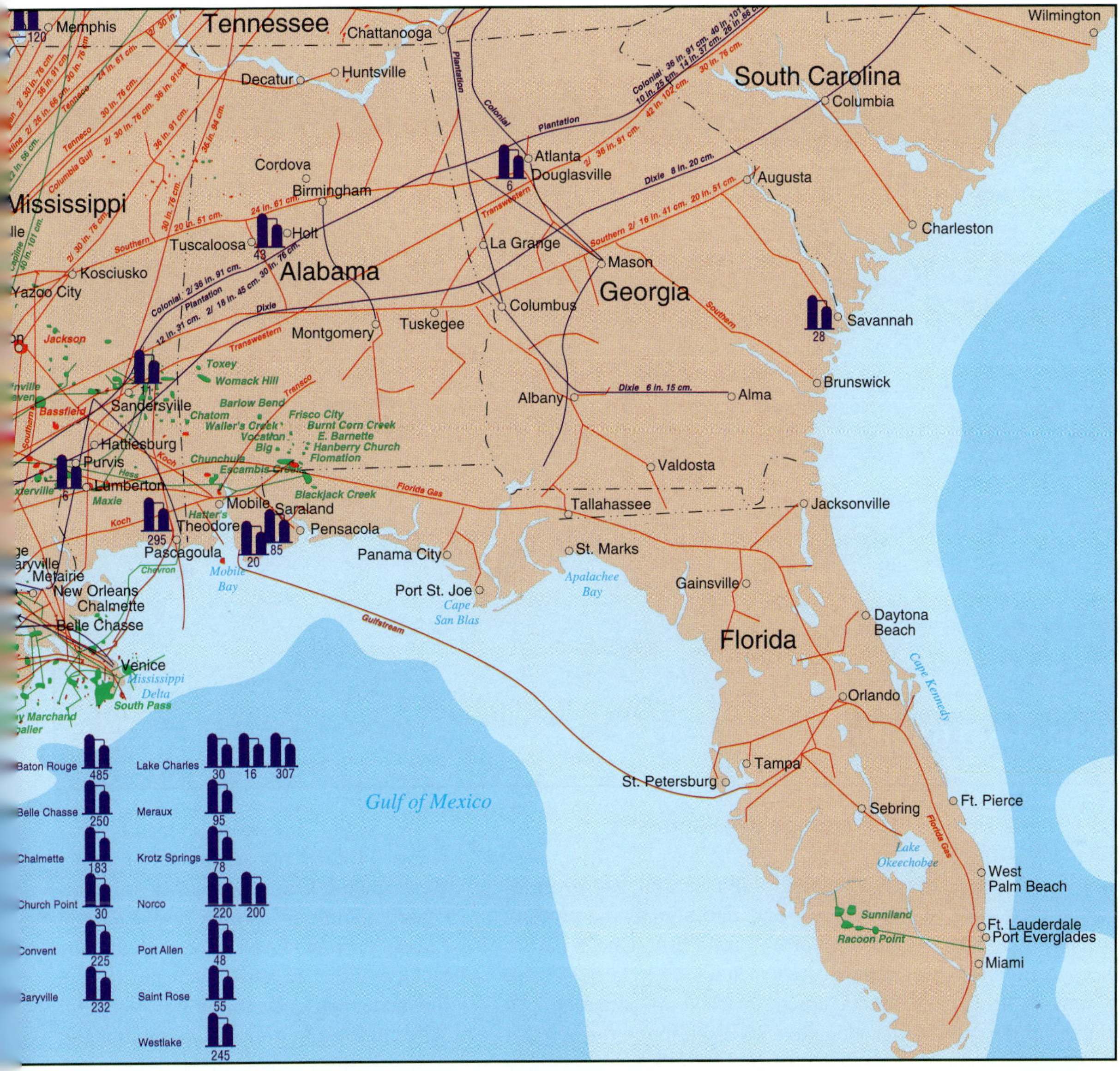

ity to 7.8 bcf from 5 bcf. Existing facilities at Cove Point included four storage tanks, three 8.45-MW gas turbine generators, an offshore LNG receiving pier with two unloading docks, and an 87-mile pipeline from the terminal to an interconnection with facilities owned by Columbia Gas Co. Plans were to build a fifth storage tank to hold 850,000 bbl of LNG, scheduled for service in 2003, bringing Cove Point's sendout capacity to 1 bcfd.

Other LNG projects in the US also gained momentum during 2001 as lower regasification costs and better transportation methods made LNG imports more attractive economically. Work was under way at Cabot Corp.'s Everett, MA, receiving terminal to double vaporization capacity to more than 1 bcfd

a faulted anticline that could cover 1,000 acres. The first well flowed 17.6 MMcfd of natural gas from a pay zone in the Middle Wilcox sands.

Nine tight-gas wells were restimulated at three sites to evaluate techniques for identifying the best candidates for such treatment. Tests were conducted in the Green River basin's Frontier formation, Piceance basin's Williams Fork formation, and East Texas basin's Cotton Valley sandstone. The treatments increased incremental reserves by 2.9 bcf at an average cost of $0.26/Mcf, reported Advanced Resources International and the Gas Technology Institute.

LNG in the spotlight

The US FERC gave the green light in late 2001 to Williams Cos.' plans to reopen Maryland's Cove Point LNG terminal in 2002 and expand its storage capac-

from 435 MMcfd. Also on the East Coast, El Paso Corp. was planning to reactivate

LNG tanker en route to CMS Trunkline LNG's Lake Charles, LA, terminal, which was being modified during 2001 to increase peak sendout capacity to 1 bcfd.

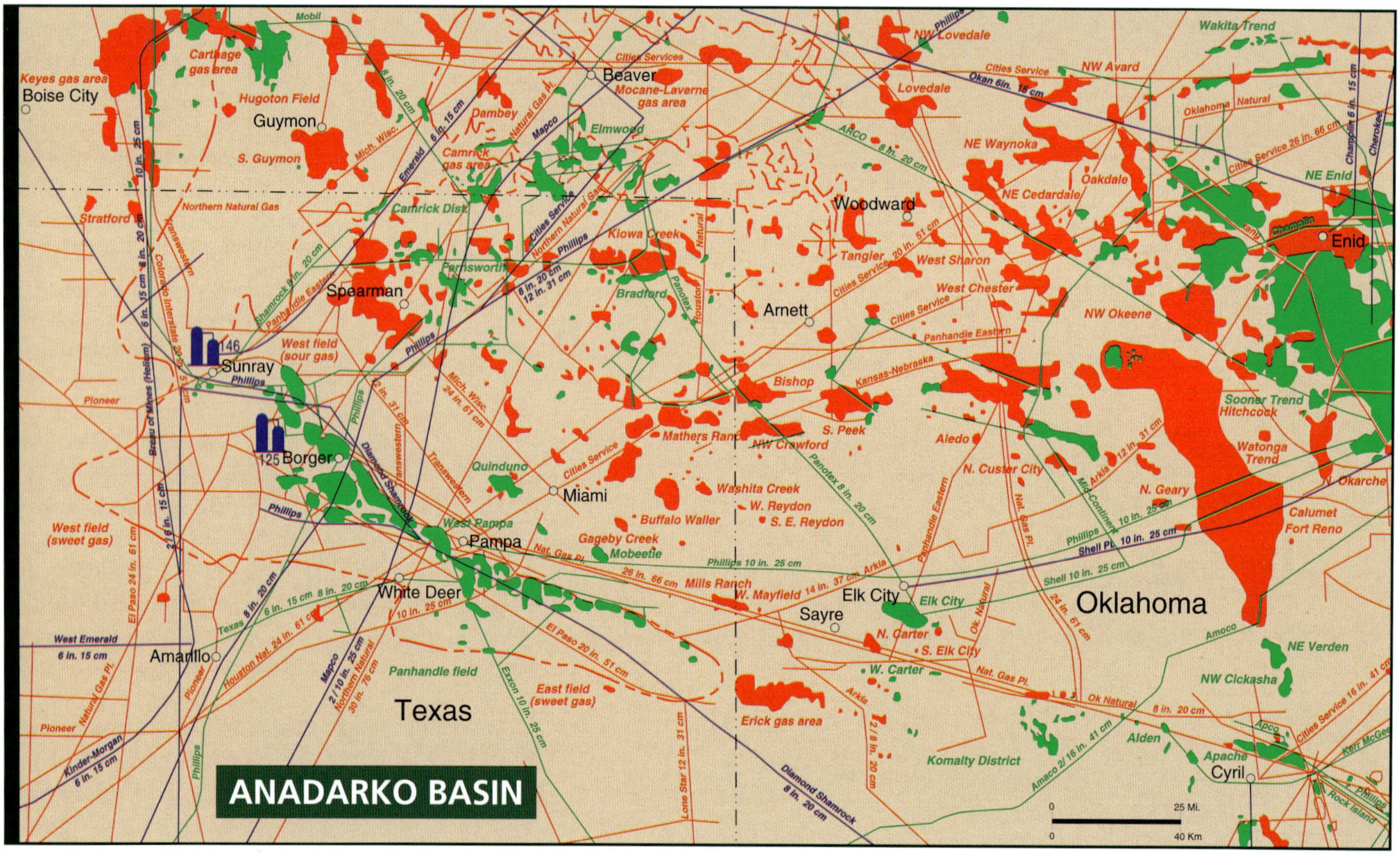

its Elba Island LNG terminal off Savannah, GA, to receive LNG from Trinidad. Southern LNG Inc. was getting quotes to add 80% capacity, increasing sendout to 440 MMcfd in 2003 and to 800 MMcfd after expansion is completed. Shell Gas & Power, a unit of Royal Dutch/Shell Group, acquired rights to all the additional storage capacity being planned.

On the Gulf of Mexico, CMS Trunkline LNG, a unit of CMS Energy Corp., was expanding peak sendout capacity of its Lake Charles, LA, terminal from 700 MMcfd to 1 bcfd. Modifications would include eliminating bottlenecks in the regasification process, said CMS. Also on the Gulf Coast, ChevronTexaco was considering building a new LNG terminal with an initial 1-bcfd capacity expandable to 2 bcfd, and Chenier Energy, Inc., announced its intent to develop three terminals on the Texas Gulf Coast, including one on Quintana Island near the Port of Freeport and one at the Port of Brownsville. Several West Coast sites in California and Mexico were also being considered for new LNG import facilities.

Gas storage developments

Unusually high natural gas inventories at year-end 2001 signalled a production dip for 2002, said EIA. Due in part to abnormally warm winter in late 2001, US gas in storage reached an estimated 2.86 tcf, which was 1.1 tcf above the level at year-end 2000 and 19% above the average for the previous five years. Earlier, EIA also announced that it will begin a weekly survey of gas storage to supplement and then replace the the one previously published by A.G.A., which will be discontinued.

An uncontrolled escape of natural gas and brine from an underground salt storage cavern occurred in January 2001, killing two residents of Hutchinson, KS. The eruptions resulted from the loss of 73 MMcf of gas from the Yaggy storage facility 7 miles away. The gas likely escaped through a fist-size hole in the casing of an injection well that was drilled out in 1992 to convert the former propane storage facility into a gas storage unit. Regulators of Kansas and other states containing hydrocarbon storage facilities began to re-examine existing rules.

A natural gas storage facility was being built in northern California by Western Hub Properties, which was purchased jointly by Aquila and ArcLight Energy Partners Fund. The Lodi storage facility, one of two merchant facilities in the state, was scheduled for year-end 2001 completion and will serve demand from gas-fired power plants coming on line in 2002 and 2003. It can store 12 bcf of natural gas with a planned injection rate of 400 MMcfd and withdrawal rate of 500 MMcfd, allowing it to cycle its inventory up to six times a year, compared with cycling rates of one or two times per year for most existing storage facilities. Independent power producer Calpine entered into a long-term firm agreement for daily deliverability at Lodi equal to about 20% of Calpine's western region peak-day gas requirements in 2002.

Capital spending and earnings

During the first quarter of 2001, strong oil and gas prices continued to fuel healthy capital spending on the heels of year-2000's record growth in exploration and production expenditures. Salomon

Smith Barney projected that worldwide exploration and production outlays would sustain near-record growth in 2001. "This increase in E&P spending marks a period of transition for the oil services sector from a state of recovery to one of steady growth."

The key factors driving E&P spending shifted during 2001. "For the past two years, cash flow was the most often cited factor. However, due to strong 1999-2000 commodity prices, most E&P companies are less concerned with their balance sheets," said Salomon. "Moreover, oil field fundamentals have improved sharply, to the point that most respondents feel confident in the sustainability of healthy oil and gas prices. Thus, we believe the focus has clearly shifted from the boardroom to the field, as operators actively seek out attractive drilling opportunities." This shift would seem to indicate another growing trend, Salomon pointed out, namely "the desire to maximize investment returns, as opposed to the more-traditional practice of maximizing investment."

Many companies made gains through 2001 in both earnings and production volumes, and capital markets responded favorably as debt was shed. However, the US petroleum industry remained cautious with its spending plans, and its long-term price forecasts were lower on average than short-term projections, especially for the major oil firms. Salomon said that this conservative outlook is due first to the industry's recognition of historical oil prices, which have traded between $18 and $20/bbl most of the time for the past 20 years.

"Moreover, after the sharp decline in oil prices during late 1998-early 1999, it would appear that E&P companies are reluctant to assume oil prices anywhere near today's historically high levels. Finally, most oil and gas projects are profitable at oil prices around $20/bbl, so higher prices are often viewed simply as 'gravy'."

In late 2001, ample energy supplies and weakening demand sent commodity prices lower, and earnings of oil and gas producing companies began to falter. Most integrated and independent E&P companies saw earnings decline by more than 35% during the third quarter compared to the same period in 2000. However, service and supply companies' earnings were up 66%, and downstream earnings improved on stronger refining and marketing margins.

The consolidation trend during 2001 shrank the OGJ 200 list of publicly traded US oil and gas producers to 197, but each company's earnings were much higher compared with 2000. Large mergers in 2001 included Chevron Corp. and Texaco Inc. (ChevronTexaco); the acquisition of Mitchell Energy & Development Corp. by Devon Energy Corp., which formed the second-largest independent natural gas producing company in the US.; and Phillips Petroleum Co.'s acquisition of Tosco Corp., creating America's second-largest refiner.

Additionally, in late 2001, Phillips and Conoco Inc. announced plans for a merger that would create a $35 billion global energy company (ConocoPhillips), which would be the third-largest integrated energy company in the US (based on market capitalization, oil and gas reserves, and production), sixth-largest energy company worldwide (based on hydrocarbon reserves), and fifth-largest global refiner. However, the merger was threatened by a bid for Conoco from Europe's TotalFinaElf SA.

Also, Devon Energy planned to purchase Calgary-based Anderson Exploration Ltd., giving Devon the highest North American oil and gas reserves of any independent. And Magnum Hunter Resources Inc. planned to merge with Prize Energy Corp. to form a $1.2 billion independent oil and gas company. IPE

Deepwater and Ultra-Deepwater Finds Rule Gulf of Mexico Plays

Ultra-deepwater plays in the Gulf of Mexico gained prominence through 2001, with large discoveries at water depths of 5,000 ft or more announced in the last two years. Previous records show that industry finds containing more than 500 million boe amounted to only nine fields. Discoveries in 2000 and 2001 in ultra-deepwater have added three or more giant fields to the roster of nine discoveries made before 1996.

The number of deepwater wells drilled in 2000 rose 35% to 40, up from 26 in 1999. Seven wells were drilled at depths of 5,000 ft or more; three were at depths exceeding 7,500 ft. Analysts estimated that deepwater finds could run as high as 40 billion boe, equaling the cumulative discovered reserves on the Gulf shelf. Twenty-five percent, or 10 billion boe, have been found to date.

Oil production from deepwater wells exceeds that from shallow areas. By late 2000, the former rose to over 750,000 b/d, while the latter dropped to around 700,000 b/d. Gas output from deepwater wells reached 2.3 bcfd by 2000. Production from wells in water depths over 1,500 ft is expected to rise during the early half of the decade as more wells come on stream.

Through the third quarter 2001, contractors had drilled 96 wells: 71 in deepwater areas, 19 in ultra-deepwater, and 6 at depths greater than 7,000 ft. Despite lower activity in shallow water due, in part, to depressed natural gas prices, the longer lead times in deepwater were not expected to significantly affect drilling activity, which was expected to remain steady. If drilling continues at its current pace, the industry could drill up to 120 wells in deepwater by the end of the year.

Exploitation of deepwater sources made up more than 50% of production in the Gulf of Mexico. This amounted to a 150% increase over production levels of less than 20% five years ago. Analysts believe that production could rise to one third of total US oil extracted by 2010, meeting more than 10% of total US supply.

Four ultra-deepwater play types identified

Geologists identified four major ultra-deepwater play types. Three occur in the Eastern Gulf thrust belt, Atwater fold belt, and the Perdido fold belt. In addition, turtle structures, found at BP's Crazy Horse reservoir on Mississippi Canyon Block 778, drew a great deal of attention. Excellent reservoir characteristics in this area implied that outstanding well productivity could be expected. Similar structures reportedly occur at Chevron's Poseidon prospect on Mississippi Canyon 727 (six blocks west of Crazy Horse), and Shell was drilling a deep test on Mississippi Canyon 687 (2½ blocks north of Crazy Horse North) targeting what was believed to be a Miocene-aged turtle structure.

Other postulated turtle structures included those imaged beneath Shell's Crosby discovery (formerly Flathead), on Mississippi Canyon 899 and below Ursa at Mississippi Canyon 809, and BP's Blind Faith discovery 20 miles northeast of Crazy Horse on Mississippi Canyon 696, as well as Dominion Exploration & Production Inc.'s Devil's Tower appraisal on Mississippi Canyon 773.

In addition to mapping 7,000 acres

Ultradeepwater Gulf of Mexico Discoveries

Discovery	Blocks	Estimated gross recoverable reserves, MMboe	Working interest ownership, %
Crazy Horse	Mississippi Canyon 778, 822	1,000+	BP 75, ExxonMobil 25
C. Horse N.	Mississippi Canyon 776	500+	BP 75, ExxonMobil 25
Atlantis	Green Canyon 698-700, 742-744	400-800	BP 56, BHP 44
Mad Dog	Green Canyon 738, 739, 781-783, 825-827	200-450	BP 60.5, BHP 23.9, Unocal 15.6
Holstein	Green Canyon 644, 645	200+	BP 50, Shell 50
K2/Timon	Green Canyon 562, 563	100	BP 39, Chevron 22.5, ENI 13.5, Unocal 12.5, Conoco 12.5
Neptune	Atwater Valley 574, 575	100-200	BP 57.5, BHP 42.5
Champlain	Atwater Valley 63, 64	100+	Texaco 75, ENI 25
Trident	Alaminos Canyon 903, 947	340-800	Unocal 59.5, Chevron 15, Ocean Energy 12.75, AGIP 8.55, Phillips 4.2
Total		**2,940-4,150**	

Sources: Industry publications, company reports, and Petrie Parkman estimates

ON OUR WALLS YOU'LL FIND OVER 6000 PATENTS.
IN OUR HALLS, PEOPLE FROM OVER 100 COUNTRIES.
IN OUR MINDS, A CONNECTION.
The connection is this: diversity equals success. Real, bottom-line success. For our clients. And ourselves. At least, it has been our experience that great ideas can come from anyone, anywhere. And for the world's leading technical services company, the best ideas are everything.
From oilfield services to smart cards to IT systems integration, and real-time energy management systems, Schlumberger is only as good as its next idea. Which explains why we've found ideas like diversity to be very good indeed.
www.slb.com
Schlumberger

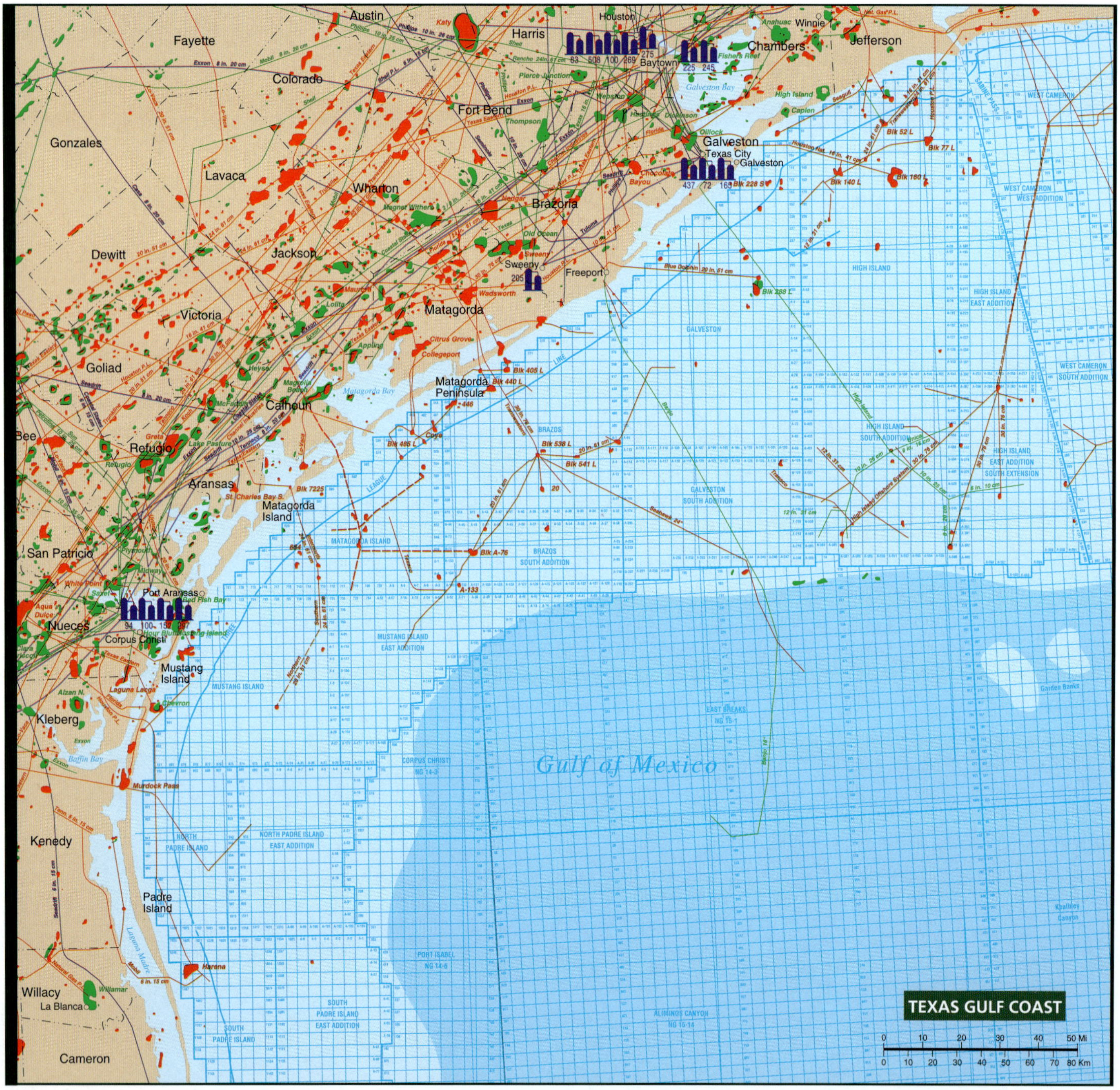

with anticipated pay intervals below 21,000 ft at its Eiger Sanction prospect on Mississippi Canyon 667 and three adjacent blocks, Anadarko Petroleum Corp. was drilling an exploratory well on the prospect. ExxonMobil Corp. was pursuing its Hawkes prospect on Mississippi Canyon 465, 508, and 509 and at Timber Wolf on Mississippi Canyon 555. Murphy Oil Corp. planned to drill its Sport of Kings prospect on the block north of Crazy Horse in 2002.

Few results surfaced on the Eastern Gulf thrust belt play. Exploratory wells at Gamera (Atwater Valley 118), Wind River (Atwater Valley 378), Phoenix (Mississippi Canyon 1002), and Showboat (Atwater Valley 336) prospects did not reveal commercial hydrocarbon deposits.

Discovered resources in the Atwater (Eastern) fold belt amount to 0.9-1.7 billion boe in the southeastern corner of Green Canyon. This area was the focus of activity following a series of large discoveries, notably Mad Dog and Atlantis. BP with co-owner BHP drilled three successful appraisal wells in the Mad Dog field (Green Canyon Blocks 738, 739, 781, 782, 783, 825, 826, and 827) and announced a revised reserve estimate of 200-450 million boe.

BP planned to start production from Atlantis on Green Canyon 699 in 2005. Production from its Holstein discovery on Green Canyon 644-645 could begin in 2004. Marathon Oil Co. had plans to drill Redwood on Green Canyon 1001 and the Flathead prospect west of Redwood in southern Green Canyon. Another prospective area was west of the Atwater fold belt in the East Breaks and Alaminos Canyon. Shell drilled 2 BAHA on Alaminos Canyon 557 in 2001, but released no data. Unocal drilled the first successful well in the Perdido fold belt, encountering more than 300 ft of net pay in its Trident well on Alaminos Canyon 903 in 9,687 ft of water.

Brisk bidding at lease sales

In late March 2001, a US Minerals Management Service (MMS) central gulf lease sale generated considerable attention. Royalty relief incentives stirred interest in drilling for deep gas in shallow waters as well as the lure of deepwater exploration. Blocks in the central gulf planning area attracted 780 bids from 90 companies netting $663,406,963 for 547 tracts. Bids at this sale exceeded 2000's central gulf sale by 60%. Independent oil and gas companies proved to be strong bidders at the sale.

ExxonMobil submitted the highest bid for a single tract, offering $26 million for deepwater Mississippi Canyon Block 912. Kerr-McGee Oil & Gas Corp. came in with the second highest bid, which was for Green Canyon Block 680 for $18.4 million. Mississippi Canyon Block 956 drew the greatest interest with six bids. Exxon, the third highest bidder at the sale, offered $18.3 million for the tract. Acreage in less than 660 ft (200 m) of water garnered the highest amount of total bids at $227.7 million; $171.9 million was bid on ultra-deepwater tracts in more than 5,250 ft (1,600 m) of water. Deepwater blocks drew the highest dollar amount per acre, averaging $281/acre over an average $191/acre at a previous central gulf sale held in 1998. Analysts suggested that results of this sale show that companies generally are financially sound, focusing on deepwater. Others seemed to be searching for deeper shelf prospects, particularly gas prospects in shallow water.

August 22, 2001, saw a large number of bids for predominantly deepwater tracts at a lease sale of US federal oil and natural gas leases in the western Gulf of

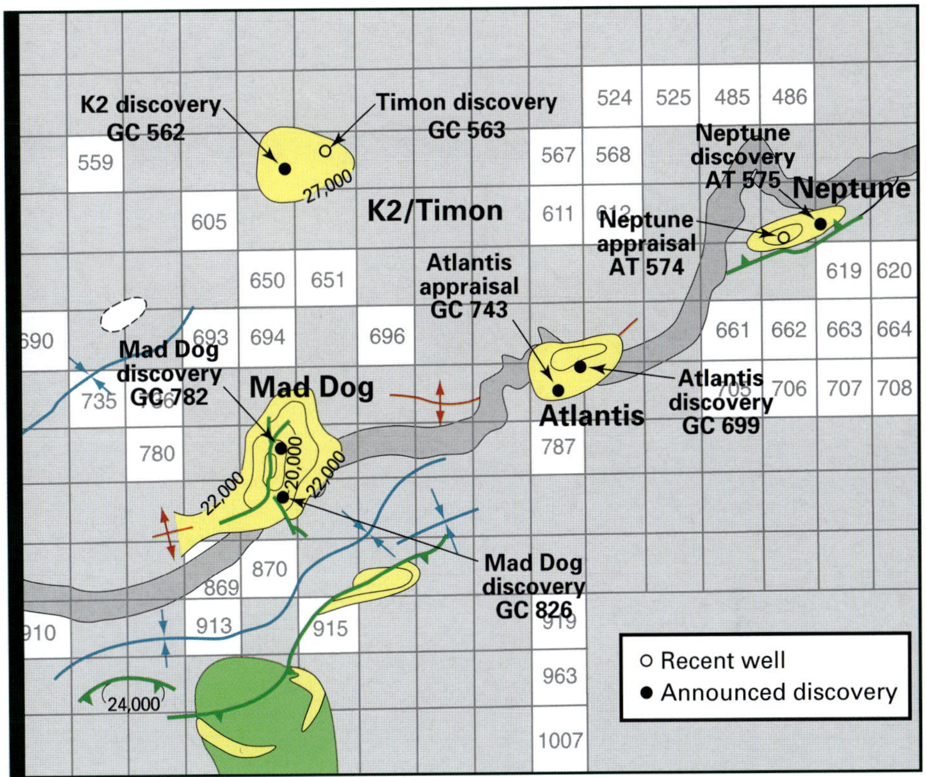

Operators in the Gulf of Mexico planned further development of the Green Canyon area's Mad Dog and Atlantis gas fields, including the Cleopatra gathering system to be built by BP.

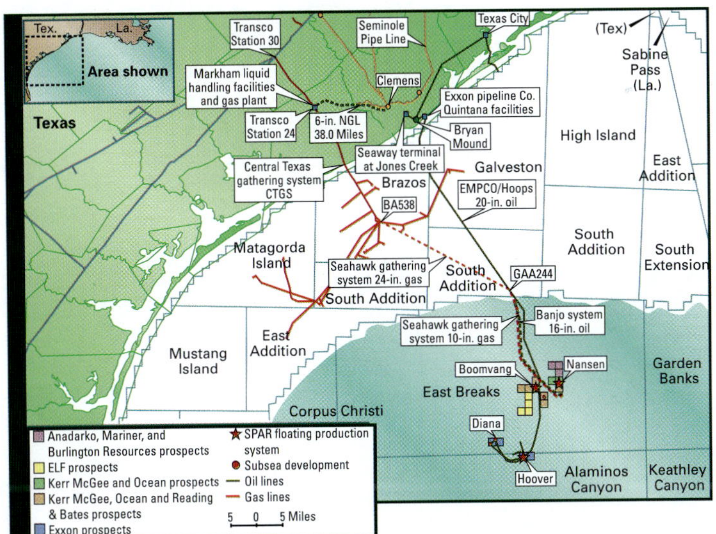

The East Breaks pipeline system, including the Seahawk gas line and Banjo oil line, was being completed in late 2001 by Williams Cos. Inc. and Stolt Offshore Inc.

Coflexip Stena Offshore was laying pipe for the East Breaks system in 3,700-foot water, and its CSO Deep Blue was the first vessel in the world to reel 18-inch OD rigid steel pipe. (Photo by Gary Langhammer, from Williams Cos. Inc.)

Texas
Louisiana
Beauregard
Allen
St. Landry
Calcasieu
Jefferson Davis
Acadia
Lake Charles
Jennings
Churchpoint
Lafayette
St. Martin
Iberville
Ascension
Livingston
Baton Rouge
Port Allen
Krotz Springs
St. John The Baptist
St. James
Convent
Reserve
Cameron
Vermilion
Abbeville
IBERIA
New Iberia
Assumption
St. Martin
St. Mary
Morgan City
Houma
Terrebonne
Port Neches
Port Arthur
Sabine Lake
Calcasieu Lake
Grand Lake
White Lake
Holly Beach
WEST CAMERON
EAST CAMERON
VERMILION
SOUTH MARSH ISLAND
EUGENE ISLAND
SHIP SHOAL
SOUTH PELTO
HIGH ISLAND EAST ADDITION
HIGH ISLAND EAST ADDITION SOUTH ADDITION
WEST CAMERON WEST ADDITION
WEST CAMERON SOUTH ADDITION
EAST CAMERON SOUTH ADDITION
VERMILION SOUTH ADDITION
SOUTH MARSH ISLAND SOUTH ADDITION
EUGENE ISLAND SOUTH ADDITION
SHIP SHOAL SOUTH ADDITION
SOUTH MARSH ISLAND NORTH ADDITION
GARDEN BANKS NG 15-2
GREEN CANYON NG 15-3
Serrano
Oragano
Magnolia
Vermilion 380
Vermilion 397
East Cameron 359
West Cameron 551
West Cameron 479
West Cameron 264
West Cameron 65
Ship Shoal 320
Tarpon
S. Marsh Is. 257
S. Marsh Is. 255
Blk 45
Blk 65
Blk 149
Blk 180
Blk 83
Blk 192
Blk 89
Blk 76
Blk 71
Blk 131
Blk 164
Blk 260
Blk 464
Blk 23
Blk 38
Blk 48
Blk 66
Blk 175
Blk 188
Blk 198
Blk 205
Blk 32
Blk 28
Blk 107
Blk 169
Blk 208
Blk 222
Blk 276
Blk 286
Blk 330
Blk 6
Blk 100
224
236
294
310
494
Tiger Shoal
Rabbit Island
Atchafalaya Bay
Vermilion Bay
Lake Sand
Lac Blanc
Recan Island
Weeks Island
Avery Island
Cote Blanche Island
Buck Point
N. Fresh Water Bayou
Transcontinental
Texas Eastern
Tennessee
Tenneco
Stingray
Trunkline
Michigan Wisconsin
Bluewater
Sea Robin
High Island Offshore
Auger Garden Banks
Koch
Shell
Mobil
Columbia Gas
Cities Service
Texas Eastern 30 in. 76 cm
Stingray 36 in. 91 cm
High Island Offshore 42 in. 107 cm
Bluewater 36 in. 91 cm
20 in. 51 cm

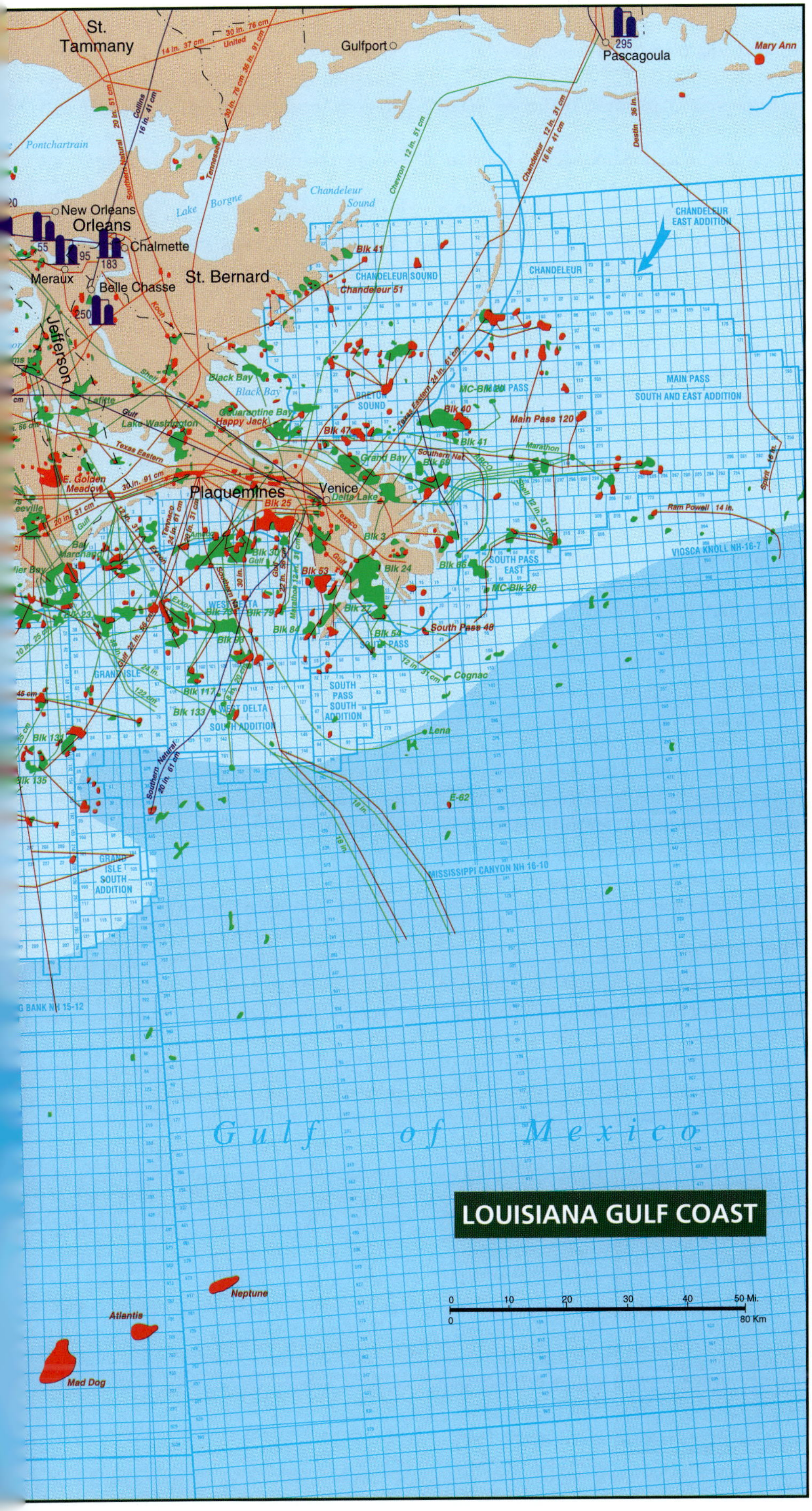

Mexico. Fifty companies submitted 386 bids, totaling approximately $190 million. Independents dominated the sale, which drew $165.6 million in high bids for 320 tracts of the 4,114 offered.

Top bidders were Kerr-McGee (42 bids for $31.9 million), Spinnaker Exploration Co. (32 bids for $17.4 million), and Petrobras America Inc. (30 bids for $16.9 million). One hundred twenty–five bids were placed for blocks in 0-199 meters of water, 18 for blocks in 200-399 m, 41 for blocks in 400-799 m, 93 for blocks in 800-1,599 m and 43 for blocks in 1,600 m or more. The highest bids went for blocks in the Garden Banks area in up to 800 m of water. Kerr-McGee ranks as the top US-based independent leaseholder in the Gulf overall as well as in the deepwater Gulf.

Oil companies bid $458.9 million for 95 of 233 tracts offered in a December lease sale in federal waters in the eastern Gulf of Mexico. Fourteen companies placed 190 bids on tracts in 1,600 m of water or deeper, subject to a 10-yr lease term at 12.5% royalty. Lloyd Ridge Block 399 and De Soto Canyon Blocks 180, 224, 269, 445, and 620 received the greatest number of offers. The sale area is about 100 miles from the Alabama-Florida state line and 285 miles from Tampa, FL. Estimates place undiscovered resources in at area at 1.25 tcf of gas and 185 million bbl of oil.

West Florida shelf and slope

An MMS study set recoverable petroleum resources on the West Florida shelf and slope and on part of the Mississippi-Alabama shelf at more than 591 billion bbl of oil and 7.1 tcf of gas. To date, the 80,000 square miles of state and federal areas covered by less than 3,300 ft of water have been only marginally explored for oil and gas. Targeting these areas for increased oil and gas exploration is in line with the Bush administration's energy plan calling for increased domestic production. However, the Florida state government maintained its policy to oppose activity within 100 miles off its shore.

The MMS recognizes four partly defined plays with potential recoverable resources. The Norphlet play contains a possible 591 million bbl of oil and 7.1 tcf of gas in a prospective, nearshore belt on the northern West Florida shelf from Alabama to Apalachicola. The Smackover play in the same belt houses an estimated 29 million bbl of oil and 48 bcf of gas.

At the perimeter of a basin on the southern West Florida shelf, the Bone Island-Pumpkin Bay-Brown dolomite-Sunniland plays have a potential 419 million bbl of oil and 31 bcf of gas. These same plays in the center of the basin could contain as much as 587 million bbl of oil and 42 bcf of gas. Based on the regional distribution of the lower Smackover oil source from onshore Alabama to Apalachicola, the MMS assessment assumed that this source continues southward from Apalachicola. It delineated areas with dry gas, condensate, and light-to-medium oil potential in Norphlet and possible Smackover plays on the northern and central West Florida shelf.

Garden Banks developments

Anadarko and BP Exploration & Production Inc. entered into a joint venture to explore 95 deepwater blocks in the Garden Banks and Keathley Canyon areas of the central Gulf of Mexico. Located in water depths ranging from 3,000 to 6,000 ft, the 95 blocks make up an area known as South Auger and are part of a larger 640-block area. The deal called for BP to assign Anadarko its rights in eight blocks in the Green Canyon near Anadarko's Marco Polo discovery in Block 608. BP holds 100% of the 95 South Auger blocks; Anadarko gained the option of earning 33-66% working interest in these blocks.

Conoco Inc. planned to invest $600 million to develop Magnolia field in 4,700 ft of water. Magnolia resides on Garden Banks Blocks 783 and 784, 180 miles south of Cameron, LA. Conoco and its partner Ocean Energy Inc. (25%) expected the field's production to reach 150 million boe. Plans called for converting three existing holes to production wells and drilling five more to bring the field to final production. The partners were considering either a tension-leg platform (TLP) or a spar for installation in summer 2004.

Shell Offshore Inc. was overseeing modification of its Mensa subsea development system to meet flow assurance requirements at its Serrano-Oregano developments. Serrano field is in Garden Banks Blocks 516 and 472 in 3,400 ft of water; Oregano is in Garden Banks Blocks 558 and 559 about 225 miles southwest of New Orleans in 3,400 ft of water. The development system includes two wells connected to a subsea flowline sled, in 2,860 ft of water, that is tied back to the Auger TLP by a single pipe-in-pipe flowline. After production startup in October 2001, Shell expected full production at 20,000 b/d in 2002 with recovery reaching 50 million boe.

Pipeline work

Late September 2001 saw on schedule installation of the shelf portion of the East Breaks pipelay project off Texas. Export pipelines will gather gas production from the deepwater Boomvang-Nansen Seahawk spars in 3,700 ft of water. Stolt Offshore Inc. installed 58 miles of 24-inch pipe from Brazos Block 538 in 97 ft of water to Galveston Block A244 in 365 ft of water. Later in November, Stolt completed riser tie-ins at the existing Brazos Block 538 platform and at Galveston Block A244 following platform installation. The Seahawk gas gathering system incorporates the deepwater and shelf installations. Target startup for the entire production system was year-end 2001.

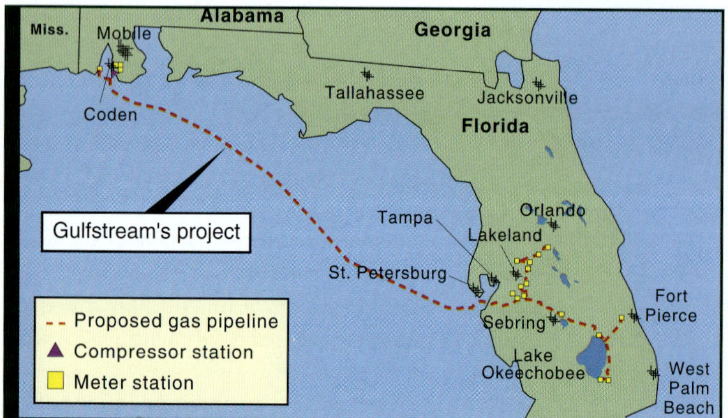

The 753-mile Gulfstream natural gas system began construction in 2001 and was scheduled to begin gas deliveries in 2002.

The Gulfstream natural gas system, a $1.6 billion pipeline project designed to deliver gas to utilities and new power generation plants throughout Florida, began construction in 2001. Late in the year, onshore pipelaying was more than 27% finished in Florida and 34% in Alabama. Compressor station construction was more than a third complete.

The 753-mile pipeline originates near Pascagoula, MS, and Mobile, AL, crossing the Gulf of Mexico with 431 miles of 36-inch OD pipe to Manatee, FL. Once ashore, 306 miles of 16-36 inch pipe will reach across southern and central Florida to Palm Beach County. The system also includes 16 miles of gathering pipeline in Mississippi and Alabama. With a capacity to move 1.1 bcfd, the system was reportedly the first new natural gas transportation system constructed to serve Florida in more than 40 years. The project is a joint interstate development by Williams Cos. Inc. and Duke Energy Corp.

Dauphin Island Gathering Partners (DIGP) announced startup of a transportation service from two Gulf projects. The service delivers offshore gas to the Mobile Bay Processing Partners' plant in Coden, LA, where it is processed for redelivery to interstate markets. In July 2001, DIGP began gathering and transport service on a new lateral from the Main Pass 265 platform, which connects four Coastal Oil & Gas Corp. Main Pass production blocks. Fourth quarter production was expected to reach 70 MMcfd of gas, up from initial levels of 25 MMcfd. DIGP also began service from Viosca Knoll Block 385 to El Paso Production Co. and to Chevron USA Production Co. IPE

ANS Gas Work Jumps Ahead While Efforts to Exploit ANWR Stall

Alaska's oil and gas reserves continued to play a central role in the Bush administration's plans to overhaul US energy policy. Focus remained on developing a scheme to transport natural gas from the Alaskan North Slope (ANS) to markets in the Lower 48 states and on opening up the Arctic National Wildlife Refuge (ANWR) coastal plain to oil and gas exploration and development.

Lured by vast deposits of natural gas estimated to amount to more than 217.9 tcf of reserves, of which 88% remain undiscovered, energy companies are eagerly seeking alternatives for bringing these stranded reserves to market. In addition, rising US gas consumption and demand for cleaner burning fuels add to the attractiveness of exploiting Alaskan resources.

Pipeline alternatives for moving North American Arctic gas to Lower 48 markets.

An industry-sponsored study revealed that exploiting ANS gas reserves is feasible at Henry Hub prices of less than $3/MMBtu. Producers were looking at two major pipeline proposals. A meeting of the National Governors Association in early 2001 approved a resolution supporting construction of a gas pipeline from Prudhoe Bay along the Trans-Alaska Pipeline System (TAPS) that would branch off and follow the Alaska Highway through the Yukon Territory into northern Alberta, where it would tie into the existing Canadian pipeline network. This route would run close to 2,000 miles and cost about $10 billion to construct.

Another option proposes construction of an 1,800-mile pipeline that would originate at Prudhoe Bay and proceed east, under the Beaufort Sea, to Canada's Mackenzie Delta. At this point it would travel south along the Mackenzie Valley to northern Alberta or British Columbia, where it would tie into the Canadian grid to the US.

The North American Natural Gas Pipeline Group – a consortium of Phillips Alaska, BP PLC, and ExxonMobil – favors developing a pipeline that would parallel the newly opened Alliance line, which extends from northeastern British Columbia 1,800 miles through Alberta and Saskatchewan to markets in Chicago.

Another option under discussion called for building a gas processing plant at Prudhoe Bay, where the product would be transported through an 800-mile pipeline that would run parallel to TAPS to a liquefaction plant and shipping terminal that would be built at Valdez. Then the LNG would be shipped by tanker to markets in Asia, the Lower 48 states, and possibly Mexico.

Risks associated with the cost and timing of ANS development include potential delays and costs due to multiple and overlapping jurisdictions, as well as opposition supporting a shorter, more economic route to the continental gas grid across the Beaufort Sea and along the Mackenzie River Valley in the Canadian Northwest Territories.

Alaska oil and gas leasing program

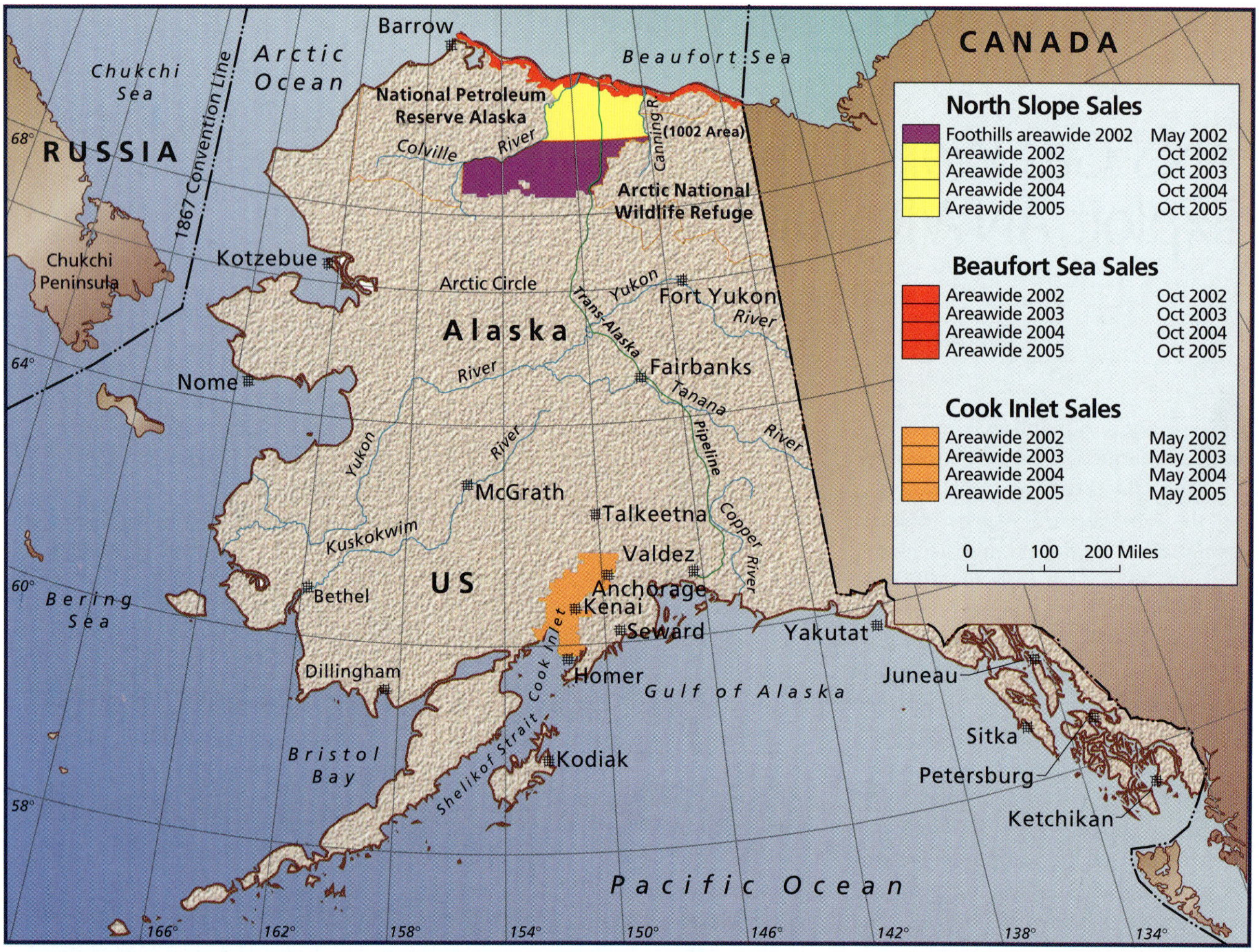

Source: Alaska Department of Natural Resources, Division of Oil and Gas

Industry analysts were concerned, too, that massive supplies brought into the Lower 48 would flood the market, thereby lowering gas prices to uneconomic levels and discouraging exploration and development in the contiguous states. Some analysts questioned whether the combination of gas from the Arctic plus multiple LNG projects in the works during 2001 would undermine the price threshold needed to initiate potential ANS developments.

A study by the Interstate Natural Gas Association of America (INGAA) concluded that it would be insufficient merely to deliver ANS gas to markets in the Lower 48. Introducing substantial deliveries into the system would mandate additional new pipeline capacity out to the East and West Coasts, and to the Midwest as well as to the South. Analysts believe that 38,000 miles of new natural gas transmission pipelines would need to be added to the existing 270,000-mile network to carry additional gas supply from the frontier system.

ANS revival

Declining oil output on the Alaskan North Slope saw production from North America's two largest oil fields, Prudhoe Bay and the 5.9 billion bbl Kuparuk River field, cut in

During 2001 the Polar Endeavour tanker, purpose-built for the Alaskan oil trade, was the first double-hulled crude carrier to transport North Slope oil. The design was mandated by the federal government after the ExxonValdez oil spill.

Production facilities at Endicott, the world's first producing offshore Arctic oil field, were being used to process output from nearby satellites. Such satellite developments were credited with rejuvenating the outlook for North Slope oil production.

half over the last dozen years. Despite this decline, in mid-2001 Alaska projected a 6.9% increase in its oil output in fiscal 2002 from 995,500 b/d in 2001 to 1.069 million b/d.

Since late 1997, operators have been exploring for and developing smaller satellite fields nearby. In May 2001, the first North Slope foothills lease sale in Anchorage represented the most acreage ever sold in a single sale. It drew bids worth $11 million for offshore tracts in the Beaufort Sea and onshore in the North Slope. Although the region shows little potential for oil production, gas shows have surfaced in successful wells drilled all along the North Slope.

Major players in the North Slope have changed dramatically. BP replaced ARCO to become the sole operator at Prudhoe Bay, North America's largest oil field. Phillips Alaska Inc. gained prominence with the startup of the newest field and production hub at Alpine. New independent players Anadarko Petroleum Corp. and Alberta Energy Co. have gained a presence alongside the majors.

Alpine field startup

This new oil field, owned by Phillips Petroleum (78%) and Anadarko (22%), is the largest North American onshore oil discovery in the last 10 years. Advances in 3D seismic technology, which reshaped thinking about the prospectiveness of the mature North Slope, facilitated discovery of the Alpine play and other finds in the National Petroleum Reserve-Alaska (NPR-A) since the latter was reopened to exploration and development.

Alpine, a fourth generation oil field development on the North Slope, reached a record output of 100,000 b/d in mid-2001, which was 25% over design expectations of 80,000 b/d. Phillips' use of long, horizontal wells, essentially barefoot openhole completions, at the reservoir demonstrated the feasibility of low-impact, environmentally sensitive development without affecting much surface area.

Company plans specified future capacity expansions at Alpine with the prospect of tying in satellite developments to a proposed new products hub on the western North Slope. Previous discoveries include the Nanuq and Fiord satellites. Phillips' drilling program calls for 12-14 exploratory wells per year, targeting fields from 50 million bbl up to larger, stand-alone prospects containing 300-400 million bbl. The operator was reviewing optimization schemes for the single production train at Alpine, utilizing the existing pipeline network to Kuparuk and Prudhoe Bay.

Additional NPR-A finds announced

In 2001 Phillips and its partner Anadarko announced several successful wells in the NPR-A 15-25 miles southwest of Alpine. Five wells and a sidetrack all encountered oil or gas and condensate in the Alpine producing horizon. These included the Spark 1A well that tested at 1,550 b/d of liquids and 26.5 MMcfd of natural gas at a flow pressure of 1,500 psig; Spark 1 was twinned to complete it for 1A.

Tests on Rendezvous A produced an unstimulated rate of 360 b/d of liquid hydrocarbons and 6.6 MMcfd of gas. A confirmation well, Rendezvous 2, was drilled in 2001, but Phillips suspended tests until next year. Testing Lookout 1 at Moose's Tooth C was delayed for a year pending selection of appropriate test methods.

Phillips and BP Exploration (Alaska) Inc. released news of a discovery that extends the Kuparuk field to the western North Slope. Palm 1, located about 3 miles west of the Greater Kuparuk Area, contains an estimated 35 million bbl of recoverable reserves. A sidetrack well, Palm 1A, tested at an unstimulated rate of 2,500 b/d of 26º API gravity oil. With production scheduled for 2003, the Palm discovery will be processed through existing Kuparuk field facilities. As the second largest oil field in North America, the Greater Kuparuk Area was producing about 225,000 boe/d from four satellite fields.

NORTHSTAR OFFSHORE PIPELINE DATA

Pipe OD	273.1 mm (10.75 in.)
Pipe WT	7.1 mm (0.594 in.)
Steel grade	API 5L Grade X-52 (358 MPa)
Pipe manufacture	Seamless, 12 m (40 ft) joint lengths
Coating	1.0 mm (40 mils) dual-layer FBE
Cathodic protection	Aluminum sacrificial anodes
Pipe SG	1.60 with respect to seawater
Length	Two pipe bundle 9.7 km (6.0 miles) long
Water depth	0–11.3 m (0–37 ft)
Maximum allowable operating pressure	10.2 MPa (1,480 psig)
Oil max. throughput	7,750 cu m/day (65,000 b/d)
Gas max. throughput	2,830,000 cu m/day (100 MMscfd)

Northstar field begins production

BP Exploration (Alaska) started operations at its Northstar Unit lease 6 miles offshore Alaska's Beaufort Sea coast. Its expansion of Seal Island, a manmade gravel production facility, and installation of the first trenched subsea oil pipeline in the Arctic preceded production by late 2001.

The subsea section of the Northstar line ties into an 11-mile elevated overland pipeline that transports oil into TAPS.

Beaufort Sea
Chukchi Sea
Point Barrow
Barrow
Smith Bay
Simpson
Prudhoe Bay
Endicott
Prudhoe Bay
Camden Bay
Mackenzie Bay
Koakoak
Amauligak
Issungnak
Pitsiulak
Atkinson
Mayogiak
Adgo
Malik
Tuk
Niglintak
Kumak
Titalik
Ya Ya
Parsons Lake
Kavik
Arctic Plains
National Petroleum Reserve
Umiat
15
15
Gubik
E. Umiat
Kavik
Brooks Range
Kotzebue Sound
Arctic Circle
Fort Yukon
Trans Alaskan
Bering Strait
Circle
Tanana
Fairbanks
North Pole
Dawson
Alaska
15
210
Yukon Territory
St. Lawrence I.
Norton Sound
US
Alaskan Range
Chugach Mountains
Whitehorse
Haines Jct.
Ivan River
Anchorage
Beluga River
Granite Point
North Cook Inlet
45
Bethel
Birch Hills
Valdez
Yakataga
Skagway
Nunivak I.
Trading Bay
72
Gravina Point
Yakutat
McArthur River
Swanson River
Montague I.
Juneau
Kuskokwim Bay
Middle Ground Shoal
Kenai Peninsula
Kenai
Sitka
Cook Inlet
Alaska Peninsula
Bering Sea
Kodiak
Kodiak I.
Gulf of Alaska
ALASKA
0
100
200
300 Mi
0
200
400 Km

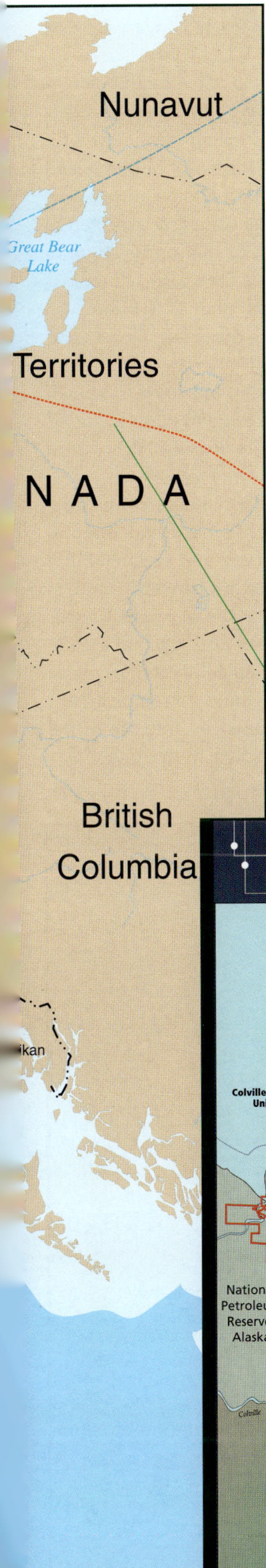

Production was expected to reach a peak rate of 65,000 b/d in 2002. Gas reinjection was being employed to maintain reservoir pressure and improve recovery. Northstar reserves amount to an estimated 175 million bbl of crude oil.

Borealis comes on stream

Borealis, a satellite field in the western Prudhoe Bay area producing from the Kuparuk sands, started operations in late 2001. BP reported that the first six wells were averaging more than 5,000 bbl each. Total November production reached a rate exceeding 30,000 b/d. The development is expected to increase North Slope production by approximately 70,000 b/d.

The operator invested $100 million to build two new gravel pads with well manifolds and pipelines that link Borealis to Prudhoe Bay facilities. Environmentally sensitive construction of the pads, which are designed to support about 50 wells, features reduced Arctic footprints with 15-ft spacing between wells. Location of the pads will allow drilling Schrader Bluff and Sadlerochit reserves.

ANWR leasing still contentious

Political furor still surrounds the Bush administration's proposed leasing in the Arctic National Wildlife Refuge, specifically in the 1.5-million acre coastal plain. Originally scheduled for Congressional consideration as early as September 2001, the issue was sidelined by the more immediate security considerations after the terrorist attack on September 11.

A 1998 US Geological Survey concluded that there was significant hydrocarbon potential in the coastal plain. Revising estimated oil potential upwards from previous figures in its 1995 survey, the agency cited a 95% chance of yields as high as 5.7 billion bbl of oil over a period of about 25 years. Success rates of finding hydrocarbon-bearing structures in the area have improved with the use of 3D seismic technology.

Alaska Power, funded mostly by Alaska state government and partly by industry sources, has started an education and lobbying campaign to open the ANWR coastal plain to drilling. Phillips' Alpine field startup was cited as a prime example of environmentally sensitive

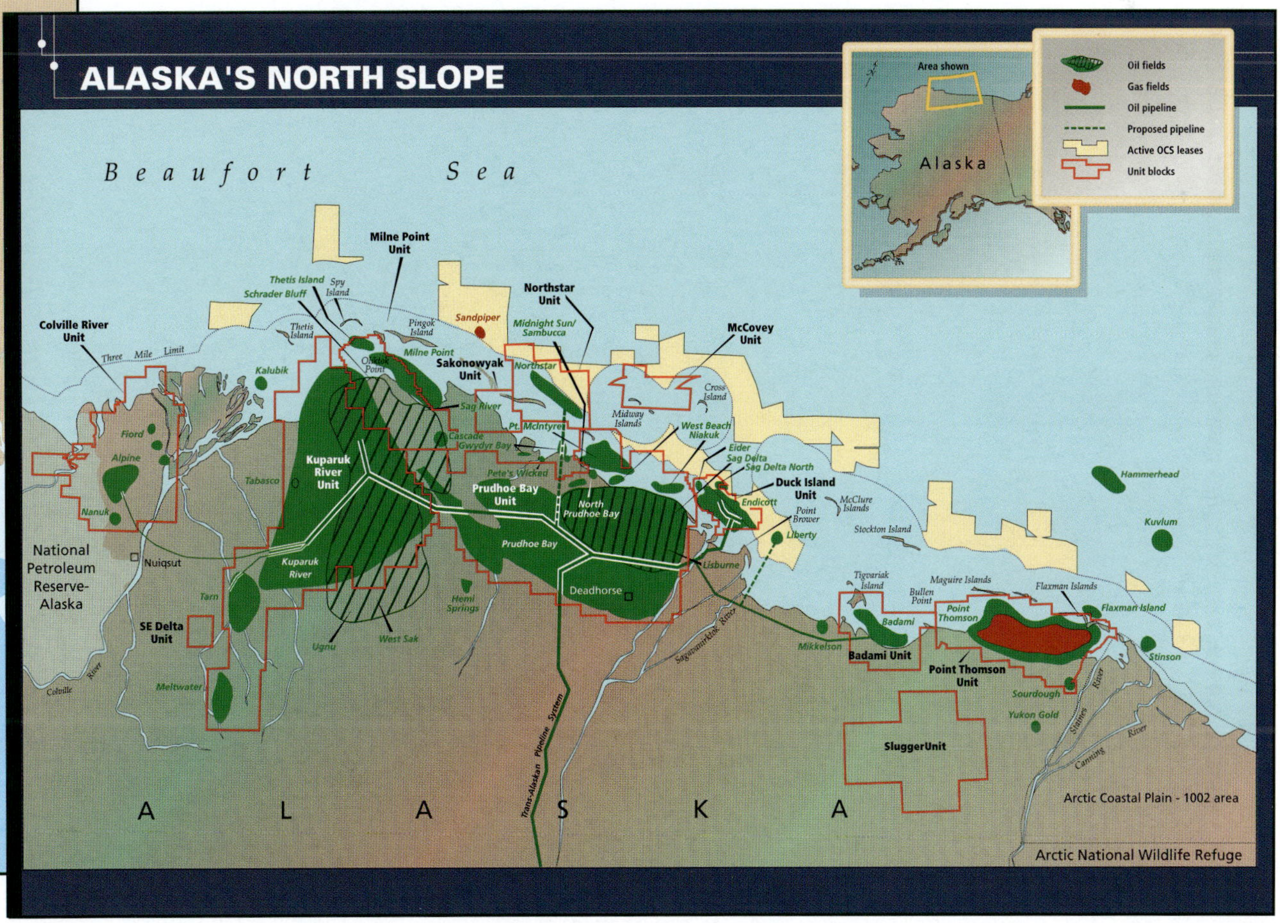

development.

First production from the coastal plain could be realized in as few as 2 or 3 years after leasing. Construction of a feeder pipe only 25 miles in length would be needed to tie into TAPS.

CANADA

CAPITAL: OTTAWA
MONETARY UNIT: CANADIAN DOLLAR
REFINING CAPACITY: 1,943,900 B/CD
OIL PRODUCTION: 2.04 MILLION B/D
OIL RESERVES: 4.9 BILLION B/D
GAS RESERVES: 59.7 TCF

The Canadian government began a 60-day public comment period in early 2002 on draft regulations that would reduce sulfur content of diesel fuel by 95% to a 15 ppm maximum. The proposals, which would align with the US's diesel requirement timeline, were part of Canada's Clean Air Agenda. After the 60-day period, the department will complete and publish the regulations.

The 15-ppm limit would be effective June 1, 2006, for producers and importers and Sept. 1, 2006, for most sellers. The delay will allow for a diesel supply turnover. In Arctic regions, the limit on sales would come into effect a year later to allow for a slower turnover and to reflect logistical difficulties.

Environmental moratoria continued to halt exploration in the West Coast offshore basins of British Columbia, including the potentially prolific Queen Charlotte basin. The provincial government initiated a review process in 2001 that was designed to open up this area to future exploration. The process assesses the basins as subjective conceptual plays with peer reviewed exploration risks assigned to each.

In late 2001, the Canada-Nova Scotia Offshore Petroleum Board accepted bids totaling $527.2 million (Can.) for nine exploration licenses off Nova Scotia, with bids representing the amount expected to be spent on each lease for the first five years of the nine-year lease. The board's decision was subject to federal and provincial ministerial veto, and any exploration work would be preceded by an environmental impact statement and other requirements. The petroleum board's next call for bids was delayed into 2002.

Capital spending

In late 2001, Shell Canada Ltd. announced plans for a $4.4 billion (Can.) capital and exploration expenditure during 2002-2007. The company expected to spend $1.2 billion of that on its share of the $5.2 billion Athabasca Oil Sands Project in Alberta and $1.3 billion on opportunities in the East Coast and Mackenzie Delta frontier areas (subject to various investment approvals). Shell Canada's 2002 investment plans included modifications to its 96,550-b/d Scotford refinery.

A Canadian Energy Research Institute (CERI) study, "2001 Canadian Natural Gas Producer Survey and Deliverability Outlook," revealed an upward trend in capital spending plans over 2000 levels by 19% in 2001, 30% in 2002, and 41% in 2003. Despite higher spending plans, producers' revenue reinvestment remained constrained by the size of the drilling fleet and a lack of personnel.

At Syncrude Canada Ltd.'s North Mine in Alberta, a hydraulic shovel loads mined oil sands into a hauler. Expansions and new projects proliferate in Canada's oil sands industry. Syncrude, the biggest producer, expected to boost productive capacity by almost a third during 2001-06. Photo courtesy of Syncrude.

Completions grew by 13% in 2001 to 10,183 wells, up from 9,042 in 2000. The largest increases occurred in the shallow gas areas of southeast Alberta and Saskatchewan. Although producers' drilling plans continued to focus on shallow gas areas, CERI analysts believed that the industry is beginning to move to deeper resources to the west and north of the Western Canada Sedimentary Basin. However, additional deliverability from the frontiers is not likely to be realized until 2005-2007.

CERI projected natural gas deliverability growing by an average 800 million cfd/yr. Under CERI's base case, production would increase by an average 730 million cfd/yr during 2002-03.

Natural gas supply

The Western Canada Sedimentary basin was judged to be the most prolific of North America's eleven producing basins and also the center of conventional gas supply in Canada. With yields that total 16.4 bcfd of gas, it sends 9.3 bcfd to US markets, meeting 16% of Lower 48 demand and one quarter of North American requirements. Analysts hold that the basin's annual output of about 6 tcf could climb above 8 tcf by 2015. The gas resource potential was estimated at 270 tcf.

The Western Canada Sedimentary basin's nominal marketable gas, however, was estimated at 142 tcf. Producers have already exploited large, high-productivity pools, and future production is expected to come from the smaller, short-lived pools. A report by the Canadian Gas Potential Committee suggested that about 10-20% of nominal undiscovered gas in the Western Canada Sedimentary basin will not yield marketable gas due to lack of exploration access and economics.

Established plays in the Near Frontier occurred on the shelf off Nova Scotia, the Mackenzie Corridor and Eagle Plain, and in the Mackenzie Beaufort basin. The Near Frontier basins – including the sedimentary basins outside the Western Canada Sedimentary basin and southern Ontario and Quebec – were estimated to contain an estimated 46 tcf, while the Remote Frontiers in the Arctic Islands, Newfoundland and Labrador, and Hudson Bay hold an estimated 44 tcf. However, remote plays in the Hudson Bay would face severe technical and economic constraints before any gas production could be realized. The Deep Water Slope play off Nova Scotia attracted a great deal of attention and might hold potential equal to that of the Gulf of Mexico. However, additional seismic tests and delineation drilling would be required in these areas before producers could commit to required production and transportation development.

A study by the Canadian National Energy Board estimated that Canada has 665 tcf of remaining recoverable reserves: Alberta, 270 tcf; British Columbia, 60 tcf; Saskatchewan, 9 tcf; the Northwest Territories and high Arctic, 175 tcf; the eastern Arctic frontiers, 89 tcf; and East Coast offshore areas, 63 tcf. The industry was under pressure to drill more wells – about 25,000 over the next three years – to offset declining output from existing wells.

CERI CANADIAN NATURAL GAS FORECAST

	2000	2001	2002	2003
Capital spending		Million $ (Can.)		
Survey, total capital spending	20,580	24,507	26,707	28,923
Survey, gas spending	13,448	16,969	18,171	19,505
CERI base case	—	17,053	18,236	19,577
CERI increased spending case	—	17,860	22,423	23,908
Well completions		No. of wells		
Survey	9,042	10,183	—	—
CERI base case	—	9,852	10,094	10,262
CERI increased spending case	—	10,387	12,669	12,880
Natural gas deliverability, production		Bcfd		
Survey deliverability	18.3	20.8	22.0	23.5
Survey production	17.0	18.7	18.9	19.4
CERI base case				
Deliverability	—	18.2	18.9	19.8
Production	—	17.8	18.5	19.2
CERI increased spending case				
Deliverability	—	18.2	19.3	20.7
Production	—	17.8	18.8	20.1

Source: Canadian Energy Research Institute

Oil sands development

Despite falling oil prices in the second half of 2001 from a high exceeding $25/bbl to around $19/bbl, Canadian petroleum industry players continued their oil sands development projects in Western Canada. The combination of a stable regulatory market over the past few years, plus the lure of potential US investment in oil sands, the east coast offshore, and the Arctic, spurred the majors into action.

Operators, determined to lower operating costs to about $8.15/bbl from initial levels of $16.30/bbl, were developing nearly $21.3 billion worth of projects to raise oil sands output to 1.65 million b/d by 2015. These projects would account for more than half of Canada's total crude oil output.

A consortium comprising Shell Canada (60%), Chevron Canada (20%), and Calgary-based Western Oil Sands Inc. (20%) scheduled a late 2002 startup date for Athabasca's Muskeg River Mine, making it the first new oil sands project to come on stream since Syncrude in 1978. With a production goal of 155,000 b/d, the project includes a 5 billion bbl lease and a 267-mile pipeline to Shell Canada's Scotford refinery near Edmonton. A low-temperature extraction process would be used to produce bitumen, targeted to start in late 2002. There was a possibility that startup of synthetic crude units could be deferred to early 2003.

Taking advantage of low-temperature extraction technology, Shell Canada ramped up production at its Peace River leases in northwest Alberta to 8,000 b/d with plans to reach daily production levels of 12,000 bbl. Thermal recovery projects also included Petro-Canada's MacKay River development, which was set to produce 30,000 b/d in 2002 and 80,000 b/d by the end of the decade, and Pan-Canadian Petroleum's Christina Lake near Fort McMurray, which had a 10,000 b/d production goal set for 2002.

A number of smaller investors were drawn to new oil sands schemes by Suncor Energy Inc.'s North American refineries, which were converted to run heavier crude, and the availability of more economic extraction technology. A joint venture between TrueNorth Energy, a subsidiary of Koch Petroleum Canada, with 78% interest, and UTS Energy (22%) would exploit Fort Hills, an estimated 2.4 billion bbl bitumen resource in the Athabasca sands. The approximately 46,000-acre Fort Hills venture encompasses oil sands Leases 5, 8, and 52. Phase one of the project targeted production levels of 95,000 b/d of bitumen, with full production scheduled for mid-2005. Phase two of the project was expected to reach a total output of 190,000 b/d in 2008.

BlackRock Ventures, 75% owner of the Orion project at Hilda Lake in northeast Alberta, invested $112.86 million to produce 2,000 b/d from 154 million bbl reserves. Its plans called for expanding the steam-assisted project to 10,000 b/d production levels in 2002 and later to 10,000 b/d.

Other new investors included OPTI Canada, Japan Canada Oil Sands, and SyEnCo Energy. The joint venture between OPTI Canada (50%) and Nexen Inc. (50%) planned to construct and operate a heavy oil in-situ recovery and field upgrader on Lease 27 of the Long Lake project, 25 miles southeast of Fort McMurray. The partnership's $282.15 million Long Lake development was expected to begin a pilot project of 3,000 b/d of bitumen steam-assisted gravity drainage (SAGD) in 2002, followed by a 35,000 b/d commercial SAGD project in 2004. OPTI would design, construct, and operate the upgrader, which was expected to process 60,000-70,000 b/d of bitumen from several oil sands projects in the area by 2006.

Japan Canada Oil Sands – a consortium of Japan National Oil (69.2%), Japan Petroleum Exploration (6.2%), and a group of seventy other Japanese firms (24.6%) – was investing $81.51 million in Hangingstone, an SAGD project 50 miles southwest of Fort McMurray in Alberta. The project was expected to raise production from levels of 1,500 b/d in 2001 to 10,000 b/d of bitumen in 2005.

SyEnCo Energy's 2.7 billion bbl lease south of Fort McMurray was the site of preliminary studies to begin commercial production in 2004. The producer would invest $940.5 billion to upgrade output

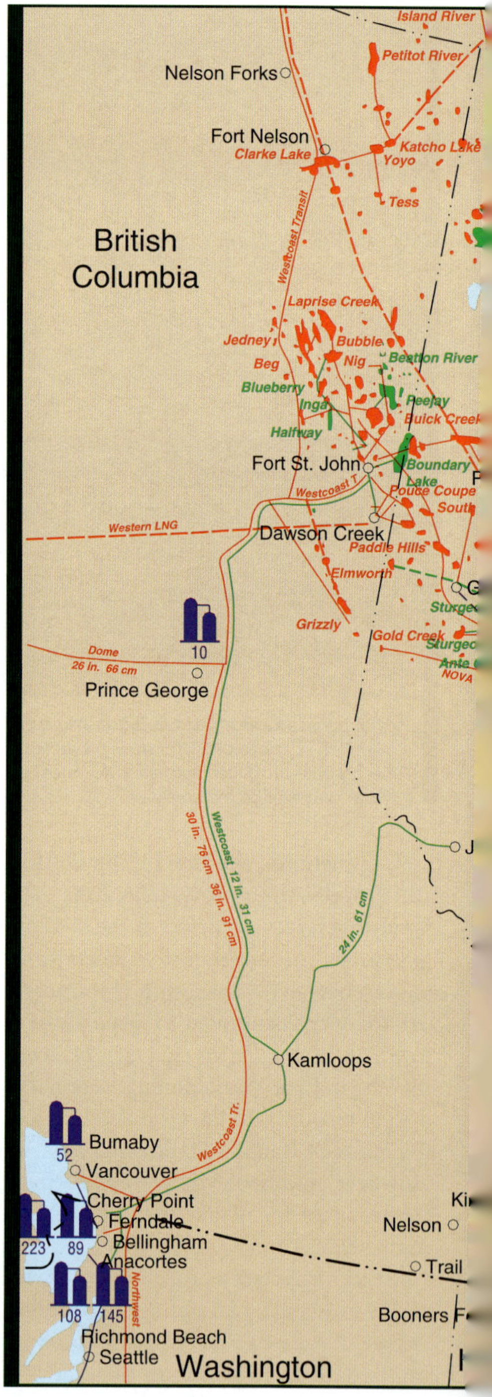

from 20,000 b/d to 80,000 b/d. Unlike the other projects that use natural gas to upgrade the heavy oil, SyEnCo would use coal to fire its upgraders.

Meadow Creek, MacKay River

Petro-Canada, a major player in oil sands development, filed for regulatory approval for a proposed major oil sands development and refinery conversion program in northeast Alberta at Meadow Creek south of Fort McMurray. The part-

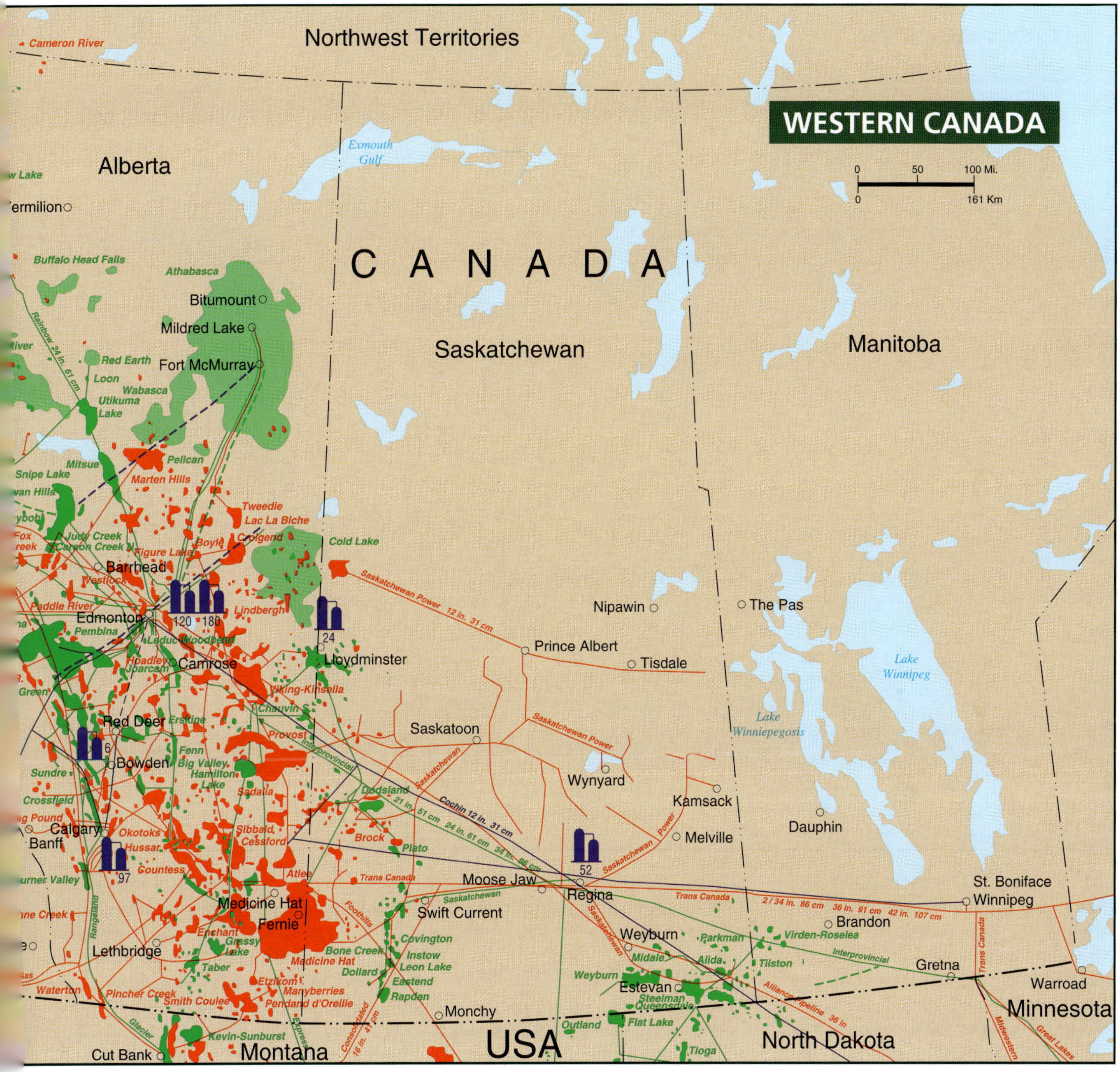

nership between Petro-Canada (75%) and Nexen (25%) would produce up to 80,000 b/d of bitumen for 25 years starting in 2007.

Integrating its upstream production with its downstream refinery, Petro-Canada's refinery conversion program would enable its Edmonton facility to produce low-sulfur gasoline, replacing its existing crude feedstock with bitumen. Phase one, set to come on stream in 2007, would convert the refinery's conventional light crude oil capacity to upgrade and refine about 85,000 b/d of bitumen, raising total bitumen handling capacity to 170,000 b/d.

Construction at Petro-Canada's MacKay River holdings, northwest of Fort McMurray, was well underway in 2001, with bitumen production scheduled to begin in late 2002. The company had 100% working interest in a total 76 sections of MacKay River, located 37 miles northwest of Fort McMurray, Alberta. The development, covering only 11.5 sections of this area, contains an estimated 230 to 300 million bbl of recoverable bitumen. This amounts to sufficient resources to produce 30,000 b/d of bitumen for 25 years.

Drilling activity

Anadarko Canada Corp. planned to invest $238 million to drill 600 wells in western Canada, primarily expanding existing wells. The company's core proj-

ects included the Moose Hills/Kehewin heavy oil program in Northeastern Alberta, the Jean Marie gas play in Northeastern British Columbia, and the Hatton shallow gas play in southwestern Saskatchewan.

Approximately $166 million of Anadarko's budget was slated for drilling development and exploratory wells. In the Moose Hills area, seismic data were to be used to delineate recent discoveries on nearly 80,000 net undeveloped acres in the area. Four rigs were planned to be operating in the area by mid-2001.

The Jean Marie play was one of the most active in northeastern British Columbia. Over 700 bcf of recoverable gas reserves had been discovered as of early 2001. Exploration was underway in the Jedney/Buckinghorse area, adjacent to Jean Marie. The 44-a Buckinghorse well tested at a stabilized rate of 5.5 MMcfd of gas.

Mackenzie Delta-Beaufort Sea frontier

Anadarko Canada acquired 400,000 acres of Mackenzie Delta-Beaufort Sea holdings and planned to extend drilling to the Arctic, where the Delta's prime exploration areas have been stalled. Decisions were expected as to whether the Delta would be linked to an anticipated pipeline from the Alaskan North Slope or whether development could support a stand-alone Canadian pipeline before the company's projects could continue.

Gas shows marked a successful exploration well completion by Petro-Canada and Anderson Exploration in the Mackenzie Delta region of the Northwest Territories. The Kurk M-15 well, located about 93 miles northwest of Inuvik, was the first of a multi-year drilling program. Petro-Canada held about 0.6 million net acres in the Delta, making it one of the largest leaseholders in the area.

The Nunavut Territory opened a large call area north of 75ºN latitude to oil and gas nominations. Discovered oil in the Arctic Islands comprised 5% and gas 15% of remaining discovered recoverable resources in Canada. The area included the entire Sverdrup basin plus a major gas field at Drake Point and Bent Horn oil field. In addition to their existing partnership on 182,851 acres, Chevron Canada and BP Canada planned to acquire 50% interest in 471,386 acres held by Chevron. Chevron and BP would take 33% interest in Burlington Canada's previous acquisition of 360,761 acres.

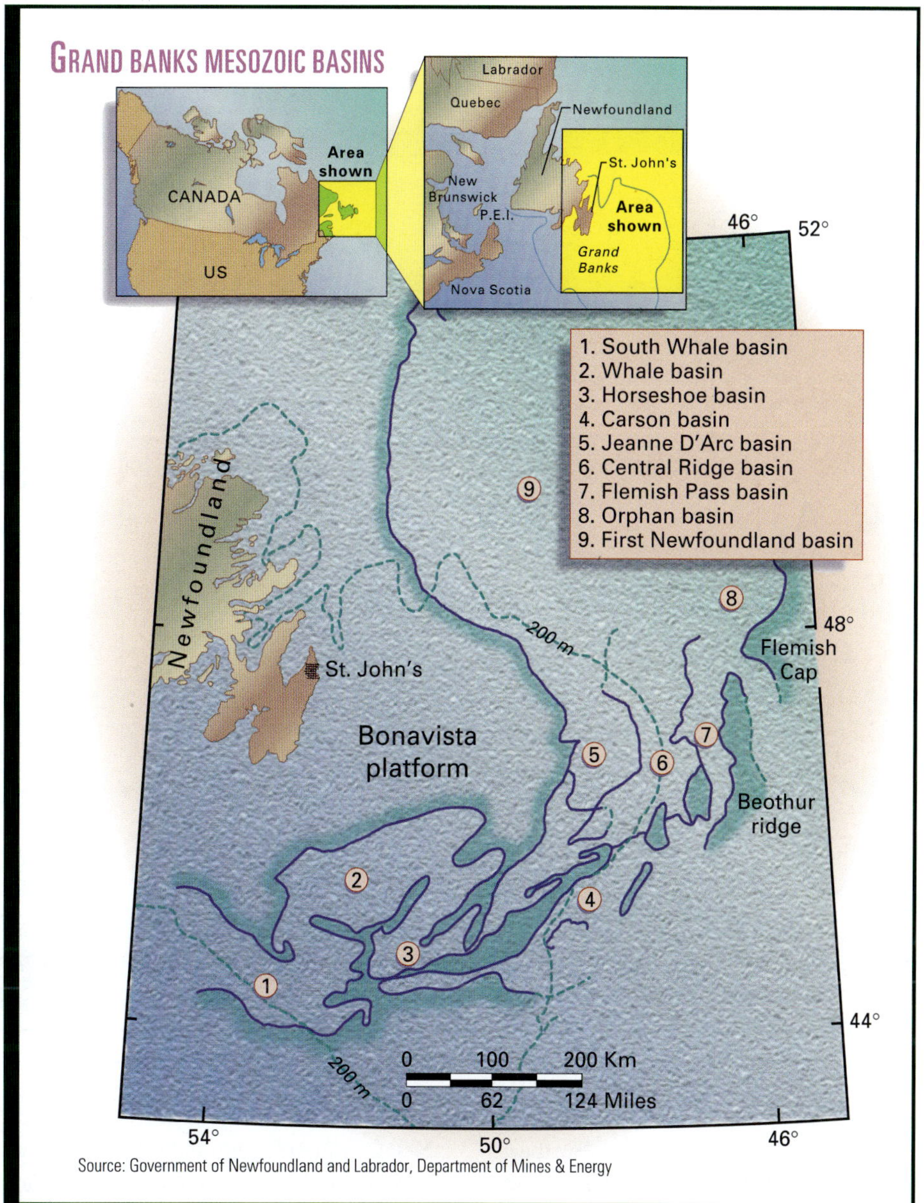

Source: Government of Newfoundland and Labrador, Department of Mines & Energy

Development on the Grand Banks off Newfoundland was gaining momentum during 2001 with Terra Nova field poised to come on stream and other Jeanne d'Arc fields becoming targets for development.

Offshore Eastern Canada

The East Coast of Canada exhibited potential for emerging as an exporter of oil, natural gas, and electric power to markets in the northeast US. Reports indicated that the Grand Banks and Nova Scotia offshore areas have reserves amounting to an estimated 5.3 billion bbl of oil and 63 tcf of gas. Eighteen exploration wells on the northern Grand Banks contain over 2 billion bbl of recoverable oil, 5 tcf of natural gas, and approximately 300 million bbl of NGLs. This area could contain more than 6 billion bbl of undiscovered oil and more than 50 tcf of recoverable gas. Discoveries offshore Labrador yield an additional 4 tcf of recoverable natural gas and more than 100 million bbl of NGL. Another 7 tcf of gas was found off Nova Scotia.

Nearly 5 million acres in 14 separate parcels off Newfoundland were up for bids. These included Parcels 1-3 in the Flemish Pass basin and Parcels 4-7 in the northern part of Jeanne d'Arc basin. Parcels 8-10 are in South Whale basin, where Husky Energy identified a prospect that could contain about 2.5 tcf of recoverable gas. Parcels 11-14 on Newfoundland's west side are in an area where there have been several oil shows.

The partnership of Husky Oil Operations Ltd. (82.5%) and Petro-Canada (17.5%) awarded a design engineering contract to Maersk-Seabay for a 100,000 b/d FPSO platform. The $1.07 billion project included plans to bring the 250 million bbl South White Rose field on stream in late 2004. Field owners filed a successful development application with the Canada-Newfoundland Offshore Petroleum Board.

Terra Nova

Terra Nova field in the Jeanne d'Arc basin on the Grand Banks, east-southeast of St. John's, Newfoundland, is the second largest oil field off the Canadian East Coast, with estimated recoverable reserves of 370 million bbl of oil. Petro-Canada, operator of the field, announced production delays due to design, engineering, and workmanship problems on its FPSO platform, which was custom built for Terra Nova. As a result, the consortium of Petro-Canada (33.99%), ExxonMobil Canada (22%), Norsk Hydro Canada Oil & Gas Inc. (15%), Husky Oil Operations (12.51%), Murphy Oil Company Ltd. (12%), Mosbacher Operating Ltd. (3.5%), and Chevron Canada Resources (1%) hoped to realize first oil production of 129,000 b/d by the end of 2002.

In late 2001, Petro-Canada announced success in drilling the first well into the separate Far East fault block of Terra Nova field. Terra Nova C-69 1 confirmed the presence of hydrocarbons in this part of the field. Previous pre-drill estimates set potential reserves at about 100 million bbl. Log and sample data suggest that the Far East sandstones are equivalent to delineated portions of the field. This discovery extended Terra Nova's ability to sustain maximum production beyond 2006.

Onshore New Brunswick gas finds

Partners Potash Corp. of Saskatchewan (50%) and Corridor Resources Inc. (50%) drilled a successful well onshore in the Canadian Maritimes known as the Magdalen basin. With 300 bcf of potential gas reserves, the McCully 1 near Sussex, New Brunswick, was

believed to be large enough to serve local markets around Sussex with future connection to the Maritimes & Northeast Pipeline to supply the Maritimes and the northeast US.

Corridor Resources and Globex Resources Ltd. spudded the Case Settlement exploration well west of Sussex and approximately 14 miles west southwest of the McCully gas field. Planned to reach a depth of 7,550 feet (2,300 m) in early 2002, the well will test prospects in the Albert formation in a sub-salt position on trend with McCully 1 and McCully 2 discovery wells.

Corridor completed drilling and logging operations at a fourth well. McCully 4 encountered gas-bearing sands in the upper Albert formation, significantly extending McCully field by at least 3 miles southwest of McCully 1.

Deer Lake

The Deer Lake basin onshore western Newfoundland, bounded on the east and west by bold granitic highlands, is an unexplored hydrocarbon bearing basin in the settled part of North America, according to Deer Lake Oil & Gas Inc. The area has a developed infrastructure including jet airport, major transmission lines, the all-weather Trans Canada Highway, and a deepwater year round seaport, as well as a university center only 10 miles to the south.

Deer Lake Oil & Gas drilled a gas discovery with the first modern well in the basin. After an 8-year search, the company affirmed that the basin has all the necessary characteristics to become a significant oil and gas producing area. However, it remained to be seen whether the gas discovery is commercial. IPE

LATIN AMERICA

ARGENTINA

CAPITAL: BUENOS AIRES
MONETARY UNIT: PESO
REFINING CAPACITY: 639,075 B/CD
OIL PRODUCTION: 760,000 B/D
OIL RESERVES: 3.0 BILLION BBL
GAS RESERVES: 27.5 TCF

Argentina's economy deteriorated sharply in mid-2001 as its recession persisted, and an infusion of cash from the International Monetary Fund (IMF) in September looked unlikely to avoid further financial problems. In late 2001, President Fernando de la Rua resigned and his government collapsed, while the IMF refused additional aid. With billions of dollars in global interest payments due in 2001, the country's debts precipitated a loan default and led to a currency devaluation. Long-term exploration and production investments were not greatly affected, but some wildcat projects were shelved. Also, currency devaluation threatened the value of earnings for companies with large Argentine investments, such as Spain's Repsol-YPF SA and Endesa.

Argentina's 27.5 tcf of proven natural gas reserves represents one of the biggest in Latin America. In 2001, the country produced about 1.3 tcf, mainly from four basins: Neuquen, Austral, Northwest, and Gulf of San Jorge. Gas production in the Austral basin was constrained by the capacity of its pipeline to Buenos Aires.

The Neuquen and San Jorge basins also yielded most of the country's oil. The largest oil-producing companies in 2001 were Repsol-YPF SA, Perez Companc SA, and Petrolera San Jorge (owned by ChevronTexaco), while Repsol-YPF was also the largest gas producer, followed by Panamerican Energy LLC, TotalFinaElf SA, Perez Companc, and Pluspetrol SA.

In 2001, a new hydrocarbons law was being considered which could make the country's oil and gas sectors more attractive to international investors by decentralizing regulatory powers, and a new fuel tax law reduced taxes on petroleum products.

In late 2001 it seemed likely that Enron Corp., as part of its bankruptcy proceedings, would sell its 35% stake in Transportadora de Gas del Sur (TGS), South America's largest pipeline company. TGS operates a 4,125-mile, 2-bcfd natural gas pipeline stretching from the southern tip of Argentina to Buenos Aires. The other major shareholder in the pipeline is Argentina's Perez Companc. Petrobras reportedly was interested in purchasing the Enron asset.

Southern Cone gas fuels power sector, exports

Since Argentina's energy sector began privatizing in 1992, gas and power markets in South America's Southern Cone region have begun to integrate. Privatization has stimulated Argentina's domestic demand and exports of natural gas and LPG. Most domestic gas was used in new combined-cycle power plants and petrochemical operations, as well other industrial facilities and residential markets and as a vehicle fuel. Argentina's electricity network has grown considerably with construction of gas-fired combined-cycle capacity, and the installed power generation base totaled 22,000 MW in 2001. The country's ratio of capacity to peak demand was 1.5:1, which could support future exports of power if the transmission network is improved.

Argentina's existing gas pipelines serve neighboring Chile (its largest customer), Brazil, and Uruguay. Gas exports totaled over 156 bcf in 2001 and close to 190 bcf in 2000, at prices ranging from \$0.90-\$1.40/MMbtu, while the average city-gate price in Buenos Aires was \$1.90/MMbtu. Four pipelines delivered

ARUBA
CURACAO
GRENADA
BARBADOS
Bridgetown
TRINIDAD and TOBAGO
Port of Spain
COSTA RICA
Pto. Limon
PANAMA
Colón
Panama
Santa Marta
Barranquilla
Cartagena
Maracaibo
Caracas
Lechoso
Jose
Puerto la Cruz
Punzan
Grico
Penal-Barrackpore-Wilson
Teak
Balata-Bouvallius
Southeast Galeota
Radix
E. Queens Beach
Quarry-Coora-Quinam
Tibu
Hato
Sinco
Palmita
Silvestre
S. Silvestre
S. Cristobol
Barrancabermeja
Ciudad Bolivar
Georgetown
Cantimplora
VENEZUELA
GUYANA
Tambaredjo/Jossikreek
Volcanera
Cupiagua
Cusiana
Apiay Complex
Chichimene
Bogota
Tulua
Cali
Castilla
Andalucia
San Francisco
Dina
Tello
SURINAME
Karanambo
Tumaco
Esmeraldas
COLOMBIA
Quito
ECUADOR
Guayaquil
Amistad
Tumbes
Talara
Bayovar
Iquitos
Concordia
Manaus
Jurua
Rio Urucu
Southwest
Leste do Urucu
Nova Olinda
BRAZIL
Maquia
Aguaytia
Pucallpa
Aguas Calientes
Camisea Fields
Lima
La Pampilla
Conchan
PERU
BOLIVIA
Lago de Titicaca
Pirin
La Paz
Surubi
Pto. Villarroel
Cochabamba
Sicasica
Colpa
Santa Cruz
Oruro
Tita
Rio Grande
Sucre
Tatarenda
Camiri
Corumba
Arica
Monteagudo
Vuelta Grande
Boyuibe
Buena Vista
Caigua
Tigre
Sanandita
Madre Jones
PARAGUAY
Toro
Campo Duran
Bermejo
Ramos
Lomitas Tranquitas
Rio Pescado
Villa Elisa
Calmancito
Jujuy
Asuncion
Salta
Antofagasta
0 100 200 300 400 Miles
0 200 400 600 Km
SOUTH AMERICA

SOUTH AMERICA
Legend
Oil field
Oil sand
Gas field
Crude oil pipeline
Natural gas pipeline
Products pipeline
Pipeline planned or under construction
Refinery in operation
Refinery capacity in 1,000 b/d
Tanker terminal
Cities
Capital
International boundary
Water depth
0 to 200 m
200 m and deeper
0 100 200 300 400 Miles
0 200 400 600 Km
Georgetown
Tambaredjo/ Jossikreek
Paramaribo
Cayenne
SURINAME
FRENCH GUYANA
GUYANA
Belem
Amazon River
Sao Luis
Manaus
Nova Olinda
Atum
Curima
1-CES-8
Xareu
Espada
Pescada
Fortaleza
Fazenda Belem
Ubarana
Canto do Amaro
Agulha
Serraria
Lorena
Guamare
Faz. Pocinhos
Estreito
B R A Z I L
Recife
Maceio
Aracaju
Salvador
Cuiaba
Brasilia
Corumba
PARAGUAY
Aquidauana
Campo Grande
Betim
Belo Horizonte
Campos
Duque de Caxias
46
6
306
151
13
355
10°
5°
0°
15°
20°
55°
45°
35°

Oruro
Arica
Sucre
Santa Cruz
Tita
Rio Grande
Tatarenda
Camiri
Boyuibe
Monteagudo
Vuelta Grande
Buena Vista
Caigua
Sanandita
Tigre
Madre Jones
Toro
Campo Duran
Bermejo
Ramos
Lomitas Tranquitas
Rio Pescado
Antofagasta
Caimancito
Jujuy
Salta
Metan
CHILE
PARAGUAY
Corumba
Aquidauana
Campo Grande
Villa Elisa
Asuncion
Formosa
Resistencia
Corrientes
Paso de los Libres
Uruguaiana
Betim
Belo Horizonte
Campos
Tremembe
Duque de Caxias
Maua
Rio de Janeiro
Sao Paulo
Cubatao
Santos
Sao Sebastiao
Apucarana
Irati
Curitiba
Merluza
Sao Francisco do Sul
Perimbo
Florianopolis
Laguna
Sao Gabriel
Canoas
Porto Alegre
Tramandai
Rio Grande do Sul
URUGUAY
Cordoba
San Lorenzo
Lumunta
Tupungato
Cacheuta
Mendoza
San Luis
Concon
Valparaiso
Santiago
Refugio
Colorados
Barrencas
El Sosneado
Campana
Buenos Aires
Montevideo
La Plata
Dock Sud
Las Flores
Cnl. Pringles
Concepcion
Colonia Catriel
El Medanito
Guanaco
Bahia Blanca
Punta Alta
Puerto Rosales
Plaza Huincul
Challaco
Cerro Bandera
Barda Gonzales
San Antonio
Golfo San Matias
Atlantic Ocean
ARGENTINA
Rivadavia
Las Flores
Comodoro Rivadavia
Anticlinal Grande
Cerro Dragon
Cerro Tortuga
Canadon Grande
Comodoro
Canadon Seco
Caleta Olivia
Pico Truncado
Santa Cruz
Falkland Islands (Islas Malvinas)
Rio Gallegos
Punta Arenas
Tierra Del Fuego
Rio Grande
Ushuaia
0 100 200 300 400 Miles
0 200 400 600 Km
SOUTH AMERICA

Argentine gas into Chile, and an existing gas pipeline was being extended from Uruguaiana, near the Argentina-Brazil border, to Porto Alegre, Brazil, slated for completion in 2002. Another line was being built to Uruguay and could later be extended into Brazil. Two additional lines to Brazil were being planned, the Trans-Iguacu and the Mercosur.

Exploration and development

In mid-2001, two operators reported successful ventures in Argentina: a ChevronTexaco San Jorge group gauged a large oil discovery in the Gulf of San Jorge basin, and Pioneer Natural Resources Co. expanded its operations in the Neuquen basin.

ChevronTexaco's discovery was at the La Yesera X1 exploratory well on the Rio Negro Norte block in Rio Negro Province. The well, 12.4 miles (20 km) southwest of giant Loma Negra complex, flowed at rates of more than 3,000 b/d of 53º gravity oil, reported partner Repsol-YPF. Total depth was 15,895 feet. Block interests were Chevron and San Jorge Energy each 18.75%, Repsol 35%, Corp. Financiera Internacional 15%, and Metro Holding 12.5%.

In the Neuquen basin, Pioneer acquired the 80,000-acre Anticlinal Campamento Block, which contains proved reserves of 31.7 bcf of gas and 170,000 bbl of oil. A 100% working interest in the block, which produces 7 MMcfd of gas and nominal oil, was acquired from Alberta Energy Co. for $12.2 million. Pioneer also bid successfully on the 169,195-acre Cerro Vagon Block, where it planned to acquire 169 sq mi (440 sq km) of 3D seismic data and drill five wildcat wells during 2001-2005. Pioneer discovered several fields on adjacent blocks during 2000-2001.

Pioneer also committed to drill two wildcats on two blocks covering 51,500 acres in the Dos Hermanas concession, also acquired in 2001. The company was evaluating a large deep gas exploration project on the 55,328-acre La Calera Block near giant Loma La Lata gas field.

Argentina's southernmost area also will be further developed by a TotalFinaElf subsidiary, which planned to work in the offshore Carina and Aries fields east of Tierra del Fuego. Two natural gas production platforms and related pipelines would be installed in water depths of about 200-250 feet. Field development contracts were awarded to units of J. Ray McDermott International in late 2001.

Exploration continues in Neuquen

Reserves in the Neuquen basin, Argentina's most prolific, totaled 7.31 billion bbl of oil and 14.9 tcf of gas in 2001. Production is from clastic and carbonate reservoirs of Jurassic, Cretaceous, and Tertiary age.

In 2001, the Neuquen provincial government launched a licensing round for 19 blocks in the basin. In addition to the Cerro Vagon Block acquired by Pioneer, blocks were won by Repsol-YPF in the Piedra Chenque and La Banda areas and by ChevronTexaco's Petrolera San Jorge in Las Tacanas; also, TotalFinaElf offered a high bid for the El Curqui area. The government planned to extend the bid round into early 2002, and a subsequent licensing round was expected to offer chancier blocks.

The areas offered for licensing are located in three sub-provinces with poorly defined limits – the fold belt, the Neuquen embayment, and the northeastern platform. The southern limit of these three areas is the Huincul Dorsal, which separates them from the Picun Leufu sub-basin. In most of the areas, geological conditions are favorable enough to contain commercial quantities of hydrocarbons. However, to establish new prospects would require further exploration.

In late 2001, Argentina's energy secretariat awarded 100% interest for five-year exploration periods in Neuquen basin Blocks CN VI A/B and CNQ 35 (Bajo Hondo) to a unit of Aventura Energy Inc. CN VI A covers 48 sq mi (124 sq km) 6 miles (10 km) north and east of Puesto Molina and Puesto Molina Norte oil fields. CN VI B is 12.7 sq mi (33 sq km) about 2 miles (3 km) northeast of Puesto Hernandez oil field. Ventura's Petrolera del Comanue planned to acquire 13.5 sq mi (35 sq km) of 3D seismic data within three years or could substitute other work. The blocks are in Mendoza Province. CNQ 35 covers 77 sq mi (200 sq km) in Rio Negro Province south of Comahue's 90%-owned Blanco de los Olivos Block. It must acquire 15 sq mi (40 sq km) of 3D seismic data or conduct similar work within three years.

BOLIVIA

CAPITAL: LA PAZ
MONETARY UNIT: BOLIVIANO
REFINING CAPACITY: 63,000 B/CD
OIL PRODUCTION: 30,000 B/D
OIL RESERVES: 440.5 MILLION BBL
GAS RESERVES: 24.0 TCF

Bolivia's economy grew less quickly than expected during 2001, hampered by the slowing world economy, especially in the US and in Bolivia's neighbors. Also, strict economic policies reining in government spending contributed to social

unrest, leading to protests and blockades in early 2001.

Improving relations with Chile during the year opened the possibility for Bolivia to export LNG from a Chilean port, while expansion of Bolivia's GTL capacity was also being considered. Although Bolivia's proven natural gas reserves stand at 24 tcf, its likely reserves are as high as 70 tcf. Greater exports of pipeline gas and LNG could improve Bolivia's economic conditions while its domestic gas demand remains small, far below production levels. President Jose Quiroga said that Bolivia intends to increase its annual gas exports more than five-fold during 2002-07. The government was considering involvement in a new pipeline that would transport Bolivian gas to Argentina, Paraguay, and Brazil.

Bolivia was also nearly self-sufficient in oil and consumed most of its production domestically during 2001. A bid round that started in May received little immediate attention, as companies requested that the government change the terms of the licenses to allow operators to drill and produce when world market conditions are favorable for new production, rather than on a predetermined schedule.

San Alberto gas

During its first-stage development in 2001, the San Alberto field was producing 246 MMcfd (6.6 million cu m/day) of gas, under a 20-year take-or-pay contract with Petrobras, as well as 1,300 b/d of condensate. Second-stage development of the field got under way during 2001 with anticipated production of 447-484 MMcfd (12-13 million cu m/day) in 2002-03, escalating to 819 MMcfd (22 million cu m/day) in 2004-05. The field's reserves were 5.6 tcf (150 billion cu m) of gas and 90 million bbl of condensate. Earlier in 2001, the San Alberto gas processing plant in Carapari, Bolivia, went on stream.

LNG exports to Mexico, US possible

In late 2001, parties agreed to begin negotiating a 20-year supply of Bolivian LNG to be delivered to northwestern Mexico and southern California. The proposed agreement, between utility holding company Sempra Energy and a joint venture led by Repsol-YPF SA, would include the output of a two-train LNG plant on South America's Pacific Coast that could export 800 MMcfd of Bolivian gas to North America. The LNG would be delivered to a Sempra Energy-CMS Energy Corp. receiving terminal with a sendout capacity of 1 bcfd, proposed to be built near Ensenada in Baja California, Mexico. The gas would serve new and existing power plants in the area. Chilean ports such as Mejillones and Tocopilla were being considered as the site for exporting the Bolivian LNG to points north.

The joint venture, JV Pacific LNG, also included BG Bolivia Corp. and Pan American Energy LLC; in early 2002 TotalFinaElf joined the consortium. Repsol-YPF and some of these companies were also partners in Bolivia's 13-tcf Margarita field, which would be the gas source and was operated by Repsol. Sempra stated that Bolivian gas is highly competitive with gas from other Pacific sources because of the shorter shipping distance. Repsol said that Margarita field's reserves are sufficient to supply four LNG trains, and several phases were envisioned for the project.

Gas processing expansion

In late 2001, BG Group PLC completed an expansion and revamp of its natural gas processing plant in La Vertiente. The 10-month project, which cost $30 million, increased the plant's capacity to 156 MMcfd from 60 MMcfd (to 4.2 million cu m/day from 1.6 million). The La Vertiente plant takes gas from Taiguati, El Escondido, and Los Suris fields. Transredes-operated pipelines carry it into Petrobras's Bolivia-Brazil pipeline to Comgas's delivery points in Sao Paulo.

In mid-2001, GTL Bolivia SA (GTLB) signed a memorandum of understanding with Rentech Inc. to use Rentech's proprietary GTL technology. After execution of a licensing agreement, GTLB planned to design a 10,000 b/d GTL plant to be built in Bolivia. Rentech said the final agreement could grant GTLB the right to operate an additional 50,000 b/d.

New pipeline to Brazil

Plans moved forward in 2001 for another Bolivia-to-Brazil gas pipeline to be built by Transierra, a consortium led by Petrobras. The 268-mile (431-km), $300 million pipeline would connect southern Bolivia's gas fields with an export pipeline in Brazil, making Petrobras the largest exporter of Bolivian gas by 2003. Objections from Transredes, an affiliate of Enron and Shell, were halted in late 2001, and other complaints by local Indian groups were likely to be dropped as well.

Petrobras, Repsol-YPF, and TotalFinaElf, operators of the Transierra consortium, received the environmental license to build the line in Bolivia and expected to begin construction in 2002. The three companies controlled large gas reserves in Bolivia's San Alberto and San Antonio fields, which hold 16 bcf of proven reserves.

In late 2001, BG Group was again prevented from using Petrobras's existing Bolivia-Brazil gas pipeline, which was transporting only 454 MMcfd (12.2 million cu m/day) but has a potential capacity of 1.12 bcfd (30 million cu m/day). Petrobras subsidiary Transportadora Brasileira Gasoduto claimed that it will use the capacity in 2002 when new electric power plants come on stream.

BRAZIL

CAPITAL: BRASILIA
MONETARY UNIT: REAL
REFINING CAPACITY: 1.786 MILLION B/CD
OIL PRODUCTION: 1.210 MILLION B/D
OIL RESERVES: 8.5 BILLION BBL
GAS RESERVES: 7.8 TCF

After strong progress during 2000, Brazil's economy faltered in 2001 as Argentina's financial crisis infected its neighbors and as growth in the US and other South American countries slowed. Brazil experienced electricity rationing and power shortages – including a major blackout hitting Rio de Janeiro, Sao Paulo, and other large cities in early 2002 – which reduced industrial output and dampened consumer confidence. However, rationing was relaxed for the year-end holidays and was scheduled to end in 2002. Also, economists predicted that Brazil would recover during 2002 more quickly than expected, expanding 2.5% after a rise of only 1.7% during 2001. Oil price controls were scheduled to end on Jan. 1, 2002, with a new fuels market opening to competition, while privatization of Petrobras remained unlikely before 2003, when President Cardoso's term concludes.

Elektro Electricidade e Servicos, Brazil's sixth largest power distributor, was up for sale in early 2002, with a shortlist of five potential bidders (Endesa, AES, Duke Energy International, Tractebel, and RWE). The company was Enron Global Services' largest Latin American asset. Earlier, the government created a new public company to supply electricity in northeast Brazil and to manage water resources.

Although Brazil continued to strive for self-sufficiency in oil production, its consumption still exceeded production by about 500,000 b/d, with imports coming mostly from Venezuela and Argentina. The offshore Campos basin, north of Rio de Janeiro, was the country's most prolific oil production area and holds the most of its natural gas reserves. Brazil planned to increase the gas share of its energy mix and reduce the country's dependence on hydropower over the next several years by building gas-fired power plants.

Also to combat hydropower dependence, Brazil looked to renewable resources as well as oil and gas. Its electricity sector regulator authorized construction of 41 wind power plants beginning in 2002. The government also reduced taxes on wind, solar, and biomass projects and continued to implement the Gerasol project to install solar generation equipment in schools and health clinics.

In late 2001, Petrobras announced that it would invest a record $8.9 billion in its refining sector between 2002 and 2010. Eight of Brazil's 13 refineries were to be revamped to absorb the growing production of heavy crude oil with low sulfur content. Petrobras planned to produce more added-value products and improve the quality of gasoline and diesel oil, the company said.

Also late in the year, Petrobras announced discovery of a natural gas reserve totaling 745 bcf (20 billion cu m) in the Camamu basin offshore Valenca in the northeastern Brazilian state of Bahia. As a result, Bahia expected to increase natural gas production by more than 50% by mid-2003.

Near year-end 2001, it seemed likely that Enron Corp., as part of its bankruptcy proceedings, would sell its stake in two Rio de Janeiro gas distributors.

Offshore E&D proceeds after accident

In March 2001, the world's largest semisubmersible production platform, Petrobras's P-36, sank in the Campos basin after suffering explosions in a supporting pillar, caused apparently by natural gas that leaked and collected there. The rig had been producing about 84,000 b/d and 48 MMcfd (1.3 million cu m/day) from Roncador field, leading Petrobras to lower its 2001 production estimate from 1.42 to 1.39 million b/d, although new wells were beginning to compensate for the loss.

Installation of three pairs of rigid flowlines and umbilicals connecting three wells to the P-36 had earlier in 2001 set a pipelaying water-depth record of 6,135 feet (1,870 m). Petrobras planned to have a provisional platform on stream in 2002 and a permanent replacement for P-36 in operation by 2007. In addition to crude oil and diesel fuel spilled from P-36, Brazil was plagued by smaller oil spills in the basin as well as an overland pipeline leak.

During 2001, Brazil held its third licensing round, offering 53 blocks in June of which 43 were offshore, mostly in deep and ultra-deep water. A wide range of aggressive bidders participated, earning Brazil's National Petroleum Agency (ANP) about $240 million. As in previous rounds, Petrobras was a big winner, receiving 19 concessions and taking a 100% stake in 7 of those blocks, as well as becoming joint venture partner in two others with operators ExxonMobil and TotalFinaElf. Companies that planned to work with Petrobras included El Paso, Enterprise, Statoil, and Petrogal.

Phillips Petroleum offered $48.7 million for Block BM-ES-11. Ocean Energy and Amerada Hess jointly bid for BM-C-15 and BM-S-22. El Paso CGP was another sizable bidder. Others offering successful bids included Repsol-YPF, Shell, Wintershall, and Amerada Hess. In late 2001, ANP disclosed that a record 55 additional blocks (15 onshore and 40 off) in 19 sedimentary basins would be offered for exploration in June 2002.

E&D action picked up slowly in the offshore Santos basin during 2001, with Petrobras announcing in May a deepwater crude oil discovery. The 1-RJS-551 well, in 4,914 feet (1,498 m) of water on Block BS-500, was drilled to 14,855 feet (4,528 m). Petrobras planned a wildcat on BS-500 and one or more delineation wells near the 1-RJS-551. Also, a Petrobras-led joint venture of four companies spudded an exploratory well on Block BS-3 with total depth of 17,060 feet (5,200 m). All told, Petrobras was active in 12 exploratory blocks in Santos basin, including partnerships with private companies on 10 blocks.

As of mid-2001, about 23% of the world's operational floating production systems were in use offshore Brazil.

Gas, power markets intertwined

Natural gas in 2001 emerged as an important alternative to hydropower in Brazil's energy policy plans, and Petrobras was poised to become a large electricity producer and distributor by

2010. Several new gas-fired power plants were planned to begin operations during 2001-02, totaling more than 9,000 MW of capacity, reported Prates & Carneiro Advogados. However, at year-end 2001, only 224 MW of new thermal plants had come on-line, reported South American Business Information.

The Brazilian government's growth targets for 2001-11 would require additional gas-fired facilities which, along with other industrial and residential demand growth, would consume enough natural gas to strain the country's production and pipeline import capacities, said Purvin & Gertz Inc. In addition to domestic exploration for gas, Brazil could rely on Bolivia for future additional gas supplies.

Brazil's official projections called for gas-fired power plants to account for 18% of the Brazilian power mix by 2005, compared with less than 7% in 2001, according to BNP Paribas, yet capacity additions were being outstripped by burgeoning demand for power. The primary hurdles to building new gas-fired power plants were governmental delays and regulatory bottlenecks, said BNP Paribas.

In late 2001, Petrobras indicated it would invite international construction companies to bid on a contract worth $500 million to build three gas pipelines in southeastern Brazil. Additional pipelines connecting Argentina to Brazil were in various stages of planning.

CHILE

CAPITAL: SANTIAGO
MONETARY UNIT: PESO
REFINING CAPACITY: 204,849 B/CD
OIL PRODUCTION: 7,000 B/D
OIL RESERVES: 150 MILLION BBL
GAS RESERVES: 3.5 TCF

Chile continued to star as one of Latin America's strongest economies, with growth of 4.9% during 2001. About a third of the country's international trade is with Asia, and copper is Chile's largest export, so weak Asian import demand and low global copper prices kept a lid on Chile's growth. Its oil company, Empresa Nacional de Petroleo (ENAP), and copper company, Codelco, remained under government control, but could be considered for privatization after year-end 2001 congressional elections.

Without large natural gas reserves, Chile produces less than half of its domestic gas consumption and has encouraged imports to help meet increasing energy demand for copper mining, power generation, and other industrial uses, as well as the country's growing urban populations. Argentina supplied all of Chile's imported gas, transported through six pipelines, two of which were under-utilized serving a handful of power plants. As Chile's power capacity increases and its power system becomes more interconnected, gas pipeline capacity could become better utilized, said Purvin & Gertz Inc.

Chilean ports such as Mejillones and Tocopilla were being considered as the site for exporting Bolivian LNG to points north.

Chile produced less than 20% of domestic oil consumption and imported the rest chiefly from Nigeria, Gabon, and Venezuela. During 2001, ENAP focused more on oil exploration and production than in the past, seeking to reactivate old fields using new technology and to satisfy 30% of Chilean demand.

In 2001, Repsol-YPF sold a 36% share of Chile's 260-mile TransAndino oil pipeline to Canada's AEC Pipelines but retained 18%; other owners were Unocal (28%) and ENAP (18.3%). The line transports half of Chile's crude supplies from Argentina's Neuquen basin to the Petrox refinery, Chile's largest of three. Petrox began work in mid-2001 to upgrade its facility, planning to spend more than $50 million to increase its crude throughput capacity 25% by early 2002.

ENAP and Advantage Resources International hired Syntroleum to study a 10,000 b/d GTL plant for Chile, proposed for installation at the ENAP-owned Cabo Negro Industrial Park near Punta Arenas. The plant would be supplied with pipeline gas from Chilean and Argentine fields, and ENAP would sell GTL products to Santiago consumers. If the study confirms expectations, the plant might be designed and engineered by Syntroleum and use its GTL technology.

COLOMBIA

CAPITAL: BOGOTA
MONETARY UNIT: PESO
REFINING CAPACITY: 285,850 B/CD
OIL PRODUCTION: 625,000 B/D
OIL RESERVES: 1.75 BILLION BBL
GAS RESERVES: 4.3 TCF

Following a deep recession in 1999, Colombia's economy continued to improve, with growth estimated at 4% in 2001 even though production of oil, the country's largest export earner on the books, continued to decline. The country could become a net oil importer as early as 2004, yet vast areas remained unexplored, so encouraging foreign investment in oil exploration and production was a top priority in 2001. Almost half of Colombia's crude oil production went to the US, and about 60% came from three fields.

In late 2001, Argosy Energy International signed an association contract with Colombian state oil company Ecopetrol for the Rio Magdalena area, said Aviva Petroleum Inc., which owns 30% of Argosy. The contract, which has 6-year exploration and 22-year development periods, covers 222 sq mi in the Upper Magdalena region.

Companies becoming involved in Colombian oil for the first time in 2001 included Spain's Cepsa, Venezuela's Tecnoil, and Texas's Lone Star; Occidental and Colombia's Geopetrocol were expanding their presence. Government officials planned to decide up to 30 contracts during 2001. Private companies started drilling seven wells in mid-2001 which could, if successful, start producing in 2002.

However, bombings and other security problems continued to plague the Colombian oil industry. In early 2001, Occidental declared force majeure on production at Cano Limon field and its 480-mile (780-km) pipeline. Occidental also failed to discover oil at the Gibraltar 1 well site, a focal point of protests and negative publicity. Production from Ecopetrol's Orito field in the Putumayo basin was stunted in 2001 due to explosions along its connecting TransAndino pipeline, while BP's Cusiana/Cupiagua operation in the eastern Andes foothills was less affected because more of its pipeline is underground.

In September 2001, Ecopetrol signed a non-binding memorandum of understanding with PDVSA Gas and ChevronTexaco to conduct a feasibility study that would evaluate the construction of a 124-mile (200-km) natural gas pipeline linking Colombia and Venezuela.

CUBA

CAPITAL: HAVANA
MONETARY UNIT: PESO
REFINING CAPACITY: 301,400 B/CD
OIL PRODUCTION: 44,000 B/D
OIL RESERVES: 750 MILLION BBL
GAS RESERVES: 2.5 TCF

A report from the Cuba Policy Foundation concluded that lifting the US trade embargo against Cuba could provide US energy firms $2-3 billion/year in new revenue, based on a projected increase in oil demand of 148,000-184,000 b/d. The study said natural gas offshore Cuba could be pipelined to Florida or converted to products in a GTL plant.

Overall production from Cuba's oil fields reportedly was up from 42,000 b/d in 2000 (still the industry's official figure) to 70,000 b/d in April 2001, according to the Cuban government. State oil company Cubapetroleo claimed that it will increase petroleum production from 52,000 b/d in 2001 to 120,000 b/d in 2005. Most domestic production consisted of heavy oil with a high sulfur content.

In October 2001, Sherritt International Corp. said its gross operated production for the second quarter averaged 30,113 b/d (19,352 b/d net), up from the previous quarter and 10% higher than in first quarter 2001. The hike was due mainly to output from new wells in Cuba's Yumuri, Canasi, and Seboruco oil fields, said Sherritt. A Canasi development well as successful, and development wells were spudded at Puerto Escondido and Varadero West. Two field wells were returned to production, a Seboruco well was deepened to 13,123 feet (4,000 m), and a Yumuri well was re-entered to remove a mechanical obstruction. In early 2002, Sherrit looked likely to set an oil production record for the fifth straight year.

Earlier in 2001, Pebercan increased its interests in Block 7 on Cuba's northern coast to 60% of the Canasi struture and 55% of the rest of the block. Gross production from Block 7 averaged 5,000 b/d in May 2001, and the government asked that it be accelerated, Pebercan said.

ECUADOR

CAPITAL: QUITO
MONETARY UNIT: US DOLLAR
REFINING CAPACITY: 176,000 B/CD
OIL PRODUCTION: 400,000 B/D
OIL RESERVES: 2.1 BILLION BBL
GAS RESERVES: 3.7 TCF

Oil exports were the Ecuador government's main source of revenue once again in 2001, and the state oil company Petroecuador worked toward attracting foreign investment in the country's five

largest oil fields: Shushufindi, Sacha, Libertador, Cononaco, and Auca. Privatization of Petroecuador was still down the road in 2001, but construction finally began on a second crude oil export pipeline.

Political instability in Colombia continued to spill over into Ecuador, whose only existing oil export pipeline was bombed in January 2001. The border region, including Ecuador's productive oil province of Sucumbios, experienced criminal activity and threats to oil operations.

Pipeline progress

The oil pipeline infrastructure in Ecuador has consisted of only one major pipeline – the 311-mile Trans-Ecuadorian line, known by its Spanish-language acronym SOTE – which was insufficient to support a substantial increase in production, operating at 380,000 b/d. In early 2001, however, Ecuador selected the Oleoducto de Crudos Pesados SA consortium (OCP Ecuador) to build a $1.2 billion oil export pipeline capable of carrying 450,000 b/d of 16-22º gravity crude. The OCP consortium, led by Repsol-YPF, also comprises four other companies that produce oil in the Oriente basin (Occidental Ecuador, Alberta Energy, Kerr-McGee, and Agip), as well as Argentina's Techint.

Ecuador's President Gustavo Noboa said that the new line could allow the country's oil output to reach 800,000 b/d by 2003. Blocks that had remained closed to exploration would be offered in new bidding rounds, and other development projects could be carried out by Petroecuador. The OCP pipeline would double the country's oil transport capacity and allow development of fields that have not been able to begin production because of lack of capacity on the existing line.

Construction of the pipeline began in July 2001 after final regulatory approval was granted in June. OCP Ecuador's Techint was building the line under a fixed-price contract allowing 25 months for completion, or until mid-2003. The line's route was planned to parallel the existing pipeline, linking Lago Agrio to Esmeraldas on the coast, except for a 62-mile deviation north of Quito to Quinindе in the coastal plain, where the line was to cross environmentally sensitive areas in the Andes' western foothills.

However, construction was delayed by environmental protests later in 2001. Fueling environmental concerns was damage to Ecuador's largest pipeline, which cracked during a mudslide and spilled 8,000 bbl of oil into a river. The project could face delays, as Petroecuador was in discussions in late 2001 with local officials regarding safety and environmental hazard precautions.

Meanwhile, talks resumed in mid-2001 between Ecuador and Peru on the possibility of pumping crude oil from Ecuador's fields near Peru's northern border through the existing 200,000 b/d North Peruvian pipeline, which had surplus capacity, to the Bayovar marine terminal on Peru's north coast.

Gas development steps up

Ecuador's Amistad offshore gas field in the Gulf of Guayaquil was due for installation of a 42-mile, 12-inch pipeline beginning in late 2001. Noble Affiliates subsidiary EDC Ecuador signed a letter of intent with Horizon Offshore to install the line, using a pipelay-bury barge. Work was scheduled for completion in early 2002. In late 2001, EDC reported that it had completed 50% of the pipeline connecting the offshore area to Ecuador, where it will supply the 312-MW Machala Power plant at Bajo Alto.

In August 2001, Noble Affiliates said the Amistad-7 well on Block 3 in the Gulf of Guayquil flowed on test at a rate of 19.4 MMcfd of gas through a 32/64-inch choke at 3,208 psi flowing tubing pressure from 172 feet of perforations. The well is part of a four-well program on the block. Noble operates the project and holds 100% working interest. It planned to use the field's production to fuel a 240-MW power plant.

In early 2002, Petroecuador gave the green light for Amazonia Gas to process and distribute natural gas produced in the Sacha field, Sucumbios province.

Other exploration, production developments

In early 2001, Petroecuador planned to begin reactivating 90 shut-in oil wells, boosting Ecuador's output by 20,000 b/d. Scheduled for reactivation were 20 wells in Auca field, 18 in Lago Agrio, 22 in Libertador, 22 in Sacha, and 8 in Shushufindi.

In mid-year, Petrobank Energy & Resources Ltd. had an agreement with Petroecuador to evaluate the commercial potential of Pungarayacu and Oglan heavy oil fields in Ecuador's Oriente basin.

In early 2002, Ecuador prepared to call its ninth international oil bidding round, and eight blocks were identified by Petroecuador: offshore Blocks 4, 5, 39, and 40 and Amazon Blocks 20, 32, 33, and 35. The company was considering adding three Amazon blocks to the sale. Petroecuador has identified 13 structures on the blocks with possible reserves of 350 million bbl of heavy crude.

GUATEMALA

CAPITAL: GUATEMALA CITY
MONETARY UNIT: QUETZAL
REFINING CAPACITY: 16,000 B/CD
OIL PRODUCTION: 20,000 B/D
OIL RESERVES: 526 MILLION BBL
GAS RESERVES: 109 BCF

Economic concerns and political uncertainty continued into 2001 after President Alfonso Portillo was inaugurated in January. Guatemala had agreed with Mexico to build a natural gas pipeline from Ciudad Pemex in southern Mexico to Escuintia in southern Guatemala, but as of year-end 2001, work had yet to begin. The $450 million, 347-mile pipeline, which was scheduled for 2004 completion, would follow the path of an existing oil line in Guatemala's remote Peten jungle region, where the country's oil reserves are concentrated. As Central America's only oil-producing country, Guatemala has enjoyed interest in these reserves because of their proximity and similarity to Mexico's productive Tabasco formations.

In mid-2001, Anadarko Petroleum Corp. sold its Basic Resources International subsidiary to Perenco, a privately held European exploration and production company, for $120.5 million. Basic Resources, Guatemala's largest oil producer, produces and refines oil from fields located in the Xan and Rubelsanto areas, which collectively produce about 20,000 b/d and account for virtually all of the country's oil production. Basic's proved and probable reserves totaled 58 million bbl. Infrastructure at the Xan field, the country's largest oil producing area, included processing plants, a 295-mile pipeline system, and storage and loading facilities at the Piedras Negras shipping terminal on the Caribbean coast. Basic also owns and operates a 6,000 b/d refinery at La Libertad.

Perenco planned to continue developing the Xan field by drilling new water disposal wells, upgrading electric submersible pumps, and drilling new production wells.

MEXICO

CAPITAL: MEXICO CITY
MONETARY UNIT: PESO
REFINING CAPACITY: 1.525 MILLION B/CD
OIL PRODUCTION: 3.1 MILLION B/D
OIL RESERVES: 26.9 BILLION BBL
GAS RESERVES: 29.5 TCF

In 2001 President Vicente Fox continued to emphasize the need for greater participation by private industry in Mexico's energy sector. In March contracts were let to drill wells in the northeastern Burgos basin, and similar contracts were expected for future years. Petroleos Mexicanos (Pemex) said in September 2001 that multiple oil service contracts would be offered in 2002 to speed exploration and production in Mexico. A program of refinery upgrades was also underway during 2001. In early 2002, President Fox inaugurated six thermoelectric plants totaling 1,519 MW.

The LNG and LPG businesses could offer investment and marketing opportunities, said Mexico Energy Intelligence. However, the prospects for further natural gas deregulation dimmed in 2001 because of a backlash against spiking gas prices in Mexico. Overall, the expectation was that the Fox government's pro-business philosophy would eventually make its way down the corridors of the state-owned oil and power companies, according to Mexico Energy Intelligence.

Boosting oil production

Mexico's government approved Pemex's 2001-06 upstream investment program, and in 2001 Pemex described several projects aimed at increasing output of crude oil and natural gas, including nitrogen injection in Bay of Campeche fields, development in the coastal Tabasco province, and production increases in the Delta del Grijalva region. Earlier, the world's largest nitrogen-generation plant was commissioned on schedule and began service for Cantarell reservoir pressure maintenance.

Cantarell oil field was undergoing a major upgrade and expansion project in 2001 that included installation of about 250 miles (more than 400 km) of pipelines. Cantarell, Mexico's largest oil

MEXICO-CENTRAL AMERICA

Texas
Coahuila
Ulua
Buena Suerte
Nuevo Laredo
Nuevo/Laredo
Laredo
Lampazos
Monclova
Corpus Christi
US
La Presa
Arcos
Sta. Rosalia
Lojitas
Culebra
Canor
Carretas
McAllen
Reynosa
235
Mision-Trevino
Brasil
Brownsville
Cadereyta
Reynosa
Fortuna
Matamoros
Primavera
Cabeza
Matamoros
Monterrey
S. Bernardo
Ramirez
Benavides
18 de Marzo
Chapul
Pobladores
Huizache
San Fernando
Nuevo Leon
Tamaulipas
Zacatecas
Lerma
Gulf of Mexico
Ciudad Victoria
San Luis Potosi
195
Barcodon
Tamaulipas-Constituciones
San Luis Potosi
Ciudad Madero
Aguas
Ebano Panuco Fields
Arenque
Tampico
Cabo Nuevo
Tuxpan
Arrecife Medio
Isla de Lobos
Guanajuato
Golden Lane Fields
Tiburon
Esturon
Marsopa
Guanajuato
Queretaro
Hidalgo
Tuxpan
Queretaro
Salamanca
Bagre
Escuala
320
La Barca
Poza Rica
Tula
Pachuca
Planche
Morelia
MEXICO
Tlax.
Veracruz
Mexico City
Jalapa
Tlaxcala
Michoacan
Toluca
D.F.
Veracruz
Remudauera
Cuernavaca
Puebla
Miralejos
Capite
M. Pancho
Mecayuman
Angastura
Mirador
Cuapo
Novillero
Veinte
Mata Verde
Bay of Campeche
Cacuite
Maculle
Tabasco
Sinán
Citam
Bolontikú
Kix
Yum
May
Dos Bocas
Ciudad del Carmen
Fontera
Cardenas
Ciudad Pemex
Villahermosa
Merida
Yucatan
Campeche
Campeche
Quintana Roo
Morelos
Puebla
Azcapotzelco
Minatitlan
200
Guerrero
Chilpancingo
Oaxaca
Oaxaca
Acapulco
Tehuantepec
Juchitan
Salina Cruz
330
Tuxtla
Chiapas
Xan
6
Belmopan
BELIZE
Caribe
1 San Diego
W. Chinaja
Rubelsanto
Yalpemech
Gulf of Tehuantepec
GUATEMALA
Guatemala
HONDURAS
Escuintla
16
21
San Salvador
Sonsanate
EL SALVADOR
100 0 100 200 Miles
100 0 100 200 300 Km
MEXICO-CENTRAL AMERICA

MEXICO-CENTRAL AMERICA

field, lies about 50 miles (80 km) off the Yucatan peninsula in the Bay of Campeche in water depths ranging from 115 to 130 feet (35 to 40 m). Its four major fields are Akal (containing the world's sixth-largest reservoir), Nohoch, Chac, and Kutz, along with reserves in the newly discovered, deeper Sihil field. The piping network that was being installed will enable Pemex not only to combat declining reservoir pressures, but also increase reserves, more than double production, and reduce flaring and utilize gas by adding sufficient compression and process capacity. Pemex expected that after the expansion project is completed, Cantarell will produce about 2.2 million b/d in 2002.

For increasing production from the Ku-Zaap-Maloob (KZM) complex – Mexico's second largest group of fields – the Pemex budget allowed for drilling 47 new wells and building 22 pipelines. From 2001 through 2014, nitrogen injection in KZM was expected to approximately double annual production, from about 1 billion bbl to more than 2 billion bbl, said Universidad Nacional Autonoma de Mexico. Nitrogen injection could also increase gas production to an estimated 450 bcf.

Pemex's program also targeted six reservoirs in Tabasco fields, where drilling of 24 wells and installation of 12 offshore platforms during 2001-06 was expected to increase production to a peak of 110,000 b/d of light and super-light crude and 288 MMcfd of gas.

The Delta del Grijalva program included Caparroso, Sen, and Luna fields, the most important in terms of proven reserves and productivity. By 2004, development could increase light crude production from 70,000 b/d to 102,000 b/d and gas production from 260 MMcfd to 360 MMcfd.

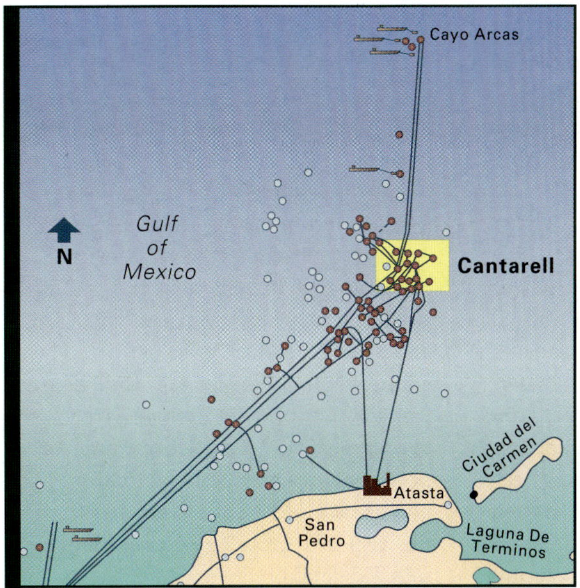

Cantarell oil field was undergoing a major upgrade and expansion project during 2001 including installation of about 250 miles (400 km) of pipelines.

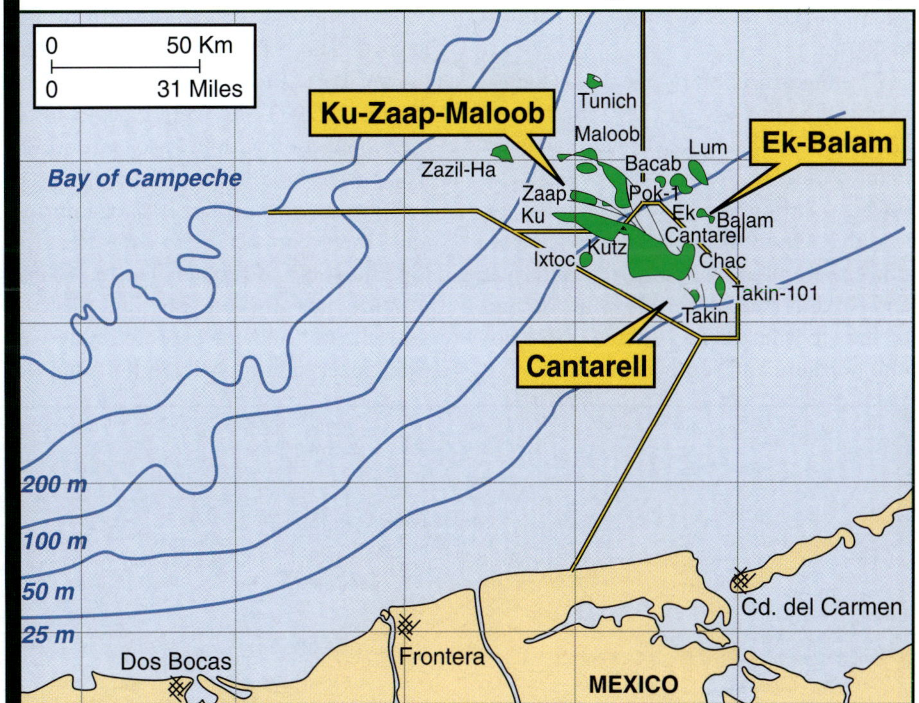

The Ku-Zaap-Maloob (KZM) complex was also undergoing expansion and upgrade, including nitrogen injection to increase production.

Burgos basin development

A March 2001 announcement revealed that Pemex awarded a contract to BJ Services Co. and Precision Drilling Corp. to drill and complete 240 wells in the Burgos basin, which covers more than 19,300 sq mi (50,000 sq km) in northeastern Mexico near Camargo. Construction of roads and drilling locations, pipeline installation, and tie-in work would be subcontracted through the partnered project. Work was to begin in mid-May 2001 and take 2 years to complete. The total value of the project was $270 million, with BJ and Precision revenues estimated at $70-$80 million each. BJ will provide cementing, fracturing, and coiled tubing services, while Precision's services will include drilling, wireline, logging, and directional drilling.

Aggressive exploration and development of the Burgos basin in northeastern Mexico could lead to substantially higher natural gas production over the next two decades, said Gas Technology Institute (GTI). Its study found that undiscovered resources in the Burgos basin might be 3-10 times current proven recovery. Estimates published by Pemex and the US Geological Survey stand at 9.7 tcf, but the GTI study suggested that another 11-65 tcf of gas could be added to this resource estimate, yielding a total undiscovered gas potential of 21-75 tcf.

The Burgos basin is an extension of the Texas Gulf Coast province with very similar geology, said GTI. Gas production has recently increased more than four-fold, to a level of 970 MMcfd in early 2001. Drilling success rates have been very high, said the study, nearly 100% for development wells and 50% for exploratory wells drilled mainly in existing fields. Although GTI expected Mexico to remain a net gas importer through 2015, exploitation of the Burgos basin could help meet growing domestic demand without greatly increasing imports from the US.

LNG terminal plans

In mid-2001, Shell Gas & Power (owned by Royal Dutch/Shell) and El Paso Global LNG announced plans to jointly develop an LNG regasification terminal at Altamira, Tamaulipas state, on Mexico's east coast. The $300 million terminal would target fast-growing gas demand in northeastern Mexico, driven largely by electric power demand growth. The initial capacity of the terminal was not revealed, but it was said to be expandable to 1.3 bcfd. The 50:50 joint venture acquired land rights at the Altamira port and began design studies in 2001.

The regasification terminal was tentatively scheduled to begin LNG imports in 2004, with gas being marketed directly to Pemex, industrial users, and power producers. Arthur Andersen LLP analysts said that Mexico could emerge as a major power supplier to southern California and Arizona during 2001-06, contributing to the division of the western US grid into northern and southern markets.

Gas pipelines

In September 2001, Tidelands Oil & Gas Corp.'s subsidiary Reef International LLC filed for construction and operating permits for a 12-inch gas pipeline, a 6-inch propane and butane pipeline, and a 6-inch standby line from Eagle Pass, Texas, into Piedras Negras, Mexico. The pipeline system will deliver product to Conagas, the local Mexican distributor, and to industrial users. Most US gas was exported to Mexico via the existing Samalayuca pipeline, which crosses the border east of El Paso, Texas, and serves power plants near Ciudad Juarez. The new pipeline crossing is needed to meet the region's growing energy demands, said Tidelands.

A consortium of Sempra, PG&E, and Mexico's Proxima Gas planned to build a 30-inch, 212-mile natural gas pipeline connecting the US and Mexican gas grids. The $230 million, 500-MMcfd North Baja Pipeline could come on stream in 2003. It would begin at an interconnection with El Paso Natural Gas Co. in Arizona, run through southeastern California and northern Baja California, and then connect with the Rosarito Pipeline south of Tijuana. The line would provide gas for an existing oil-fired power plant and for expected new gas-fired generation capacity in the region.

Baja California was experiencing rapid energy demand growth, much of which could be met with US gas, and Mexico's energy policy called for converting many oil power plants to gas by 2005. In early 2002 US regulators approved the 80-mile (129-km) US segment of the line, while the 135-mile (217-km) Mexican portion had been approved earlier.

PERU

CAPITAL: LIMA
MONETARY UNIT: NEW SOL
REFINING CAPACITY: 182,250 B/CD
OIL PRODUCTION: 93,000 B/D
OIL RESERVES: 323 MILLION BBL
GAS RESERVES: 8.7 TCF

Peru's new president Alejandro Toledo took office in July 2001, and his new energy minister signalled the government's commitment to expanding the country's upstream petroleum sector and completing privatization of the sector by year-end 2003. Peru was expected to work with Bolivia on a scheme to export natural gas to the US and Mexico through Peru.

The government continued to promote the Camisea natural gas project, aimed at North American LNG markets, and the offer of blocks near Camisea was also accelerated. Privatization of Peru's refineries remained on hold. The US Trade and Development Association granted the state company, Perupetro, $346,000 to study the feasibility of upgrading its 62,000 b/d Talara refinery to improve environmental standards.

Most of Peru's oil was produced by Argentina's Pluspetrol Peru Corp., with Perupetro and US-based PetroTech and Barrett

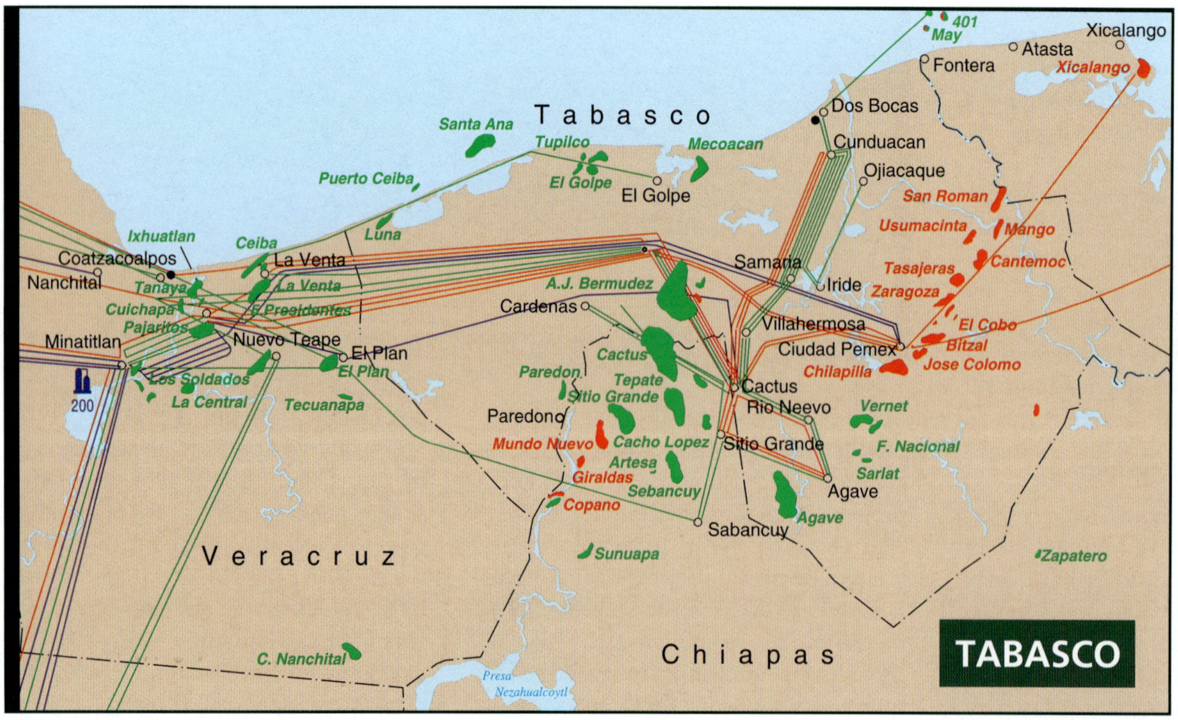

Resources Corp. accounting for most of the remaining production. The biggest producing area was Block 1-AB, operated by Pluspetrol. Perupetro reportedly reduced royalties for several companies operating in the country and planned to continue revising contract terms to stimulate investor interest.

LNG project targets deliveries to US

Peru could begin delivering LNG to US West Coast markets in 2007, according to Argentina's Pluspetrol, using gas reserves from the Camisea fields and neighboring blocks. Pluspetrol operates the field development project, and Tecgas NV operates the pipeline and transportation projects. In 2001 six companies involved in Camisea conducted a feasibility study of options including a \$1.4-\$1.5 billion, 600-MMcfd LNG plant on Peru's southern coast.

The study, led by project partner Hunt Oil Co., pegged recoverable proven and probable reserves at 7.4 tcf in San Martin and Cashiriari fields and an additional 5.8 tcf of recoverable potential in the Camisea area. It estimated Peruvian demand over 20 years at 3.5 tcf (680 MMcfd in 2022) and 3.8 tcf for exports at 550 MMcfd over 20 years. The consortium began shipping equipment to Iquitos for barging to the Camisea site 310 miles (500 km) east of Lima. The Camisea project was scheduled to begin production in 2004, and its environmental impact assessment was approved by Peruvian authorities in December 2001.

Upstream developments

In 2001, Perupetro offered a development license for Block 56 and exploration and development licenses for Blocks 57 and 58, all near the Camisea gas fields. In late 2001, four companies were considering leases of these blocks: TotalFinaElf SA, Repsol-YPF SA, Occidental Petroleum Corp., and Hunt Oil, which represents the Camisea consortium.

Block 56 holds the Pagoreni gas structure discovered by a Royal Dutch/Shell Group subsidiary in 1998; this structure and the small Mipaya structure opened previously by Shell have estimated resources of 4 tcf. Earlier in 2001, Peru had offered four unexplored blocks in deep water off the north coast, which are near Occidental's Block Z-3.

Perez Companc del Peru in October 2001 signed an exploration and production license with Perupetro for Block 99 in Peru's Ucayali basin. At minimum Perez Companc planned to reprocess 560 miles (900 km) of 2D seismic lines and drill four exploration wells. Block 99 lies directly north of Blocks 34 and 35, which Perez Companc is exploring in a 50:50 agreement with Repsol-YPF, operator of both blocks. In mid-2001, Repsol drilled its first wildcat in Block 35 on the bank of the Urubamba River with Parker Drilling Co. as contractor.

In mid-2001, Pluspetrol announced plans to hike production from Block 1-AB, which was producing 38,000 b/d, by 25% over three years under a new development contract. Plans included drilling 14 development wells and reworking 30 wells.

Maple Production del Peru, a unit of US-based Maple Gas Co., in early 2001 planned to begin a \$10 million exploration project in Block 31-E in northern Ucayali basin. Occidental also planned to evaluate Blocks 21 and 22 in the Pachitea and Ucayali basins.

Conex SAC conducted modeling of geological factors indicating that Peru's offshore Tumbes-Talara area has certain similarities to other productive basins. Exploration of Blocks Z-4 and Z-5 in the area could signal a new play in offshore Peru, said Conex.

NGL-GTL venture

In 2001, Syntroleum Corp. and BPZ Energy were planning an integrated NGL/GTL project in northwestern Peru. Syntroleum held 95% interest in the project and was discussing other companies' participation. In phase one, Syntroleum planned to build a 2,000 b/d NGL plant to upgrade and replace an existing plant. The new facility would process gas and extract NGLs from 30 MMscfd of associated gas from oil fields around Talara as well as other associated gas not being produced.

At full capacity in phase two, the plant's capacity would be expanded to process 100 MMscfd of gas from Talara and extract 10,000-11,000 b/d total of NGLs and GTLs. Also, a 5,000 b/d GTL plant would be built to produce clean synfuels from residue gas. The last phase of the project would expand the GTL plant facility by 20,000-40,000 b/d as the local gas supply increases. Additional subsequent work might include exploitation of undeveloped discoveries and new exploratory drilling on Block Z-1 off Mancora to Peru's border with Ecuador. The Z-1 exploration stage would involve three exploratory wells and seismic work, estimated by Perupetro to cost at least \$26 million.

Pipeline planning

Argentina's Pluspetrol in 2001 revealed plans to build an 78-mile (125-km), 8-inch oil pipeline between two Peruvian assets, Nueva Esperanza in Block 8 and the Jibaro Marshalling area in Block 1-AB, both operated by Pluspetrol. The pipeline, which could begin operating in 2002, would carry a mix of light oil from Block 8 and heavy oil from the eastern area of Block 1-AB. Together the

blocks average 60,000 b/d of production.

During 2001, Peru and Ecuador resumed talks on the possibility of linking their crude oil pipeline systems. In late 2001, Peru's government agreed with the Canadian government to conduct feasibility studies for building a gas pipeline extending into Camisea's network, possibly connecting to the southern port of Ilo.

TRINIDAD & TOBAGO

CAPITAL: PORT OF SPAIN
MONETARY UNIT: DOLLAR
REFINING CAPACITY: 160,000 B/CD
OIL PRODUCTION: 112,000 B/D
OIL RESERVES: 716 MILLION BBL
GAS RESERVES: 23.5 TCF

As the largest LNG exporter to the US, Trinidad and Tobago has also become one of the major gas development centers in the world. Although oil continued in 2001 to earn more revenues, gas was expected to surpass oil's position in the near future. However, the Trinidad government continued to encourage investment in oil exploration as well.

In late 2001, BHP Billiton Ltd. claimed a significant oil discovery off Trinidad's east coast, as indicated by the results of its Canteen-1 delineation well, which was drilled to a total depth of 7,070 feet in Block 2c in 164 feet of water. The well encountered 700 gross feet of hydrocarbon-bearing sands including 200 feet of net oil pay and 179 feet of net gas pay. It was tested at a rate of 3,700 b/d of 33º gravity oil from a 60-foot interval through a 1-1/8 inch choke. BHP said that its data indicated a high-quality

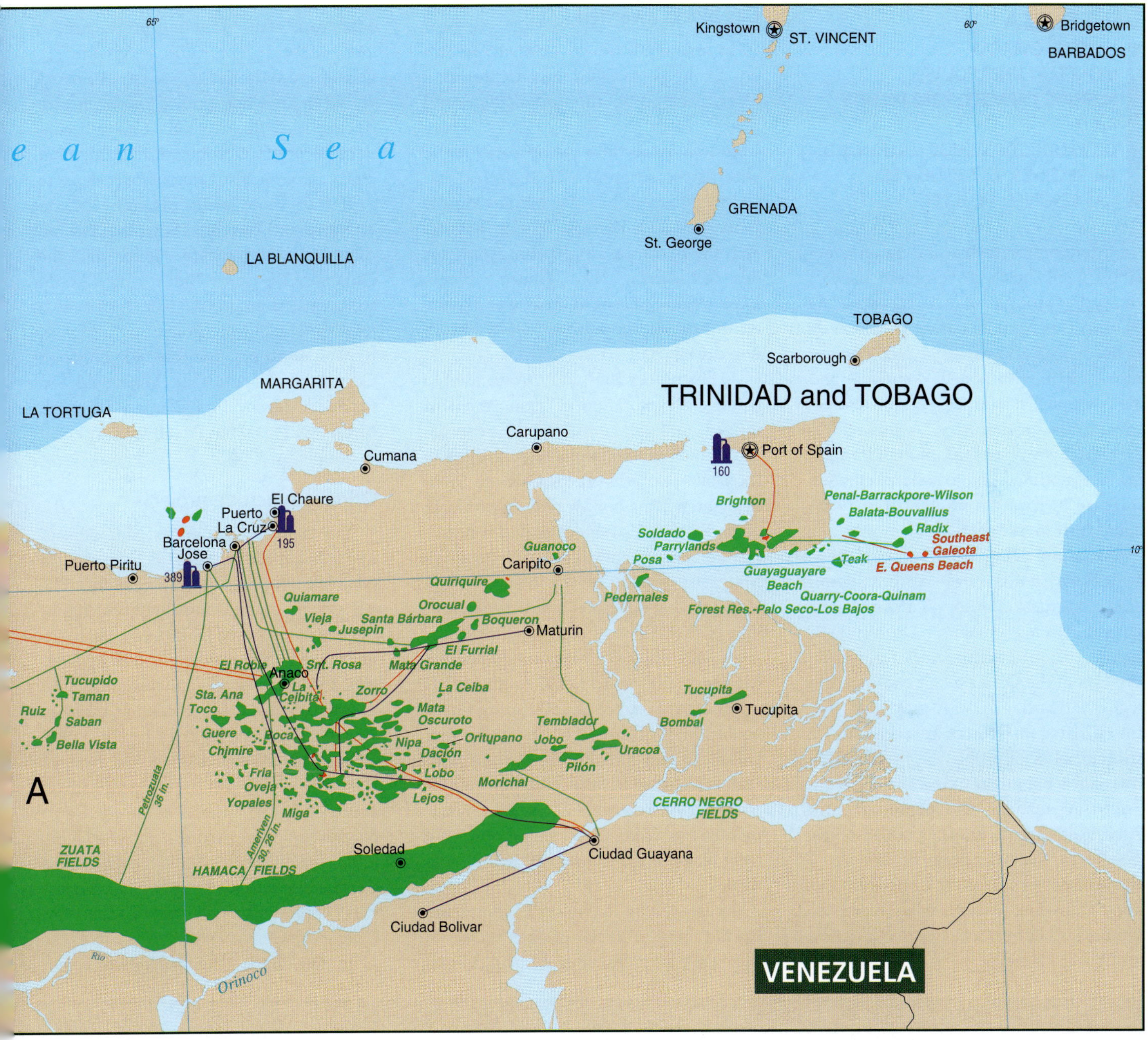

reservoir formation similar to the Kairi-1 well.

BHP operated the drilling program with a 45% interest; its partners were Elf Petroleum Trinidad BV (30%) and Talisman Energy Inc. (25%). Earlier in 2001, offshore Block 3a was awarded to a group of companies led by BHP Petroleum (Americas) Inc. Partner Talisman said that Block 3a could also contain an extension of the Kairi-1 play.

Vintage Petroleum Inc. in 2001 tested a high gas flow on its Carapal Ridge No. 1 wildcat on Trinidad's Central block. The well flowed more than 50 MMcfd of gas and 1,500 b/d of condensate.

Following two discoveries in 2000, BP PLC affiliate BP Trinidad and Tobago in 2001 said that its Cashima-1 well off Trinidad's east coast found more than 1 tcf of gas. Cashima, in 263 feet of water, was drilled to 13,160 feet and found 520 feet of gas-bearing sand in four pay horizons. BP planned to drill several more wells during 2001, including two more exploration wells on the shelf, one in deep water in Block 27, and an appraisal well in Red Mango field.

Petrobras Trinidad reached a farmout agreement with Norsk Hydro for deepwater Block 27 covering 471 sq mi (1,220 sq km). The first of two exploration wells was scheduled to be spudded in 2001.

Earlier in 2001, Atlantic LNG Co. of Trinidad and Tobago awarded the design contract for a fourth process train at its LNG plant in Point Fertin. Atlantic LNG was already building its second and third trains, expected to come on stream in 2002 and 2003, respectively.

In mid-2001, Methanex Corp. and BP decided to proceed with construction of a \$400 million, 1.7 million tonne/yr methanol plant in Trinidad. The Atlas facility was expected to begin market delivery in 2004.

VENEZUELA

CAPITAL: CARACAS
MONETARY UNIT: BOLIVAR
REFINING CAPACITY: 1.282 MILLION B/CD
OIL PRODUCTION: 2.830 MILLION B/D
OIL RESERVES: 77.7 BILLION BBL
GAS RESERVES: 147.6 TCF

After political debate and union opposition, Venezuela's government in 2001 pushed through plans to separate PDVSA Gas from its parent Petroleos de Venezuela (PDVSA), as a step toward replacing it with a new company, and also supported private sector involvement in the gas industry. A new hydrocarbons law was passed, though it might later be modified due to concerns that an increase in royalties to 30% from 16.6% could hurt investment.

As a founding member of OPEC, Venezuela holds the western hemisphere's largest oil reserves and one of its largest refining systems. During 2001, heavy oil production and upgrading projects in the Orinoco belt continued to move forward. PDVSA estimated that the 54,000 sq mi Orinoco area contains 1.2 trillion bbl of oil in place, of which 100-300 billion bbl might ultimately be recovered. Venezuela also contains Latin America's largest natural gas reserves, mostly in the Maracaibo basin. As a result of gas sector reforms, opportunities abound in four main areas: Norte de Ambrosio, Barrancas, Guarico-Cojedes, and Yucal-Placer.

Gas blocks awarded

In June 2001, the government awarded six nonassociated gas exploration blocks, including the promising Yucal Placer North and South Blocks, which went to a consortium led by TotalFinaElf with partners Repsol-YPF, Inelectra, and Otepi. Pluspetrol of Argentina won the Barbacoas and Tiznado Blocks, Repsol-YPF the Barrancas, and Perez Companc de Venezuela the Tinaco Block. Exploration of licensed areas was expected to generate investments of $700 million during 2001-2009.

In September 2001, PDVSA decided to reactivate exploration of offshore basins for gas and lighter crude oil. The company planned to start exploratory drilling in the Delta platform off the Orinoco River near the marine border with Trinidad in 2002. PDVSA also said it would run geophysical surveys north of La Tortuga and Margarita islands, to be completed in 2002 and costing a total of nearly $22 million. The work was to include the recording of 435 miles (700 km) of 2D seismic data and 1,200 sq mi (3,100 sq km) of 3D seismic, over a total of 1,930 sq mi (5,000 sq km) in 320-3,200 feet (100-1,000 m) of water. In late 2001 PDVSA agreed with Statoil to increase oil production, which was 18,000 b/d, from Ceuta Area 2 Sur field in Lake Maracaibo, using new well technology in the technically challenging field.

In late 2001, Repsol-YPF completed the last stage of its natural gas project in Venezuela's Quiriquire field in Monangas state, doubling production to 261 MMcfd (7 million cu m/d). Quiriquire field reserves were estimated at 145 million boe, and it had been producing 13,000 b/d of oil. Gas and condensate production began in September 2001. Repsol also announced plans to drill four more wells through 2002 to appraise and develop its world-class oil discovery in Tacata field in the Quiamare-La Ceiba Block. Its TAG-14 well, drilled to 16,000 feet, tested at a total of 21,000 b/d from four intervals. Repsol partners include ExxonMobil Corp. and Tecpetrol de Venezuela SA. Production from the Quiamare-La Ceiba Block was 14,000 boe/d but was expected to double by 2004.

Orinoco syncrude projects

In August 2001, the first shipment of upgraded crude oil from the Cerro Negro complex left port for a US refinery. The 8.5º gravity crude was diluted with naphtha and transported by pipeline 190 miles to the port of Jose, where it was upgraded into a 16º gravity synthetic crude. The upgrader's installed processing capacity was 120,000 b/d of extra-heavy crude, which it converts to 108,000 b/d of upgraded crude and byproducts such as sulfur and coke. The $790 million Cerro Negro project's partners were PDVSA and ExxonMobil (42% each) and Germany's Veba Oel (16%). The project was operated by ExxonMobil-led Operadora Cerro Negro.

In late 2001, ChevronTexaco began

Venezuela began exploring offshore basins for gas and lighter crude in 2001, including exploratory drilling off the Orinoco River Delta and seismic work north of La Tortuga and Margarita islands.

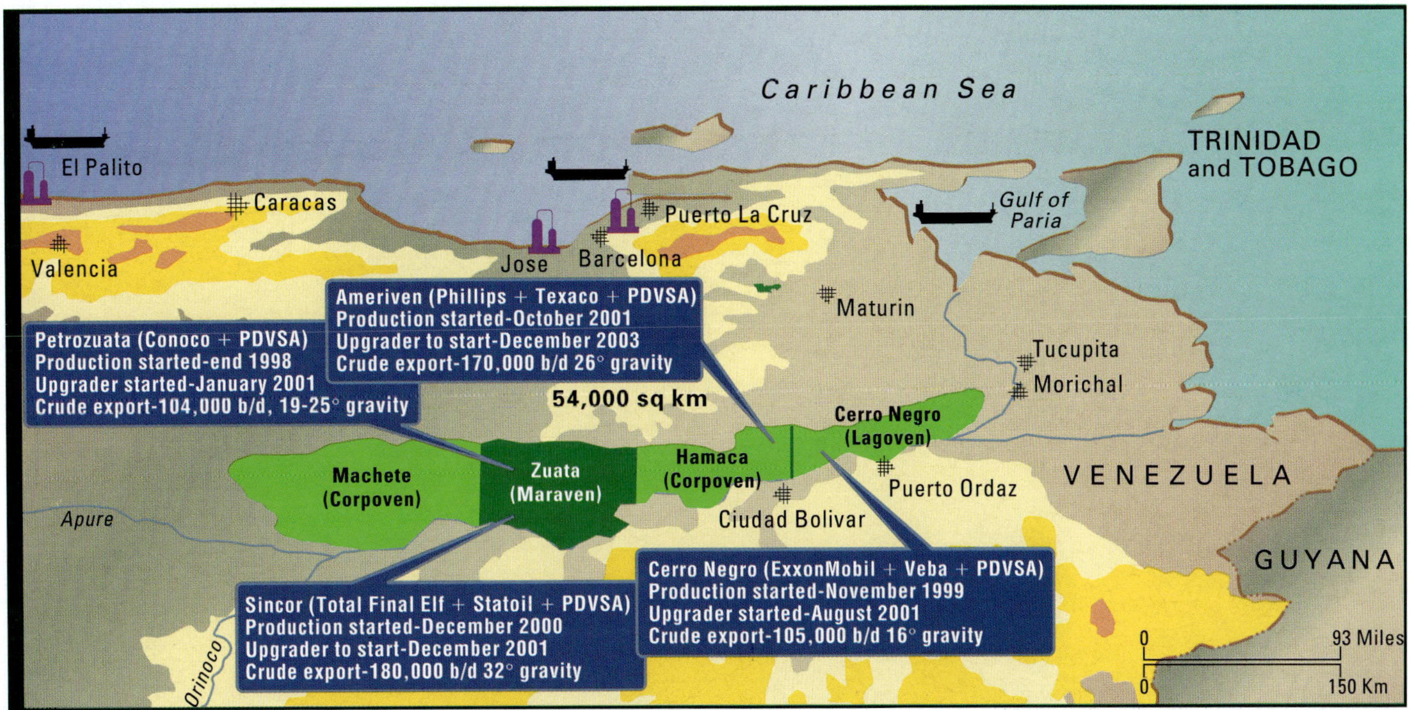

Four projects were exploiting portions of Venezuela's Orinoco belt by investing in production facilities and upgraders to transform the heavy oil into lighter syncrudes.

production from Hamaca field, also in Venezuela's Orinoco belt. Hamaca was producing 30,000 b/d of 8.5º gravity oil, which was being blended with lighter crudes and sold internationally. The 253-sq mi (657-sq km) Hamaca area contains more than 30 billion bbl of oil. Over the $1.1 billion development project's 34-year life, 2.1 billion bbl could be produced, the partners estimated. The project includes construction of a heavy crude upgrade unit at the Jose industrial complex, to be brought on-stream in 2004, which will upgrade the field's oil production to 26º gravity crude. At peak field production of 190,000 b/d, the upgrading process will yield 180,000 b/d of 26º gravity oil. Other partners in the project are PDVSA and Phillips Petroleum Co.

Another heavy crude upgrader at the Jose complex was nearing completion in late 2001 and on schedule for commercial lifting starting in February 2002, as part of a $4.2 billion integrated upstream and downstream project being operated by Sincor (formerly Sincrudos de Oriente Sincor CA). One-third of the investment was for the initial 250 wells and main production station near San Diego de Cabrutica in the Orinoco heavy-oil belt, while the other two-thirds was earmarked for the 180,000 b/d upgrader, storage, and export facilities at Jose. The project encompasses transportation via 124-mile (200-km) pipeline and upgrading of the heavy crude to a 32º gravity crude for export. Sincor calls this synthetic crude "Zuata Sweet" and was planning to market it in early 2002. Sincor partners were TotalFinaElf (47%), PDVSA (38%), and Statoil ASA (15%).

In June 2001, PDVSA signed an agreement for operation of the oil storage and loading terminal at the Jose complex by a consortium of Williams, Enbridge Inc., and Northville Industries Inc. The port complex at Anzoategui can handle up to 800,000 b/d of oil. The facilities' platform has a berthing capacity for up to 34 ships per month.

PDVSA also began engineering, procurement and construction in 2001 of an aquaconversion unit at its Isla refinery, which was scheduled to begin commercial operation in mid-2002.

Along with ChevronTexaco and Ecopetrol, PDVSA planned to study a 124-mile (200-km) gas pipeline from northeast Colombian fields in the Guajira region to markets near Venezuela's Lake Maracaibo. IPE

The world trusts us.
Pipelines bring water, oil and natural gas right where they are needed. Over the years, EUROPIPE and its predecessor companies have manufactured over 30 million tonnes of large-diameter pipe – enough to circle the earth twice. Thanks for counting on us.
EUROPIPE GmbH Germany · Formerstraße 49 · D-40878 Ratingen · www.europipe.com

EUROPE

ALBANIA

CAPITAL: TIRANA
MONETARY UNIT: LEK
REFINING CAPACITY: 26,300 B/CD
OIL PRODUCTION: 6,000 B/D
OIL RESERVES: 165 MILLION BBL
GAS RESERVES: 100 BCF

In the Balkan region, Albania holds the second-largest oil reserves (after Romania), and oil is its most important industry, with state company Albpetrol being the major player. Domestic production in 2001 accounted for roughly half of oil consumption. Albania produced enough natural gas in 2001 to satisfy its small domestic demand, though gas imports were expected to rise and new pipelines were being planned to deliver more gas.

In mid-2001, Lundin Albania Ltd., a unit of Sweden's Lundin Oil AB, withdrew from Albania's onshore Block 2, citing disappointing drilling results on the Shpiragu-1 well.

Albania was poised to become more important as part of a Balkan transit center than an oil producer. The US-based Albanian Macedonian Bulgarian Oil Co. (AMBO) continued plans in 2001 to construct a 560-mile (900-km), $1.13 billion crude oil pipeline from the Bulgarian Black Sea port of Burgas through Macedonia to the Albanian port of Vlore. Estimated capacity would be 750,000 b/d, which would help reduce tanker traffic through the Bosporus Straits and Aegean Sea. The line might also accommodate oil coming indirectly from the Caspian region. AMBO began to assemble financing for the project in 2001, and construction was expected to take three years, with the pipeline scheduled for 2005 operation.

AUSTRIA

CAPITAL: VIENNA
MONETARY UNIT: SCHILLING*
REFINING CAPACITY: 208,600 B/CD
OIL PRODUCTION: 18,500 B/D
OIL RESERVES: 85.7 MILLION BBL
GAS RESERVES: 915 BCF
* Switching to euro in 2002.

Austria's largest oil and gas company, OMV AG, made little progress during 2001 in its attempt to merge with other leading regional players such as Poland's PKN Orlen and Hungary's MOL Rt. A geographically dominant, unified company would be well-positioned in the region's oil products market, but financing and political issues continued to thwart consolidation.

Work continued in 2001 on the looping of the TransAustria gas pipeline, which moves Russian natural gas into Austria from Slovakia and across the country into Italy, Slovenia, and Croatia. Construction of the 234-mile (377-km) segment, which began in 2000, aims to increase the line's capacity to 1.23 tcf/yr (33 billion cu m/yr).

Austria gets around 5 million bbl/yr from fields that started production in 1938. The reason these fields can remain in economic production is attributable to bacteria and research into oil spill cleanup at Norway's University of Bergen by scientist Egil Sunde, now with Statoil.

BELGIUM

CAPITAL: BRUSSELS
MONETARY UNIT: FRANC*
REFINING CAPACITY: 791,013 B/CD
OIL PRODUCTION: N/A
OIL RESERVES: N/A
GAS RESERVES: N/A
* Switching to Euro in 2002.

In late 2001 Distrigaz SA, Belgium's monopoly gas distributor, received approval from the Belgian government to

EUROPE
Legend
Oil field
Oil sand
Gas field
Crude oil pipeline
Natural gas pipeline
Products pipeline
Pipeline planned or under construction
Refinery in operation Refinery capacity in 1,000 b/d
Tanker terminal
Cities
Capital
International boundary
Water depth
0 to 200 m
200 m and deeper
0 50 100 150 Miles
0 100 200 Kilometers
IRELAND
N. IRELAND
ENGLAND
WALES
NETHERLANDS
BELGIUM
LUX.
GERMANY
DENMARK
FRANCE
SWITZERLAND
CORSICA
SARDINIA
PORTUGAL
SPAIN
MINORCA
MAJORCA
IBIZA
MOROCCO
ALGERIA
TUNISIA
North Sea
Irish Sea
Celtic Sea
Bristol Channel
Lyme Bay
Bay of Biscay
Gulf of Cadiz
Mediterranean Sea
Londonderry
Belfast
Sligo
Ballina
Galway
Dublin
Foynes
Limerick
Tralee
Cork
Waterford
Whitegate
Kinsale Head
Morecambe
Newcastle
Carlisle
Darlington
Tees
Middlesbrough
Barrow
Lancaster
Heysham
Bradford
Manchester
Liverpool
Sheffield
Killingholme
Grimsby
Lincoln
Theddlethorpe
Bacton
Birmingham
Cambridge
Ipswich
Coryton
Milford Haven
Pembroke
Neath
Gloucester
Oxford
London
Bath
Exeter
Plymouth
Southampton
Fawley
Portsmouth
Weymouth
Amsterdam
Rotterdam
Groningen
Emden
Antwerp
Brussels
Dunkerque
Gelsenkirchen
Duisburg
Essen
Dusseldorf
Cologne
Wesseling
Bonn
Dortmund
Hamburg
Kiel
Brunsbuttel
Wilhemshaven
Cuxhaven
Heide
Bremen
Frankfurt-am-Main
Karlsruhe
Ingolstadt
Munich
Vohburg
Luxembourg
Metz
Le Havre
Gonfreville L'Orcher
Port Jerome
Notre Dame de Gravenchon
Paris
Cherbourg
Brest
Quimper
Lorient
Vannes
Donges
Nantes
Angers
Orleans
Blois
Laval
St. Malo
Les Sables
La Rochelle
Royan
Le Verdon
Pauillac
Ambes
Bordeaux
Bayonne
Toulouse
Lacq
Carcassonne
Narbonne
Marseille
Toulon
Lyon
Feyzin
Grenoble
Geneva
Zurich
Bern
Turin
Genoa
Milan
Nice
Monaco
Bastia
Ferrol
La Coruna
Aviles
Santander
Bilbao
Vitoria
Burgos
Logrono
Opporto
Zamora
Vallodolid
Zargoza
Madrid
Barcelona
Tarragona
Castellon de la Plana
Vallencia
Puertollano
Cordoba
Sevilla
Huelva
Cadiz
Malaga
Algeciras
Almeria
Cartegena
Murcia
Lisbon
Setubal
Sines
Cabo Ruivo
Algiers
Bejaia
Skikda
Arzew
Oran
Tunis

transfer its international trading activities to a new company that will take over the Distrigaz name. Operation of the gas transport grid, storage facilities, and LNG terminal remained under the original company, which was renamed Fluxys. Distrigaz also was awarded a five-year contract by Rhodia Energy to supply a total volume of 335 bcf (3 billion cu m) of gas to France's Chalampe chemical complex, beginning in 2002.

A new 380,000 ton per year (tpy) polypropylene plant was scheduled to come on line by year-end 2001 in Feluy, Belgium. Atofina, TotalFinaElf's chemical branch, was planning to produce bimodal high-density polyethylene by the end of 2002 using metallocene technology.

The NC2 ethylene plant in Antwerp was being upgraded and expanded during 2001, with completion scheduled for late 2002. Under an engineering, procurement, and construction contract with Fina Antwerp Olefins, M.W. Kellogg Ltd. was to install a new pyrolysis furnace and modify the recovery section of the facility. The upgrade project would improve the plant's energy efficiency.

In late 2001, a unit at Belgium's 300,000 b/d Antwerp refinery was prepared for shutdown by Camom and partner GTI Electromechanical NV. Also, the reformer unit at the 65,000 b/d refinery in Antwerp was shut down by Petroplus International NV following a brief fire in June 2001, which caused no injuries or danger to the environment.

Belgium's Zeebrugge gas terminal was shut down for three weeks in April-May 2001 following the discovery of a minor leak into the trunkline transporting natural gas from the Norwegian North Sea. It was later determined that a valve controlling gas flow from the Sleipner East field into the Zeepipe was leaking.

BULGARIA

CAPITAL: SOFIA
MONETARY UNIT: LEV
REFINING CAPACITY: 115,240 B/CD
OIL PRODUCTION: 1,000 B/D
OIL RESERVES: 15 MILLION BBL
GAS RESERVES: 210 BCF

Although the pace of economic reform in Bulgaria picked up during 2001, the country's transition to democracy and a market-based economy continued to lag behind other parts of Europe. However, Bulgaria could play a strategic role in the European transport corridor to bring Caspian oil exports to European markets.

In mid-2001, Bulgaria's new prime minister Simeon Saxe-Colburg pledged to attract foreign investors and pursue European Union (EU) membership. The country's gas market took early steps toward partial liberalization, and the government planned to allow large industrial gas consumers and gas distributors, starting in early 2002, to negotiate imports of gas directly from external suppliers, circumventing Bulgargaz, which owns and operates the country's 1,554-mile pipeline network (including more than 400 miles of transit pipelines).

Gas exploration and production

In May 2001, Bulgaria signed natural gas concessions to the UK's Petreco Sarl, a subsidiary of Melrose Resources PLC. Petreco planned to explore the shallow-water Kaliakra 99 concession in the Bulgarian sector of the Black Sea and also extract gas from the Galata field, which has estimated reserves of 49-53 bcf and probable reserves of 80 bcf.

The Galata concession was carved out of a 652 sq mi (1,690 sq km) area of exploration Block 91-III, where Petreco had earlier wrapped up a 453-mile (729-km) 2D seismic survey over a number of leads in the northern part of the block. Kaliakra 99 lies southeast of Block 91-III and crosses over parts of the old Blocks II and IV. The new area of 1,004 sq mi (2,601 sq km), said Melrose, includes a number of attractive prospects in water depths of 66-1,640 feet (20-500 m), along with the 1986 Samotino More R-1 discovery, which was tested at rates up to 9.5 MMcfd of gas and 204 b/d of condensate.

Melrose announced in late 2001 that Petreco contracted for sale of Galata gas to Bulgargaz, which will purchase 14.9 bcf (400 million cu m) per year. At a gas price linked to area gas prices, a net price in the range of $2.60-$2.75/Mcf ($92-$97/thousand cu m) was anticipated. First production under the contract was set for mid-2003. Petreco also entered into a gas transportation contract with Bulgargaz that will let Petreco sell gas directly to Bulgarian consumers. Petreco was discussing another gas sales contract with a second purchaser.

Ramco Energy PLC agreed in 2001 to farm into Anschutz Bulgaria Ltd.'s A-Lovech license area in northwestern Bulgaria, subject to government approvals. A-Lovech lies about 50 miles northeast of Sofia and covers 1,832 sq mi (4,744 sq km). The acreage is a thrust belt play similar to that in Ramco's Carpathian Block in southern Poland. Anschutz, which specializes in thrust belts, has identified several deep structures around 13,100 feet (4,000 m). Minimum obligations of the agreement, which gives Ramco a 20% interest and Anschutz 80%, included a two-year geological work program and reprocessing of existing seismic data by April 2002, then acquisition of 62 miles (100 km) of seismic in 2003.

Boosting gas pipeline capacity

Bulgaria continued during 2001 to increase its natural gas transit capacity by widening existing facilities and building compressor stations. With a 2001 budget of $45 million, Bulgargaz enlarged the network to pump more Russian gas to Balkan neighbors Greece, Macedonia, and Turkey. Volumes were expected to increase from 423 bcf in 2000 to 494 bcf after 2002 and to 670 bcf by 2010.

Bulgaria and Greece during 2001 continued to plan an oil pipeline from the Bulgarian Black Sea port of Burgas to Alexandropoulis on the Aegean Sea, but Russian backing to supply the oil was considered critical to the project going forward. The three countries were conducting a $2.2 million feasibility study and continued to negotiate launch of a Trans-Balkan Oil Company for development and construction. The pipeline would offer another alternative to transport through the Bosporus, in addition to the AMBO pipeline also being planned. Several Bulgarian companies have indicated an interest in investing in the Burgas pipeline.

CROATIA

CAPITAL: ZAGREB
MONETARY UNIT: KUNA
REFINING CAPACITY: 292,990 B/CD
OIL PRODUCTION: 21,300 B/D
OIL RESERVES: 92.2 MILLION BBL
GAS RESERVES: 1.2 TCF

Croatia's government hired advice during 2001 on how to reform its oil and gas markets. PricewaterhouseCoopers PLC and Deutsche Bank AG were to advise the Balkan country regarding a restructuring of the state-owned petroleum company, Industrija Nafte DD Zagreb (INA), an integrated firm involved in E&P, refining, and marketing of oil and gas. The joint report was to include an examination of the regulatory framework for Croatia's oil products market and natural gas energy sector. In July 2001, laws were passed to liberalize the energy sector in preparation for privatization and to comply with EU regulations.

Croatia continued to seek to increase its domestic gas production, and by early 2001, four exploration wells had been completed, including INAGIP's offshore gas field discovery at Marica 1, with unproven reserves of 106 bcf, in the northern Adriatic Sea's Aiza Laura Block. INA, through INA's joint venture with ENI SPA subsidiary Agip, also made progress in 2001 on its scheme to tie in another four satellite gas fields to the Ivana A platform. It was expected to have production of 2.5 bcf (67 million cu m). The Ivana area was estimated to contain 279 bcf.

The platform tie-in is integral to larger plans, the Gas Energy Adria (GEA) project, which is aimed at delivering gas to Croatia and exporting it to central Europe through a 200-mile (321-km) pipeline running from Casal Borsetti in Italy via Pula and on into the Balkans. The 24-inch line would consist of an 81-mile (130-km) offshore section and a 119-mile (191-km) onshore section from Pula to Rijeka and then to Karlovac, where it would connect to Croatia's existing gas pipeline. Total capacity was planned to be 186 bcf/yr (5 billion cu m/yr), though the pipeline would probably operate initially at lower volumes prior to installation of compressors. Construction was expected to take three years.

INA was reportedly very interested in connecting its fields with the greater European network. The company calculated that it had proved and probable gas reserves of 1.34 bcf (36 million cu m) in Croatia and another 328 MMcf (8.8 million cu m) abroad. INA said that the GEA development could become one of the largest foreign investment projects in Croatia.

CZECH REPUBLIC

CAPITAL: PRAGUE
MONETARY UNIT: KORUNA
REFINING CAPACITY: 198,000 B/CD
OIL PRODUCTION: 3,500 B/D
OIL RESERVES: 15 MILLION BBL
GAS RESERVES: 140 BCF

The Czech Republic's Vienna basin was the site of an oil discovery in early 2002 at the Postorna-1 wildcat drilled by Carpathian Resources Ltd. The well intersected its primary target, the Miocene-aged shallow marine deltaic sand identified as the Lab horizon, at 4,757-4,793 feet (1,450-1,461 m). The well found strong hydrocarbon indications, and the sand was thicker than expected. After perforation, the hole flowed 626 cfd (16.8 cu m/d). Carpathian prepared for a long-term production test and permanent production facilities.

Postorna-1, the company's first hole drilled on the Breclav license, reached 5,420 feet (1,652 m) total depth in November 2001. Carpathian as operator held 17.65% interest, with partners Unigeo AS, Ceska Naftarska Spolecnost AS, and Geocan Energy Inc. each holding 27.45%. The 11 sq mi (28.5 sq km) Breclav license is flanked by producing fields – to the east Hrusky, the Vienna basin's largest, which has produced 11 million bbl of oil and 64 bcf of gas, and Lednice to the west, which has produced 2.1 bcf.

Earlier in 2001, Unigeo and Geocan Energy acquired the 12,355-acre Bucovice Block, in addition to their previous acquisition of the Rostin and Breclav Blocks. This brings the companies' land holdings in the Czech Republic to 56,647 acres.

In late 2001, the Czech Republic sold its 63% stake in the oil and petrochemical refining company, Unipetrol AS, to Agrofert SA, a Czech-Swiss chemical company, in a bid backed by Conoco Inc., which was to acquire Agrofert for $168.3 million in 2002. Conoco will take over Ceska Rafinerska AS, the largest refiner in the country, and Benzina AS, a service station network. Agrofert will keep the rest of Unipetrol, including petrochemical makers Chemopetrol AS and Kaucuk AS.

Also late in the year, the government sold 97% of the national gas utility Transgas and 46-58% of each of its eight regional gas utilities. Germany's RWE AG acquired Transgas and gained the majority in six of the eight utilities. Earlier in 2001, the government had postponed privatization of the Czech electricity sector after rejecting several bids from Western European companies for the state's majority shares in electrical engineering firm CEZ. Czech officials promised to evaluate the future of the sector in 2002.

DENMARK

CAPITAL: COPENHAGEN
MONETARY UNIT: KRONE
REFINING CAPACITY: 176,400 B/CD
OIL PRODUCTION: 335,000 B/D
OIL RESERVES: 1.11 BILLION BBL
GAS RESERVES: 2.7 TCF

Denmark's national oil and gas company agreed to deliver 74 bcf/yr (2 billion cu m/yr) of natural gas to Poland beginning in 2003. The export deal was signed in mid-2001 by Dansk Olie ag Naturgas AS (DONG) and Polskie Gomictwo Naftowe I Gazownictow SA (POGC). The deal requires that DONG also reach agreement with Denmark's Dansk Undergrunds Consortium (DUC) to increase gas deliveries from Danish North Sea gas fields, bringing export volumes to Poland up to 596 bcf (16 billion cu m) by 2011. At year-end 2001, DUC reported that production of natural gas from its offshore fields rose 7.1% compared with December 2000, to 27.0 bcf (725 million cu m).

Denmark's efforts to open up its gas market were expected to proceed more quickly than in most other European countries, as legislative amendments to unbundle DONG's operations were considered.

NORTHERN EUROPE
0 50 100 150 Miles
0 100 200 Kilometers
Norwegian Sea
North Sea
Irish Sea
Celtic Sea
Donegal Bay
Galway Bay
Moray Firth
Firth of Forth
Firth of Clyde
North Channel
FAEROE I.
SHETLAND I.
ORKNEY I.
NORWAY
SCOTLAND
N. IRELAND
IRELAND
WALES
ENGLAND
DENMARK
NETHERLANDS
BELGIUM
LUX.
GERMANY
SWITZERLAND
Midgard
Åsgard
Tyrihans
Draugen
Njord
Åsgard Transport
Agat
Troll
Tjeldbergodden
Trondheim
Sullom Voe
Lerwick
Mongstad
Kollsnes
Sotra
Bergen
Oslo
Slagentangen
Vall oy Tonsberg
Karsto
Kopervik
Stavanger
Sola
Arendal
Mandal
Goteborg
Skagen
Kirkwall
Flotta
Wick
Stornoway
Nigg Bay
Inverness
St. Fergus
Mallaig
Cruden Bay
Aberdeen
Dundee
Finnart
Ardyne Point
Grangemouth
Edinburgh
Glasgow
Ardrossan
Corrib
Londonderry
Moffat
Belfast
Carlisle
Newcastle
Port Clarence
Sligo
Ballina
Darlington
Tees
Middlesbrough
Barrow
Lancaster
Scarborough
Heysham
Morecambe
Bradford
Ballough
Galway
Dublin
Formby
Leeds
Hull
Manchester
Liverpool
Eastham
Grimsby
Foynes
Limerick
Arklow
Tralee
Ellesmere Port
Theddlethorpe
Lincoln
Nottingham
Wexford
Waterford
Bacton
Aberystwyth
Birmingham
Leicester
Cork
Whitegate
Bantry
Galley Head
Seven Heads
Helvick
Ardmore
Kinsale Head
Milford Haven
Pembroke
Llandarcy
Coventry
Northampton
Cambridge
Neath
Gloucester
Ipswich
Oxford
London
Bath
Reading
Exeter
Bournemouth
Fawley
Southampton
Portsmouth
Plymouth
Weymouth
Arhus
Fredericia
Kaergard
Esbjerg
Kalundborg
Skaelskor
Kiel
Heide
Brunsbuttel
Cuxhaven
Wilhelmshaven
Schwerin
Harburg
Hamburg
Groningen
Emden
Grasbrook
Amsterdam
Salzbergen
Meerdf
Roxforrde
Rotterdam
Gelsenkirchen
Dortmund
Antwerp
Duisburg
Essen
Dunkerque
Calais
Dusseldorf
Cologne
Wesseling
Eisenach
Lille
Brussels
Liege
Verviers
Bonn
Suhl
Frankfurt-am-Main
Maubeuge
Notre Dame de Gravenchon
Port Jerome
Cherbourg
Le Havre
Rouen
Grandpuits
Luxembourg
Gonfreville L'Orcher
Gargenville
Vert-La-Gravelle
Reims
Velye-Germinon
Bussey-Lettree
Metz
Petite Couronne
Plestin Les Greves
St. Malo
Brest
Paris
Villeperdue
Dommartin-Lettre
Courdemanges
Karlsruhe
Marolles
Valence en Brie
Soudron
Grandville
Trois Fontaines
Vohburg
Quimper
Rennes
Laval
Villemar
Aufferville
Donnemarie
Forcelles
Ingolstadt
Lorient
Flacy
St. Martin-de-Bossenay
Baden-Wurttemberg
Vannes
Gisy-les-Nobles
Joigny
Bourbonne
Reichstett-Vendenheim
Orleans
Chateaurenard
Blois
Donges
Nantes
Zurich

EASTERN EUROPE
Copenhagen
Malmo
BORNHOLM
RUGEN
Kaliningrad
Krasnoborskoye
Ushakovskoye
Vilnius
Gdansk
Grodno
Baranovich
Unecha
Kursk
BELARUS
Davydovskoye
Minsk
Vishanskoye
Mozyr
Kharkov
Reinkenhagen
Kolberg
Rostock
Stettin
Schwedt
3 Lines
2 Lines
Pinsk
Kobrin
Plock
Warsaw
Brest
Ovruch
Berlin
Vestrup
Rudersdorf
Poznan
Frankfut-an-der-Oder
Lodz
Pulawy
Kovel
Kiev
Magdeburg
Staakow
Grunberg
Radom
Lutsk
Rovno
Cottbus
Guben
POLAND
Berdichev
UKRAINE
Dubno
Sandomierce
Rozwadow
Staro-Konstantinov
Kielce
Brody
Kremenchug
Dresden
Grobla
Partinia
Lubaczow
Vinnitsa
Zeitz
Jaroslaw
Ternopol
Trzebina
Jasto
Rudki
Litvinov
Kralupy
Lvov
Pardubice
Dashava
Kolin
Czechowice
Kradobnensk-Kalusz-
Waidhaus
Prague
Pribor
Zukow
Stryy
Gorlice
Lugskoye
Starynj
Mikova
Strazske
Dolina
Bitkov
Kosov
MOLDOVA
CZECH REP.
Brno
Uzhgorod
Nadvornaya
Chernortsy
Kosmachskoye
Krasnoilsk
SLOVAKIA
Trebio
Vacenovice
Breclav
Hodonin
Salgahsk
Miskolo
Kherson
Ugeny
Znojmo
Stefanov
Malacky
ROMANIA
Kishinev
Odessa
Linz
Vienna
Bratislava
Wysok
Hajduszoboszlo
Biharnagybajom
Sarmasel
Teleac
Stanesti
Bacau
Victorovca
Baimaclia
Valeni
Munich
Budapest
Nadudvar
Zemes
Cucadesti
Salzburg
HUNGARY
Szandaszolos
Zau-de-Cimpie
Miercurea
Saros
Dramanesti
Izmail
Bogota
Singeorgiu
AUSTRIA
Lovaszi
Budafap
Nagylengyel
Pusztafoldvar
Battonya
Taunt
Bazna
Noul-Sasesc
Boldesti
Moreni
Frasinu-Mislea
Black Sea
Graz
Ulles
Szeged
Arad
Guru-Ocnitej
Lebada
Midia
Navodari
Constanta
Paluzza
Maribor
Petisovci
Legrad
Nagylanizsa
Algyo
Knezevac
Timisoara
Ploiesti
SLOVENIA
Ljubljana
Zagreb
Pecs
Mokrin
Srpska Crnja
Resita
Cimpina
Belluno
Dugo Selo
Zutica
Babosca
Gakovo
Bizovac
Velebit
Ada
Boka
Medja
Gospodjinci
Ticleni
Bilteni
Pitesti
Bucharest
Trieste
Jezero
Ivanic Grad
Janja Cipa
Pepelana
Benicanci
Obed
Srbobran
Jamarica
Nova Gradiska
Nicolinki
Mramorak Selo
Leonardi
Budisterni
Tjulenovo
Kavarna
Porto Marghera
Koper
Rijeka
Sisak
Bosanski Brod
Pancevo
Belgrade
Turnu-Severin
Giurgiu
Tolloukhin
Varna
Staro Orjachovo
Gulf of Venice
Pula
BOSNIA &
HERZEGOVINA
YUGOSLAVIA
Gigen
Dolni Dabnik
Gorni Dabnik
Chiren
Ferrara
Bologna
Ravenna
Ivana
Gospic
Devetak
Vraca
BULGARIA
Otmanli
Burgas
Aleksinac
Nis
Sarajevo
Pristina
CROATIA
Split
Sofia
Stara Zagora
Florence
San Marino
Falconara
Ancona
Adriatic Sea
Plovdiv
MONTENEGRO
Istanbul
Izmit
ITALY
Dubrovnik
Shkoder
Skopje
Bellante
Cellino
Marta
Allano
Chieuti
Rospo
Lanciano
San Stefano
Cupello
Vasto
ALBANIA
MACEDONIA
Kavala
Alexandroupolis
Bandirma
Bursa
Valle Capo
S. Salvo
Portocannone
Monte Stillo
Tirana
Prinou
S. Kavalla
Thessaloniki
Rome
Ballaj
Divjaka
Seman
Marinze
Panaja
Kucove
Bubullima
Cakran
Gorisht
Drashovice
Petroliard-Farnesina
Terracina
Candela
Giovinazzo
Bari
Burhaniye
TURKEY
Gaeta
Naples
Pietralunga
Grottole
Ferrandina
Cerro Falcone
Monte Alpi/Enoc
Tempa Rossa
Taranto
Brindisi
Vlore
GREECE
Pomarico
Pisticci
Rotondella
Finiq
Larissa
Agean Sea
Izmir
Tyrrhenian Sea
Nova Siri
Arta
Crotone
Luna
San Leonardo
C. Cimiti
Delfi
Aspropyrgos
Soke
Ionian Sea
Gulf of Corinth
Eleusis
Athens
Argos
Tripolis
Vibo Valentia
Rosarno
Lavinia
Bagnara
Messina
SICILY
Palermo
Milazzo
Gaglioano
Rodhos
Mazara del Vallo
Marinella
Bronte
Clara
San Nicola
Catania
Ardizzone
Marsala
Porta Enpedocle
Gela
Commarata
Pone
Drillo
Perla
Ragusa
Vega
Augusta
Mediterranean Sea
Khania
Iraklion
Rethimnon
CRETE
MALTA
150 Miles
200 Kilometers

Pipeline agreements move forward

DONG also agreed with POGC to form a consortium to build the 143-mile (230-km) BaltPipe gas pipeline, which would run from Danish Zealand to Poland's Baltic coast. The two companies expected Norwegian gas producers, who were negotiating gas sales to Poland, to join the consortium in 2002. Another pipeline being evaluated was a 684-mile (1,100-km) line between Denmark and Norway through the Skagerrak Channel, extending to Niechorze on the Baltic coast, where it would link with the Polish gas network. Over the longer term, the pipeline could also supply Denmark.

Offshore E&P

In late 2001, Amerada Hess reported its first oil discovery off the self-governing Danish Faroe Islands in the northeast Atlantic. Islanders have hoped that offshore oil might enable them to win political and economic independence from Denmark. Amerada Hess planned an appraisal program and report to be completed by mid-2002.

In concert with BP PLC's earlier decision to develop its Clair discovery between the Faroes and the Shetland Islands, Amerada Hess's discovery could further rekindle interest in exploration in Faroe waters. The well, designated 6004/16-1, was located in 3,117 feet (950 m) of water in Faroe License 001 close to the UK boundary. After reaching the planned depth of 12,565 feet (3,830 m), the well was deepened to 14,025 feet (4,275 m) total depth. Amerada Hess said that light oil and gas were found in a gross interval of 558 feet (170 m), but a flow test was prevented by weather conditions. Amerada Hess as operator had 71.547% of the license; other holdings were DONG AS (28.286%) and Atlantic Petroleum Inc. (0.167%).

Earlier in 2001, Statoil Faerøyene spudded a wildcat to test the deepwater Longan prospect off the Faroes. The well was in 3,087 feet (941 m) of water on License 003 in the Atlantic Margin near the UK continental shelf. Transocean Sedco Forex's semisubmersible Sovereign Explorer was drilling the hole during 2001.

Development continued in the Danish North Sea Siri and Stine fields, even as ownership interest changed hands in mid-2001. Enterprise Oil Denmark Ltd., which owns Enterprise Oil PLC's interest in the producing Siri oil field and the nearby Stine discovery, was acquired by Paladin Resources PLC for $35 million plus interest. Enterprise said that the Siri interest represents 4 million bbl of oil reserves and had produced 5,900 b/d during the first half of 2001. Paladin acquired a 20% interest in License 6/95 and 60% interest in and operatorship of Block 7/98.

The Stine Segment II discovery was drilled in record time by rig Noble George Sauvageau, reported Norway's Statoil in mid-2001. Including coring and logging, the rig finished the well in less than 10 days. The discovery well was located in 197 feet (60 m) of water 4 miles (7 km) east of Siri field, which Statoil operates on the Danish continental shelf. It was drilled almost vertically to 6,860 feet (2,091 m) below sea level, where oil was found in the Paleocene sand. The wildcat was seeking to increase oil volumes to the Siri platform, which has a production capacity of 50,000 b/d but was producing about 30,000 b/d. The rig was moved to the platform to drill a 23,622-foot (7,200-m) horizontal production well to the discovery, with output scheduled to begin in late 2001.

Earlier in the year, DUC announced plans to spend $520 million on development of Denmark's offshore Halfdan field. The program would include drilling of 11 wells with 9,800-16,400 foot (3,000-5,000 m) horizontal sections for production of oil and 11 related water injection wells. Also, the central platform would be extended with facilities for processing and compression of gas and processing of produced water. Accommodation and flaring platforms were to be bridge-connected to the central platform. The development was expected to increase overall production from the Halfdan field to 100,000 b/d in 2005. DUC partners included A.P. Møller company Mærsk Olie og Gas AS, Royal Dutch/Shell Group, and ChevronTexaco.

In mid-2001, a natural gas explosion disrupted oil and gas production at Gorm field and its satellite fields in the Danish North Sea, resulting in no oil pollution and only minor injuries. A ruptured high-pressure pipe from a gas compressor triggered the explosion, and Module 5 on the Gorm C platform sustained damage. Production at Dan field was resumed within a day, while Gorm's production was suspended for a longer period.

Although corrosion problems in Gorm field reduced production rates, increasing development of Halfdan field and the possible development of the Cecilie and Nini finds should make the Danish sector more self-sufficient for longer than previously predicted and also generate a surplus of gas for export, said an analysis by Infield Database and Douglas-Westwood Associates. In addition to Denmark's gas supply agreement with Poland, a connection to the UK gas pipeline system was in early stages of evaluation.

Onshore developments

In mid-2001, an onshore exploration well found oil on License 5/98 in the northern part of the Danish Central Graben. Phillips Petroleum International Corp. Denmark said that well Hejre-1 encountered the oil in Late Jurassic sandstones. Phillips' extensive evaluation program rendered test production unnecessary. The discovery size was being evaluated, and the rig was moved to the eastern part of the graben to drill exploration well Svane-1 in License 4/98.

A group led by UAB Minijos Nafta in early 2001 was awarded the 900,000-acre License 1/01 in the Zechstein basin of Denmark's southern Jutland. Initial work in the six-year licensing period was to involve evaluation of existing data, seismic reprocessing, a pilot geochemical survey, and prospect-lead definition. Other group partners were Sterling Resources Ltd., DONG, and Dansk Venture Olieefterforskning Aps. The Zechstein basin is geologically similar to Sterling's acreage in the UK Cleveland basin but has not had the benefit of modern exploration techniques.

FINLAND

CAPITAL: HELSINKI
MONETARY UNIT: MARKKA*
REFINING CAPACITY: 238,800 B/CD
OIL PRODUCTION: N/A
OIL RESERVES: N/A
GAS RESERVES: N/A
* Switching to euro in 2002.

During 2001, progress continued on the Baltic Pipeline System linking Finland and other countries to Russia's oil supplies. The 162-mile line to Finland's port at Pimorsk from Kirishi, Russia, was completed, and an oil terminal was built on the Gulf of Finland at Primorsk. The first tanker was scheduled to be loaded in Primorsk in late 2001. Meanwhile, work began on the second stage of the project, which would include construction of a parallel pipeline and additional storage facilities in Primorsk that would increase the capacity of the system to 600,000 b/d.

The 51,800-b/d refinery in Naantali, Finland, owned by Fortum Corp., was being expanded with a low-sulfur gasoline unit during 2001. The unit, which will use Institut de Francais du Petrole's technology, was scheduled to come on line in 2002.

An ethylene cracking facility in Porvoo, Finland, was also being expanded in capacity by 10% to 310,000 tonnes. Borealis SA planned to debottleneck the cracker by modifying the compressor and installing an ethylene splitter. Borealis expected that the expansion would help reduce emissions of volatile organic chemicals by 5%. The modifications were expected to be completed in 2002.

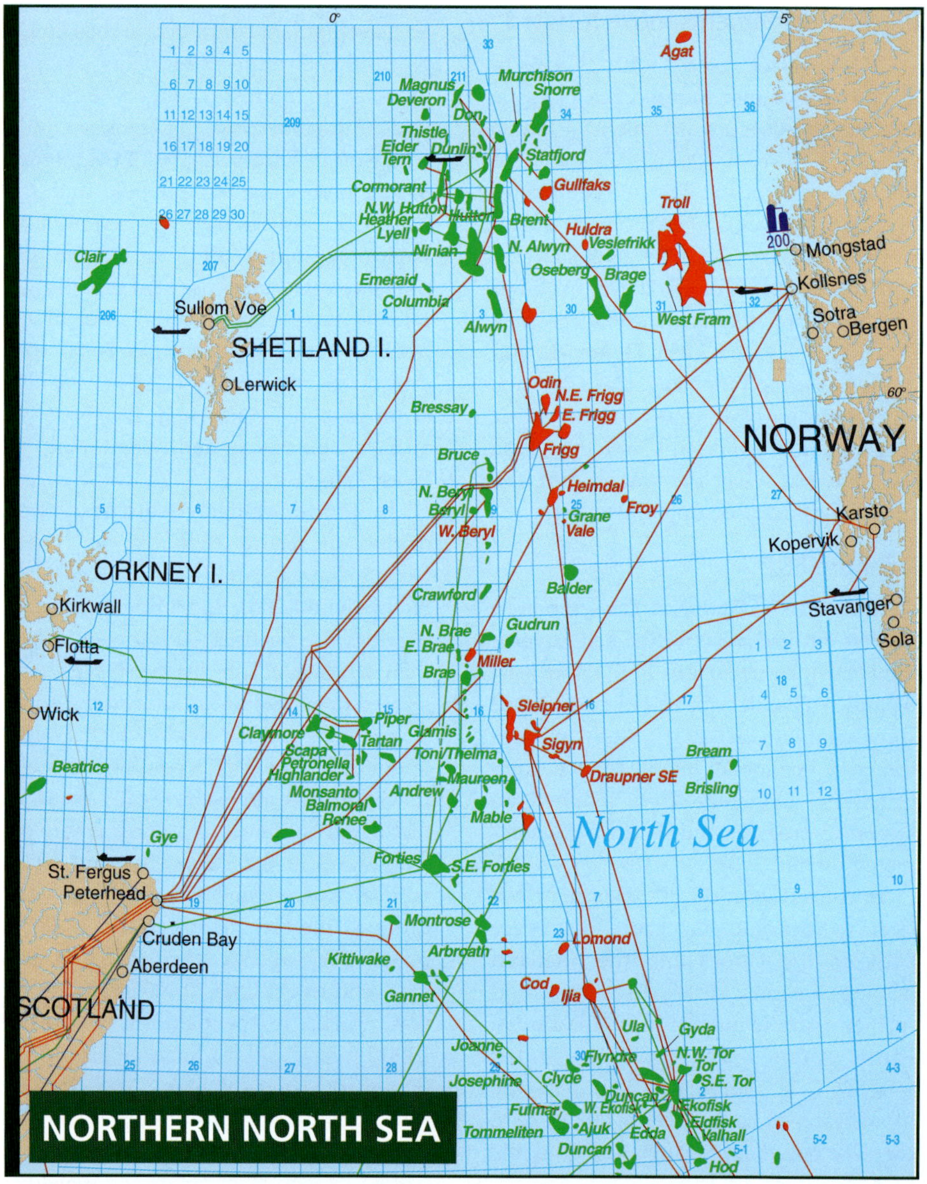

NORTHERN NORTH SEA

FRANCE

CAPITAL: PARIS
MONETARY UNIT: FRANCE*
REFINING CAPACITY: 1,896,493 B/CD
OIL PRODUCTION: 27,500 B/D
OIL RESERVES: 140 MILLION BBL
GAS RESERVES: 403 BCF
* Switching to euro in 2002.

During 2001 France dragged its heels on deregulating its natural gas and electricity businesses. The monopolistic home markets of Gaz de France SA and Electricité de France SA had been opened by only 20% and 30%, respectively, as of mid-2001. The French government decided not to transcribe the EU gas deregulation into French law as scheduled, and the European Commission (EC) planned to bring France before the European Court of Justice for its failure to import the directive, stating that France's refusal to meet the EC timetable would handicap the target of liberalizing all European gas markets by 2005. In early 2002, the EU gave France a deadline of July 2002 to endorse liberalization, and EC President Romano Prodi threatened to use a special single-market directive on reluctant states to force them to make progress on the liberalization issue.

Gaz de France moved ahead with investment plans that would at least double its share of the European gas market, to 20%, and Electricité de France affiliate EDF Trading Ltd. agreed to sell 745 bcf (20 billion cu m) of gas over 10 years to Wingas GMBG of Germany, which previously was to purchase the gas from BG Group PLC's subsidiary BG International Ltd. Electricité de France also announced a restructuring plan including independent subsidiaries for US operations and other units. The company hoped to generate half of its annual sales internationally in the future, compared with about one third currently.

French and Spanish companies were considering building a natural gas pipeline to improve gas transmission between the two countries. Gaz du Sud-Ouest was evaluating the 18-mile (29-

km), 18.6-bcf (500 million cu m) pipeline along the Basque coastline in partnership with Ente Vasco de Energia and its affiliate Gas de Euskadi. Gaz du Sud-Ouest was owned 30% by Gaz de France and 70% by TotalFinaElf SA. The line would run between Bidart, south of the French coastal town of Biarritz, to Spain's Irun, adding a link to the countries' gas networks. If undertaken, the pipeline would be on stream in winter 2002-03. The French and Spanish partners were also conducting a feasibility study for construction by 2010 of a high-capacity link between the underground gas storage reservoir in Lussagnet, France, to Bilbao, Spain.

During late 2001, French courts were considering bringing formal charges against TotalFinaElf SA for the loss of the oil tanker Erika off Brittany in December 1999.

Boom in French downstream

High refining margins in France during 2000 led companies to continue expanding the country's refining and petrochemical industry in 2001. Atofina, the chemical division of TotalFinaElf SA, focused on capacity expansions and integration of large-scale olefins plants with existing refining assets.

The company completed a 245-mile (395-km) ethylene pipeline in 2001 linking the Carling petrochemical complex near Metz with the Viriat underground storage facilities near Lyon. The line constitutes part of Europe's largest ethylene transportation network and provides Atofina with a transport connection between northern and southern France.

Atofina said the line would help Carling optimize its capacities with those of the Solvay Atofina cracker in Feyzin, near Lyon, as well as with Atofina-BP Chemical's Lavéra cracker and Royal Dutch/Shell Group's Berre cracker on the French Riviera. Exchanges between the petrochemical sites would have a total ethylene capacity of more than 2 million tpy. Atofina was scheduled to complete a 90,000-tpy expansion of one of its naphtha cracker trains at Carling in 2001. With the pipeline on stream, Carling's ethylene capacity would rise to 570,000 tpy by late 2001, from 480,000 tpy previously.

The Lavéra cracker sits close to BP PLC's 210,000-b/d refinery at Lavéra and TotalFinaElf's 145,150-b/d refinery at La Mede. Atofina planned to increase the integration of the 740,000-tpy Lavéra ethylene unit with the two refineries, and BP invested heavily in maintenance and modernization of its facility, where a cracker was debottlenecked, increasing its capacity to 270,000 tpy from 235,000, and an amines unit was increased in capacity to 50,000 tpy from 30,000. Also, Fina Antwerp Olefins, a joint venture of Atofina and ExxonMobil Corp., planned to increase its ethylene production capacity to 1.4 million tpy by year-end 2002 by upgrading its No. 2 naphtha cracker.

GERMANY

CAPITAL: BERLIN
MONETARY UNIT: DEUTSCHE MARK*
REFINING CAPACITY: 2,258,800 B/CD
OIL PRODUCTION: 67,500 B/D
OIL RESERVES: 364 MILLION BBL
GAS RESERVES: 12.1 TCF
* Switching to euro in 2002.

Germany had not yet set a date, as of mid-2001, for liberalization of its gas and electricity markets. The EU requires members to import its liberalization directives into their national laws as part of the deregulation process, which is to culminate in 2005. The German government said that it would address the issue during 2001.

Germany's RWE AG acquired the Czech Republic's national gas utility Transgas and gained the majority in six of the eight Czech regional gas utilities put on the auction block in 2001. The deal, to be consummated in early 2002, would make RWE Europe's second-largest pipeline operator, with 54,400 miles (94,000 km) of grid length, and would raise the company to the number four position in the European natural gas business. RWE would also gain access to natural gas from Russia, Europe's largest supplier, through Trangas's major transit pipeline links.

RWE also acquired in late 2001 British gas producer Highland Energy Holdings Ltd., adding 484 bcf (13 billion cu m) to RWE's 1.27 tcf (34 billion cu m) of gas reserves. Long-term supply agreements were in place with customers in the UK and the Netherlands for the existing gas production volumes, giving RWE access to new market links.

A new German joint venture, Shell & DEA Oil GMBH, looked likely to receive regulatory approval, creating a new leading company in the German petroleum business. The combination of RWE-DEA and Shell would include the petrochemical operations, refineries, logistics, and distribution and marketing of both companies.

The first commercial unit to use BP's proprietary OATS gasoline desulfurization technology came on stream in late 2001 at the Bayernoil refinery at Neustadt, Bavaria. The process is based on catalytic conversion of thiophenes into high boiling compounds that can be readily removed from the gasoline stream by fractionation, said BP. Bayernoil's desulfurization unit has the capacity to produce 15,000 b/d of low-sulfur gasoline to meet German regulations, which will restrict maximum sulfur concentrations to 10 ppm by 2003. The 260,000-b/d refinery complex is a joint venture of BP, Agip, and Ruhroel (a joint venture of Veba Oil Refining and Petrochemicals and PDVSA). BP expected that two other plants using the process would come on stream in 2002.

The Schmierstoff Raffinerie Salzbergen (SRS) lube oil refinery reported results in 2001 of a two-year maintenance program that had been outsourced for greater profitability. As a result of the maintenance outsourcing strategy and alliance, the refinery exceeded industry benchmarks for reliability, plant availability was increased, and the cost structure in maintenance and spare-parts management was optimized. Partners in the program were SRS, Ferrostaal Maintenance Nord GMBH, and Ferrostaal Industrial Plant Services GMBH.

Basell Polyolefine GMBH planned to build a 320,000-tpy high-density polyethylene plant at Wesseling, Germany, to produce pipe grades in black, blue, and orange-yellow, using Basell-owned technology. The plant, scheduled for startup early in 2004, would be Basell's largest capacity facility based on the Hostalen technology.

The SRS lube oil refinery in Salzbergen, with a total capacity of 350,000 tpy, makes more than 840 different products including technical and medical white oils, paraffins, ink, base oils, and lubricants for industrial and automotive applications.

GREECE

CAPITAL: ATHENS
MONETARY UNIT: DRACHMA*
REFINING CAPACITY: 406,500 B/CD
OIL PRODUCTION: 4,000 B/D
OIL RESERVES: 9 MILLION BBL
GAS RESERVES: 18 BCF
* Switching to euro in 2002.

Greece was one of four offshore areas identified in early 2001 by Wood Mackenzie Consultants Ltd., in a survey of southern Europe's upstream sector, as likely contributors to future E&P activity. However, ongoing political wrangling between Greece and Turkey over continental shelf rights in the Aegean Sea has restricted Greece's ability to award acreage. This was particularly frustrating, suggested Wood Mackenzie, because the area has arguable prospectivity in the post-orogenic Paleogene to younger Neogene basins, where Tertiary sediments up to 9,840 feet (3,000 m) in thickness can occur, and favorable geological structures have been identified.

Enterprise planned to drill an exploration well in the Ioannina license in northwestern Greece, where it holds a 63.3% interest. Previous exploration onshore had been unsuccessful, as Triton Energy Corp. drilled four dry wildcats in western Greece. A delineation well was also drilled.

Greece was one of ten LNG importing countries during 2001. Although Greece is a small purchaser, its activity as an Atlantic basin LNG buyer represents a break from the old dedicated model of LNG projects. Rather, Greece is supplied with what could be considered surplus capacity from existing projects. Liberalization of Greece's gas sector continued to be delayed and was scheduled for 2006.

Bosporus bypass, other pipelines

Greece and Bulgaria during 2001 continued to plan an oil pipeline from the Bulgarian Black Sea port of Burgas to Alexandropoulis on the Aegean Sea, but Russian backing to supply the oil was considered critical to the project going forward. The three countries were conducting a $2.2 million feasibility study and continued to negotiate launch of a Trans-Balkan Oil Company for development and construction. The pipeline would offer another alternative to transport through the Bosporus. Several Greek companies have indicated an interest in investing in the pipeline, but no construction date was set. Meanwhile, Greece will begin receiving greater transit volumes of gas after 2002, under an agreement reached between Russia and Bulgaria.

Greece's state-owned Hellenic Petroleum (HP) was scheduled to complete in mid-2001 a $100 million, 143-mile pipeline to carry crude oil from the northern port city of Thessaloniki to HP's newly acquired Okta refinery near Skopje in Macedonia. The pipeline was to be managed in partnership with Macedonia and will carry crude that had been shipped by rail.

In early 2001, Greece joined a tripartite committee with Armenia and Iran to seek foreign capital for completing a pipeline to transport natural gas from Iran into Armenia and surrounding areas. The IGAT-3 pipeline would run north from South Pars field in the Persian Gulf.

HUNGARY

CAPITAL: BUDAPEST
MONETARY UNIT: FORINT
REFINING CAPACITY: 161,000 B/CD
OIL PRODUCTION: 23,500 B/D
OIL RESERVES: 111 MILLION BBL
GAS RESERVES: 1.28 TCF

The 25% state-owned Hungarian Oil and Gas Co. (MOL Rt.) is one of central Europe's largest integrated oil and gas companies and the largest in Hungary in terms of sales. In early 2001, MOL bought a 36% share in the Slovakian oil and gas company, Slovnaft, with an option to buy a majority stake.

To gain accession to the EU by 2004, Hungary was working toward adopting EU directives during 2001, and the Hungarian government remained committed to liberalizing the natural gas and electricity markets. A new law was enacted in September 2001 to begin opening the power markets in 2002-03, but natural gas market liberalization was not expected to be fully accomplished before EU membership.

The country's gas pricing scheme was due to change at year-end 2001. It was expected that in the new formula, retail gas prices would not be able to fluctuate with the price of oil, but would be related instead to a long-term benchmark import gas price based on oil price forecasts from market information services and on estimates from Hungary's finance ministry. The formula would take into account capital costs and currency exchange rates. Also, a limited element of risk management would be built into the formula, which would apply until the market is liberalized and would remain in place for regulated prices to captive customers. However, analysts foresaw no special risk management for MOL's long-term supply contract with Russia's Gazprom, which provides the vast majority of Hungary's gas imports.

Like Croatia, Hungary also was becoming increasingly important as a transit center for natural gas. Hungary's Barsvo (Brotherhood) pipeline carries Russian gas through Horgos, Hungary, into Serbia, Bosnia, and central Europe. Imports through this Hungarian pipeline were expected to increase in 2001 to 53 bcf from only 9 bcf in 2000.

IRELAND

CAPITAL: DUBLIN
MONETARY UNIT: IRISH POUND*
REFINING CAPACITY: 71,250 B/CD
OIL PRODUCTION: 0
OIL RESERVES: 0
GAS RESERVES: 700 BCF
* Switching to euro in 2002.

Although Ireland's energy resources are limited, the country made progress during 2001 on developing its gas fields and diversifying its energy supplies. Average field size for Ireland increased greatly, from only 10 million boe/yr to 63 million boe/yr, with the start-up of the Corrib North field, according to an analysis by Infield Database and Douglas-Westwood Associates. Further prospects included the Seven Heads and Galley Head fields, each of which are gas developments with long subsea tiebacks to shore as the preferred development scheme.

Corrib field development underway

In mid-2001, owners of Corrib gas field off western Ireland withdrew an application to jointly market their production, due to the EC's concerns about competition. Enterprise Energy Ireland Ltd. was operator of Corrib with 45%; Statoil Exploration (Ireland) Ltd. had 36.5%, and Marathon International Petroleum Hibernia Ltd. 18.5%. The decision was expected to offer gas consumers in Ireland, especially power generators and energy-intensive industrial users, a wider choice among gas suppliers. Later in 2001, Enterprise agreed to sell 60% of its share of gas from Corrib to a trading arm of Bord Gáis Eirann, Ireland's state-owned gas company.

Enterprise Energy Ireland opened the field, which was estimated to hold 931 bcf (25 billion cu m), in 1996 on Block 18/20. The field's fifth appraisal well was drilled in mid-2001 and tested at 30 MMcfd; earlier tests showed flow rates of up to 60 MMcfd. Development of Corrib, which began in 2001 and was scheduled for completion in 2003, was to include seven subsea wells in more than 1,100 feet of water, tied back to a central gathering manifold. A pipeline would move the gas to a landfall at Broadhaven, County Mayo, in northwestern Ireland, and from there to the processing terminal, permitted in August to be built at Bellanaboy bridge. Gas was to be exported from the terminal via pipeline to a tie-in at Craughwell near Galway. Corrib was the first significant find off Ireland since Kinsale Head gas field in 1973. First gas delivery was anticipated for the second half of 2003 and plateau of 8.9 MMcfd the following year. Enterprise estimated the field life between 15 and 20 years.

Exploration continues

In mid-2001, operator Ramco Energy PLC awarded a contract to Pride Foramer SAS to use the Pride North Sea rig to drill in Seven Heads accumulation off Ireland. The rig was to drill one firm well in summer 2001 on Block 48/24 and one optional well. The first well was expected to reach 4,000 feet subsea in 335 feet of water. If enough gas is found and regulatory approval forthcoming, the group could bring Seven Heads on stream in 2003.

Ramco increased its interest in the Seven Heads project to 49% from 37.5% when Ireland converted the consortium's licensing option to a lease. Other partners were Duke Engineering & Services Ltd. (26%), Island Petroleum Developments Ltd. (20%), Northern Petroleum PLC (4%), and Sunningdale Oils (Ireland) Ltd. (1%).

The group was permitted to carry out the evaluation of four wells previously drilled in the structure, which covers the southern halves of Blocks 48/23 and 48/24 and the northern halves of Blocks 48/28 and 48/29. The group acquired 64 miles (103 km) of new 2D seismic over the area. Seven Heads is 29 miles (47 km) from landfall on the Cork coast and 42 miles (68 km) from the landfall of Marathon's gas pipeline.

Also in Irish waters, Statoil AS drilled the first exploration well on the Sarsfield prospect during 2001. Sarsfield is on Block 35/21 in the Porcupine basin west of Ireland in 2,133 feet (650 m) of water. Statoil-contracted drilling rig Sovereign Explorer, owned by Transocean Sedco Forex Inc., spudded the wildcat. Other partners in the exploration license included Conoco UK, Enterprise Energy Ireland, and Dana Petroleum.

Onshore, Evergreen Resources Inc. agreed to acquire a 100% interest in 605,000 acres of prospective tight gas sand properties in Northern Ireland's Northwest Carboniferous basin. Evergreen planned to drill four exploratory wells during 2001-02. The company identified seven drillsites for permitting and other locations for exploratory and delineation drilling.

ITALY

CAPITAL: ROME
MONETARY UNIT: LIRA*
REFINING CAPACITY: 2,282,800 B/CD
OIL PRODUCTION: 74,000 B/D
OIL RESERVES: 622 MILLION BBL
GAS RESERVES: 8.1 TCF
* Switching to euro in 2002.

ENI SPA of Rome reported record earnings as of mid-2001, attributed to a 15% jump in operating income from increased hydrocarbon production and sales outside of Italy, as well as other gains. ENI's foreign production included startups of gas fields in Egypt, Norway, Nigeria, Tunisia, and Croatia, while its domestic production from mature gas fields declined by 13.5%. ENI said that Italgas SpA is a strategic asset, and the group does not intend to sell its controlling 41.15% stake.

Italy was one of the countries to which the US International Trade Commission extended in mid-2001 antidumping measures on oil country tubular goods other than drill pipe.

Exploration and development

In early 2001, ENI's Agip Division gauged 1,000 b/d of 34º gravity oil and 1.1 MMcfd of associated gas at Miglianico 1 on the Bucchianico license 100 miles east of Rome in a lightly explored area north of Ortona on the Abruzzo coast. Geological and geophysical studies identified hydrocarbons in carbonate rock of unspecified age, and the well cut 328 feet (100 m) of net pay at 15,748 feet (4,800 m). ENI's data suggested 40-45 million bbl of reserves.

Italy's largest home-based project in the Adriatic Sea being planned during 2001 was the Alto Adriatico, which would consist of 19 platforms on 15 fields and could require more than 250 miles (400 km) of pipeline to fully develop. However, the fields' proximity to the shore city of Venice raised environmental concerns about possible subsidence effects should the natural gas, estimated at about 1 tcf, be withdrawn from the reservoirs. Meanwhile, development projects underway offshore Libya were expected to feed large volumes of gas to the Italian domestic market over the next several years, to 2005.

Pipeline for Val d'Agri production

In late 2001, the first oil began flowing through a pipeline linking the Val d'Agri oil center at Viggiano in Italy's Southern Appenines to Taranto. Construction of the 20-inch, 85-mile (136-km) line, with a capacity of 150,000 b/d, was completed earlier in the year. The line moves crude from the Monte Alpi/Monte Enoc fields in the Val d'Agri area of the Basilicata region and would enable its production to reach around 45,000 b/d by year-end 2001. Previously, oil center production had been limited by trucking logistics to 9,500 b/d.

AgipPeteroli SpA has a refinery at Taranto with an 84,000-b/d capacity. The crude could also be exported to the Mediterranean market. The Val d'Agri oil center was developed by Enterprise Oil PLC (29% share), which said that it would focus on increasing output from Monte Alpi, developing Cerro Falcone field, and moving Tempa Rossa field toward sanction. All three fields are operated by ENI in partnership with Enterprise.

Italy's Enel SPA, in a joint venture with Sonatrach and Wintershall AG, studied the feasibility during late 2001 of a second gas pipeline linking Italy to Algeria. The proposed 930-mile (1,500-km) pipeline would be built in four sections and cost $2 billion. The first phase would carry 372 bcf/yr (10 billion cu m/yr) of natural gas, and a later extension could supply gas to Germany. The existing Trans-Mediterranean line from Algeria supplies Italy with about 745 bcf/yr (20 billion cu m/yr) of gas.

Desulfurization unit added

An 11,000-b/d desulfurization unit was planned for a refinery near Genoa, to be built by SulphCo Inc. and partners. The unit, being installed at the 35,000-b/d IPLOM refinery at Busalla during 2001, would be able to reduce sulfur content in diesel fuel and fuel oil products by more than 95%, to 5-10 ppm, using SulphCo's proprietary technology. IPLOM, a refiner and fuel distributor in northern Italy, was purchasing the desulfurization unit and paying installation costs. Operation was scheduled for late 2001.

Enichem SPA planned to revamp instrumentation and automation at its ethylene plant at Porto Torres, Sardinia, with completion and startup scheduled for 2002. The turnkey contract was awarded to Technip SA.

Agip's refinery in Priolo reduced the sulfur content in its full-range gasoline by 35% by using a new fluid catalytic cracking catalyst technology, reported AgipPetroli in early 2001. The unit also increased its gasoline selectivity and reduced gas and coke yields.

NETHERLANDS

CAPITAL: AMSTERDAM
MONETARY UNIT: GUILDER*
REFINING CAPACITY: 1,205,842 B/CD
OIL PRODUCTION: 27,000 B/D
OIL RESERVES: 107 MILLION BBL
GAS RESERVES: 62.5 TCF
* Switching to euro in 2002.

Continued exploration during 2001 in the Dutch sector of the North Sea moved forward a number of potential fields for development. Clyde Petroleum Exploratie BV (formerly owned by Gulf Canada Resources Ltd. and acquired in 2001 by Conoco Inc.) discovered commercial gas and brought other existing fields on stream, all within a short period of time.

Castricum-Zee, P6-D

In mid-year, Clyde began production from Castricum-Zee gas field off the Netherlands. The field, opened 4 miles off the coast in 2000, has potential reserves of 35 bcf, or 70 bcf gross, said Clyde. The field was to be brought into production via the existing Q8-A platform 3 miles away. The extended reach well was drilled at a high angle to 17,067 feet MD (5,807 feet vertical depth) in 56 days. Castricum-Zee was Clyde's second field to be brought on stream in less than a year, having been preceded by the Q4-9 in late 2000. Operator Clyde holds 50% interest in the field and partner Dyas BV the other 50%.

Also in 2001, Clyde brought on stream P6-D field off the Dutch coast, beginning production 19 months after its discovery in 2000. The discovery well tested at 45 MMcfd of gas with wellhead pressure of 2,450 psi from a 164-foot (50-m) perforated interval in the Triassic and 9 MMcfd from the Permian. The well's expected initial production rate was 70 MMscfd. The gas was being transported via a 12-inch pipeline 6 miles to Clyde's P6 main platform for processing.

Capping this string of successful developments, Clyde made yet another gas discovery in offshore Netherlands, its seventh within three years, in late 2001. The Q4-10 well flowed at an unstimulated rate of 43 MMcfd from a lower 184-foot (56-m) perforated interval and at an unstimulated rate of 39 MMcfd from an upper 131-foot (40-m) perforated interval. Both tests were within longer reservoir sections.

The Q4-10 was 6 miles (10 km) from Clyde-operated pipeline and production facilities. The discovery was described as commercial but was to be appraised to determine optimal development. The discovery extends from Block Q4 into Block Q1 and was to be named Q1-A. Operator Clyde held 45.46% with partners Energie Beheer Nederland BV (40%), DSM Energie BV (10.32%), CLAM Petroleum BV (2.77%), and Dyas Nederland BV (1.45%).

The capacity of Dow Chemical Co.'s ethylene plant at Terneuzen, the Netherlands, was increased by 600,000 tonnes in 2001, making the facility the largest naphtha-cracking ethylene site in the world, with annual production of 1.7 million tonnes.

Hanze field

Another operator active in offshore Netherlands in 2001 was Veba Oil Nederland, which brought on stream Hanze oil field in the North Sea. The initial flow rate was 11,000 b/d, but peak production of 30,000 b/d was anticipated by year-end 2001. The peak was expected to account for two thirds of overall oil production in the Dutch North Sea.

The development included a production storage platform, dedicated oil offloading systems, and gas export pipelines. The oil was being offloaded via a single-anchor loading system and shuttle tanker. In mid-year, dedicated shuttle tanker Betty Knutsen was to offload the first crude. Associated gas was being used to power the platform, with any excess exported via the A6-F3 pipeline and the NOGAT pipeline to Den Helder on the Dutch coast.

Groningen renovation

Gas production was being enhanced in late 2001 on the Groningen acreage in the northern Netherlands. Nederlandse Aardolie Maatschappij (NAM) BV awarded a three-year contract for renovation of three production sites – Zuiderpolder, Schaapbulten, and Oudeweg – at Slochteren gas field. The contract went to the Stork GLT consortium.

The first phase of the renovation project was to cover 11 of the 29 production installations. The Stork consortium, which has performed similar work for NAM at other production sites, had an option for remaining installations. NAM and Stork had earlier signed a 25-year contract for maintenance of all the renovated production installations.

To keep production volumes up, both Clyde and NAM were planning to install reusable platforms that can be moved rapidly from one small accumulation to another, reported Infield Database and Douglas-Westwood Associates in mid-2001. The Dutch were confident enough of their production prospects to consider a new gas export line to the UK if needed.

Ethylene plant upgrade

The capacity of Dow Chemical Co.'s ethylene plant at Terneuzen, the Netherlands, was increased by 600,000 tonnes in 2001, making the facility the largest naphtha-cracking ethylene site in the world, with annual production of 1.7 million tonnes. Dow invested $500 million in the major ethylene expansion, northwestern Europe's first since mid-1990. The expansion also enhanced the environmental performance of existing units, decreasing their carbon dioxide and nitrogen oxides emissions, and increased the production of polymer-grade polypropylene, heavy and light gasoline, pyrolysis gas, crude C4, carbon-black feedstock, high-pressure methane, hydrogen, tail gas, and ethane and propane recycle.

The additional ethylene production supports growth in Dow's derivative plants. Dow expected that by 2003, all of the ethylene produced in Terneuzen would be used captively. The capacity will also support several other planned expansions in Terneuzen, including a cumene plant scheduled for startup in late 2002 with a 300,000-tonne expansion. Also in late 2002, production of magnum acrylonitrile-butadiene-styrene (ABS) resins would increase by 75,000 tonnes.

In a processing development, Shell Nederland Chemie BV planned to invest more than $25 million in a benzene extraction unit at its petrochemical facility at Moerdijk, the Netherlands. The 500,000-tpy unit was scheduled to come on stream in late 2002 and will be based on Krupp Uhde's proprietary extractive distillation technology. The Moerdijk cracker and other Shell sites will provide feedstock for the new unit, and the benzene product will be consumed by the local ethylbenzene unit. Overall, shipping movements in and out of Moerdijk were to be reduced.

NORWAY

CAPITAL: OSLO
MONETARY UNIT: KRONE
REFINING CAPACITY: 310,000 B/CD
OIL PRODUCTION: 3.133 MILLION B/D
OIL RESERVES: 9.45 BILLION BBL
GAS RESERVES: 44 TCF

In May 2001 the Norwegian Parliament, the Storting, sold 15% of the country's SDFI (state's direct financial interest) assets on the Norwegian continental shelf to state-owned Statoil ASA for $4.2 billion (38.6 billion kroner). Norway's SDFI assets included offshore oil and gas licenses for 10 of the largest producing fields on the shelf – Troll, Gullfaks, Snorre, Vigdis, Tordis, Viscund, Statfjord East, Asgard, Sleipner East, and Sleipner West. Another 6.5% of SDFI assets will be sold to Norsk Hydro AS and other companies. The sale increased Statoil's reserves base from 3 billion boe to 4.5 billion boe and its production from 700,000 boe/d to 1 million boe/d. SDFI's name was changed to Petoro.

In June 2001, Norway sold 17.5% of state-owned Statoil for $2.9 billion through an IPO on the Oslo and New York stock exchanges. The government retained 33% of Statoil to ensure that the company preserves its character as a "Norway business."

The 68 blocks offered in Norway's 17th licensing round were in areas near the Norwegian continental shelf's existing infrastructure to encourage maximum production from mature and declining areas. Statoil, Royal Dutch/Shell Group, and Enterprise Oil Ltd. entered into long-term area cooperation agreements to take part in the late 2001 licensing round.

Snorre, Gullfaks

Idemitsu of Japan produced 70,000 bbl/d crude oil from the northern part of Snorre (B) starting in July 2001. The production was expected to increase to 110,000 bbl/d in 2003. Although Idemitsu was field operator, the company held just a 9.6% share.

In 2001 Statoil started the 22-mile (35-km) pipeline linking the Gullfaks platforms A and C with the Statpipe trunkline. Gullfaks field is located on Block 34/10 in the North Sea's Norwegian sector. The Gullfaks production facility was also being expanded to include three platforms and two additional gas-processing modules to serve satellite developments during 2001.

As of September 2001, the Statoil-operated Statfjord C platform had produced for 231 days without stopping, which set a production record on the Norwegian continental shelf. However, corrosion damage was later discovered, resulting in loss of oil production. Shown is the Statfjord complex. Photo courtesy of Statoil.

Gas deliveries from Gullfaks, previously exported via the Statfjord field to Statpipe, were expected to rise from 494 MMcfd (13 million cu m/day) to 745 MMcfd (20 million cu m/day) with added production from Gullfaks South. Eventually, 931 MMcfd (25 million cu m/day) would be transported from the Gullfaks area to Statoil's gas treatment plant at Kårstø (capacity of 2.31 bcfd, or 62 million cu m/day). Gullfaks operator Statoil had an 18% stake, Petoro (formerly SDFI) 73%, and Norsk Hydro Produksjon 9%.

The pipeline link was in preparation for gas production at Gullfaks South field via the Gullfaks C platform as part of the Gullfaks satellite phase II development. This development program focused on Gullfaks South's reserves, believed to stand at 1.28 tcf (34.4 billion cu m), and associated gas from Rimfaks field, where reserves are 107 million bbl and production was 13,000 b/d. Rimfaks field is on Block 33/12. The platform produced at 170,000 b/d, about half of which was from Vidgis and Viscund fields operated by Norsk Hydro.

Kvaerner Oil and Gas was awarded a $43.1 million (380 million kroner) contract to boost gas treatment capacity on platform A. In 2003 the modified Gullfaks A platform will receive gas from both Gullfaks South and Rimfaks. The A platform will then process about 149 MMcfd (4 million cu m/day) from the combined gas output of the satellite fields.

Statoil found hydrocarbons at 1,608 feet (490 m) in its exploration well 34/10-J-4H in Rimfaks field. Statoil, operator, held 61% interest, Petoro 30%, and Norsk Hydro 9%.

Fram West, Vale, Heimdal

In spring 2001, Norsk Hydro received approval of its development and operation plans for the Fram West and Vale fields in the North Sea. At a cost of $440 million (4 billion kroner), Fram West will have four wells on subsea templates tied back to Hydro's nearby Troll C platform for processing. Aker Maritime AS was selected to provide a production module, modify Troll C platform, and install the module on the platform. The contract was for $55.1 million (500 million kroner). Reserves were 100 million bbl of oil and 298 bcf (8 billion cu m) of natural gas. Fram West was expected to flow in 2003 and to produce at a plateau rate of at least 60,000 b/d. Operator Norsk Hydro owned a 25% stake, Petoro 30%, Statoil 20%, and ExxonMobil 25%.

The Vale gas-condensate field (Block 25/4) was planned to be developed, at a cost of $100 million (880 million kroner), as a subsea satellite tied back to the Heimdal riser platform. As part of the development program, Norsk Hydro awarded Stolt Offshore and Fabricom Contracting the project to coat, lay, and trench a 10-mile (16.5-km), 8-inch flowline between the Vale subsea production system and the Heimdal riser platform. Fabricom will make abovesea modifications on the Heimdal platform for the Vale and Vesterled field projects. Vale was expected to flow in 2002 from reserves of 93 bcf (2.5 billion cu m) of gas and 21 million bbl of condensate. Hydro held a 28.5% share, Marathon 46.9%, and TotalFinaElf 24.2%.

The Vesterled North Sea pipeline, linking the existing Heimdal gas network and the Frigg Norwegian Association (FNA) pipeline in the Norwegian North Sea, became operational. The project involved laying a 21-mile (33-km) pipeline from the Heimdal platform to FNA's pipeline and from Frigg to the St. Fergus, Scotland, gas terminal. The pipeline has a capacity of 410 bcf/yr (11 billion cu m/yr). Statoil may use this pipeline to deliver the 59.6 bcf/yr (1.6 billion cu m/yr) Norwegian gas that was contracted to BP in the UK from 2001 to 2016.

Grane, Sigyn

Development work continued on the Grane field on Block 25/11, northwest of Stavanger, in 417 feet (127 m) of water. Grane was the largest proven but undeveloped oil field in the region, with estimated reserves of 700 million bbl of heavy oil. Its development involved an integrated platform with drilling facilities, living quarters, and production facilities. First production from the field was due in 2003; the field's expected maximum output of 214,000 b/d was estimated for 2005. Norsk Hydro held a 24.4% interest in Grane, Petoro 43.6%, ExxonMobil Corp. 25.6%, and Conoco Inc. 6.4%.

Development of Sigyn gas-condensate field on Block 16/7-4 in the Norwegian North Sea involved a subsea template tied back to Statoil's Sleipner East field. Sigyn is 7.5 miles (12 km) southeast of Sleipner A. (Sleipner is on Blocks 15/6 and 15/9.) Oil and gas production would be moved to Sleipner A for processing and transportation to shore. The Sigyn East and Sigyn West deposits would be produced by means of one and two wells, respectively, via the same template. The wells were due on stream in 2003. The field has an estimated lifespan of 10 years. Sigyn has reserves of 209 bcf (5.6 billion cu m) of gas as well as condensate.

ExxonMobil was Sigyn operator, but Statoil was to execute Sigyn drilling and subsea work, in addition to topsides modifications on Sleipner platform. ExxonMobil held 40% interest, Statoil 50%, and Hydro 10% interest in Sigyn field.

Huldra, Skuld, Norne

Statoil's Huldra field, on Blocks 30/2 and 30/3 in the North Sea, began production of gas and condensate. After separation, the gas was being sent to the Norsk-Hydro-operated Heimdal field for further transport. The condensate was piped to Veslefrikk for processing. Huldra will eventually produce from six wells, building up to a daily plateau of 372 MMcf (10 million cu m). The field's production life was planned for 7 years. Statoil was operator with a 51.62% stake, TotalFinaElf held 24.33%, ChevronTexaco Norway AS 23.34%, Paladin Resources Norway AS 0.5%, and Svenska Petroleum Exploration AS 0.21%.

In other North Sea activity, Statoil began drilling wildcat 16/2-2 on the Skuld prospect south of Balder (ExxonMobil) and Grane (Norsk Hydro) fields. The wildcat was to go to 6,234 feet (1,900 m) or the Tertiary. Separately, Statoil drilled wildcat 21/12-4 between the Sleipner fields, operated by Statoil, and the Frigg field, operated by TotalFinaElf. The prospect was a dry hole.

In late 2001, Statoil announced a "promising" discovery on the Falk structure northeast of its Norne field. Exploration well 6608/11-2, drilled from the West Navion drillship in 1,150 feet (350 m) of water, was terminated 7,218 feet (2,200 m) below sea level. The well proved an "uncertain" size find of oil in early Jurassic sands. Statoil was looking to confirm a connection between the Falk area and its earlier Svale discovery to the south.

Oseberg, Viscund

Norsk Hydro announced two oil and gas finds on the Vesterflank section of the Oseberg field of the North Sea that were estimated to represent recoverable reserves of about 120 million bbl of oil and 127 bcf (3.4 billion cu m) of gas.

Norsk Hydro found oil in two zones with exploration well 34/8-12 S on the Viscund South field in the Norwegian North Sea. The well, drilled to a total vertical depth of 10,367 feet (3,160 m) and completing in the Triassic, was not tested and was plugged. Partners included Norsk Hydro, TotalFinaElf SA, Conoco Inc., Petoro, and Statoil ASA.

Mikkel, Kristin

The Mikkel gas-condensate field straddles Blocks 6407/6 and 6407/7 on the Halten Bank in the Norwegian Sea, 22 miles (35 km) south of the Midgard deposit on the Åsgard license. The development and operation plan, approved in 2001, specified production to start in 2003 and continue until 2017. Late in 2001, Statoil awarded Transocean Sedco Forex Inc. a contract to drill and complete three wells in Mikkel field. The Transocean Searcher was to begin the 140-day contract in 2002. Mikkel field had estimated recoverable reserves of 138 million boe and 760 bcf (20.4 billion cu m) of gas. Statoil was the operator and held a 50% interest in the field. ExxonMobil had 40% and Norsk Hydro 10%. The field was being developed as a subsea system involving two subsea templates, with two production wells each. Development costs were estimated at $255 million (2.4 billion kroner).

The well stream was to be transferred via Midgard to Statoil's Åsgard B platform for processing. Field operator Statoil awarded European Marine Contractors a contract to lay a 25-mile (40-km), 18-inch pipeline between Mikkel gas-condensate field and the subsea template, with completion expected in 2002. The processing facilities and the control room on Åsgard B would be modified.

Gas was expected to stream through

the new 439-mile (707-km), 42-inch Åsgard transport trunkline to the processing plant at Kårstø; condensate was to flow through an existing line to the Åsgard C storage ship for export. Statoil was moving rich gas from the Åsgard offshore field outside mid-Norway to the Kårstø gas terminal in a 42-inch deepwater gas gathering system.

In early 2001, Statoil spudded an exploration well on License 134B about 6 miles (10 km) north of its high-temperature, high-pressure Kristin field to determine if the gas/condensate field will be developed. Saipem SPA's Scarabeo 5 rig drilled 16,214 feet (4,942 m) below sea level in 1,168 feet (356 m) of water and found oil under high pressure in Jurassic sands. The well was in Block 6506/11 in the Norwegian Sea.

Statoil also spudded an exploration well on Production License 257 in the Halten Bank area of the Norwegian Sea to determine natural gas and condensate potential in the northern part of the Erlend structure. The well is 3 miles (5 km) west of Kristin field. Transocean Sedco Forex's Transocean Arctic semisubmersible drill rig was expected to take 76 days to drill to a total depth of some 16,400 feet (5,000 m).

Results from these two wells will be critical in determining long-term development in this Halten Bank area. Kristin is one of three fields (with Sigyn and Oseberg Delta) expected to supply gas to continental Europe over the next 15 years. Kristin alone was thought to hold approximately 1.49 tcf (40 billion cu m) of gas and 200 million bbl of condensate. Reserves in Erlend, Kristin, and two other adjacent Halten Bank fields were estimated at 5.21 tcf (140 billion cu m) of gas and 440 million bbl of oil and condensate.

Kristin might be developed via a floating production platform equipped with gas and condensate processing facilities in order to build in maximum flexibility to accommodate potential tiebacks from other reservoirs in the area. Gas produced would flow through the Åsgard transport system to the Kårstø treatment facilities north of Stavanger for further processing and then to continental Europe. A storage ship would export the stabilized condensate. Partners in the licenses were Statoil 46.6%, Agip SPA 9%, Norsk Hydro AS 12%, TotalFinaElf SA 3%, and ExxonMobil Corp 10.5%. Petoro had an 18.9% stake in the Kristin field.

Goliath, Snohvit

Confirmation tests by Norsk Agip on Goliath field discovery well 7122/7-1, drilled in 2000 in License PL229, found a 230-foot (70-m) oil column, and production tests yielded 4,300 b/d of 33º gravity crude, reported ENI in late 2001. Unofficial estimates suggested that Goliath field could contain 175-300 million boe, or recoverable reserves of 35-40 million boe.

Licensees in the Snohvit field in the Barents Sea ordered the first of three specialized tankers to transport LNG to the US market. The tanker, to be built by Mitsubishi Industries in Japan, would be operated by Leif Hoegh & Co. of Norway and Mitsui OSK Lines Inc. of Japan on behalf of Statoil (22.29% interest), TotalFinaElf SA (18.4%), Gaz de France SA (12%), Norsk Hydro (10%), Amerada Hess (3.26%), RWE-DEA AG (2.81%), and Svenska Petroleum BV (1.24%). The Norwegian government had a 30% interest in the project.

Kawasaki Kisen Kaisa (K-line) and Mitsui & Co. were contracted for two additional tankers on behalf of Statoil, Amerada Hess, Norsk Hydro, RWE-DEA, Svenska Petroleum, and Petoro. All three vehicles would be fully committed to Snohvit for 20 years, starting with production in 2006, and would be adapted to North Atlantic climatic and environmental conditions. The licensees expect to ship 70 LNG cargoes/yr to meet existing gas sale agreements of 89 bcf (2.4 billion cu m) of gas to customers in the US and 59.6 bcf (1.6 billion cu m) to Spain, all from the field's planned land terminal at Melkoya near Hammerdest in northern Norway.

Processing

Statoil awarded M.W. Kellogg Ltd. a $7.75 million (70 million kroner) contract to provide engineering, procurement, and construction assistance for an upgrade to the Kårstø complex north of Stavanger. Statoil recommended the Statpipe partners build an extraction plant there to process rich gas from Mikkel field, slated to come on stream 2003. Statoil had 25% interest in Statpipe. Other partners were Petoro with 33.25%, TotalFinaElf 12%, ExxonMobil 12%, Norsk Hydro 10%, Royal Dutch/Shell 5%, and Conoco 2.75%. The Statpipe system links fields in the northern North Sea to the continental Europe gas market. In late 2001, Statoil also awarded ABB Group a $36 million, 5-year contract for maintenance and modification of the Kårstø gas treatment plant north of Stavanger and at the Draupner platforms in the North Sea. Three 2-year extensions were optional.

Also late in 2001, Statoil began testing a process to convert methanol that was previously flared into propylene gas. The gas was flared because the infrastructure for export was lacking. The objective was to achieve a yield of 70% propylene, a raw material for plastics production, from the methanol. The demonstration plant at Tjeldbergodden in mid-Norway will show if the technology, developed by Lurgi AS of Germany, can be used at a large plant. Tjeldbergodden provided a site for landfall gas from the Ormen Lange field in the Norwegian Sea, Norway's second largest gas discovery.

Statoil was installing a desulfurization plant at the 200,000 b/d Monstad refinery near Bergen. The plant would allow Statoil to produce 10-ppm sulfur gasoline and diesel oil. It should be on stream in 2003, 2 years before such fuel becomes compulsory in the EU.

Statoil planned to build a new liquid propane storage cavern at the Monstad refinery to replace the existing cavern that was damaged by a rock fall leading to severe ice formation and reduced storage capacity. The Vestprocess partners, which operate the refinery, determined the damage could not be repaired. The new cavern should be completed by late 2003.

Almost 259 million bbl of crude were shipped from the Monstad refinery in 2001, making it Europe's second busiest oil port, after Rotterdam, in terms of volume. The partners were Statoil with 17%, Petoro 41%, Norsk Hydro 17%, ExxonMobil 10%, Norske Shell 8%, TotalFinaElf 5%, and Norske Conoco 2%.

POLAND

CAPITAL: WARSAW
MONETARY UNIT: ZLOTY
REFINING CAPACITY: 382,000 B/CD
OIL PRODUCTION: 17,000 B/D
OIL RESERVES: 115 MILLION BBL
GAS RESERVES: 5.1 TCF

During 2001 Poland discussed trade agreements with Russia, Norway, and Denmark for imports of natural gas. Reportedly, Poland did confirm in early 2002 its commitment to financing the first unit of the Yamal-Western Europe gas pipeline in the originally expected amount. In 2000, Russia supplied approximately 257 bcf (6.9 billion cu m) of gas to Poland. Of its total natural gas consumption of 428 bcf (11.5 billion cu m), Poland was annually importing 279 bcf (7.5 billion cu m).

Norway's Gas Negotiating Committee (GFU) and state-owned Polish Oil & Gas Co. (POGC) agreed on terms for delivery of nearly 2.8 tcf (74 billion cu m) of Norwegian gas to Poland over 16 years. Government authorities must approve the deal. Starting in 2008, deliveries would increase over 3 years to 186 bcf/yr (5 billion cu m/yr) and remain at that level until 2024. Norwegian gas producers involved in the deal were Statoil, Norsk Hydro, and Norske Shell, along with ExxonMobil and TotalFinaElf. It had not yet been decided which fields would supply the gas. The contract replaces one for 18.6 bcf/yr (500 million cu m/yr) from 2000-06 that POGC and GFU negotiated in 2000.

A gas pipeline would be required for export. Statoil proposed the option of a 684-mile (1,100-km) line through the Skagerrak Channel between Norway and Denmark, extending to Niechorze on the Baltic coast, where it would link with the Polish gas network. This pipeline could supply Sweden and Norway's Grenland industrial region south of Oslo and, longer term, Denmark. If the export volumes are insufficient to justify the proposed pipeline, the Norwegian gas producers can withdraw from the agreement.

Danish state oil and gas firm DONG signed a deal with POGC to deliver more than 74 bcf/yr (2 billion cu m/yr) of gas to Poland starting in 2003, culminating in 2011 with total Danish exports of 596 bcf (16 billion cu m) of gas. For delivery, DONG and POGC proposed the 143-mile (230-km) BaltPipe natural gas trunkline, to run from Danish Zealand to Poland's Baltic coast. The two companies further proposed that a consortium of gas producers with interest in exporting to Poland could fund BaltPipe. The deal could fall through if Denmark's Dansk Undergrunds Consortium and DONG cannot reach agreement on increased gas deliveries from Danish North Sea fields.

The amine regeneration unit at Polski Koncern Naftowy Orlen SA in Plock, Poland, was part of a hydrodesulfurization project under construction during 2001 to produce ultralow sulfur diesel. Photo courtesy of ABB Lummus Global.

Gas exploration and production

Gas flow started in early 2001 at 2 MMcfd in the Kleka-11 well located near Radlin. FX Energy Inc. held 49% interest; POGC, operator, held 51%. The flow rate was to be raised to 4 MMcfd and maintained for 2 years, then evaluated. The well was a step-out to 39-bcfd Kleka field. POGC agreed to purchase the gas produced from Kleka-11. FX Energy claimed to be the first western company to produce and sell gas from a discovery in Poland.

Apache Corp. reported gas flow of 9.5 MMcfd from a preliminary open-hole test of its Tuchola 108-2 wildcat on its Pomeranian concession in northwestern Poland. This was the first discovery in the Pomeranian area since the mid-1980s. The flow was the limit of the surface equipment, and the gas was from the secondary objective, a 200-foot gross interval in the Permian main dolomite at 8,318-8,514 feet. The well was Apache's first on the concession, but drilling was to continue toward the primary objective, the Devonian sand, and other leads would be explored in the area if the discovery proves commercial. Tuchola 108-2 was 60 miles from the nearest production and 3 miles from a gas pipeline. Apache, the operator, held 42.5% of the concession, FX Energy had 52.5%, and Polish Oil & Gas Co. had 15%.

RWE-DEA AG acquired a 58.8% share of Ramco Energy PLC subsidiary of Medusa Oil & Gas, Poland. Medusa Oil & Gas (Europe) BV was to hold the balance. RWE-DEA essentially took a farmout of Ramco's interest in Carpathian Blocks 434, 435, 454, and 455 in southern Poland. The German company would fund the next $12 million of an agreed work program, including seismic surveys and the drilling of two exploration wells before late 2002. The blocks covers 802 sq mi (2,077 sq km) along Poland's southern border with Slovakia. Medusa Poland has completed three seismic surveys totaling 370 miles (596 km) of 2D data and has identified target structures with proven gas indications on trend.

Refining

Polski Koncern Naftowy (PKN) operated the 260,000-b/d refinery in Plock where a hydrodesulfurization project to produce ultralow sulfur diesel was under construction.

A proposed integrated gasification combined cycle (IGCC) project at the Rafineria Gdanska SA refinery in Gdansk would allow Poland to meet 5-6% of its CO_2-abatement target set by the Kyoto protocol and enable the refinery to produce higher value-added products. Installation of IGCC would mean that the refinery can produce electric power in an environmentally cleaner way and probably allow shutting down of older, polluting coal-fired power plants. A CO_2 emission reduction of between 1.2 and 1.6 million tpy was possible. Production of 800,000 tpy of heavy sulfur fuel oil for export would be halted, which would yield a reduction of about 50,000 tpy of SO_X emissions. This is a reduction of 5% of the national annual SO_X emissions.

In late 2001, Rotch Energy paid about $250 million to buy 75% of the Gdansk refinery. Rotch planned to invest $600-$700 million in expansion and bolster the refinery's capacity to about 150,000 b/d over the next few years.

PORTUGAL

CAPITAL: LISBON
MONETARY UNIT: ESCUDO*
REFINING CAPACITY: 304,172 B/D
OIL PRODUCTION: 0
OIL RESERVES: 0
GAS RESERVES: 0
* Switching to euro in 2002.

In early 2001, Portugal was cited by the EC for failing to open its internal markets to competition, and Portuguese authorities responded to the EC that implementation is imminent.

Petrogal (Petroleos de Portugal, the country's state oil company and part of the larger shareholder-owned GALP energy conglomerate) awarded a 23 million euro contract to Flour Daniel, a unit of Fluor Corp., for the engineering, procurement, and construction of a hydrogen desulfurization facility at the Sines refinery. This was Fluor's first contract with Petrogal. The Sines refinery was granted an extension until early 2002 to meet the EC's directive to limit sulfur in its gasoline, and this upgrade would enable the plant to meet the EC's regulations. Petrogal is based in Lisbon and owned by Galp Energia shareholders.

Mohave Oil and Gas planned to drill onshore in the Lusitanian basin area north of Lisbon. Two wells in less than 328 feet (100 m) of water each, one targeting primarily gas and the other oil, were expected to be spudded to meet the work requirements for the San Pedro de Muel and Sao Mamede concessions.

Mohave and Desire Petroleum PLC plugged and abandoned the Alijubarrota 2 sidetrack in the Lusitanian basin because open hole production rates, up to 100 Mcfd, indicated very low permeability. Previously, Mohave suspended drilling at Alijubarrota 3 because of geological problems. Mohave and Desire will take more onshore seismic data on several concessions in the area.

Wood Mackenzie Consultants Ltd. identified the virgin deepwater acreage off Portugal as one of four offshore areas to be likely contributors to future E&P. Based on direct hydrocarbon indications in the deepwater outer continental shelf, the Portuguese government planned to open exploration license bidding in 2002, and drilling could begin in 2004. Interested parties included Swedish independent Taurus Petroleum, with hope of finding oil reserves in the Galacian area offshore.

ROMANIA

CAPITAL: BUCHAREST
MONETARY UNIT: LEU
REFINING CAPACITY: 504,037 B/CD
OIL PRODUCTION: 120,000 B/D
OIL RESERVES: 956 MILLION BBL
GAS RESERVES: 3.6 TCF

In mid-2001, Romania's Industry Minister announced that partial privatization of SNP Petrom, the nation's vertically integrated national oil company, would be launched in 2002. SNP Petrom was undergoing restructuring to streamline its managment and operations.

Transylvania, Doina

Wintershall AG grew its exploration and production business into activities in Romania. Wintershall planned to join with state-owned gas company Exprogaz SA to explore the Transylvania South Block. Before deciding to spud one or two exploration wells at a cost of $4 million, Wintershall was acquiring 311 miles (500 km) of 2D seismic in the Saros-Sighisoara region, as part of a $4 million initial phase.

Sterling Resources Ltd. and Paladin Expro Ltd., partners in Block XV in the Black Sea, awarded a contract to Petrom to drill an appraisal well off Romania. The well was a follow-up to the Doina discovery.

Romania protested Ukraine's discovery of commercially viable oil and gas deposits near Zmiynyy Island, an object of a territorial dispute between Romania and Ukraine.

Gas transport and distribution

Ruhrgas Energie Beteilingungs AG signed a participation agreement with Congaz for gas distribution in Romania. Ruhrgas was the first international oil and gas company to have a stake in Romania's gas distribution. Congaz was newly established in Romania and expected to develop sales of more than 11.2 bcf (300 million cu m) as more homes and industrial customers in the Constantza region are connected to the national gas network. According to Ruhrgas, Romania has the largest gas market in central Europe, consuming 633 bcf (17 billion cu m) yearly. Three-quarters of that supply was from indigenous sources and one-quarter from Russia.

To make natural gas consumption more efficient, Romania has begun to modernize its 9,000-mile natural gas distribution network to cut down on leakage and upgrade its measuring stations. In early 2001, the Romanian natural gas trader Itera and Romanian and Moldovan natural gas companies agreed to build a 72-mile pipeline connecting Dorchia and Ungheni, Moldova, with Iasi, Romania. The pipeline was expected to cost $60 million and have an annual capacity of 141 bcf of natural gas per year. The pipeline would allow Romania to import more Russian natural gas via Moldova.

Romania continued to be a major transit point for oil and gas supplies. Recent studies on the feasibility of major new pipeline options to bring Caspian Sea oil to European refineries included a route from Constantza, Romania, to Trieste, Italy. The pipeline was estimated to cost $900 million and would incorporate existing pipelines to connect Constantza with 10 refiners. There were several alternatives for the route, all beginning at Constantza and ending at Trieste.

SLOVAKIA

CAPITAL: BRATISLAVA
MONETARY UNIT: KORUNA
REFINING CAPACITY: 115,000 B/CD
OIL PRODUCTION: 1,000 B/D
OIL RESERVES: 9 MILLION BBL
GAS RESERVES: 530 BCF

Among other countries, Slovakia is capable of meeting European specifications for cleaner fuels and is in a position to provide exports of diesel fuel. Slovakian refiner Slovnaft sold a 32% stake in its business to Hungary's MOL Rt. in 2000.

Slovakia would benefit from Ukraine's project to modernize the gas transit system. The improvements will make it possible to transfer up to 1.08 tcf/yr (29 billion cu m/yr) of gas from Russia to the Ukrainian borders of Slovakia and Hungary.

The Slovak government in 2001 launched the privatization of 49% of the natural gas transport company Slovensky plynarensky priemysel (SPP). However, a decision on new owners was not expected until early 2002. SPP was considered a crucial part of the European pipeline network, carrying gas to western and southern Europe.

SPAIN

CAPITAL: MADRID
MONETARY UNIT: PESETA*
REFINING CAPACITY: 1.294 MILLION B/CD
OIL PRODUCTION: 7,000 B/D
OIL RESERVES: 21 MILLION BBL
GAS RESERVES: 18 BCF
* Switching to euro in 2002.

Restructuring in the energy industry, in some cases to meet Spain's internal policies and EU requirements, has driven consolidation and divestiture. Compania Logistic de Hidrocarburos CLH SA, Spain's largest pipeline network and storage business, agreed to sell 25% of CLH to Enbridge, Inc. CLH had been majority owned by Spanish refiners Repsol-YPF SA, BP Oil Espana SA, and CEP.

Repsol-YPF planned to invest 23.1 billion euros over the next five years, with 8.4 billion euros invested in Spain and the balance in Latin America.

Exploration and development

Near the end of 2001, the government of Spain announced it is ready to grant Respol-YPF nine licenses to explore for oil in waters north of the Canary Islands. Repsol had applied for permission to explore the nine blocks, consisting of 2,378 sq mi (6,160 sq km) in the Atlantic Ocean surrounding the Canary Islands, but clearance was denied until the government determined whether the territory belonged to Spain. Two of the nine areas to be licensed to Repsol were also claimed by Morocco. Separately, GS-NOPEC Geophysical Co. ASA and Fugro Geoteam AS began a nonexclusive 2D seismic survey in the southern part of the Gulf of Valencia off Spain, with data to become available in early 2002.

Repsol-YPF announced an oil discovery off Spain near the Casablanca complex of fields. The Chipron Northeast discovery flowed 5,000 b/d in production tests from each of two wells. If developed, production would be connected via an existing platform and pipeline to Repsol's refinery in Tarragona.

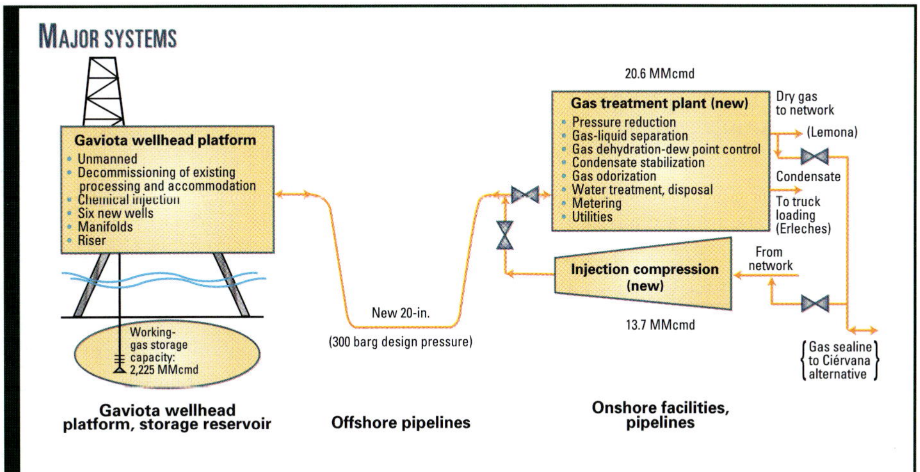

The Gaviota platform's gas processing and compression facilities will be moved to an expanded onshore plant, while the offshore underground gas storage capacity will be almost tripled.

Transportation

Gaz du Sud-Ouest (owned 70% by TotalFinaElf SA and 30% by Gaz de France SA) and Ente Vasco de Energia of Spain and its affiliate Gas de Euskadi planned an 18-mile (29-km), 18.6-bcf (500 million cu m) line along the Basque coast to improve gas transmission between the two countries. The 10.7 million euro cost would be subsidized by the EU as part of a policy to strengthen gas links between member countries. The proposed line, from Bidart in France to Irun in Spain, could be on stream in the winter of 2002-2003. The companies have also begun a feasibility study on the prospect of building, by 2010, a high-capacity link between Bilbao and the underground gas storage reservoir in Lussagnet, France. The 204-mile (328-km) pipeline would cost 137 million euros.

LNG

LNG played an increasing role in Spain during 2001, with significant LNG receiving terminal and regasification projects underway. LNG will become more important for power generation in Spain, especially as uneconomic coal use declines. Union Fenosa awarded M.W. Kellogg Ltd. a contract to design and prepare engineering, procurement, and construction enquiry packages for a $240-million LNG receiving terminal at Sagunto Harbour, Valencia. The project included an estimated storage capacity of 15.9 MMcf (450,000 cu m) and a nominal LNG vaporization capacity of about 180 bcf/yr (5 billion cu m/yr) in 2004.

Also, Enagas SA awarded a joint venture to build an LNG tank at Palos de Frontera, Spain. The 5.3-MMcf (150,000 cu m) tank would expand the LNG import terminal in Huelva harbor. The 40 million euro construction project was to be led by Technigaz (subsidiary of Bouygues Offshore SA) with 30% of the contract held by Initec Corp.

Seven Spanish companies and Sonatrach formed the Reganosa group to build the Ferrol import terminal in northwest Spain and about 56 miles (90 km) of pipeline connecting the regasification terminal with power plants. Also, a regasification plant with associated power plant, scheduled for startup in 2004, was underway in northern Spain with EVE (Basque Energy Authority) as part of the consortium. Separately, the Gas Natural Group planned to expand three regasification terminals and build a new terminal to begin regasification in 2004.

Gas storage and processing

To help accommodate increasing demand for seasonal use of natural gas, Repsol-YPF and Murphy Spain Oil Co. planned to increase the working gas capacity of their Gaviota natural gas storage reservoir nearly threefold, from 27.5 to 78.5 bcfd (780 to 2,225 million cu m/day). The Gaviota reservoir is in Spain's Bay of Biscay at 7,218 feet (2,200 m) below sea level. A peak gas withdrawal rate of 727 MMcfd (20.6 million cu m/day) was targeted. The multi-year project was scheduled for start-up of the injection plant and production facilities in 2005.

SWEDEN

CAPITAL: STOCKHOLM
MONETARY UNIT: KRONA
REFINING CAPACITY: 423,500 B/CD
OIL PRODUCTION: 0
OIL RESERVES: 0
GAS RESERVES: 0

Refining continued to be a major activity in Sweden. Preem Raffinaderi awarded Foster Wheeler a $34 million contract for engineering and procurement to revamp the existing Penex isomerization unit at its Gothenburg refinery. Also, a new Molex unit would be added and associated interconnected services put in place. These investments would debottleneck the existing isomerization process with the aim of improving the octane level of gasoline to meet requirements of future EU gasoline regulations. Work was due to be completed in 2002. The refinery has a capacity of 106,000 b/d.

Vattenfall AB sold its 51% share in the Vattenfall Naturgas subsidiary to the other shareholders, leaving Ruhrgas Energie Beteiligungs AG (30%), Statoil AS (30%), Fortum (20%), and DONG AS (20%) as owners of the company. Vattenfall Naturgas annually sells 37 bcf (1 billion cu m) of natural gas in Sweden, where the fuel accounts for only 2% of primary energy consumption. Most of the gas was imported from Denmark and sold to local distribution companies and industrial customers. Vattenfall was concentrating more on the electricity market.

To reinforce the gas distribution network, Sydgas of Sweden awarded a $5.96 million (65 million Swedish kroner) contract to a consortium to lay a 28-mile (45-km) gas line near Halmstad in southern Sweden. The consortium contractors were Bohlen & Doyen and Ludwig Freytag, both of Germany, Puhl Aarsleff AB of Sweden, and Per Aarsleff AS of Denmark. The installation would involve welding high-pressure 16 and 8-inch (40 and 20-cm) pipe. Work was planned to start and finish during 2002.

Sweden could benefit from a pipeline proposed to run 684 miles (1,100 km) through the Skagerrak Channel between Norway and Denmark, extending to Niechorze on the Baltic coast, where it would link with the Polish gas network. Although Statoil proposed this option to deliver an agreed-upon 2.76 tcf (74 billion cu m) of Norwegian gas over 16 years to Poland, the line could also supply Sweden and Norway's Greenland industrial region south of Oslo.

Vattenfall AB agreed to buy Mirant Corp.'s 44.8% ownership position in Bewag AG for $1.63 million. Vattenfall would have a 89.6% stake in Bewag and 92% of voting rights. Mirant is one of Europe's top 10 gas marketers.

SWITZERLAND

CAPITAL: BERN
MONETARY UNIT: FRANC
REFINING CAPACITY: 132,000 B/CD
OIL PRODUCTION: 0
OIL RESERVES: 0
GAS RESERVES: 0

At a meeting in Bonn, Switzerland joined with Canada, the EU, and Japan in saying they will spend a total of $420 million to finance CO_2-absorbing projects, such as tree planting or clean-power conversions, to meet the Kyoto climate change treaty. To advance its environmental agenda, Switzerland gives subsidies for photovoltaic users.

Switzerland seldom participates in oil and gas exploration. In 2001, Switzerland had one exploration well: Forest Oil Corp. tested Weiach 2 with disappointing results.

TURKEY

CAPITAL: ANKARA
MONETARY UNIT: LIRA
REFINING CAPACITY: 719,275 B/CD
OIL PRODUCTION: 56,000 B/D
OIL RESERVES: 295.8 MILLION BBL
GAS RESERVES: 310 BCF

Just as Turkey had been the crossroads of world trade between East and West in historic times, it will continue to be that crossroads for energy transportation from Central Asia and Middle East petroleum producing countries to global consuming countries. Turkey's unique geographical position has earned it the sobriquet, "energy bridge between east and west."

Safety, security, cost, political partners, and engineering feasibility were seen as the primary constraints for emerging oil and gas transportation from production facilities in Iraq, Iran, Turkmenistan, Azerbaijan, Russia, and other countries seeking indirect access to the Mediterranean Sea and international markets. Turkey would be the terminal country for several oil and gas pipeline proposals. Some were moving forward, while others required more consideration pending economic studies and future oil and gas production.

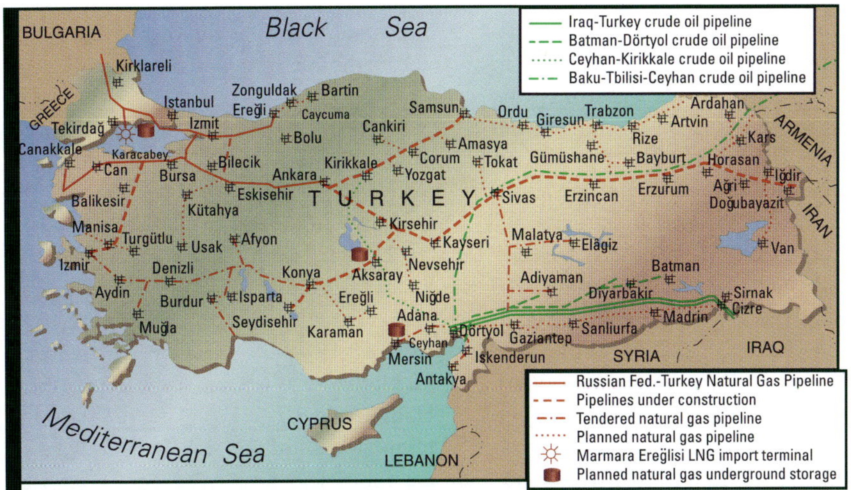

Determined to remain at the center of the East-West energy corridor, Turkey continued during 2001 to expand its existing oil and gas pipeline network, with the Iran-Turkey gas pipeline coming on stream, the Blue Stream gas pipeline approaching completion, and the Baku-Tbilisi-Ceyhan oil pipeline negotiators reaching key agreements.

Pipeline planning moves forward

The Baku-Tbilisi-Ceyhan oil pipeline, planned to extend from Azerbaijan through Georgia and into Turkey, made steady progress during 2001. First, Turkish President Ahmet Necdet Sezer and Azerbaijani President Haidar Aliyev signed a 15-year contract for 233 bcf/yr of gas, beginning in 2004, to be exported from Azerbaijan's Shah Deniz field. Then this $3 billion, 1 million b/d pipeline – after being agreed upon by heads of state and pending approval by parliaments in Azerbaijan, Georgia, and Turkey – was confirmed by US officials in early 2002 as being on time for completion in 2005 and on budget. Talks were being held to develop financing. The US has supported a Baku-to-Ceyhan route to mitigate Russia's and Iran's influence over Caspian oil exports. The pipeline would also prevent additional tanker traffic through the crowded Bosporus Straits, since oil would be delivered to Turkey's Mediterranean port.

Turkey's agreement with Russia to import gas via the Blue Stream pipeline was near startup in 2001. The 758-mile pipeline, including a 247-mile length under the Black Sea, was scheduled to be operational in 2002. Gas was expected to cost $3.40/cubic foot or about half again as much as Turkmeni gas. The Saipem 7000 lay barge was laying the subsea section in late 2001, after more than 18 months of retrofitting in preparation for work under more than 6,500 feet (2,000 m) of water.

Separately, the 160-mile gas pipeline from Iran to Turkey, tied to a 23-year import agreement, was complete and intended to be operational by the end of 2001, according to the Turkish Commercial Counselor Osman Bakaroglu, but the project had been delayed prior to completion of the metering station in Turkey. The line will initially supply 112 bcf/yr (3 billion cu m/yr), increasing deliveries gradually to 372 bcf/yr (10 billion cu m/yr) in 2007.

Russia had proposed a new oil pipeline route across the Caucasus to the Russian port of Novorossiysk (frozen port for two months/year), then through the Bosporus by tanker or pipeline to Bulgaria and Greece. Many alternative oil pipelines had been proposed for Caspian and Central Asia oil, with some terminating in Turkey. One plan was an oil pipeline parallel to the Baku-Tbilisi-Ceyhan gas pipeline.

Turkey indicated that it does not want more tanker traffic in the Bosporus Straits. With some 5,500 oil tankers passing through the straits annually, the country worries about safety and environmental risks. Turkey wants more control of the size and age of vessels passing through the straits. The International Energy Agency estimates that the traffic through the straits is near the maximum recommended.

Privatization and growth

Turkey's energy demand has outpaced its production, and billions of dollars of investment will be needed in the coming years. In order to move closer to the growth plans of the government and requirements of the EU, Turkey was moving toward privatization of government energy monopolies. One goal has been to partially privatize Botas, the state-owned company that controls gas imports.

Toward meeting energy goals, Turkey ratified the Energy Charter Treaty in April 2001 that provides the legal framework for international investment in the energy sector.

Production

Turkey's oil production generally comes from small fields in scattered locations. The country's oil prospects were primarily in the Hakkari basin, European provinces, the Black Sea shelf region, and the Aegean Sea. The latter has not been explored because of conflicting claims with Greece. Turkey has licensed new acreage off its southern Mediterranean coast. State-owned Turkish Petroleum AO, partnering with Romania's Petrom and Houston-based El Paso Corp., was exploring the area.

UNITED KINGDOM

CAPITAL: LONDON
MONETARY UNIT: POUND
REFINING CAPACITY: 1.771 MILLION B/CD
OIL PRODUCTION: 2.537 MILLION B/D
OIL RESERVES: 5 BILLION BBL
GAS RESERVES: 26.8 TCF

Early in 2002, the UK government offered 300 blocks in its 20th licensing round and announced that Argyll, the first oil field to be developed in the UK North Sea, would return to production 11 years after it had been closed as uneconomic. Argyll was expected back on stream in 2003. The government also said that three small North Sea gas developments would proceed.

UK North Sea production was shifting focus from a small number of very large projects to a larger number of smaller projects. The Pilot task force, which is made up of 23 government representatives and oil industry chiefs, announced in mid-2001 the next fields for prospective development under the leading oil and gas industry competitiveness (Logic) scheme: the North Sea Don (Northeast, Southwest, South) and Don West fields. Redevelopment of these fields, located over four blocks in around 500 feet of water in the UK Continental Shelf, could produce around 35 million bbl of reserves from the Brent sandstone reservoir and possibly further volumes from the Statfjord sandstone reservoir in the Don fields, from Don West, and from a westerly expansion of Don. Altogether, about 180 million bbl could be available.

The Logic satellite accelerator scheme allows for industry investors to gain a share in a field's reserves or production in exchange for the capital expenditure required to bring the redevelopment on stream, thus prolonging the UK Continental Shelf unexploited reserves. In February 2001, Logic chose the Nisus consortium (Global Marine Integrated Services Reservoir Management, Stolt Offshore, and Wood Group Engineering) to create development design for BP's marginal Wood field, the first discovery to advance under the satellite accelerator program. Wood field, which has estimated recoverable reserves of 10 million boe, is on Block 22/18-6 between Arbroath and Marnock fields in the central North Sea. BP considered the prospect "subeconomic under current development scenarios." Wood was the first of five undeveloped UK North Sea discoveries that Logic has brought to market. The other four are Solan, Strathmore, Kestrel, and Kessog.

Beauly, Kyle

In early 2001, Talisman Energy Inc. announced production at the rate of 12,000 b/d from Beauly field, Block 16/21 in the UK North Sea. The field has reserves of 3 million bbl. It was developed with a single horizontal well tied back 3 miles (5 km) to the Balmoral floating production vessel. Talisman, operator, held 60% stake in the field, and North Sea Oil Ltd. had 40%.

Kyle field in the UK North Sea came on stream in early 2001 through two wells at 7,000 b/d, in addition to gas production. Ranger Oil UK Ltd. was operating the development on Block 29/2c in the central North Sea. Output was expected to stabilize at around 20,000 b/d of oil. The oil was being processed via the Maersk Curlew FPSO 18 km south of Kyle, and the gas was to be exported through the SEGAL pipeline system. Kyle partners were Premier Oil Exploration Ltd 35%, Ranger Oil UK Ltd. 20%, Ranger Oil PC Ltd. 20%, Roc Oil UK Ltd. 12.5%, and Bow Valley Petroleum UK Ltd. 12.5%. A third production well was to be drilled. In 2000 at Kyle, 1.5 million bbl was produced during an extended well test.

Blake, Ross

BG Group and its partners brought Blake oil field on stream 18 months after project sanction – two months ahead of target and 10% below budget. It cost £145 million to develop. Blake field, in the Outer Moray Firth, is located on Blocks 13/24 and 13/29b in the northern North Sea. Initial production of 40,000 b/d was expected to increase to 60,000 b/d later in 2001. Oil was produced through the Bleo Holm FPSO vessel at Talisman Energy UK's nearby Ross field. BG estimated gross field reserves at 70 million bbl. BG operated Blake, with 44% interest; Talisman owned 53.6%, and Paladin Resources owned 2.4%.

In mid-2001 Coflexip Stena Offshore Ltd. (CSOL) installed in the Blake field what was claimed to be the largest and heaviest reeled pipe. The Blake field is located in 312 feet (95 m) of water about 6 miles (9 km) north of the Ross field in the outer Moray Firth (northern North Sea). Production from Blake into the existing Ross floating production, storage, and offloading (FPSO) vessel Bleo Holm would add 44,000 b/d to the continued Ross field production of 25,000 b/d.

A sixth development well in the Ross field, well 13/29a-A4, flowed oil at 19,000 b/d in mid-2001. This was the fourth of five production wells planned by Talisman Energy Ltd. and partners for Ross. Production would be tied into the Bleo Holm FPSO vessel. Oil was also found in exploration well 13/29b-7 in May. That well flowed at 2,200 b/d. Talisman held 80% interest in the block and Paladin Expro Ltd. owned 20%.

Captain, Leadon

One of the largest subsea templates installed in the North Sea lay in the Captain field, UK North Sea Block 12/22A, in about 490 feet (150 m) of water. Designed and engineered in a joint venture between Cameron, a division of Cooper Cameron Corp., and Halliburton Subsea, a division of Halliburton Co., the 18 slot, 740-tonne template manifold was installed about 2.2 miles (3.5 km) from two platforms (one new) that were connected by a bridge. The template manifold serves as a drilling template for 18 wells and as a pipe manifold for production, water injection, and future polymer injection. Texaco North Sea UK Co., 85% interest, operated the field, and Korea Captain Co., Ltd., held the remaining working interest.

In late 2001, Kerr-McGee brought Leadon field, on Blocks 9/14a and 9/14b about 220 miles northeast of Aberdeen, on stream at production of 10,000 b/d. Subsea horizontal wells were tied back to FPSO vessel Global Producer II. Production was expected to peak at 45,000 b/d in 2002. Production began less than 2 years after discovery and less than 1 year after government approval of the development. The field cost $650 million.

Discoveries

Talisman Energy Ltd. reported in spring 2001 an oil discovery at the Lucy prospect, 5 miles (8 km) northeast of the Piper Bravo platform in the UK central North Sea. The discovery, believed to have reserves ranging from 20 to 50 million bbl, is adjacent to the Talisman-operated Piper field. Talisman held 76.7% interest in the 15/12b Block, which contains the discovery well, and 60.4% in the adjacent 15/17 Block into which the discovery extends. Lasmo PLC held 20% in each block. Intrepid Energy North Sea Ltd. held 3.3% in 15/12b and 19.6% in 15/17 Block.

BP also announced a "significant" oil discovery in the first well drilled on Buzzard exploration prospect in the central UK North Sea. The well, 62 miles (100 km) northeast of Aberdeen, flowed at 6,547 b/d of light oil and 0.97 cfd of gas in mid-2001. The partners spudded appraisal wells 19-10A in November; and planned to spud well 20/6-D in 2001 as well. Partners in the Buzzard license were PanCanadian 45.01%, BG 19.99%, Intrepid Energy North Sea 30%, and Edinburgh Oil and Gas 5%.

Kerr-McGee Corp. named three discoveries on its Block 9/23 in the UK North Sea: Tullich field (reserves of 35-40 million boe, 100% owned by Kerr-McGee), Blue Sky, and Blue Sky 2. The block is immediately south of Kerr-McGee-operated Gryphon field and east of the Harding field, in which Kerr-McGee held 25% interest. Plans were to develop Tullich as a subsea tieback to the Gryphon FPSO facility, with first production in 2003. In that same area, development sanctioned for the Maclure field was expected to lead to another subsea tieback to the Gryphon facility. A gas pipeline linking the Gryphon, Leadon, and Beryl fields was planned.

Elgin, Franklin

Both Elgin and Franklin gas-condensate fields in the UK North Sea came on stream in 2001, four years after development began. Elgin, on Blocks 22/30b, 22/30c, and 29/5b in about 300 feet (92 m) of water, came on stream early in the year. Franklin, on Block 29/5b, was 3.4 miles (5.5 km) from Elgin in 305 feet (93 m) of water. It came on stream in late 2001. Franklin and Elgin each had unmanned wellhead platforms, and their productions were being processed through a central platform at Elgin. Condensate was to be exported by a 24-inch, 24-mile line to BP's Marnock platform and moved ashore via the Forties pipeline. Gas was to exported via the new Shearwater Elgin Area Line to the Bacton terminal in eastern England, where it would be tied into the onshore gas transmission systems.

First production from the jointly developed Elgin-Franklin fields was 13,500 b/d of condensate and 42 MMcfd of natural gas. This was expected to plateau at 140,000 b/d of condensate and 455 MMcfd (13 million cu m/day) of natural gas. TotalFinaElf was operator of both fields through unit Elf Franklin Oil & Gas, in which TotalFinaElf had 77.5% interest. The operator had a 46.2% stake in the fields, and Gaz de France had 22.5% interest. Remaining equity in the fields was held by Agip, BG, Ruhrgas, ExxonMobil, Texaco, Dyas, Oranje-Nassau and BP. TotalFinaElf said the project is the largest high-pressure, high-temperature development in the world (1,100 bar and 200°C) and the deepest in the UK North Sea at 18,044 feet (5,500 m).

Nuggets, Murdoch

Debottlenecking the gas treatment plant of the Alwyn North platform in the northern North Sea has increased the processing capability and enabled TotalFinaElf Exploration UK PLC to develop and operate the Nuggets field, south of the platform, via a tieback to the platform. Nuggets gas would be processed on the Alwyn North platform, controlled from the Dunbar platform, and moved through the Frigg pipeline to the St. Fergus terminal in Scotland. All three fields were expected on stream by year-end 2001.

The Nuggets field is to the east of the Shetlands Islands in 361-410 feet (110-125 m) of water. The gas-bearing accumulations were designated Nuggets N1, N2, N3, and N4. In 2001, the UK government approved TotalFinaElf's development of the Nuggets N2 and N3 gas fields, which are 12 miles (20 km) south of Alwyn satellite field Dunbar. The Nuggets N1 development plan was approved in mid-2000. Nuggets N4 was to be developed later. Combined production from N, N2, and N3 would be 165 MMcfd.

Downstream activities

Late in 2001, Conoco Ltd. announced it was selling 175 company-operated Jet service stations to Fuelforce Ltd., a new independent retailer. The stations would continue to operate under the Jet brand, and Conoco would supply the Fuelforce stations. With the deal, Concoco's Jet-branded network increased from 388 to more than 560 outlets. IPE

The most thoroughly tested drilling control system is also the most reliable.

MIDDLE EAST

BAHRAIN

CAPITAL: MANAMA
MONETARY UNIT: DINAR
REFINING CAPACITY: 248,900 B/CD
OIL PRODUCTION: 102,500 B/D
OIL RESERVES: 124.6 MILLION BBL
GAS RESERVES: 3.3 TCF

Interest in exploration in Bahrain was stimulated in 2001 when the country was awarded sovereignty over the Hawar Islands, which are adjacent to Qatar's main onshore oil field. Bahrain and Qatar had been in a territorial dispute over the area. However, initial survey data were reported to be disappointing.

Bahrain's Ministry of Oil awarded Texaco UK and Petronas new offshore exploration blocks southeast of Bahrain in September 2001, which had been unawarded previously due to the boundary dispute with Qatar. Texaco planned to begin drilling in 2002 in the Eastern Offshore Area in the Persian Gulf on Block V. Bahrain's existing reserves are contained in a single oil field, Awali, which was nearing depletion.

Most of Bahrain's natural gas is associated with the Awali field, with gas production and processing controlled by the state's Bahrain National Gas Co. (Banagas). Given Bahrain's demand for power generation fuel, the country could become a gas importer unless offshore reserves are located in the next few years.

In early 2002, Bahrain signed a memorandum to purchase Qatari gas as part of a 1.75-bcfd enhanced gas utilization project for Qatar's supergiant North field. The agreement supports efforts toward a regional gas pipeline network promoted by the Gulf Cooperation Council (GCC). In mid-2001, a study was commissioned to examine the feasibility of a 1-bcfd gas import pipeline connection to a proposed Qatar-Kuwait pipeline.

The Bahraini government continued efforts to consolidate the corporate structure of its state-owned petroleum sector, merging the upstream Bahrain National Oil Co. (Banco) with the refining operations of Bahrain Petroleum Co. (Bapco). The merger was scheduled for completion in 2002.

Bapco proceeded with modernization of its vintage refinery south of Manama. The program was scheduled for completion in 2004. In addition to Bahrain's oil output of 40,000 b/d, the refinery processes Saudi crude, of which 140,000-b/d from Abu Safa offshore field is provided free of charge.

Bahrain was among the six Arab oil-producing nations, which together control half of the world's known oil reserves, that signed a trade agreement at year-end 2001. The agreement, expected to help stabilize the area, will accelerate startup of a region-wide customs schedule, imposing a 5% tariff on imported goods beginning in 2003 instead of 2005.

Other members of the GCC signing the agreement were Kuwait, Oman, Qatar, Saudi Arabia, and the United Arab Emirates (UAE). The countries also will establish a single currency by 2010 and eventually establish a free-trade zone with the European Union (EU), the Middle East's biggest trading partner. The GCC's action to liberalize trade comes at a critical time. Oil revenues, which make up more than 80% of the income of GCC states, were expected to drop 25% in 2002.

IRAN

CAPITAL: TEHRAN
MONETARY UNIT: RIAL
REFINING CAPACITY: 1,484,000 B/CD
OIL PRODUCTION: 3,130,000 B/D
OIL RESERVES: 89.7 BILLION BBL
GAS RESERVES: 812.3 TCF
IRAN

Iran retained its importance as an oil and gas province with 9%, or 90 billion bbl, of the world's proven oil reserves

and estimated gas reserves of 812 tcf. Additionally, the country ranks second behind Saudi Arabia as the largest producer in OPEC. President Mohammad Khatami's reelection in June 2001 bolstered international hope for domestic economic reform that would impact the petroleum sector. Stringent measures were adopted that aimed at increasing growth, initiating structural growth and privatization, encouraging internal and foreign investment, and liberalizing trade.

Iranian crude is produced mainly from five onshore fields in the Khuzestan region near the Iraqi border and Persian Gulf: Ahwaz-Bangestan, Marun, Gachsaran, Agha Jari, and Bibi Hakimeh. Plans for Ahwaz-Bangestan call for production increases from 250,000 b/d to 600,000 b/d over the next eight years, financed by a $2.5 billion investment. Domestic crude consumption was about 1.2 million b/d and net exports about 2.7 million b/d.

Oil Minister Bijan Namdar Zanganeh revealed that Iran's oil fields are depleting at a rate of 250,000-300,000 b/d annually, showing an obvious need for upgrading and modernization. This has not affected the country's plans to double foreign investment to $24 billion and to double national oil production to around 8 million b/d by 2020.

Foreign investment

In July 2001, the National Iranian Oil Company (NIOC) signed a contract with Japex and Indonesia Petroleum (both majority owned by the Japan National Oil Co.) to develop the Azedegan field, a giant onshore field in southwestern Khuzestan east of the Iraqi border. Early estimates set potential oil reserves at up to 24 billion bbl, with production potential of 300,000-400,000 billion bbl. The contract is based on the "buyback" model in which the contractor funds all investments, receives compensation from NIOC in the form of an allocated production share, and then transfers operations to NIOC once the contract is completed.

In June 2001, NIOC and Italy's ENI signed a $1 billion 5½ year buyback contract to develop the Darkhovin field near Abadan. The field was estimated to contain 3-5 billion bbl of low-sulfur crude. ENI has a 60% share in the project, NIOC 40%. Estimated production could peak at 160,000 b/d.

Spain's Cepsa signed a $300 million buyback contract to develop Iran's Cheshmeh-Kosh field. Cepsa hopes to effect a 167% increase in oil production from 30,000 b/d to 80,000 b/d during 2001-05. TotalFinaElf, Shell, and BP placed bids to develop Bangestan field in 2002.

Australia's BHP Billiton Ltd. appeared to be close to signing a $359 million buyback contract to develop Foroozan-Esfandiar, expecting to raise production levels at the fields by 200% from 50,000 b/d to 150,000 b/d. Meanwhile Shell was reportedly expediting its $800 million Soroush-Nowrooz development to raise output to 500,000 b/d by 2003. First production was expected by late 2001.

Additional oil fields that received development bids include Salman, Sa'databad-Sarvestan, and Dehloran. NIOC has plans to develop five oil and gas fields in the Hormuz area, which include Henjam A (HA) and HB, HC, HD, and HE, in addition to the A field near Lavan Island, Esfandir field near Kharg Island, and two areas near South Pars. NIOC estimates that the five Henjam fields could contain 400 million bbl of oil and could produce up to 80,000 b/d. The company plans production increases at Doroud, Nosrat, Farzam, and Salman from today's levels of 105,000 b/d to 130,000 b/d in the next 3 years.

Iran's field development plans

Iran's five-year plan revolved around boosting domestic natural gas consumption in order to free up more oil for export. The government oil ministry had plans to expand its gas export links over the wider area, focusing on South Pars, its largest nonassociated gas field with estimated reserves of 280 tcf of gas. Russian and local contractors started construction of the 56-inch IGAT-3 pipeline and were planning a possible IGAT-4 line to ship the gas north, where it would be reinjected to boost oil output from the declining Agha Jari field and possibly the Ahwaz and Mansouri fields that make up part of the Bangestan reservoir in southwest Khuzestan. Gas from South Pars could also be exported via pipeline or as LNG by tanker.

Phase one of the South Pars project involves production of 900 MMcfd of natural gas and 40,000 b/d of condensate. First gas production from the field was due in early 2002. Phase two and subsequent phases would include construction of twin undersea pipelines from South Pars to onshore facilities at Assulyeh and a major terminal and onshore treatment facilities at the port. Iran's South Pars natural gas field in the Gulf was expected to generate $300 billion in revenues for the country over 30 years, said Oil Minister Bijan Namdar Zanganeh.

Development of the 48 tcf North Pars field to produce 3.6 bcfd of natural gas comprises another part of Iran's long-term gas utilization plans. North Pars gas would be slated for reinjection into mature southern fields: 1.2 bcfd into the onshore Gachsaran, Bibi Hakimeh, and Binak oil fields, plus 2.4 bcfd to the Agha Jari field.

Significant gas finds underscored Iran's natural gas development plans. Large discoveries occurred in the Zireh field (800 bcf) in Bushehr province, Homa field (4 tcf) in southern Pars province, and Tabnak field (14 tcf) in southern Iran. The country's other large nonassociated gas reserves include an offshore North Pars field (47 tcf), the onshore Nar-Kangan fields, and the Aghar (13 tcf) and Dalan fields in Fars province, as well as the Sarkhoun and Mand fields.

Oil swaps

Iran continued to focus on expansion of its oil and gas export markets for production from Kazakhstan and Turkmenistan in terms of oil and gas swap arrangements between the Caspian Sea countries. As a result, plans called for infrastructure development and construction of several oil swap pipelines.

The first of these was a new pipeline linking Neka, a south Caspian seaport to its refinery in Rey south of Tehran. The 243-mile 32-inch pipeline and three pumping stations were said to cost about $240 million. Infrastructure was developed to accommodate a swap capacity of 350,000 b/d at a cost of $360 million.

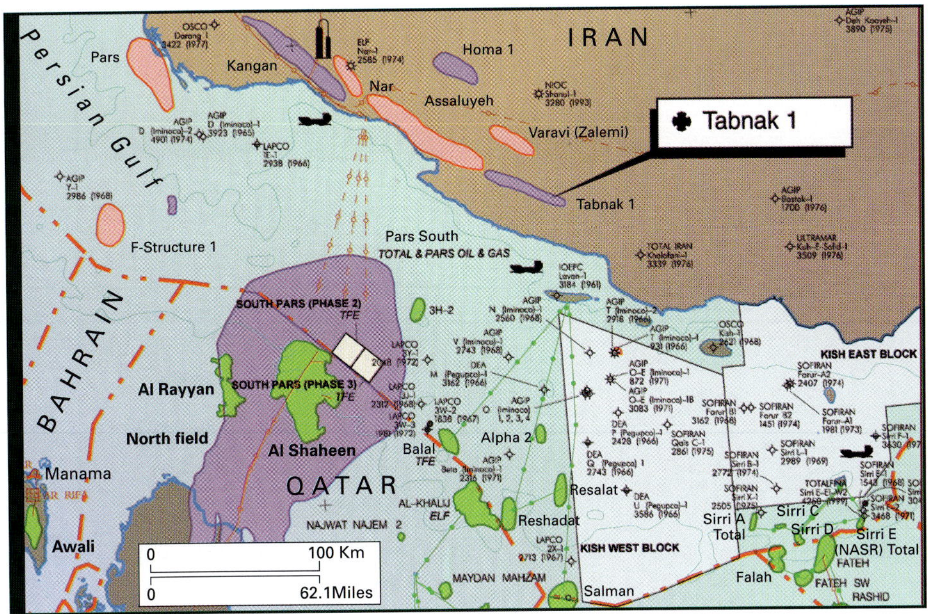

Iran's supergiant Tabnak sweet gas field.

About $260 million would be invested in refinery modification and for storage and blending facilities in Neka. Turkmenistan and Kazakhstan were shipping about 50,000 b/d from the ports of Turkmenbashy and Aqtau. Future shipments were expected to reach more than 425,000 b/d with pipeline upgrades. There remained the possibility of adding three new routes to expand the swap network to 700,000 b/d. The first would begin at Anzali port and connect by means of a 100-mile trunkline to the Tabriz and Rey pipelines; the second route would involve constructing a line from the Port of Noshahr north of Tehran. The third option would convert an existing Soviet-built pipeline that runs from Astara on the coast at the Azerbaijan border to a refinery at Tabriz. NIOC operates five export terminals in the Persian Gulf that could accommodate future swap arrangements.

Other developments in the swap arena include increased shipments of 85,000 b/d (up from 6,000 b/d) by Dragon Oil PLC from its Turkmenistan Cheleken operation to Iran. Larmag Energy Associates initiated a trial shipment of 23,000 b/d to Noshahr port from Cheleken, Turkmenistan.

Crude swaps with Kazakhstan were limited by problems at Iranian refineries, which initially were unable to process Kazakhstan crudes containing high sulfur, mercaptans, and salt. Refineries at Tabriz and Rey were upgrading their facilities to process about 337,000 b/d of Kazakh crude in the future. Refineries at Isfahan and Arak could also be upgraded and additional capacity added. Kazakhstan and Iran planned to resume swaps up to 20,000 b/d as soon as possible with increases to 120,000 b/d by 2006.

Gas-oil swaps

Iran and Turkmenistan were conducting a number of gas-oil swaps. Northern Iran needs Turkmenistan gas to supply its domestic markets. A swap arrangement sends Iranian oil to oil-poor Turkmenistan as partial payment for the gas. Both countries were considering an arrangement that would send equal volumes of gas to Iran for sale in markets in Turkey and Europe. Turkmenistan was planning on selling Iran 186 bcf/yr (5 billion cu m/yr) of gas by 2003.

Gas exports

Iran was affirming its place in the region as a gas exporter. The 1,590-mile (2,557-km) Iran-Turkey pipeline was already in place to accommodate a 25-year agreement to deliver 8.5 tcf (228 billion cu m) of gas annually. The line was being inspected, and startup was scheduled for year-end 2001. Initially, the line would supply 112 bcf/yr (3 billion cu m/yr), increasing to 370 bcf/yr (10 billion cu m/yr) by 2007. It was expected to supply 57 Turkish villages by 2003.

The IGAT-3 pipeline was under construction north from South Pars through the Caucasus region to Turkish or European destinations. IGAT-3 would supply Armenia and could ship gas to European markets via Georgia-Russia-Ukraine. Additionally, Iran was considering shipping gas through its IGAT-1 line into Azerbaijan.

Plans under consideration included building a 58-inch, 1,660-mile (2,670-km) pipeline into India through Pakistan at a cost of $5 billion. Late in 2001, Iran awarded Australian consultant BHP Kinhill a contract, along with Snamprogetti of Italy, to study the feasibility of the proposed $6 billion gas pipeline. The line would start at NIOC's

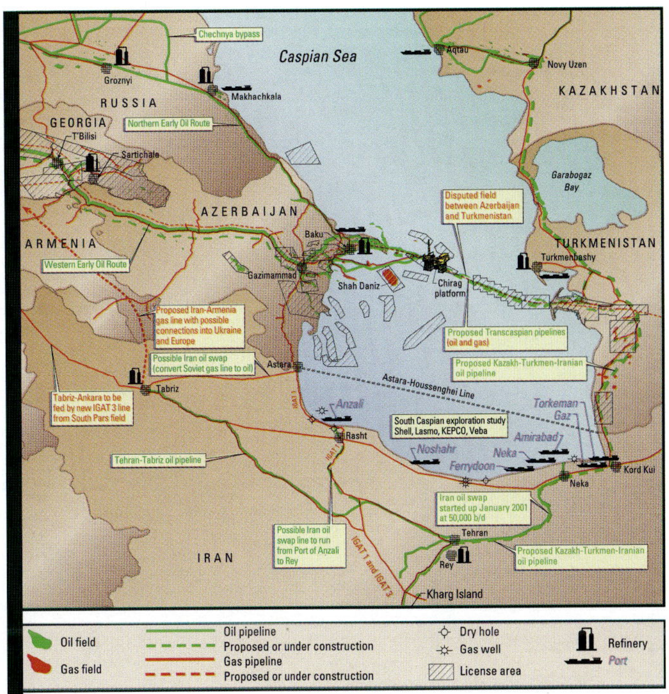

Iran finished linking up a new oil pipeline at the south Caspian Sea port of Neka with its refinery at Rey, sparking the possibility of several "oil swap" pipelines in the area

PERSIAN GULF
Legend
Oil field
Oil sand
Gas field
Crude oil pipeline
Natural gas pipeline
Products pipeline
Pipeline planned or under construction
Refinery in operation Refinery capacity in 1,000 b/d
Tanker terminal
Cities
Capital
International boundary
Water depth
0 to 200 m
200 m and deeper
IRAN
IRAQ
KUWAIT
SAUDI ARABIA
BAHRAIN
QATAR
UNITED ARAB EMIRATES
OMAN
TURKMENISTAN
AFGHANISTAN
Caspian Sea
Persian Gulf
Gulf of Oman
Strait of Hormuz

gas fields in southern Iran and would transport 2.4 bcf/yr (66 million cu m/yr) of gas. The report was expected to be ready by March 2002. It was proposed that 70% of the line's capacity be contracted to India and the rest to Pakistan. Planners have suggested that a consortium of multinationals develop the project in order to moderate potential conflicts between the countries involved.

LNG

BP and India's Reliance Group signed a memorandum of understanding with NIOC to develop an LNG plant in Iran. The two-train plant, to be located in Asaluyeh, could produce 8 million tonnes per year (tpy) of LNG for export to India and Europe. Output from phases one and two of the South Pars field would source the new plant.

IRAQ

CAPITAL: BAGHDAD
MONETARY UNIT: DINAR
REFINING CAPACITY: 417,500 B/CD
OIL PRODUCTION: 1,960,000 B/D
OIL RESERVES: 112.5 BILLION BBL
GAS RESERVES: 109.8 TCF

Iraq's trade status continued to be debated by nations seeking to restrict the country's ability to amass weapons and military goods. In early 2001 US Secretary of State Colin Powell proposed so-called "smart sanctions," which would allow freer movement of civilian goods, but the United Nations (UN) Security Council postponed consideration of the plan until May 2002. Russia's opposition to smart sanctions weakened in late 2001 in light of the terrorist attacks on New York and Washington, DC, and subsequent military action in Afghanistan.

In late 2001, the UN Security Council again extended Iraq's oil-for-aid program, this time through May 2002. The council also continued to wrangle over a UK proposal, supported by the US, to change the Iraqi oil pricing scheme. The UK and US were concerned that Iraq was using oil price fluctuations to impose a de facto surcharge on oil purchasers and that the extra money was going directly to the Iraqi government free of UN control. Earlier in the year, Iraq had suspended oil exports for a month to protest the US/UK attempts to change the terms of its oil program. In September 2001, Iraq denied any connection or involvement with the terrorist attacks.

A UN report in June 2001 said that Iraqi oil production capacity would fall sharply unless technical and infrastructure problems were addressed. The report estimated, for example, that output from the Kirkuk region, one of Iraq's two largest active oil fields, could fall by 50% over 12 months to 500,000 b/d, and production from southern Rumaila, the other main field, could also be reduced substantially.

Oil field development and rehab

Iraq hoped to counter this by a large-scale program to drill new wells; 417 were planned, mostly to be drilled by Russian, Chinese, Iraqi, and Romanian companies. If sanctions were lifted, Iraq claimed that it could increase its oil production capacity to at least 6 million b/d and perhaps 10 million. Reportedly, Iraq had signed several multi-billion dollar deals with foreign oil companies, mainly from China, France, and Russia, but was frustrated with their failure to start work, due to UN sanctions.

Russia was pursuing its interest in rehabilitating West Qurna field, where two more degassing stations were planned for 2001, potentially raising production to around 400,000 b/d. Russian companies also received UN approval to begin drilling Saddam and other fields. JSC Slavneft, a joint Russian-Belarus company, signed a contract to develop Luhais oil field in southern Iraq west of giant Rumaila field. The contract with the oil ministry would allow Slavneft to drill 25 wells and export the oil via Persian Gulf ports. With full development, Luhais could produce 100,000 b/d. Other contract provisions called for drilling wells in Nahr Umr and Zubair oil fields. Production sharing contracts (PSCs) for other fields were also in various stages of negotiation. TotalFinaElf was close to an agreement to develop Nahr Umar field, with an expected initial output of 440,000 b/d.

In May 2001, Iraq announced separate oil and gas discoveries in its Western Desert, reported the Organization of Petroleum Exporting Countries (OPEC) news agency. North Oil Co. was planning to drill more wells in the new oil area. No details were released about the gas field.

Pipelines and transport

Iraq's oil minister announced in August 2001 that an agreement was reached with Turkey to build a $2.5 billion natural gas pipeline capable of increasing exports to Europe, with Turkey functioning as a key transit center. Work to rehabilitate and expand capacity on the parallel Kirkuk-Ceyhan pipelines carrying Iraq's oil into Turkey was reportedly well behind schedule during 2001. Iraq claimed that the crucial IT-2 pumping station, which was damaged and was being bypassed, would be restored to operation.

In early 2001, Iraq and Syria were planning to build a 1.4 million b/d oil pipeline to replace a 20-year-old line having leakage and corrosion problems. The pipeline's capacity was unknown, but it had been transporting 200,000 b/d. The tender for the Iraqi section was being delayed until Iraq's financial situation improves. Iraqi Oil Minister Amer Mohammed Rasheed said another oil pipeline was planned through Jordan. The Jordanian government has issued a tender for a section extending from the Iraqi border to the 90,400 b/d refinery at Zerka. Iraq would build another section, from Haditha, 162 miles (260 km) northwest of Baghdad, to the Syrian border.

Iraq also made progress toward repairing its Khor al-Amaya tanker terminal, which had been virtually destroyed during the war with Iran. Reports in March 2001 indicated that two berths had been completely repaired and that the terminal would "soon be ready to receive oil tankers," according to the Oil Ministry. Gulf War damage to Iraq's largest terminal, Mina al-Bakr, appeared to have been largely repaired, and the terminal was capable of handling up to 1.3-1.4 million b/d, but restoration to nameplate capacity would require more extensive infrastructure repairs.

In April 2001 an Iraqi-owned vessel,

the Zainab, sank off the Dubai coast, leaking more than 1,000 tons of smuggled diesel oil and polluting Gulf waters and UAE beaches.

ISRAEL

CAPITAL: JERUSALEM
MONETARY UNIT: SHEKEL
REFINING CAPACITY: 220,000 B/CD
OIL PRODUCTION: 100 B/D
OIL RESERVES: 3.8 MILLION BBL
GAS RESERVES: 1.47 BCF

Samedan Oil Corp., a unit of Houston's Noble Affiliates Inc., continued to develop its Mari-B and Noa fields off Israel. Samedan was part of Israel's Yam Thetis gas consortium (also including Avner Oil, Delek Drilling). Mari-B field is in 787 feet (240 m) of water 15 miles (25 km) from Ashkelon, and Noa is in 2,600 feet (790 m) of water 26 miles (42 km) from shore. Samedan awarded an engineering and project management services contract to CSO Aker Engineering Inc. for subsea development, flowlines, and an export pipeline at Mari and Noa. An electrohydraulic multiplex control system for the two fields, along with the platform and subsea equipment, would be supplied by Kværner ASA unit Kværner Oilfield Products under another contract.

Spudding of the Har Sedom No. 1 well in the Dead Sea was rescheduled again for August 2001 by Ness International Inc., after being delayed twice.

MIDOR (Middle East Oil Refinery Ltd.), a joint venture of Israel and Egypt, commissioned its $1.3 billion, 100,000-b/d refinery in Alexandria in April 2001, but in June, Israel's Merhav confirmed that it had sold its 20% share in MIDOR to the National Bank of Egypt, making the refinery 100% Egyptian-owned. The sale effectively marked the end of what had been the largest Arab-Israeli joint venture to date. Earlier in the year, Bazan Oil Refineries Lt. selected the SCANfining process for its Ashdod, Israel, refinery. ExxonMobil Research & Engineering Co.' proprietary process was also selected by Bazan for its Haifa refinery in 2000.

Several gas suppliers were negotiating

Israel gas development in Mari-B and Noa fields.

sales to Israel Electric Co. , including Yam Thetis, Isramco/BG, the East Mediterranean Gas Co. consortium, and local group Sdot Yam.

KUWAIT

CAPITAL: KUWAIT CITY
MONETARY UNIT: KUWAITI DINAR
REFINING CAPACITY: 773,300 B/CD
OIL PRODUCTION: 1,440,000 B/D
OIL RESERVES: 94 BILLION BBL
GAS RESERVES: 52.2 TCF

Kuwait decided early in 2002 to purchase Qatari gas for 25 years, under a preliminary protocol and terms agreement, said deal participant ExxonMobil Corp. The arrangement would be part of a 1.75-bcfd enhanced gas utilization project in the works for Qatar's supergiant North field and would support efforts toward a GCC gas pipeline network.

Kuwait was also among the six GCC Arab oil-producing nations that signed a trade agreement at year-end 2001. The agreement, expected to help stabilize the area and liberalize trade, comes at a critical time. Oil revenues, which make up more than 80% of the income of GCC states, were expected to drop 25% in 2002.

To help prop up its oil production, Kuwait continued building gathering centers in its producing fields during 2001. Kuwait Oil Co. and China Petroleum Engineering Construction Corp. completed two centers, one at Umm Gudair field with a design capacity of 190,000 b/d and one at Minagish with 210,000 b/d. As of mid-year, the centers were handling 80,000 and 100,000 b/d, respectively. A water injection system was also being installed at Minagish.

Khafji concession

Kuwait extended the concession to Arabian Oil Co. to continue operating and drilling Khafji offshore oil field in the Kuwaiti sector of the Neutral Zone. Arabian Oil, Japan's largest crude oil producer, said that renewal of the contract was crucial since the company had already lost its rights in the Saudi Arabian-controlled side of the same field. Kuwait and Saudi Arabia each had a 10.9% share of the Japanese company, with other shareholders including Tokyo Electric Power Co., Kansai Electric Power Co., and Nippon Steel Co. The new agreement would allow Arabian Oil to keep exploring the field beyond the January 2003 expiration date. Japan, heavily dependent on oil imports from the Middle East, was trying during 2001 to maintain a strong foothold in overseas oil-producing areas, especially the Middle East.

Petrochemicals construction, refining

In mid-2001, Kuwait announced plans to build two petrochemical plants by the end of 2005 to produce materials for the plastics industry. As part of efforts to liberalize the Kuwaiti economy, private companies would be allowed to own some share of both projects.

One proposed plant, projected to cost $2 billion, would produce 850,000 tpy of ethylene, 450,000 tpy of polyethylene, 650,000 tpy of ethylene glycol, and 70,000 tpy of propylene, reported the Kuwait News Agency. Kuwait's Petrochemical Industries Co. (PIC) would be the local partner, with a leading international company expected to invest in the plant. The second facility, also to be partly owned by PIC and costing $1.4 billion, would produce 650,000 tpy of paraxylene and 500,000 tpy of styrene.

Kuwait's most modern refinery, al Shuaiba, was reportedly operating at around 300,000 b/d, after Iraq claimed the plant had been sabotaged by "national resistance" groups in Kuwait. In August 2001, it was reported that Kuwait was planning to spend more than $3 billion to upgrade its three refineries.

OMAN

CAPITAL: MUSCAT
MONETARY UNIT: OMANI RIYAL
REFINING CAPACITY: 85,000 B/CD
OIL PRODUCTION: 960,000 B/D
OIL RESERVES: 5.5 BILLION BBL
GAS RESERVES: 29.28 TCF

Oman continued to focus on reducing its dependence on oil and expanding the use of natural gas to the domestic sector and export market. The country's oil fields are smaller and less productive than those in other Persian Gulf countries, and they cost $3-4/bbl more to produce. Oman was among six GCC Arab oil-producing nations that signed a trade agreement at year-end 2001 to help stabilize the area and liberalize trade. The agreement comes at a critical time, since oil revenues, which make up more than 80% of the income of GCC states, were expected to drop 25% in 2002.

Oman's proven reserves are mainly in the north in Yibal, Natih, Fahud, Al-Huwaisah, and Lekhwair fields. Their combined production amounts to almost half of Oman's. Yibal is the country's largest, producing about 180,000 b/d. Nimr in the southern part of the country is second largest with production of approximately 178,000 b/d from more than 307 wells. Petroleum Development Oman (PDO) – a consortium of the Omani government (60%), Shell (34%), TotalFinaElf (4%), and Partex (2%) – controlled over 90% of oil reserves and produced about 94% of the country's output.

PDO announced discovery of two major oil fields in southern Oman at Al-Noor and Al-Shomou that contain estimated combined reserves of 340 million bbl. If successful, the consortium planned to increase reserves from both to 1.8 billion bbl by 2003 and to 2.7 billion bbl by 2011.

In March 2001, Maersk Oil of Denmark was awarded a concession for Blocks 45 and 48. The company will invest $18 million for exploration in the

area over a period of four years. Oman exports most of its crude to markets in East Asia, mainly China, Japan, and South Korea; India also imports significant amounts.

Oman Refinery Co. operates an 85,000 b/d refinery at Mina Al-Fahal to supply local markets. A second refinery was planned to be built near the northern city of Sohar. Construction of the 75,000 b/d facility could begin in 2002, with completion set for 2004.

Diversification into petrochemical production comprised part of Oman's energy plans. A joint venture between BP (40-49%) and Muscat Securities Market (40%) was planning to build an ethane cracker and polyethylene unit in Sohar. Due to come on stream between 2001 and 2002, the facility would produce fertilizer and polyethylene with a total output of about 450,000 tpy. An ethane pipeline that was under construction would supply the project.

Natural gas

Oman's proven gas reserves of 29.3 tcf in 2001 were up 138% from 1992 levels of 12.3 tcf. PDO reported a large discovery in Dakhiliya, which could add 1-2 tcf of reserves.

The country's existing pipeline network was being extended during 2001. A contract was awarded to India's Dodsal Co. for a $124 million line that will connect gas deposits in central Oman to the coastal city of Sohar in the north. A consortium of Italy's Snamprogetti and Saipem received a second contract to build a $180 million pipeline to Salalah on the southern coast. In April 2001, Canada's Enbridge and BC Gas were contracted to operate Oman's gas transportation and distribution network for the next five years.

Oman Liquefied Natural Gas Co. (OLNGC) was considering addition of a third train to its 6.6-million tpy liquefaction plant at Qalhat. The plant comprises two 3.3 million tpy trains that were sourced with nonassociated gas from the Saih Nihayda, Saih Rawl, and Barik gas fields. The plant's customers included Korea Gas Corp. (4.1 million tpy over 25 years) and Japan's Osaka Gas Co. (700,000 tpy over 25 years). Shell agreed to buy 700,000 tpy over five years from OLNGC. The gas, shipped via tanker chartered by Shell, will supply its customers in Spain. Oman was reported to be considering building a fleet of tankers to export LNG produced at Qalhat. In November 2001, Oman purchased Enron's 40% share in the new vessel Lakshmi. LNG production from the plant would also be destined for sale on the spot market.

QATAR

CAPITAL: DOHA
MONETARY UNIT: RIYAL
REFINING CAPACITY: 57,500 B/CD
OIL PRODUCTION: 574,000 B/D
OIL RESERVES: 15.2 BILLION BBL
GAS RESERVES: 509 TCF

Qatar contains the world's third largest natural gas reserves and the largest nonassociated gas field. During 2001, the country continued to expand its LNG exporting operations and develop its gas fields. Early in 2002, Qatar agreed to sell natural gas to Kuwait for 25 years, under a preliminary protocol and terms agreement, said deal participant ExxonMobil Corp. The arrangement would be part of the 1.75-bcfd enhanced gas utilization project for Qatar's super-giant North field, which includes the Dolphin project, and would support Qatar's efforts toward developing a GCC regional gas pipeline network. Qatar was also among the six GCC Arab oil-producing nations that signed a trade agreement at year-end 2001.

Crescent Petroleum completed studies in early 2002 for a proposed $3 billion Qatar-to-Pakistan offshore gas pipeline traveling mainly through Iranian waters, reported OPEC News Agency. It would extend from Ras Laffan through Hamriyah, then Dibba on the Gulf of Oman, terminating at Gadani, Pakistan. The section between Hamriyah and Dibba would be onshore.

A territorial dispute was resolved in court between Qatar and Bahrain about sovereignty over the Hawar Islands, which are adjacent to Qatar's main onshore oil field. Bahrain was awarded control of the area, while Qatar retained the neighboring islands of Zubarah and Janan.

Dolphin gas development

In late 2001, Qatar signed a development and production contract with the Dolphin Energy Ltd. joint venture for the $3.5 billion Dolphin project, which is designed to provide export outlets for natural gas in the region. The project is key to the first cross-border gas pipeline agreement in the Middle East. The partners in Dolphin Energy, United Arab Emirates Offsets Group (UOG) with 75.5% interest and TotalFinaElf SA with 24.5%, signed a 25-year agreement for development of the Khuff formation in Qatar's 500-tcf North gas field.

Enron Corp. previously held 24.5% of the project and would have operated the pipelines; its stake, acquired by UOG, was up for bid. Five companies – BP PLC, ChevronTexaco Corp., ExxonMobil Corp., Occidental Petroleum Corp., and Royal Dutch/Shell Group – had until year-end 2001 to replace Enron's position in the development. UOG was anticipating close to $1 billion for the project stake.

TotalFinaElf, which will operate the upstream section of the scheme, said that two unmanned platforms will produce the offshore reserves. Production should plateau at 2.5 bcfd of gas, which will be delivered to Qatar's Ras Laffan terminal, where the wet gas will be treated to strip out condensate, ethane, sulfur, and LPG. Then 2 bcfd of methane-rich lean, sweet gas will travel through a 48-inch subsea line to Taweelah in Abu Dhabi, UAE, serving the markets of Abu Dhabi, Dubai, and Fujairah. The line will be either 236 or 273 miles long (380 or 440 km). The pipeline construction agreement was also signed but no details revealed.

Qatar will own all the condensate, sulfur, NGL, and LPG extracted from the wet gas and will charge $1.30/MMBtu for the dry gas, which Dolphin Energy will export, aiming at delivering first gas to its customers in the UAE in 2005. Dolphin's initial phase will require an investment of $3.5 billion, followed by a further "multi-billion dollar investment" in downstream activities over 6-7 years. In the second phase, the volume of gas could be increased by 50%.

Also for work in Qatar's North field, Ras Laffan Liquefied Natural Gas (II), a joint venture of ExxonMobil and Qatar Petroleum Co., awarded ABB Group a contract for eight surface wellheads over two years. The requirement could rise to 40 systems over 4 years.

Oil production

Early in 2001, Qatar approved plans for further development of Al Shaheen field on Block 5 off Qatar, said Maersk Oil Qatar. The program would increase production to 200,000 b/d of oil in 2004 from 112,000 b/d. During 2001 Maersk planned to drill 40 production wells, 20 water injection wells, and a number of appraisal wells. The company also was to convert 14 existing wells to water injection and install production platforms interconnected with pipelines, facilities for gas compression, and a gas export pipeline to Qatar Petroleum's North field Alpha platform. Once pipelines are installed, some production could begin from the drilling rigs before permanent facilities are in place in 2003.

LNG developments

In late 2001, Qatar's Ras Laffan Liquefied Natural Gas Co. Ltd. (RasGas, a joint venture between Qatar Petroleum and an ExxonMobil Corp. affiliate) awarded a Technip-Chiyoda joint venture a 100 million euro contract to expand the Ras Laffan LNG plant on the northeastern coast of Qatar. The contract calls for debottlenecking three LNG trains by replacing or upgrading compressors and turbines, which would expand the capacity of each train to 3 million tpy from 2 million. Work will be performed during scheduled shutdowns in 2002-05.

RasGas was also expanding its LNG facilities at Ras Laffan in Qatar by adding a fourth train, bringing total capacity to 16 million tpy. RasGas II signed a contract with a joint venture of Chiyoda and Mitsui, along with Snamprogetti, to build the LNG train and related onshore facilities. Another RasGas contract hired J. Ray McDermott Middle East (Indian Ocean) to build the offshore facilities and a 60-mile (90-km) pipeline.

The project consists of offshore production, transportation, and liquefaction facilities to produce 4.7 million tpy for the Indian LNG consortium Petronet LNG under an existing agreement

MIDDLE EAST
TURKEY
SYRIA
IRAQ
IRAN
LEBANON
ISRAEL
JORDAN
SAUDI ARABIA
KUWAIT
WEST BANK
GAZA STRIP
Red Sea
Lake Van
Lake Uromiyeh
Aksaray
Mersin
Ceyhan
Dortyol
Batman
Mosul
Irbil
Kirkuk
Baghdad
Tabriz
Bostanabad
Shanpur
Uromiyeh
Naqadeh
Halabia
Kermanshah
Sar-e Pol-e Zahab
Baniyas
Tartus
Homs
Tripoli
Beirut
Sidon
Damascus
Haifa
Nazareth
Tel Aviv-Yafo
Jerusalem
Amman
Hebron
Elat
Tadmur
Tibni
Mayadin
Al Hadr (Hatra)
Ash Sharqat
Qayyarah
Al Qawayr
Bayli
Samarra
Al Haditha
Hit
Ar Rutbah
Karbala
Al Hashimiyah
Al Kut
Ad Diwaniyah
Abu
Sukhayr
As Samawah
Ar Rifai
Ash Shatrah
Suq ash Shuyukh
Jalibar
Safwan
Ar Rawdatayn
Badanah
Al Jawf
Sakakah
Rafha
Al Qaysumah
Al Bad
Ash Sharmah
Duba
Al Waih
Marsa Alam
Umm Cajj
Medina
Mina Baranis
Yanbu al Bahr
Badr Hunayn
Riyadh
Dumah
As Salamiyah
Ad Dilam
Al Hillah
Kifri
Khanaqin
Naft-e Shah
Sumar
Ilam
Mehran
Jelow Gir
Al Hayy
As Suwayrah
0 100 200 Miles
0 100 200 300 Km

involving deliveries of 5 million tpy to a proposed import terminal at Dahej, Gujarat, and another 2.5 million tpy to one at Cochin, Kerala. Deliveries to Dahaj were slated to begin in late 2003 using existing LNG trains at the RasGas plant until the RasGas II train is completed. Work has already started on the new train, with production slated to begin in 2004.

In June 2001, Edison Gas SPA agreed to buy 3.5 million tpy of LNG from RasGas for 25 years. RasGas would begin delivering the LNG in 2005 to a receiving-regasification facility in the North Adriatic off Italy.

GTL projects

In mid-2001, Qatar Petroleum and ExxonMobil announced plans to conduct a technical feasibility study for a GTL plant in Qatar. The proposed plant would convert North field gas into low-sulfur diesel fuel, naphtha, and lubricant basestocks for export to world markets. The study will provide Qatar Petroleum with information on the use of ExxonMobil's proprietary AGC-21 technology in a GTL plant. The study will also identify and determine potential synergy opportunities with other projects and infrastructure in Qatar.

In July 2001, Qatar Petroleum and Sasol Synfuels International Co. awarded Foster Wheeler Corp. a $30 million contract for the front-end engineering design of a GTL plant to be built at Ras Laffan Industrial City. Foster Wheeler was the first contractor to be involved in a Sasol GTL technology plant outside South Africa. The $800 million GTL project was scheduled for startup in 2005 and would convert 330 MMscfd of natural gas from North field into 34,000 b/d of liquids – 24,000 b/d of fuel oil, 9,000 b/d of naphtha, and 1,000 b/d of LPG. Qatar Petroleum had 51% and Sasol 49% of the project. Financing was being sought.

Petrochemicals developments

In mid-2001, Qatar Chemical Co. Ltd. was in the process of building a grassroots petrochemical facility that could almost double ethylene production capacity at the Mesaieed industrial area, which was producing 525,000 tpy. Startup was planned for late 2002 to coincide with several other large ethylene capacity expansions in the Middle East and Asia. The new facility was designed to produce 500,000 tpy of ethylene, 460,000 tpy of polyethylene, and 47,000 tpy of hexene-1, a key feedstock in manufacturing high-density polyethylene and linear low-density polyethylene. Qatar Chemical is a joint venture between Qatar Petroleum (51%) and Chevron Phillips Chemical Co. LLC (49%).

Chevron Phillips and Qatar Petroleum were also planning to build a petrochemical plant at Ras Laffan which would involve a 1.2 million tpy ethylene cracker, a 750,000 tpy polyethylene plant, and associated utilities and offsite facilities. Slated for 2006 completion, the plant would use ethane feedstock from Qatar's North field and Chevron Phillips' proprietary loop process technology.

In July 2001, Qatar Vinyl Co. officially opened a $6 million petrochemical plant at Mesaieed with a nameplate capacity of 230,000 tpy of vinyl chloride monomer, 175,000 tpy of ethylene dichloride, and 290,000 tpy of caustic soda.

In late 2001, Qatar Petroleum signed a preliminary deal for construction of what could eventually become the world's largest ethane cracker. The 1 million tpy plant at Ras Laffan Industrial City would come on stream in the second half of 2006. If expanded to 1.5 million tpy in the second phase, sponsors said it would be the largest such unit in the world. The deal brings Atofina and Qatar Petrochemical Co. into the project as equity participants in the ethane cracker. The cracker will provide feedstock via pipeline to new polyethylene plants at Ras Laffan and at Mesaieed Industrial City.

SAUDI ARABIA

CAPITAL: RIYADH
MONETARY UNIT: RIYAL
REFINING CAPACITY: 1,745,00 B/CD
OIL PRODUCTION: 6,470,000 B/D
OIL RESERVES: 259 BILLION BBL
GAS RESERVES: 219 TCF

Faced with low crude prices and production cutbacks, Saudi Arabia ended 2001 with the prospect of a sharp deterioration in its finances and economic situation during 2002. The country's real economic growth has fallen behind its population growth, and its movement toward economic reform has been uneven at best. In addition, Saudi Arabia has made only slow progress on another of its main domestic goals, attracting foreign direct investment.

Gas deals

At year-end 2001, US companies hoping to invest in Saudi Arabian natural gas projects were worried that diplomatic tensions between the White House and the Kingdom would delay finalization of the deals, which had been scheduled for March 2002. ExxonMobil Corp., the largest single foreign investor in Saudi Arabia, won most of the contracts in May 2001. Conoco Inc., Phillips Petroleum Co., Occidental Petroleum Corp., and Marathon Oil Co. also received interests in various ventures the Saudis are planning.

ExxonMobil was set to lead two of the three available "core ventures." The first project, which the Saudis were calling the most attractive investment opportunity, would include drilling in eastern Saudi Arabia at South Ghawar and using the gas to fuel two planned petrochemical plants, two power stations, and four desalination plants. Initial investment was estimated at $12-16 billion. ExxonMobil was also tapped to lead a Red Sea project on the West Coast estimated to require at least a $5 billion investment. A third project at the Shaybah field in the southeast "Empty Quarter" near the UAE border would be led by Royal Dutch/Shell Group and would require an investment of $5-10 billion, the Saudis said.

A sticking point between the Saudis and foreign investors had been the issue of how much gas the Saudis, through the state-controlled oil and gas company, Saudi Aramco Co., want to earmark for the various infrastructure projects on the table. Also, foreign investors wanted more specific details over production costs.

However, industry sources said neither of those issues should stop the proj-

OPEC skirts year-end price war threat

After making three production quota cuts totaling 3.5 million b/d in 2001, OPEC members were reluctant to further lose market share. In November 2001, the group indicated to non-OPEC oil suppliers that a price collapse could result if they didn't participate in another round of supply cuts. After much negotiation, OPEC announced in late December pledges of supply reductions from Angola, Mexico, Norway, Oman, and the Russian Federation totaling 462,500 b/d. OPEC said that it would cut an additional 1.5 million b/d from its production quota beginning January 1, 2002, bringing total cuts to almost 2 million b/d.

OPEC's successful call for non-OPEC support averted a price collapse in the near term, even if some of the pledged non-OPEC reductions do not result in real supply cuts, as was suspected, said the US Energy Information Administration (EIA). Although non-OPEC supply reductions were expected to be relatively small (conservatively, 100,000 b/d), OPEC's quota cuts could result in real reductions of over 1 million b/d, said EIA. A cut of this size was projected to largely offset the buildup in oil stocks that took place in 2001. Commercial stocks were projected to tighten in 2002 even after allowing for further OPEC leakage over the course of the year.

Energy Security Analysis Inc. (ESAI) agreed that OPEC did indeed need to take action to prevent a price collapse, but that the organization has overreached with its production accord. "If OPEC initially makes good on the production cut, crude supplies will be depleted, and prices will be headed to the $25/bbl OPEC basket target easily by mid-year," said ESAI. "As prices rise, the pressure to cheat, however, will become unbearable, attenuating the price increase at a lower level. In short, the spare capacity OPEC is creating is far too heavy a burden to carry if prices are rising and non-OPEC output is taking market share."

But if OPEC does not make good on its production cut at the outset, then the group will be faced with having to awkwardly put a good face on poor discipline, ESAI contended: "The biggest fallout from this would be a commensurate loss of discipline in non-OPEC countries. As discipline deteriorates among all producers, the $25 OPEC basket price target is out of reach."

In ESAI's view, OPEC has yet to figure out that the long-term sustainable crude oil price is not above $25/bbl, but above $20/bbl. "OPEC would have been better off making a marginal adjustment. It would have been credible, effective, and have a greater chance of enduring."

ects from moving forward. Instead, a more important concern was whether the Saudis still have the political will to allow foreign investors in, given recent tensions between the US and Arab nations. In an early 2002 meeting with top ExxonMobil officials, Saudi Arabian representatives gave no indication whether they will postpone the foreign gas investment projects.

Oil field development

In late 2001, Saudi Aramco awarded Halliburton's Energy Services Group a contract for $140 million of development work at Qatif field in eastern Saudi Arabia. The initial scope of the three-year project will be to drill oil and water wells, using drilling optimization software, extended-reach drilling, and installation of multilateral systems. In addition to Halliburton's well construction, reservoir evaluation, and completion services, the contract will include Landmark Graphics's geophysical, geological, and drilling software. Multiple wells will be directionally drilled from a single drilling pad, and, to minimize personnel required at the wellsite, the company will use information systems to transmit real-time data.

Qatif field is 25 miles long (40 km) and extends 3.7 miles (6 km) offshore. It has produced from six zones. The main reservoirs have been the Upper Jurassic Arab C (31° gravity oil) and Upper Jurassic Arab D (38° gravity). It also produces from the Arab A carbonate, Arab B carbonate, Middle Jurassic Fadhi carbonate, and Upper Jurassic Hanifa carbonate.

Petrochemicals

Saudi Arabia has ambitious plans to expand its petrochemical production using natural gas as a feedstock. State-owned (70%) SABIC, the Middle East's largest non-oil industrial company, was poised to become one of the world's top five ethylene producers by 2005. In February 2001, SABIC completed a $1 billion expansion at the Yanbu petrochemical facility, making it the largest polyethylene plant in the world. In early 2002, SABIC agreed to a $1.15 billion loan to fund a new petrochemicals plant in Jubail, an industrial city in eastern Saudi Arabia. The complex was scheduled to come on line in 2004, producing 1 million tpy of ethylene as well as olefins, polyethylene, and glycol ethylene.

SYRIA

CAPITAL: DAMASCUS
MONETARY UNIT: SYRIAN POUND
REFINING CAPACITY: 242,140 B/CD
OIL PRODUCTION: 515,000 B/D
OIL RESERVES: 2.5 BILLION BBL
GAS RESERVES: 8.5 TCF

Syria's proven oil reserves were expected to last only about 10 more years, meaning that the country could become a net oil importer within that timeframe. Consequently, the government regarded exploration for oil and natural gas as a top priority. Syria's economy was expected to grow 4.2% during 2001, in terms of GDP.

Gas project

TotalFinaElf SA announced in early 2002 that the $400 million Desgas associated gas project in eastern Syria had reached its full capacity of 450 MMcfd (13 million cu m/day). Conoco Inc. operates the project, which includes collection, treatment, and export of associated gas from the Deir Ez Zor region to the Syrian gas transmission network and production of condensate from Tabiyeh field (operated by TotalFinaElf) through gas reinjection. The project puts an end to associated gas flaring at Deir Ez Zor.

The contract involved construction of a two-train gas processing complex, six compressor stations, a 100-mile gathering system, and a 150-mile trunk pipeline. The gathering system transports up to 173 MMcfd (5 million cu m/day) of gas to the 450-MMcfd processing plant, which also processes 280 MMcfd (8 million cu m/day) from Tabiyeh field before its reinjection into the reservoir. The processed gas not reinjected at Tabiyeh is shipped via a 155-mile (250-km) line to the Syrian grid for use at fuel power plants and industrial facilities in the west of the country. The LPG associated with production is exported via rail while the condensate is transported through the Syrian Co. for Oil Transport line to the Banias terminal.

In early 2001, Syria and Iraq were planning to build a 1.4 million b/d oil pipeline to replace a 20-year-old line having leakage and corrosion problems. The pipeline's capacity was unknown, but it had been transporting 200,000 b/d. Syria planned to issue a tender for construction of the section within its borders, while the tender for the Iraqi section would be delayed until Iraq's financial situation improves.

Syria was expected to receive $65 million from Kuwait for a 300-MW expansion of its Naseriya power station located northeast of Damascus.

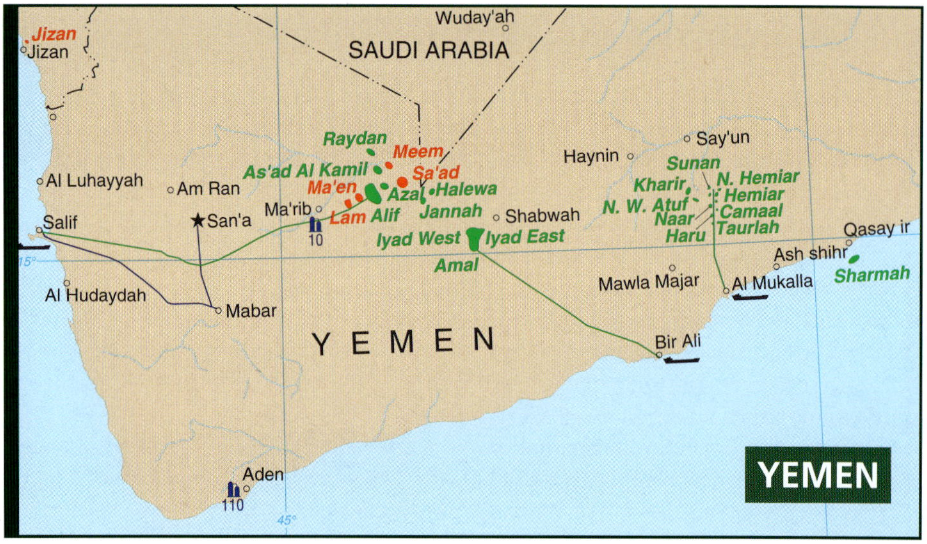

UNITED ARAB EMIRATES

CAPITAL: ABU DHABI
MONETARY UNIT: DIRHAM
REFINING CAPACITY: 514,750 B/CD
OIL PRODUCTION: 443,500 B/D
OIL RESERVES: 104.8 BILLION BBL
GAS RESERVES: 212.1 TCF

Abu Dhabi Marine Operations Co. (ADMA-OPCO) awarded in October 2001 a major gas injection project to expand the output of its primary offshore oil field, Zakum. ADMA-OPCO is an operating arm of Abu Dhabi National Oil Co. (ADNOC). Gas would be moved from giant Umm Shaif field to the Zakum West complex for injection. The project was thought to be one of the largest offshore operations in ADMA-OPCO's history.

A 200-MMcfd compressor to be installed at the gas injection platform would compress gas from Umm Shaif and deliver it to an existing wellhead tower for injection. The multimillion dollar project, due for completion by 2004, includes a year of engineering, design, and procurement. The UAE has spent more than $15 billion over the past 10 years to increase oil production capacity, according to the OPEC news agency.

UAE was due to receive natural gas from Qatar's Dolphin project beginning in 2005. Two bcfd of methane-rich lean, sweet gas would be transported from Qatar's Ras Laffan treatment terminal via a 48-inch subsea line to Taweelah in Abu Dhabi, UAE, serving the markets of Abu Dhabi, Dubai, and Fujairah.

A $1.2 billion petrochemical complex at Ruwais, UAE, began its first shipments in early 2002, said the OPEC news agency. The complex, owned by the Borouge joint venture of ADNOC and Borealis AS, Copenhagen, achieved full ethylene capacity of 600,000 tpy and 70% of polyethylene capacity of 450,000 tpy, with full polyethylene capacity to be reached within a month. Borouge contracted in late 2001 with the Abu Dhabi National Tanker Co. to undertake shipments of ethylene from Ruwais to the hub port of Mina Zayed, as well as to Jebel Ali in Dubai. Borouge also signed an accord with Mina Zayed, Mussafah port, and Jebel Ali for the export of polyethylene.

In April 2001 two ADNOC gas subsidiaries – Abu Dhabi Gas Co. (ATHEER) and Abu Dhabi Gas Industries Ltd. (GASCO) – merged and combined their UAE gas production, reported the OPEC's news agency.

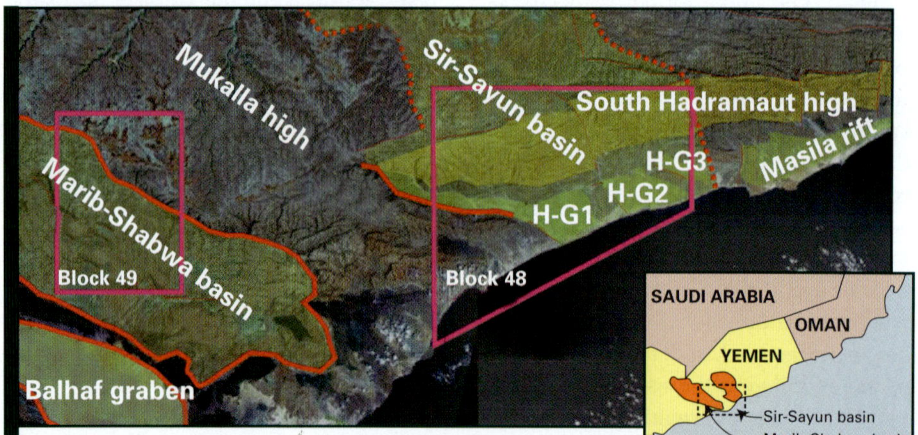

Satellite photo of Yemen showing North Mukalla Block 48 and SE Al Maber Block 49 being explored by Hungary's MOL Rt.

YEMEN

CAPITAL: SANAA
MONETARY UNIT: RIAL
REFINING CAPACITY: 130,000 B/CD
OIL PRODUCTION: 350,000 B/D
OIL RESERVES: 4 BILLION BBL
GAS RESERVES: 16.9 TCF

Although Yemen is one of the Middle East's smallest countries in terms of oil reserves, it has taken strides toward increasing its crude and condensate production and still has more potential in store. Yemen's premier producing area is the Masila Block operated by Nexen Inc. (52% working interest), followed by the Marib-Jawf area operated by Hunt Oil Co., ExxonMobil Corp., and others, and the Shabwa area led by TotalFinaElf.

In early 2001, Masila was averaging 230,000 b/d, and Nexen expected production to grow for the eighth straight year. Partner Occidental Petroleum Corp. said that Masila contained more than 126 producing wells in 16 fields. Much of Nexen's exploration acreage lay in northeastern Yemen along the border with Saudi Arabia. Nexen planned to boost overall E&D spending in Yemen in 2001 90% to $216 million.

DNO, Oslo, agreed in late 2001 to farm into and assume operatorship of First Calgary Petroleum's Block 43 in Yemen. Previously, Nexen had agreed to farm into the block, but certain closing conditions were not satisfied, said First Calgary. DNO planned to fund second-phase exploration work through either declaration of a commercial discovery or the expenditure of $7.5 million.

The commitment included drilling two wells and acquiring miles 310 miles (500 km) of seismic data. DNO would also pay lease bonuses of $600,000/yr until earning obligations are met. First Calgary retained 11.7% of foreign contractors' interest in the block, which is near Masila Block 14. DNO operates Block 32 immediately north of Block 43 and holds an interest in Block 53 to the west. First Calgary said both Blocks 32 and 53 were under development.

MOL Hungarian Oil & Gas Co. was also a Yemen player in 2001, having had a presence there since 1996. Operating in North Mukalla Block 48 and SE Al Maber Block 49, MOL developed a tectono-stratigraphic model and conducted new geochemical investigations on source rocks, leading to a revised play concept for the Marib-Shabwa and Sir-Sayun basins in Yemen. In late 2001, MOL subsidiary MOL Yemen Oil & Gas (Cyprus) Ltd. formed a joint venture with Pakistan Oilfields Ltd. and its parent Attock Oil Co. Ltd. The joint venture planned to explore two onshore blocks in the southern part of the areas studied earlier by MOL. The block sizes were 1,953 and 1,076 sq mi (5,055 and 2,788 sq km). IPE

Technology
Production-boosting Technologies
▪ Underbalanced Drilling Systems
▪ Expandable Technologies
▪ Intelligent Completions
▪ Production Optimization
Technologies that provide results. Technologies that are redefining the way operators drill, complete and produce wells. Technologies that are setting the pace in the oilfield for the 21st century.
www.weatherford.com
© 2002 Weatherford International, Inc. All rights reserved.
See where technology can take you.
Weatherford®

AFRICA

ALGERIA

CAPITAL: ALGIERS
MONETARY UNIT: DINAR
REFINING CAPACITY: 450,000 B/CD
OIL PRODUCTION: 820,000 B/D
OIL RESERVES: 9.2 BILLION BBL
GAS RESERVES: 159.7 TCF

Algeria enjoyed a sharp increase in oil export revenues during 2001, which helped reduce its external debt and improve its current account balance. Late in the year these positive trends began to reverse as oil prices slumped. The country remained highly dependent on oil and natural gas earnings, which accounted for more than 90% of its export revenues and 30% of its gross domestic product (GDP).

The International Monetary Fund (IMF) urged Algeria in late 2001 to proceed with privatization and other economic reforms, and despite the firing of privatization minister Hamid Tenmar, some progress was made. An important new hydrocarbons law moved forward, with passage possible in early 2002, though it faced labor union opposition.

Even though Algeria's first oil was discovered in the 1950s, the country is considered under-explored. In October 2001 Algeria's second licensing round resulted in Sonatrach's awarding of five new exploration contracts, all in the Berkine basin: Anadarko (Block 406b), Burlington Resources (402d), First Calgary Petroleum (406b), TotalFinaElf (432, 444s, 403n), and Repsol-YPF (401d). In November a tender for development of five oil and gas projects in the Illizi basin received 14 bids from foreign companies.

Berkine, Rhourde Oulad Djemma

Algerian oil and gas company Sonatrach and its partners continued work in early 2002 to develop oil reserves in the Berkine area, with a $257 million contract award to a joint venture of Saipem SPA and Bouygues Group. The contract involved development of Rhourde Oulad Djemma oil field and five other bordering oil fields in southeastern Algeria that are held jointly by Sonatrach, Agip SPA of Italy, and BHP Billiton of Australia. In late 2001 Billiton said it would invest $190 million for development of Rhourde Oulad Djemma field, which holds estimated reserves of 300 million boe and extends into blocks operated by Agip and Sonatrach. The project's various installations would include a central treatment unit, a gathering network, and oil storage tanks. The companies planned to to drill 20 production wells. The six fields were scheduled to come on stream at 80,000 b/d in 2003-04.

In early 2002, nearly two months ahead of schedule, Anadarko Petroleum Corp. began oil production from Hassi Berkine field on Blocks 404 and 403 in Algeria's Sahara Desert. The oil from the 75,000 b/d field will be processed through the third production train of the central processing facility at Hassi Berkine South field. A gathering system, injection lines, crude oil storage, and export facilities were installed as part of the facility expansion. Hassi Berkine occupies two blocks: 403, operated by Sonatrach and Agip (Algeria); and 404, operated by Sonatrach and Anadarko. Anadarko had 74.5% and Agip 25.5% of the unitized field.

The second oil production train at Hassi Berkine South at the central production facility was started in September 2001, and a fourth 75,000 b/d train, due for completion in mid-2002, was slated to process oil from four satellite fields on Block 404. Anadarko anticipated its gross production capacity in Algeria would exceed 500,000 b/d when all of the new equipment was up and operating. Partners on the project were Sonatrach, Agip, and Maersk Olie Ageriet. The

AFRICA
TURKEY
SYRIA
IRAN
SICILY
GREECE
CRETE
CYPRUS
LEBANON
ISRAEL
JORDAN
IRAQ
TUNISIA
Mediterranean Sea
LIBYA
EGYPT
SAUDI ARABIA
OMAN
YEMEN REPUBLIC
NIGER
CHAD
SUDAN
ERITREA
DJIBOUTI
ETHIOPIA
SOMALIA
NIGERIA
CAMEROON
CENTRAL AFRICAN REPUBLIC
CONGO (FORMER ZAIRE)
UGANDA
KENYA
GABON
CONGO
RWANDA
BURUNDI
TANZANIA
CABINDA
MALAWI
ANGOLA
ZAMBIA
ZIMBABWE
MOZAMBIQUE
MADAGASCAR
NAMIBIA
BOTSWANA
SWAZILAND
LESOTHO
REPUBLIC OF SOUTH AFRICA
Red Sea
Gulf of Aden
Indian Ocean
Mozambique Channel
0 300 600 Miles
0 200 400 600 Km

Atlantic Ocean
Mediterranean Sea
SICILY
GREECE
CRETE
Athens
Tunis
Bizerte
Skikda
Bejaia
Algiers
Arzew
Oran
Rabat
Mohammedia
Casablanca
Kechoula
Essaouira
Safi
Djebel-Jeer
Sidi Rhalem
Sidi-Kacem
MOROCCO
Las Palmas
Sidi Ifni
Tarfaya
El Golea
ALGERIA
Bel Rhazi
Touat
Ilatou
SBAA
Azzene
Reggane
In-Salah
Hassi Messaoud
2 Lines
3 Lines
TUNISIA
Tripoli
Sider
Tobruk
Farrud
Umm Farud
Amal
Nafoora
Augila
Defa
Messla
Sarir
In Amenas
Murzuk
Amasralad
Le Camp
LIBYA
WESTERN SAHARA
MAURITANIA
Nouakchott
MALI
NIGER
SENEGAL
GAMBIA
GUINEA-BISSAU
Bissau
GUINEA
Conakry
SIERRA LEONE
Freetown
LIBERIA
Monrovia
Bamako
BURKINA FASO
Ouagadougou
Niamey
Lake Chad
CHAD
N'Djamena
Kano
Kaduna
NIGERIA
Makurdi
IVORY COAST
GHANA
TOGO
BENIN
Porto Novo
Lagos
Lome
Accra
Abidjan
Espoir
Panthere
Bellier
Foxtrot
Saltpond
Gulf of Guinea
Bight of Biafra
CAMEROON
Yaounde
CENTRAL AFRICAN REPUBLIC
Bangui
EQUATORIAL GUINEA
Sao Tome
Libreville
Cape Lopez
GABON
Port Gentil
Tschengue
Rembo-Kotto
Asswe
Sette-Came
Gamba-Ivinga
CONGO (FORMER ZAIRE)
Mossaka
Brazzaville
Kinshasa
Pointe Noire
Nkossa
CABINDA
Emeraude
GCO
Essungo
Mavanga
Sulele
Guntala
Maleva North
Luanda
Benfica
Tobias
Lobito
ANGOLA
NAMIBIA
Windhoek
Walvis Bay
BOTSWANA
Gaborone
Luderitz
Kudu
Johannesburg
REPUBLIC OF SOUTH AFRICA
Cape Town
Cape of Good Hope
Mossel Bay
F-A-
Oribl
Legend
Oil field
Oil sand
Gas field
Crude oil pipeline
Natural gas pipeline
Products pipeline
Pipeline planned or under construction
Refinery in operation Refinery capacity in 1,000 b/d
Tanker terminal
Cities
Capital
International boundary
Water depth
0 to 200 m
200 m and deeper
AFRICA

Organization of Petroleum Exporting Countries (OPEC) News Agency said that production from the complex should increase Algeria's production by 285,000 b/d in 2002 and 2003. The complex has 1 million b/d of capacity, which was expected to grow 50% by 2005.

Anadarko and partners had resumed exploration in east-central Algeria under an amended production-sharing agreement (PSA) with Sonatrach concerning Blocks 404, 208, and 211 after Anadarko's original PSA ended in 1998. Altogether, the partners have drilled 24 production wells, 12 water injection wells, and 3 gas injection wells. Development drilling was to continue until reaching a total of 53 wells.

Other exploration and development

In late 2001 BHP Billiton announced plans to invest $430 million in the Ohanet gas project, with Woodside Petroleum holding 15% interest and Sonatrach and others owning the rest. Total investment would reach $1 billion, with production expected in 2004. Reserve estimates for Ohanet field exceed 3.2 tcf of pipeline-quality gas, 97 million bbl of condensate, and 115 million bbl of NGLs.

Algeria's In Salah gas fields development reached a milestone in mid-2001 when BP PLC and Sonatrach awarded three contracts worth a total of $2.5 billion to an international consortium. BP and Sonatrach agreed in 1997 to develop the seven In Salah fields, while an additional field, Boutraa, was earmarked for later work. The consortium included Halliburton unit Kellogg Brown & Root, Bechtel, Japanese company JGC, and Algerian drilling firm Enafor.

Kellogg and JGC were to provide treatment units and a gas gathering network. Bechtel would build gathering lines and a 286-mile (460-km) gas pipeline to link the project to Hassi R'mel. Enafor was to drill 71 producing wells and complete well workovers. The project was scheduled to come on stream in 2003 from three of the fields – Krechba, Teguentour, and Reg – and eventually would peak at 335 bcf/yr (9 billion cu m/yr). Development costs were shared between BP (65%) and Sonatrach (35%).

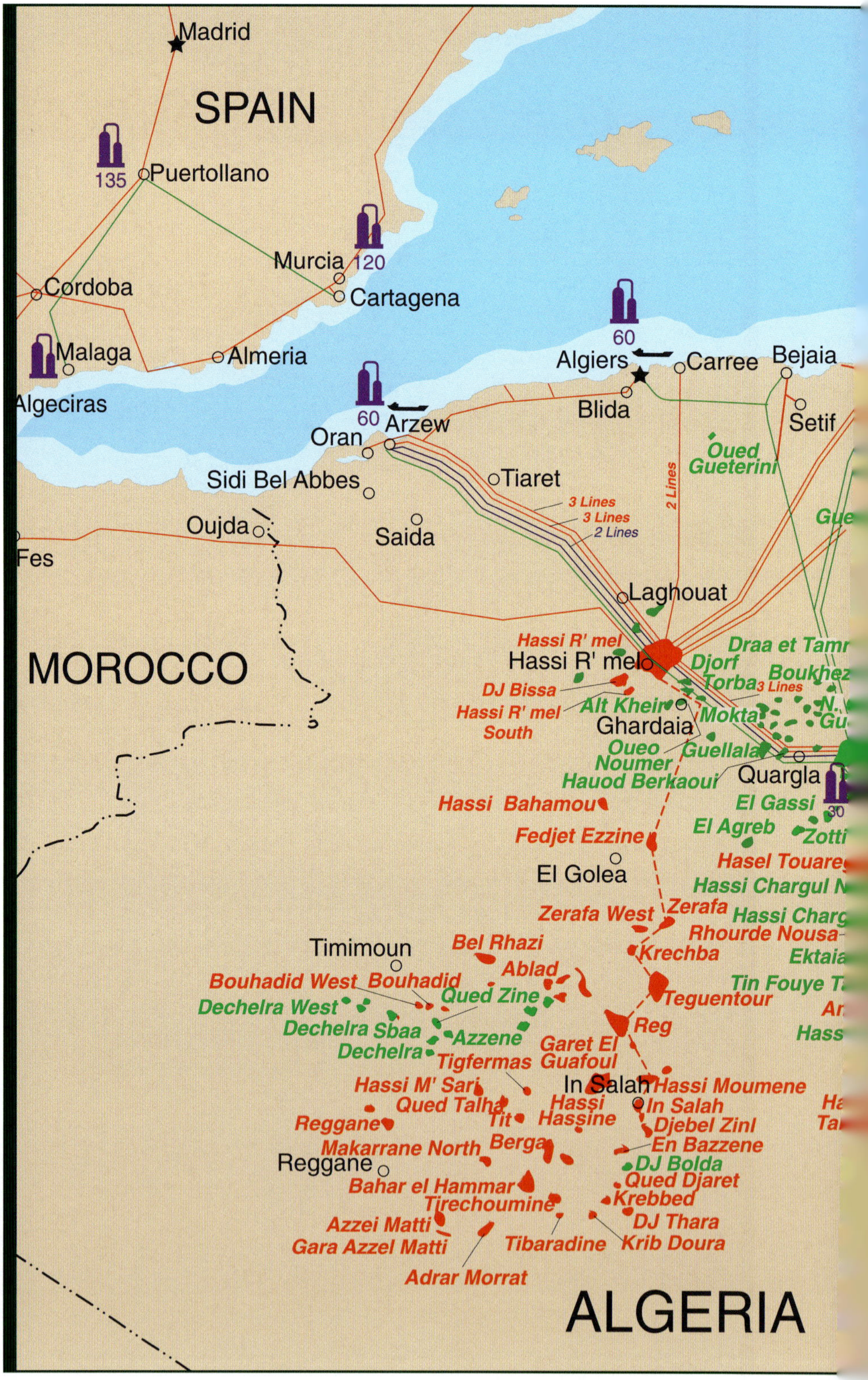

BP completed the Semhari East 1 wildcat on the Hassi Messaoud high as an oil and gas discovery, and Sonatrach's wildcat Azrafil Sud Est 1 in the West Region tested 14.4 MMcfd. Sonatrach also made an oil and gas discovery at Hassi Terfa 1 in the East Region. The well flowed 1,440 b/d of oil with some gas. Sonatrach's Djebel Heirane Kahla Tablbala 1 wildcat in the West Region tested gas.

In early 2001 Sonatrach awarded Gaz de France and Petronas a three-year exploration contract covering 6,560 sq mi (17,000 sq km) in the Ahnet basin south of In Salah in Algeria's Saharan desert. Gaz de France anticipated the exploration phase would cost 25-41 million euros. If

development is viable, production could begin in 2008. The venture falls under a cooperation agreement between Sonatrach and Gaz de France, signed in 2000, to bolster their commercial collaboration and widen it to the development of gas reserves and sales. As part of that agreement, the two companies formed a 50-50 joint venture in mid-2001 called Med LNG & Gas, with plans to market LNG in the Mediterranean and North America.

Gulf Keystone Petroleum signed an agreement in early 2001 to explore Block 126 in the Ferkane region in northeastern Algeria. A two-phase work program would involve $15 million for acquisition of 232 sq mi (600 sq km) of 3D seismic and the drilling of three wells over a five-year period, followed by $9 million to shoot another 373 miles (600 km) of seismic and drill another well.

Pipeline projects

Algeria would play a role in Nigeria's aims to revive a long-dormant project to build a natural gas trunkline to Algeria. The 2,484-mile (4,000-km) line would cost $7.5 billion and would move Nigerian gas from Abuja fields to the northwest Algerian port of Beni Saf for export to Europe. Both countries pledged to support the project with an extension of the trans-Saharan road and construction of a fiber optic line paralleling the pipeline to improve the area's telecommunications. The World Bank planned a feasibility study for 2002 and also would seek additional financing.

Algeria was exporting 745 bcf/yr (20 billion cu m/yr) of gas to Italy through the Trans-Mediterranean pipeline. In 2001 Sonatrach was considering building a second pipeline to Italy, which would carry 370 bcf/yr (10 billion cu m/year) of natural gas and would cost an estimated $2 billion. A joint venture was established by Sonatrach with Italy's Enel SPA and German company Wintershall AG to carry out feasibility studies, reported OPEC News Agency in late 2001.

The proposed 930-mile (1,500-km) pipeline would be built in four sections: 1) from Hassi R'mel gas field to the northeast part of El Kala, 398 miles (640 km); 2) subsea from El Kala to Cagliari, Sardinia, 193 miles (310 km); 3) Cagliari to Olbia on Sardinia's northeast shore, more than 124 miles (200 km); and 4) Olbia to Pascaia, southeast of Florence, about 186 miles (300 km) with a good portion offshore. A later extension of the pipeline could supply gas to Germany. Joint venture holdings were Sonatrach (50%), Enel (35%), Wintershall (15%).

Algeria also transports 298 bcf/yr (8 billion cu m/yr) of gas to Spain and 93 bcf/yr (2.5 billion cu m/yr) to Portugal through the Maghreb-Europe pipeline. During 2001 Algeria was considering construction of another pipeline to Spain. Energy Technology International analyzed the risks of building the pipeline across the Alboran Sea, where potential routes would have to cross through deep water with an uneven subsea profile and seismic activity.

Sonatrach continued efforts toward increasing Algeria's crude oil transportation capacity to 1.5 million b/d by 2004, compared with less than 1 million b/d in 2001. In mid-2001 a $70 million contract was awarded to ABB Lummus Group to extend and renovate pump stations on one of Sonatrach's oil pipelines, with the work scheduled for completion within 30 months.

ABB Group was awarded another $70 million pipeline upgrade contract earlier in 2001. The upgrade would increase the capacity of the 180-mile (290-km) OH3 pipeline to 63,000 b/d from 46,575 b/d. The line links Hassi Berkine oil field with the Haoud El Hamra terminal miles 620 miles (1,000 km) southeast of Algiers. ABB was to engineer and build an oil stocking and dispatching center at Hassi Berkine and a pumping station at Nezla. The project was to be completed in less than two years, with construction to be carried out in Algeria by Sarpi, an ABB-Sonatrach joint venture.

Refinery expansion

Sonatrach also began work to expand the Mesdar oil terminal near the giant Hassi Messaoud fields complex, awarding the 42 million euro contract to Bouygues Offshore subsidiary Sofresid in late 2001. Work was to include a new pumping station that would enable transport of 12 million tonnes per year (tpy) of oil and would be expandable to as much as 18 million tpy. The Mesdar terminal would receive, store, and dispatch the crude coming from the fields through the 26-inch OH2 pipeline.

ANGOLA

CAPITAL: LUANDA
MONETARY UNIT: KWANZA
REFINING CAPACITY: 39,000 B/CD
OIL PRODUCTION: 685,000 B/D
OIL RESERVES: 5.4 BILLION BBL
GAS RESERVES: 1.62 TCF

As sub-Saharan Africa's second largest oil producer, Angola ranks ninth in supplying crude to the US and also exports oil to Europe, Latin America, and Asia. Much of Angola's crude is high quality, with API gravity of 32-39.5º and sulfur content of 0.12-0.14%. Trade figures suggest that China's oil imports from Angola grew by more than 400% from 2000 to 2001, resulting from environmental standards that place a premium on lower sulfur crudes. Exports to South Korea have also increased.

Most of Angola's oil was produced from Block 0 offshore the enclave of Cabinda, which contains the country's largest oil fields, Takula and Numbi in Area A and Kokongo in Area B. In 2001 operator Cabinda Gulf Oil Co., a subsidiary of ChevronTexaco, completed installation of the Nemba North platform and began production in August. Nemba North was initially producing 20,000 b/d but was expected to reach peak production of about 50,000 b/d in 2002 and to average 40,000 b/d over the year. Cabinda Gulf was planning to invest heavily over the next few years in maintaining Block 0's oil production at 600,000 b/d or more through further field development and enhanced oil recovery.

Angola announced that it will host the Oil and Gas Exhibition in Luanda in December 2002.

Kuito development

During 2001 Angola's remarkable string of discoveries and field development continued. The third phase of development was completed in Block 14 on Kuito, Angola's first deepwater field put into production. Beginning in November 2001, additional development increased Kuito's output by 30,000 b/d, bringing its total to 85,000 b/d. Production in 2002 was expected to average 66,000 b/d, with peak levels targeted at 100,000 b/d. As of late 2001, Kuito had produced a total of 39 million bbl. Cabinda Gulf Oil was operator of Block 14 with partners state oil and gas company Sonangol (20%), Agip Angola Exploration B.V. (20%), TotalFinaElf Angola (20%), and Petrogal (9%). ChevronTexaco was Angola's largest oil producer, with about half of output going to the US.

The 2001 Kuito development, called Phase 1C or third phase, added a subsea production center with seven wells producing through a 12-slot manifold and an additional four water injection wells connected to the field's floating production, storage, and offloading (FPSO) vessel. Kuito produces with a gas/oil ratio of

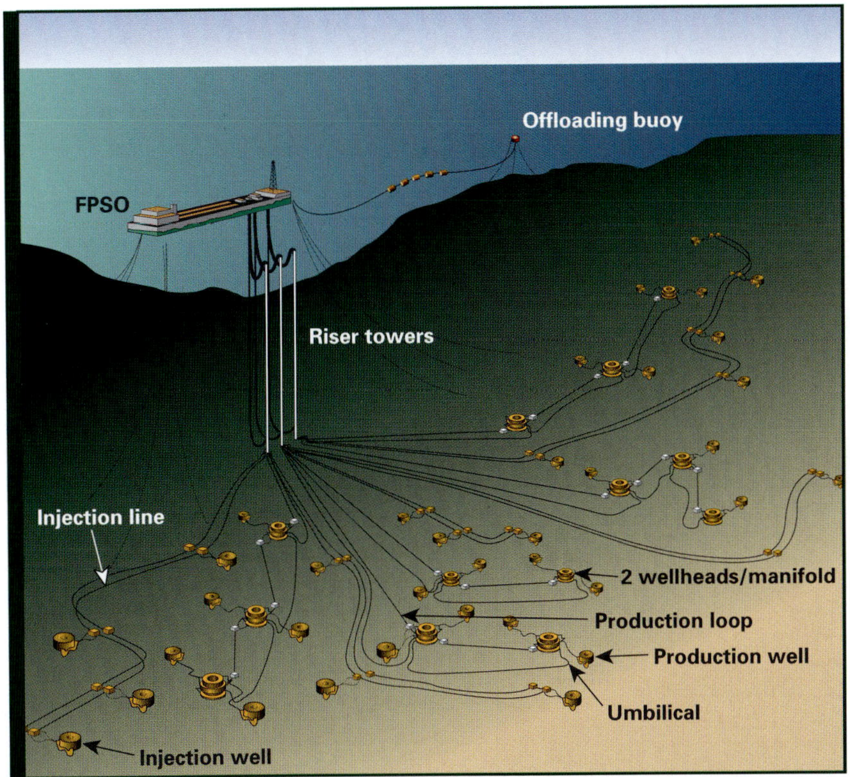

Girassol field was being developed using the world's largest FPSO, according to operator TotalFinaElf. The 50,000-tonne hull supports a 23,500-tonne topsides that processes 200,000 bo/d, treats 300,000 b/d of fluid, and injects back into the reservoir 390,000 bw/d and 8 MMcfd of gas.

200 scf/bbl. Third phase development employed a subsurface gas lift system designed for the life of the well to eliminate well intervention costs and gas flaring. Kuito became Cabinda Gulf's third "zero flare" field, after Nemba and Lomba in Block 0.

Development plans for four other fields on Block 14 were ongoing in 2001, with Benguela and Belize in an advanced stage and Lobito and Tomboco to follow, reported ChevronTexaco in October. The operator was analyzing possible tie-in synergies among the fields, and bid packages were being evaluated in late 2001.

Girassol startup

In mid-2001 Girassol field, estimated to contain 725 million bbl of oil, received what operator TotalFinaElf was calling the world's largest FPSO, and oil began flowing in December. Girassol's production in January 2002 reached 140,000 b/d, and the field's output was projected to be maintained at 200,000 b/d during 2002-06 with a peak potential level of 230,000 b/d. Girassol, located in Block 17, produces a low-sulfur crude similar to North Sea oil. Instead of flaring, associated gas was being injected but could ultimately be produced for export as LNG. TotalFinaElf held 40% of the block interest, with partners ExxonMobil's Esso Exploration Angola Ltd. having 20%, BP PLC 16.67%, Statoil 13.33%, and Norsk Hydro 10%.

The FPSO Mar Profundo Girassol was built by Bouygues Offshore SA/Stolt Offshore SA (a 50-50 joint venture called AMG). The 1,181-ft (360-m) vessel can process 200,000 b/d of oil, treat 300,000 b/d of fluid, and inject 390,000 b/d of water and 8 MMscfd of gas. It also has a storage capacity of 2 million bbl and a desulfurization unit that treats seawater for injection. The FPSO was spread moored in 4,430 ft (1,350 m) of water. Africa's relatively mild offshore environment allows lighter spread mooring instead of the heavier turret type. Conditions include an average wind speed of 82 ft/sec (25 m/sec), maximum wave height of 23.3 ft (7.1 m), and no hurricanes.

Girassol was planned to have 39 wells by 2003 – 23 oil producers, 14 water injectors, and 2 gas injectors. In early 2002, the first phase of 24 wells was near completion, and flow of 40,000 b/d per well was expected. Girassol's subsea system connects two wells to each subsea manifold. Two 1.5-mile, 16-inch oil export flowlines were being installed in 2001-02 to connect the FPSO to the unloading buoy. The flowlines were suspended at water depths of 150-1,800 ft (45-550 m) with centrally positioned buoys in a W-shaped configuration. Field development also included 43 miles (70 km) of infield pipelines linked to three 4,100-ft (1,250-m) innovative hybrid riser towers comprised of bundles of production, injection, and other flowlines encased in syntactic foam.

FPSO vessels were being considered for developing several other fields on Block 17 including Dalia and Rosa, which could start production by 2004. Girassol satellite field Jasmin was also targeted for development.

Kizomba A

Development also proceeded during 2001 on the Kizomba A project, which will combine production from Chocalho and Hungo fields. Recoverable resources expected from Kizomba A were 1 billion bbl. The two fields, located on Block 15, lie in water depths of 3,300-4,200 ft (1,000-1,300 m). Production was expected to begin in 2004 and was targeted to reach 250,000 b/d. The development project reportedly would cost $3 billion, making it the largest deepwater development project in West Africa. Operator ExxonMobil held 40% interest in the block, with partners BP Exploration (Angola) Ltd. (26.67%), Agip Exploration Angola BV (20%), and Statoil (13.33%).

The FPSO destined for Kizomba A was being built by a consortium of Hyundai Heavy Industries Ltd. and a Fluor Daniel-AMEC alliance called AFD under an $800 million contract. The FPSO, to be installed in 2004, would eclipse Girassol's as the world's largest such facility, capable of processing 250,000 b/d of oil and storing 2.2 million bbl. The deep draft caisson vessel would support about 34 oil wells completed with dry trees, eliminating the need for insulated flowlines. A tension leg plat-

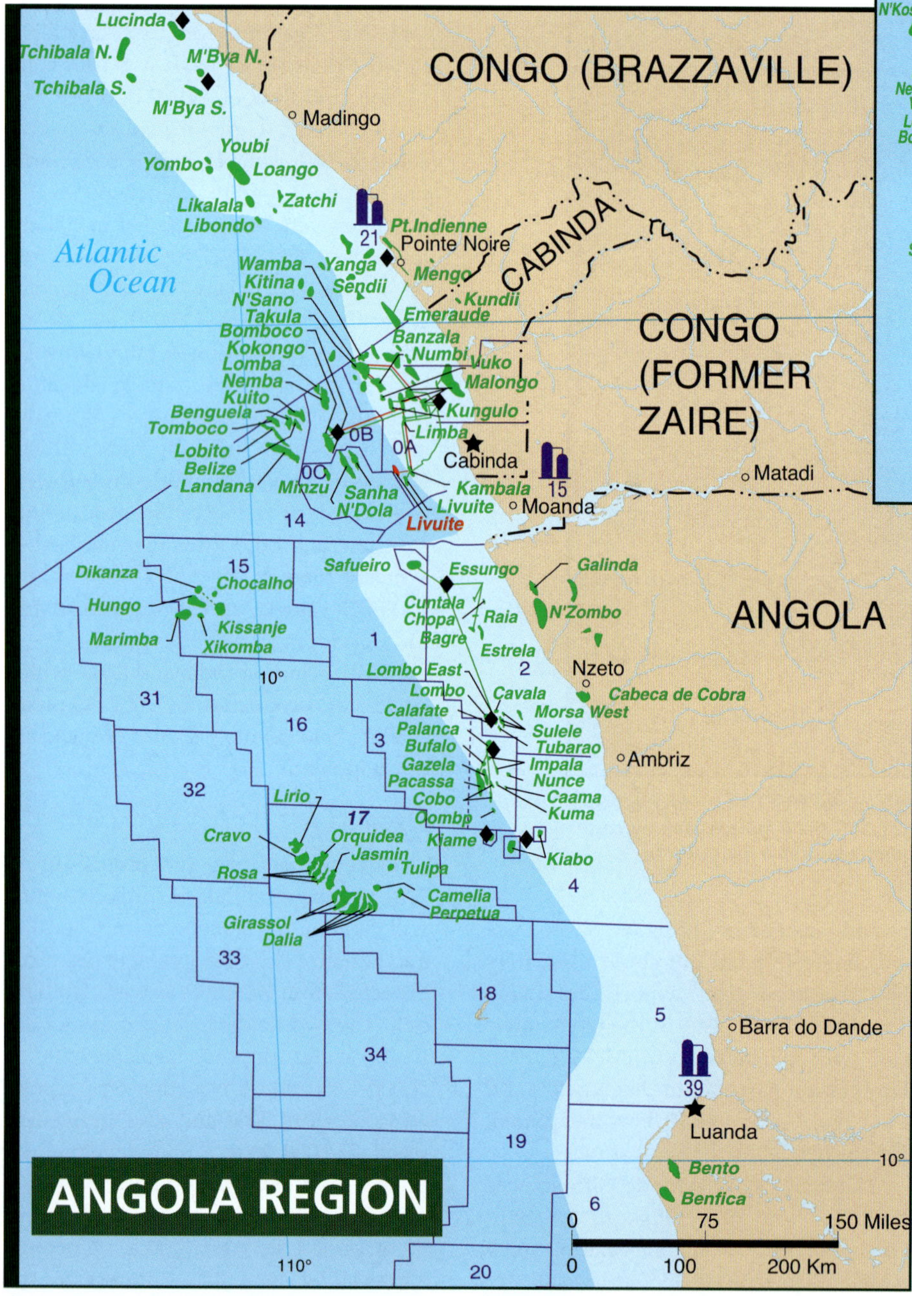

form was to be provided by ABB and partner Hereema and anchored in 4,100 ft (1,250 m) of water in 2003. Oil would be offloaded to tankers through a catenary anchor-leg mooring buoy.

Discoveries

In 2001 ExxonMobil announced three more discovery wells on Block 15, bringing its total to 12. M'blumbumba-1, revealed in April, was drilled in water depth of 2,790 ft (850 m); Vicango-1, announced in May, was drilled in 3,199 ft (975 m) of water to 6,890 ft (2,100 m); and Mavacola-1, also announced in May, was drilled in 3,790 ft (1,155 m) of water to a total depth of 9,060 ft (2,760 m) and found an oil-bearing reservoir that flowed at 3,280 b/d. Block 15's recoverable hydrocarbon potential was estimated at over 3.5 billion boe.

Block 18 operator BP PLC and 50-50 partner Royal Dutch/Shell Group continued their record of success, where the fifth and sixth consecutive successful wells were drilled – Cobalto and Cromio. All six wells are within a 12½-mile (20-km) radius collectively called the Greater Plutonio Prospect.

Another discovery occurred in Block 24, operated by ExxonMobil with partners Sonangol (20%), Ocean Energy's Ocean Angola Corp. (15%), and Petronas Carigali Overseas Sdn. Bhd. (15%). The Semba-1 wildcat, announced in June, encountered two oil-bearing reservoirs that flowed at a combined rate of 3,039-4,400 b/d. The well was the first on the block, spudded in February 2001 and drilled in water depth of 3,840 ft (1,170 m).

The success in offshore discoveries spurred even more interest in Angola's exploration blocks, and a PSA was signed in 2001 for Block 34, with Sonangol as operator (20%) assisted by Norsk Hydro (30%) and partners Phillips (20%), Petrobras (15%), and Shell (15%). Also Ocean Energy acquired a 35% interest in Block 10, which lies in up to 1,500 ft of water in the Benguela subbasin of the Kwanza basin. Angola could offer additional blocks in the very deep offshore Kwanza basin.

Gas, LNG developments

The Angolan government signed a protocol agreement in late 2001 aiming to develop up to 4 tcf of gas reserves. Sonangol signed the deal with TotalFinaElf for a $3 billion gas venture. Previously, Angola's associated gas production had been flared or reinjected. The venture will be broadened to include other companies operating in Angola. The first stage is to reevaluate reserves and choose development options. TotalFinaElf's LNG expertise was considered desirable, and some of the gas would also be used in fertilizer manufacturing and power generation.

ChevronTexaco expected to monetize

about 100,000 b/d of oil, condensate, and LPG and eliminate associated gas flaring in Areas B and C of Block 0. The Sanha gas condensate/Bomboco project on Block 0 was the cornerstone of this gas initiative.

Plans also moved forward to build an LNG plant in Luanda, Angola's capital, with ChevronTexaco as operator. Engineering contract bids for the $2 billion project were tendered during 2001, with construction to start in 2002 and startup envisioned for 2005. The plant would produce 4 million tpy of LNG from one train, but the site could accommodate additional trains for expanding capacity. Gas would come from eight offshore blocks that contain about 9 tcf. Partners in the project were Sonangol, BP PLC, ExxonMobil, and TotalFinaElf.

Refinery plans

Angola continued working on plans to build the country's second refinery, to be located in the southern coastal city of Lobito. In 2001 several companies signed a memorandum of understanding that called for construction to start in 2002 and the plant to come on stream by 2006.

CAMEROON

CAPITAL: YAOUNDE
MONETARY UNIT: CFA FRANCE
REFINING CAPACITY: 42,000 B/CD
OIL PRODUCTION: 75,000 B/D
OIL RESERVES: 400 MILLION BBL
GAS RESERVES: 3.9 TCF

Cameroon's oil production continued its steady decline, but the country's refinery capacity and its position as the terminus of an oil export pipeline from Chad could propel Cameroon into a significant transport center. During 2001 a dispute continued among Cameroon, Nigeria, and Equatorial Guinea over control of the Bakassi Peninsula in the Gulf of Guinea, site of several oil discoveries.

Major operators of oil and gas fields in Cameroon were TotalFinaElf, Perenco, and Pecten International (a Shell subsidiary). Bidding for exploration blocks during 2001 attracted relatively low interest, causing industry observers to speculate that major finds are unlikely following years of heavy exploration.

Still, UK energy analyst Douglas-Westwood Ltd. projected in late 2001 that oil firms would spend $7.3 billion in "second tier" West Africa projects before 2005. About 13% of the total, or $950 million, could go to Cameroon.

In early 2001 Cameroon's state company SNH signed a petroleum concession contract with Fusion Oil & Gas of Australia for exploration of the deepwater Ntem permit in the Southern Douala/Kribi-Campo basin. The prospect – an area of 803 sq mi (2,080 sq km) where water depths are 3,200-5,900 ft (1,000-1,800 m) – neighbors the northern border of the Equatorial Guinea territory of Rio Muni, where Triton Energy discovered the 500 million bbl deepwater Ceiba field in 1999. Fusion held 100% equity in the Ntem contract, which had three exploration terms of two years each. The company said it would shoot 2D and possibly 3D seismic over the permit area during the first term to size up several prospective features, noting that existing seismic data already suggests certain structures with possible direct hydrocarbon indicators.

Oil export facilities were under construction in 2001 near Kribi on the Cameroon coast. The facilities were being built to handle the first oil production from neighboring Chad, where a 650-mile, 24-30 inch pipeline connection was being constructed to move oil from Doba basin fields beginning in 2003-04. Partners in the pipeline and oil field development were ExxonMobil Corp. subsidiary Esso Chad (also the operator, with 40%), Malaysia's Petronas (35%), and ChevronTexaco (25%).

Cameroon's refinery, the 42,000 b/d Ste. Nationale de Raffinage in the port city of Limbe, was due for upgrading in 2001 under a 10 million euro contract awarded by Commission Nationale des Marchés Financiers du Cameroun with Bouygues Offshore subsidiary Sofresid. The debottlenecking project, which was expected to require 19 months, would increase the refinery's distillation capacity of light crude and allow it to obtain a new distillate fraction.

CHAD

CAPITAL: N'DJAMENA
MONETARY UNIT: CFA FRANC
REFINING CAPACITY: N/A
OIL PRODUCTION: N/A
OIL RESERVES: N/A
GAS RESERVES: N/A

ExxonMobil Corp. and its partners in late 2001 began drilling the first test wells in Doba basin oil fields in southern Chad, as part of a $3.5 billion project to produce the country's first oil. With its large undeveloped oil reserves, Chad could become a significant energy producer. The country is one of the poorest in the world, ranked near the bottom on the United Nations' human development index. The Doba oil project alone could double Chad's GDP.

The oil would be transported beginning in 2003 through a 650-mile, 24-30 inch pipeline connection to Cameroon's coast, where export facilities were under construction. Partners in the pipeline and oil field development were ExxonMobil subsidiary Esso Chad (also the operator, with 40%), Malaysia's Petronas Carigali Sdn. Bhd. (35%), and ChevronTexaco (25%).

The drilling project has been backed by the World Bank with a $200 million initial loan and was planned to have a 25-year life, with 300 wells producing a peak of 250,000 b/d in middle of the decade. Esso Chad's statement announcing the project start put an end to rumors that environmental groups had raised issues that were delaying progress. The Chad government said that it has in place adequate restrictions on how revenues can be spent as well as measures to protect the environment and people living near the pipeline.

The World Bank program for the Chad project included its own financing, support for private financing, petroleum revenue management, and environmental management. In addition to field construction and well drilling, the project will include a central treatment facility, an electric power generation plant, an airfield and office buildings, and a pipeline and pump station.

CONGO (BRAZZAVILLE)

CAPITAL: BRAZZAVILLE
MONETARY UNIT: CFA FRANC
REFINING CAPACITY: 21,000 B/CD
OIL PRODUCTION: 260,000 B/D
OIL RESERVES: 1.5 BILLION BBL
GAS RESERVES: 3.2 TCF

Congo (Brazzaville) could receive a generous portion of the $7.3 billion projected to be invested by oil firms in "second tier" West Africa projects before 2005, said UK energy analyst Douglas-Westwood Ltd. About 22% of the total, or $1.6 billion, could target development of Congo's reserves, the largest reserve base targeted for development over the period.

Foukanda field off Congo began production in record time for that area, said Eni SPA in mid-2001. Oil production came on stream within three years and three months of the field's discovery in March 1998, primarily as a result of contractual incentives. Production, expected to peak at 12,000 b/d, was routed via pipeline through the Kitina platform to the Djeno terminal for export. The additional volumes would boost Eni's average production in the area to 75,000 b/d.

Foukanda field's platform was located in 344 ft (105 m) of water. Its jacket was built in Congo at Pointe Noire, then joined offshore to the deck, which was built in Italy. The field, which cost 150 million euros to develop, contains total recoverable reserves estimated at 40 million bbl of oil. Eni was operator through Agip Congo with 65% interest; the State Oil Co. of Congo held the rest.

Agip was also planning to start up Mwafi field in 320 ft (100 m) of water near the northern end of Marine VI. Initial production was estimated at 4,000 b/d with a peak of 9,000 b/d in 2002. Six wells were planned to produce through a satellite platform with oil shipped via pipeline to the Loango plantform and on into Djeno. Mwafi reserves are 26 million boe.

Development of Yombo field off Congo proceeded during 2001 as two 2-mile, 18-inch pipelines were laid in 380 ft of water by Global Offshore International Ltd. (a unit of Global Industries Ltd., Houston) for CMS Oil & Gas (Congo) Ltd. Global was to use the pipelay-derrick barge Comanche to install the lines from the Yombo A and B platforms to an FPSO vessel.

In mid-2001 a 410-ft (125-m) oil column was discovered in the primary target, Lower Cretaceous Vandji sands, on the 850,000-acre Kouilou license offshore, said Heritage Oil Corp. (Calgary), partner with France's Maurel & Prom on the license. The M-Boundi-1 well, located 12½ miles (20 km) north of Kouakouala field, was drilled to 6,968 ft (2,124 m). Interests were Maurel & Prom (30%), Heritage (30%), and Tacoma Resources Ltd. (35%).

In late 2001 Eni announced a new oil discovery offshore Congo at a depth of 8,957 ft (2,730 m). The well flowed at 3,800 b/d of 32º API gravity. A second well was planned.

CONGO (FORMER ZAIRE)

CAPITAL: KINSHASA
MONETARY UNIT: ZAIRES
REFINING CAPACITY: 15,000 B/CD
OIL PRODUCTION: 24,000 B/D
OIL RESERVES: 187 MILLION BBL
GAS RESERVES: 35 BCF

According to a political risk index developed by IHS Energy, the Democratic Republic of Congo (formerly known as Zaire) is one of the five riskiest countries for oil and gas investment. During 2001 President Laurent Kabila was assassinated and his son Joseph became Congo's leader. Still, oil firms were expected to invest in the country, said UK energy analyst Douglas-Westwood Ltd. The analyst projected in late 2001 that $7.3 billion would be spent in "second tier" West Africa projects before 2005, with about 1% of the total, or $7.3 million, going to Congo (former Zaire). Also, in addition to its oil resources, the country contains large proven gas and associated gas reserves offshore, said Vanco Energy Co.

EGYPT

CAPITAL: CAIRO
MONETARY UNIT: POUND
REFINING CAPACITY: 726,250 B/CD
OIL PRODUCTION: 755,000 B/D
OIL RESERVES: 2.95 BILLION BBL
GAS RESERVES: 35.2 TCF

Natural gas discoveries by companies including BG Group PLC, BP PLC, and Apache Corp. continued during 2001, giving Egypt a stronger springboard for increasing its gas production and becoming an exporter of LNG. The country's rapidly expanding gas sector was expected to nearly double production by the end of 2002, to about 3 bcfd. The Egyptian government created state-owned Egyptian Natural Holding Gas Co. (EGAS), which was separated from Egyptian General Petroleum Corp. (EGPC).

BG gas development, LNG project

In November 2001, BG Group (UK) and Italy's Edison International confirmed the largest gas column discovered in Egypt. The two companies drilled the Sapphire 3, the second Sapphire appraisal well, on the West Delta Deep Marine concession – home to the deepwater Scarab-Saffron fields, Egypt's largest gas development – 37 miles (60 km) off the Nile Delta. The well, in the south central area of the concession 12½ miles (20 km) north of Rosetta field, was drilled to 9,514 ft (2,900 m) total depth in 1,476 ft (450 m) of water. The first target was the previously undrilled Saffron Channel C in the Scarab-Saffron complex. The well proved pressure communication between Channel C and the Saffron discovery, indicating a gas column in excess of 1,886 ft (575 m).

A second target was the Sapphire sands. The well penetrated all sands present in Sapphire 1 and 2, located 9 miles (15 km) west of Sapphire 3. Pressure data and geochemical analysis indicated that all three wells are in communication. Reserves could be greater than previously modeled, said BG. Burullus Gas Co., a joint venture of EGPC 50%, BG Egypt 25%, and Edison International 25%, was responsible for drilling and field development on the West Delta concession. Sapphire 3 was the 12th successful appraisal well drilled on the concession and the 15th well overall. Sapphire 1 and 2 flowed on test at rates of 35 MMcfd and 1,100 b/d of condensate and 33 MMcfd and 805 b/d of condensate, respectively.

The Sapphire gas and further West

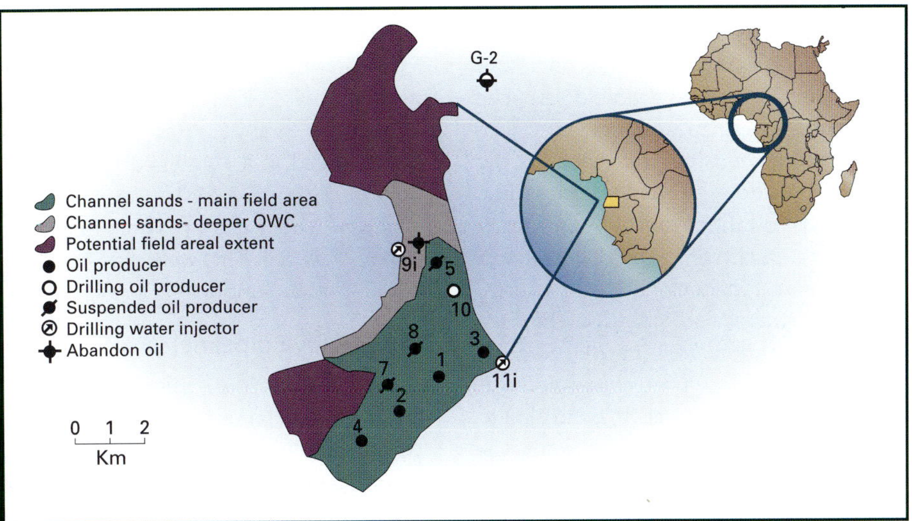

A deeper pool discovery and other recent drilling at Ceiba field off Equatorial Guinea extended the field and increased its reserves.

Delta Deep Marine reserves were being targeted for Burullus' proposed Egyptian LNG project. In April 2001, BG agreed with EGPC and Edison to advance plans for an integrated LNG export project in Egypt which would involve development of West Delta gas reserves and construction of a $900 million onshore liquefaction plant at Idku, which is near Alexandria and adjacent to the Rosetta and Scarab-Saffron gas processing terminals on the Mediterranean coast. BG envisaged that the Idku facilities would form an Egyptian gas hub with its own port facilities.

Scarab and Saffron fields were being developed for the domestic market. In early 2001 Burullus awarded a $56 million contract to the Egyptian Deepwater Development Consortium for the Scarab-Saffron deepwater subsea production system. First gas from Scarab-Saffron was scheduled for January 2003. Rosetta field, which was producing 275 MMscfd, began supplying the Egyptian national grid in early 2001. The associated Simian, Sapphire, Sienna, and Serpent discoveries would support the LNG project. As part of the integrated development of the West Delta concession, BG was having higher capacity offshore pipelines installed in order to support a potential two-train LNG development while supplying the domestic market from Scarab-Saffron.

In early 2002, Egypt signed an $8 billion, 20-year agreement to supply Gaz de France SA (GdF) with 3.6 million tpy of LNG – about 10% of France's yearly gas demand – further paving the way for the LNG project. BG, Edison, EGAS, and EGPC formed the Egyptian LNG (ELNG) joint venture to build and operate the Idku LNG export plant. Under the deal, GdF would take a 5% stake in ELNG, which was to be formally restructured.

The first train from ELNG would have a capacity of 3.6 million tpy of LNG with first production scheduled for mid-2005. ELNG partners would operate the LNG export plant as a tolling facility. The commercial structure would also facilitate future plant expansions and processing of gas reserves by ELNG from other fields and operators.

Based on its successful drilling record and discoveries since 1997 totaling more than 10 tcf, BG was counting on sufficient reserves to support imminent marketing of a second train. The contract with GdF was expected to lead to final project sanction, following completion of the front-end engineering design studies by Bechtel Group Inc., the deepwater managing contractor on the Scarab-Saffron development. Bechtel was involved in the early definition and engineering work for the LNG plant, which could use the Phillips optimized cascade process installed by Bechtel at Atlantic LNG in Trinidad.

In addition to BG's upstream activities, it was a 37.5% shareholder in the Nile Valley Gas Co., which had a 25-year franchise agreement to develop a gas transmission system and market gas in Upper Egypt. First gas was delivered in 1999.

Also, Shell Gas & Power and EGPC were considering plans to build a second-generation 75,000 b/d GTL conversion plant with at least one LNG train on the Mediterranean coast.

More gas finds

In mid-2001 BP PLC made its fifth gas find in two years in the western Nile Delta with the Libra K-1x well in the North Alexandrian concession area off Egypt's north coast. The well, in 1,476 ft (450 m) of water, discovered gas in five separate zones, one of which was tested at a rate of 22.3 MMscfd with a condensate-gas ratio of 7.6 bbl/MMscf. BP noted that the flow rate was constrained by testing equipment. First estimates were that the well encountered around 164 ft (50 m) of conventional pay, proving combined reserves of 500 bcf.

BP's five gas finds off the delta have proved reserves of more than 2 tcf. The concession partners – BP with 60% and Germany's RWE-DEA with 40% – planned to drill at least another three wells in the western Nile Delta over the next year, including a first well in the West Mediterranean deepwater concession area.

Earlier in 2001, BP Egypt announced another natural gas discovery in the North Alexandria concession with the Fayoum (L-1x) well, in 1,312 ft (400 m) of water 22 miles (35 km) from the northern coast. The well flowed on test 21 MMcfd of gas from the main target zone and 6 MMcfd from a secondary zone with more silt. BP estimated that the well proved reserves of 600 bcf.

Apache discoveries

In early 2002 Apache announced two discoveries in Egypt. The Ozoris-1X wildcat in the Western Desert found both oil and natural gas on the company's Khalda concession. The well flowed 2,504 b/d of 38.5° gravity oil and 750 Mcfd of gas through a 1-inch choke at 161 psi of flowing pressure from one of three zones. The tested zone – the Cretaceous Alam el Bueib 3D sand – is a prolific producing horizon at Khalda Ridge, 12 miles east of Ozoris-1X. Apache owned a 100% contractor's interest in Khalda.

The discovery was 6 miles from the nearest well and 9 miles from the nearest field on the 2 million acre concession. Apache said that the pay zone tested was more than 1,500 ft downdip from the crest of the Khalda Ridge, one of the largest hydrocarbon accumulations in the Western Desert, yet still had good porosity and permeability. Another 65 ft of potential pay remained to be tested in two additional uphole zones.

Also in the Western Desert, Apache completed another successful well on the South Umbarka development lease, in which it holds a 100% interest. On test, the Khepri 9 well flowed 29.5 MMcfd of gas and 220 b/d of condensate and was completed in the Alam el Bueib D-7 sand through a 1-inch choke with 1,640 psi flowing wellhead pressure from perforations at 13,340-13,368 ft. Apache was planning to test the D-1 sand. Khepri 9 was 1.2 miles southeast of the Khepri 1 discovery well. Natural gas production from the 106,253-acre development lease was transported 30 miles via pipeline to the company's Salam gas plant on the adjacent Khalda concession.

In late 2001 Apache said the JG-1X discovery in Egypt's Northeast Abu Gharadig concession flowed on test at a rate of 4,190 b/d of oil and 5 MMcfd of gas. The Western Desert wildcat was tested through a 48/64-inch choke with 1,171 psi flowing wellhead pressure. It was perforated at 10,497-10,517 ft in the Jurassic Khatatba, making it the first Jurassic sand oil discovery in this part of the Abu Gharadig basin. Operator Shell Egypt identified several potential offset locations. Apache said the JG-1X well will be put on production quickly because field facilities operated by a Shell joint venture are nearby. The discovery well was less than 2 miles from existing oil pipelines.

Apache was also drilling another well on a nearby concession, said concession partner Novus Petroleum. Ras El Hekma 2X well, on Ras El Hekma concession in the Western Desert, flowed at a stabilized rate of 11.5 MMcfd of gas and 2,120 b/d of condensate from two zones in the Lower Cretaceous AEB 5B formation. The discovery was 5.3 miles (8.5 km) from Tarek North East gas field. Apache and Novus were to submit a development plan proposing a tieback to the Tarek NE well on the Khalda concession, then transportation through the Tarek processing facility to Ameriya through the Northern Pipeline.

The partners also spudded the East Neith exploration well in 2001 on the adjoining Khalda Offset concession. It reached 12,246 ft and was to be drilled to a total depth of 14,700 ft. East Neith was 2.1 miles (3.3 km) northeast of the 1,170 b/d Neith South 3X appraisal, which confirmed the 2,634 b/d 1X discovery well. The partners have submitted a development plan for Neith that involves a tieback to facilities on the South Umbarka concession, 8 miles (13.2 km) from Neith South 1X. Production was to begin in 2001.

Other finds

In October 2001 Ocean Energy Inc. (Houston) said a well on its East Zeit concession in the Gulf of Suez might add 40-80 million bbl of reserves to the 87 million bbl already produced from the field. East Zeit A-21 reached 16,300 ft total depth and found 745 ft of oil pay. Ocean Energy, which owned 100% working interest, planned to bring the well on stream in late 2001 and was planning delineation and further development of East Zeit field for 2001. Ocean Energy held working interests in five concessions in Egypt, three onshore.

In July 2001 Centurion Energy International Inc., Calgary, spudded the Gelgel-1 exploration well on the El Manzala concession in Egypt. The target was gas in the Pliocene Kafr El Sheikh formation. Total depth was to be 3,937 ft (1,200 m). Centurion was operator and held 100% working interest in the 420,000-acre concession.

In April 2001 Devon Energy began production from its North July field in the Gulf of Suez off Egypt. North July No. 1 flowed 5,500 b/d of oil and 3 MMcfd of natural gas from Rudeis at 9,790-9,894 ft. The well was 5 miles offshore in 169 ft of water and was producing to a three-slot platform. Devon Energy also had a discovery in the southern Gulf of Suez with a well in the southwest Gebel El-Zeit concession, 6 miles offshore in 86 ft of water. The SWGEZ-2 well flowed at a combined rate of 6,700 b/d of oil and 2.4 MMcfd of natural gas from two intervals in Nubia-Matulla at 7,786-7,894 ft. The well was drilled to 8,040 ft total depth.

Egypt-Libya pipeline

In August 2001, after more than 5 years of negotiations, the Arab Co. for Oil & Gas Pipelines was created to transport crude to Egypt and gas to Libya. Egypt's General Authority for Investments approved formation of the joint company. Work was to begin immediately on fixing the route of the proposed pipeline. Egyptian Petroleum Minister Sameh Fahmy said Libya's National Oil Co. and EGPC would each provide 50% of the capital for the new firm. Fahmy said initial investment would be around $20 million. Libyan oil would be refined in Cairo and Alexandria.

Refinery commissioned

Egypt, as a majority partner in joint venture MIDOR (Middle East Oil Refinery Ltd.), commissioned a $1.3 billion, 100,000-b/d refinery in Alexandria, Egypt, in April 2001. The ultra-modern, environmentally advanced refinery includes a 33,500-b/d hydrocracker. Israel's Merhav sold its 20% share in MIDOR to the National Bank of Egypt. The refinery, built in the Ameryia Free Zone of Alexandria, was intended to produce high-quality products for both export and domestic consumption, especially local markets requiring LPG and diesel fuel.

In addition to the hydrocracker licensed by UOP LLC, the high-conversion refinery has a 100,000-b/d crude and vacuum unit and a 21,700-b/d catalytic reformer, also UOP-licensed. The Conoco-Bechtel alliance provided its delayed coking technology in a 22,800-b/d unit. A hydrogen reformer and the catalytic unit provide the refinery's hydrogen needs. The processes have a slant towards diesel production.

The feed to the new refinery was a wide spectrum of Arabian crude, ranging from 100% Arabian Light to 50% Arabian Light/50% Arabian heavy crude (or an equivalent mix). Refinery products will meet European Union standards for 2005. Unleaded high octane gasoline and low-sulfur diesel make up the largest part of the intended products. The MIDOR refinery will produce an average of 1.3 million tpy of gasoline and 2.2 million tpy of diesel. Generally, all of the refinery's LPG and diesel production will head for domestic consumption, and the remaining products (gasoline, kerosene/jet fuel, and coke) will be exported to international markets. MIDOR designed the refinery to be well within local Egyptian environmental standards. Air and water emissions will be regularly monitored.

EQUATORIAL GUINEA

CAPITAL: MALABO
MONETARY UNIT: CFA FRANC
REFINING CAPACITY: N/A
OIL PRODUCTION: 170,000 B/D
OIL RESERVES: 12 MILLION BBL
GAS RESERVES: 1.3 TCF

Equatorial Guinea continued to lead the parade of smaller West African countries attracting exploration and development capital. According to UK energy analyst Douglas-Westwood Ltd., oil firms could spend $7.3 billion in "second tier" West Africa projects before 2005. Equatorial Guinea would receive the lion's share – about 24% of the total, or $1.75 billion. In addition to the region's first deepwater production (ExxonMobil Corp.'s Topacio field) and its productive offshore La Ceiba field in Block G, six more fields in Equatorial Guinea were identified for development through 2005. The analyst said that Okume field, also on Block G, was likely to be the most significant future development and was a candidate for a floating production ship.

In mid-2001 Amerada Hess Corp. announced acquisition of Triton Energy Ltd. for $3.2 billion, making Amerada operator of Block G with 85% interest; Energy Africa Ltd. (Cape Town) held 15%.

La Ceiba development

La Ceiba was slated to enter a second development phase calling for 14 wells to be drilled and completed by early 2002 (10 producers and 4 water injectors), in addition to the previous five wells producing 50,000 b/d. In spring 2001 a deeper pool discovery and drilling at Ceiba extended the field to the northwest and increased its reserves, partly from a known zone that had not been expected to contribute significant oil. Four wells averaged a combined 45,000 b/d of 30º gravity oil, leading Triton to refine its geologic model.

In late 2001 Ceiba got a new FPSO, Sendje Ceiba, to replace the Sendje Berge vessel that had been working the field. Sendje Ceiba used to be a very large crude carrier (Berge Hus) but was upgraded into an FPSO by Bergesen DY Offshore (Oslo). The new vessel has larger onboard capacities, 160,000 b/d of liquids and 135,000 b/d of water injection, to accommodate Triton's plans to expand Block G's output in 2002. The Sendje Ceiba was to begin taking Ceiba production in early 2002.

Discoveries nearby

Exploration and appraisal of Okume and Oveng fields in Block G near La Ceiba continued during 2001. In mid-year before being acquired, Triton announced success in Okume field, 18 miles offshore, where the discovery well found gross oil pay of 407 ft, with 200 net ft of oil-bearing pay in one pool. Okume-1 was drilled to 7,064 ft total depth in 1,604 ft of water at a cost of $5.5 million. Okume's reservoirs appeared similar to those of the 45,000 b/d La Ceiba field nearby, with 28% average porosity and 37º gravity oil. Later in the year Amerada Hess continued appraisal of Okume with the drilling of Okume-4.

Also on Block G, the Oveng-1 discovery well was drilled in June 2001 and found 168 ft of net oil pay. This discovery was confirmed by Amerada Hess Corp. in early 2002. The Oveng-2 appraisal well, drilled with Global SantaFe's Glomar R.F. Bauer drillship, found 199 ft of net oil pay and 145 ft of net gas pay. The well was drilled to 6,989 ft in 696 ft of water. Oveng-2 flowed 8,457 b/d of 34º gravity oil from an interval of 124 ft.

Amerada Hess also found oil at its Ebano prospect in the Rio Muni basin on Blocks F and G in 2,052 ft of water. The 11,011-ft F-2 discovery well and a 9,446-ft sidetrack on adjacent Block G confirmed a 300 ft gross pay interval with 181 ft total vertical depth of net oil-bearing pay at 7,469-7,756 ft. The oil was 36.5º gravity. The discovery, 10 miles northeast of Ceiba field, tested reservoirs similar to the Ceiba, Okume, and Oveng discoveries. Once operations at the sidetrack well are completed, Amerada Hess planned to spud a well on Block G about 3 miles southeast of Ebano. As with Block G, Amerada Hess operated Block F with 85% working interest and Energy Africa 15%. The government of Equatorial Guinea would have a carried 5% participating interest in any Ebano-area commercial production.

Alba reserves increase

Two wells drilled in Equatorial Guinea's offshore Alba gas condensate field contributed to nearly doubling the field's proven and probable reserves, which were re-estimated at more than 300 million bbl of liquid hydrocarbons associated with 4.6-5.0 tcf of natural gas. Alba No. 9, 2.4 miles from the nearest producing well, found 1,000 ft of gross pay and 732 ft of net pay. The well flowed 37.5 MMcfd of gas and 2,400 b/d of 46º gravity condensate through a 1-inch choke with 2,473 psi flowing tubing pressure and 197 psi bottomhole pressure drawdown. The flow came from 120 ft of performations. Partner Noble Affiliates Inc. estimated the well could produce 100 MMcfd of gas and 6,200 b/d of condensate.

Following this success, Marathon Oil Co. in late 2001 paid $993 million for CMS Energy Corp.'s Equatorial Guinea assets including its operatorship of and 52.4% interest in Alba, a 37.6% interest in Block D, and substantial shares in three onshore plants (gas processing, methanol, and LPG). In late 2001 Alba field was producing 240 MMcfd of wet gas from which 17,000 b/d of condensate and 2,400 b/d of LPG were recovered. Plans had been submitted to increase gross production to 800 MMcfd and expand the condensate recovery and LPG facilities.

The methanol plant, located on Bioko Island, is run by Atlantic Methanol Production Co. LLC (AMPCO) and converts gas from Alba field. In May 2001 the operation launched its first methanol shipment, with the ship Noble Spirit to deliver the 41,000-tonne cargo to the European market in mid-year. The plant was expected to average five shipments/quarter. Methanol prices were averaging $240/tonne, which would generate revenues of $200 million/yr. The plant was producing more than its 2,500-tonne/day rated capacity. Marathon and Noble each owned 45% of AMPCO, and the government of Equatorial Guinea held 10%.

GABON

CAPITAL: LIBREVILLE
MONETARY UNIT: CFA FRANC
REFINING CAPACITY: 17,300 B/CD
OIL PRODUCTION: 300,000 B/D
OIL RESERVES: 2.5 BILLION BBL
GAS RESERVES: 1.2 TCF

Gabon's main export commodity is oil, causing the government to be concerned with declining oil production and a longer term trend of diminishing oil reserves. Gabon sought to attract new investors by revising production-sharing contracts (PSCs) and increasing the number of exploration permits.

Of $7.3 billion projected to be invested in "second tier" West Africa projects before 2005, about 20%, or $1.46 billion, could go to Gabon, said UK energy analyst Douglas-Westwood Ltd. in late 2001. The country has 21 offshore fields identified for development over the period, targeting 296 million boe of reserves, said the analyst, and Pioneer Natural Resources Co. expected production of 50,000 b/d from Bigorneau South-Olowi Marin prospect.

Atora field came on stream in March 2001, producing 20,000 b/d. Atora, which could contain 50-100 million bbl of oil, was operated by Elf Gabon with Amerada Hess Gabon holding 40% and Shell Gabon 20%. The additional production made Elf the country's largest oil producer. Elf Gabon signed a concessional agreement for Olonga Marin covering 4,000 sq mi and was planning to

drill in the ultra-deep Astrid marin concession.

One focus of development in 2001 was the Etame Marin concession off Gabon, where a large commercial discovery was confirmed by delineation well Etame 4V in mid-2001. Vaalco Energy Inc. (Houston) said it would fast-track development of the field and projected that 12,000 b/d of oil would begin tanker loading in early 2002. Vaalco awarded the FPSO Petróleo Nautipa a two-year, $15 million contract to work off Gabon in the Etame field, with a three-year extension option. Vaalco subsidiary Vaalco Gabon (Etame) Inc. operated the Etame consortium with 30.35% interest; other partners were UK's PanAfrican Energy Corp. Ltd. (32.5%), Sasol Petroleum International (Pty.) (30%), PetroEnergy Resources Corp. (4.5%), and Nissho Iwai Corp. (2.65%).

The Etame 4V well, spudded in May 2001, was drilled 1.5 miles from the discovery well to a depth of 7,290 ft (2,222 m), and log analysis indicated a gross oil column of 105 ft (32 m) with net pay of 79 ft (24 m). The well was cored and recovered 56 ft (17 m) of oil sand in the Gamba and Dentale formations, including 46 ft (14 m) in the Gamba – 30% more than other wells had found. In late 2001 Gabon gave Vaalco's subsidiary a five-year extension to its PSC for a 19.3 sq mi (50 sq km) production area on the Etame Block. The 1,158 sq mi (3,000 sq km) Etame block also holds the North and South Tchibala discoveries, which Vaalco was considering for development.

In early 2001 PanAfrican Energy signed a concession agreement to reactivate the Remboui field, which had been shut in due to low oil prices in 1998. Production from the first of 11 wells was to begin in mid-2001 and peak at 1,800 b/d in 2002. PanAfrican held 92% of the concession. Later in the year PanAfrican secured $30.6 million in joint venture financing for its projects in sub-Saharan Africa.

Also early in 2001, a 10½-mile (17-km), 6-inch, $2.2 million pipeline commissioned by PanAfrican began taking oil from Gabon's onshore Obangue field to nearby Avocette field. Previously, Obangue's 1,000 b/d of oil had been carried by river barges. As part of its plans to develop Obangue field, PanAfrican completed and began testing Obangue 5-H well and had started on Obangue 6-H in mid-year. PanAfrican Energy commissioned the pipeline and was planning development drilling of Obangue field.

In early 2001 Nuevo Energy Co. (Houston) spudded NAK No. 1 wildcat in the 1.9 million acre Accra-Keta permit off Gabon in 1,100 ft of water. A 25% interest in the permit was acquired from Nuevo by two South Korean companies, state-owned Korea National Oil Corp. and international conglomerate SK Corp. Nuevo retained operatorship and 50% interest.

A water depth record of 9,157 ft (2,791 m) was briefly held by Saipem SPA drillship S10000 working a wildcat off southern Gabon. The well, being drilled for Total Astrid Marin Gabon (acting for the Vanco Gabon Group), was the first assignment for the 97,000-tonne, $300 million vessel, which can store 140,000 bbl of oil and process 20,000 b/d. The ship was designed to operate in water as deep as 9,850 ft (3,000 m) and drill to 32,800 ft (10,000 m). Vanco Group partners were TotalFinaElf SA (28%), Unocal Corp. (25%), Vanco Energy Co. (22%), Kerr-McGee Corp. (14%), and Reading & Bates Development Co. (11%).

Energy Africa Gabon SA was to spud an exploration well in mid-2001 on its Ofoubou Ankani Block 25 miles (40 km) northeast of the TotalFinaElf SA-owned producing field, Rabi Kounga. A rig had been secured to drill the well starting in August.

King Resources, which acquired Nescor Energy Co. in mid-2001, was planning to develop the first field in the onshore Salsich Permit G4-187, with estimated reserves of 80 million bbl.

IVORY COAST

CAPITAL: ABIDJAN
MONETARY UNIT: CFA FRANC
REFINING CAPACITY: 65,200 B/CD
OIL PRODUCTION: 10,000 B/D
OIL RESERVES: 100 MILLION BBL
GAS RESERVES: 1.05 TCF

With substantial natural gas reserves and excess power generating capacity, Ivory Coast could become a significant regional energy supplier in sub-Saharan Africa. Oil firms were planning to spend \$7.3 billion in "second tier" West Africa projects before 2005, according to UK energy analyst Douglas-Westwood Ltd. About 10% of the total, or \$730 million, could go to Ivory Coast.

In early 2001, Ranger Oil Côte d'Ivoire SARL announced an oil find off Ivory Coast on Block CI-40. The Baobab 1X discovery well, drilled to 10,085 ft (3,074 m) total depth in 4,869 ft (1,484 m) of water, found oil in the Albian. Drill stem tests on selected intervals indicated a combined stabilized flow rate exceeding 6,700 b/d of 22-23º gravity crude with a gas/oil ratio of 290-320 scf/bbl. Partners in the block's PSC were operator Ranger (61%), Svenska Petroleum Exploration Production AB (29%), and Ivory Coast national oil company Petroci Holding (10%). Ranger is a wholly owned subsidiary of Canadian Natural Resources Ltd. (CNR), Calgary.

The Baobab discovery is 5 miles (8 km) south of Espoir field on Block CI-26, another of the four blocks Ranger operates off Ivory Coast. During 2001 Ranger/CNR began developing Espoir field, which had been shut down in 1988 after several years of production due to high operating costs. The Espoir Ivoirien FPSO vessel, owned byNorwegian company Prosafe ASA, was set to begin production from Espoir field off the Ivory Coast in first quarter 2002. The FPSO has capacities of 40,000 b/d oil production and 60,000 b/d water injection.

Espoir's oil would be exported by shuttle tanker and the gas piped onshore for power generation. Initial production was expected to be about 38,000 b/d of oil and 34 MMcfd of gas, with projected peak production of 28,000 b/d and 35 MMcfd. Espoir's recoverable reserves were estimated at 93 million bbl of oil and 180 bcf of gas, with a field life of 20-25 years.

CNR increased its equity interest in Block CI-26 in May 2001 to 58.67% by purchasing the assets from Addax Petroleum Côte d'Ivoire Ltd. (a wholly owned subsidiary of Addax Petroleum NV). Espoir field development was estimated to cost \$265 million, and the West African Development Bank had earlier approved a \$9.35 million loan to Petroci for the Block CI-26 project.

Additionally, Vanco Energy Co., Houston, announced plans to drill at least six deepwater exploration wells off Ivory Coast and other West African countries, starting in late 2002.

LIBYA

CAPITAL: TRIPOLI
MONETARY UNIT: DINAR
REFINING CAPACITY: 343,400 B/CD
OIL PRODUCTION: 1.36 MILLION B/D
OIL RESERVES: 29.5 BILLION BBL
GAS RESERVES: 46.4 TCF

As of early 2002, the US government was not prepared to lift its economic sanctions against Libya, despite talks between American, British, and Libyan officials. The US said Libya has not met its obligations to compensate families of the Lockerbie bombing victims and has not done enough to stop state-supported terrorism.

During 2001 US President George W. Bush signed a renewal of the Iran-Libya Sanctions Act, which seeks to punish foreign companies that invest in either country. Following the September terrorist attacks, US companies were again optimistic that new Libyan cooperation against terrorism might result in a diplomatic quid pro quo, but none materialized.

Occidental Petroleum Corp., Marathon Oil Co., and Conoco Inc. have Libyan oil fields. Earlier in 2001 the Libyan government threatened to sell those fields unless US policy changed, but it appeared to back down from that threat. The US Office of Foreign Assets Control in late 2001 said it would allow geologists from US oil companies to travel to Libya to monitor their concessions. Industry officials said production at the fields has collectively fallen to 300,000 b/d.

Murzuq basin discoveries

In March 2001 Repsol YPF SA said it had a second oil discovery at its B-1 wildcat in the Murzuq basin 500 miles (800 km) south of Tripoli, Libya. The well, the second Repsol drilled on the 1,660 sq mi (4,300 sq km) Block NC-186 in the Sahara Desert, produced 1,300 b/d of 40º gravity oil. Repsol was operator on Blocks NC-186, NC-187, and NC-190 on behalf of a consortium comprising Libya's National Oil Co. and OMV of Austria, TotalFinaElf, and Norsk Hydro affiliate Saga Petroleum AS. The wildcat was only 25 miles (40 km) from 160,000-b/d El Sharara field, which Repsol operates on Block NC-115.

In May 2001 Repsol-YPF confirmed its A-1 discovery, drilled in November 2000 on Block NC-186. The A-2 well found an oil column in the middle Ordovician sand of the Hawaz. On test, the well flowed unassisted at a rate of 2,670 b/d of 41º gravity oil through a 1-inch choke. The A-1 well had flowed 2,500 b/d of 41° gravity oil through a 3/4-inch choke. Production from the Murzuq basin started in December 1996 from El-Sharara field, which was producing 165,000 b/d of 44º gravity oil.

Late in 2001 Repsol announced its fourth discovery on the NC-186 block, a wildcat that found a significant oil column in the Ordovician Hawaz sandstone and tested flow rates up to 2,286 b/d of

41º gravity oil through a 1-inch choke. Four of five Repsol exploration wells drilled in different structures have been commercial oil discoveries, said Repsol. Preliminary evaluations indicated more than 250 million bbl of reserves. Development plans were filed with Libya's National Oil Co. for two of the discoveries. Like earlier wells, this discovery was close to the El-Sharara field, which has enough processing and pipeline capacity to handle additional crude. In early 2002 Repsol said that Blocks NC-187 and NC-190, covering 7,336 and 4,748 sq mi, respectively (19,000 and 12,300 sq km), have potential similar to NC-186.

Field development contracts

In early 2001 TotalFinaElf SA began development work on Field B of Block 137, an oil field in the Mediterranean 62 miles (100 km) off Libya in 295 ft (90 m) of water. The field was discovered in 1975 and confirmed in 1998. TotalFinaElf was operator with 37.5%; Libyan National Oil Corp. held 50% and Wintershall AG 12.5%. The development plan called for oil production to start in early 2003 from a platform connected to an FPSO vessel. Production was expected to plateau at 35,000 b/d.

In mid-2001 Compagnie des Pétroles Total Libye awarded two turnkey contracts for Field B development. Under one of the contracts, worth $54 million, a consortium would build and install a 1,700-tonne, 35,000-b/d oil production platform with a three-story deck and 6,200-tonne jacket. The consortium consisted of Bouygues Offshore, ROSBOS (a Rosetti Marino SPA-Bouygues joint venture) and Stolt Offshore, part of the Stolt Group. Delivery was expected in late 2002.

The second contract was for a cable drum, mooring system, and topsides for the FPSO vessel to develop the field. Exmar Offshore Co. awarded that contract to Doris Engineering SA, which is 51% owned by Bouygues. Equipment delivery was scheduled for mid-2002.

Also in mid-year Exmar awarded Spanish company IZAR Construcciones Navales SA a contract to build a 900,000 bbl FPSO vessel to be used at Aquitaine field in the Mediterranean Sea off Libya. The vessel would be delivered in January 2003. IZAR was created by the merger of Bazan and Astilleros Españoles.

In early 2001 International Petroleum Libya Ltd., a wholly owned subsidiary of Swedish company Lundin Oil AB, awarded contracts for processing facilities and a 58.7-mile (94.5-km) pipeline for En Naga North and West oil fields in area NC177 of Libya. The En Naga North and West field contains proven and probable reserves of more than 100 million bbl of oil.

Libya-Egypt pipeline

In August 2001, after more than 5 years of negotiations, the Arab Co. for Oil & Gas Pipelines was created to transport gas to Libya and crude to Egypt. Work was to begin immediately on fixing the route of the proposed pipeline. Libya's National Oil Co. would provide 50% of the capital for the new firm, with the other half from Egypt's oil company. Initial investment was said to be around $20 million. Libyan oil would be refined in Cairo and Alexandria.

MOROCCO

CAPITAL: RABAT
MONETARY UNIT: DIRHAM
REFINING CAPACITY: 154,901 B/CD
OIL PRODUCTION: 400 B/D
OIL RESERVES: 1.8 MILLION BBL
GAS RESERVES: 47 BCF

Morocco's insignificant energy reserves could grow with exploration being planned in 2001. In early 2002 King Mohammed VI announced economic and administrative plans aimed at promoting foreign investment. Morocco also made impressive gains in Robertson Research International's E&P ventures survey, indicating healthy spending increases ahead.

In October 2001, Morocco signed oil exploration deals with TotalFinaElf for the Dakhla offshore zone and with Kerr-McGee du Maroc Ltd. for the 44,000 sq mi Boujdour block lying along the Atlantic margin in water depths up to 10,000 ft or more. The Polisario Liberation Front disputed Morocco's rights to these Western Sahara areas and protested the contracts to the United Nations.

In early 2001 the Cap Draa Haute Mer exploration license was awarded to Enterprise Oil Co. The four-year license covers 4,635 sq mi (12,000 sq km) off western Morocco in water depths of 490-6,900 ft (150-2,100 m). Partners were operator Enterprise (30%), Energy Africa, Kerr-McGee, and Onarep. In March 2001 this group and Anschutz Corp. as operator signed two onshore reconnaissance contracts for 4,942 sq mi (12,800 sq km) in northern Morocco. Under the one-year contracts, the partners planned to perform field work and seismic reprocessing. A 48-inch gas pipeline runs through the area connecting Algeria and Spain.

In late 2001 Morocco claimed that an offshore exploration area granted by Spain to Repsol under nine, six-year licenses was in Moroccan territorial waters. The area lies close to Spain's Canary Islands and about 56 miles northwest of Morocco's coastal Tarfaya region.

Vanco Energy Co.'s proposed gas pipeline from West Africa to Europe would include first laying the pipe to potential gas fields off Morocco before continuing construction to the south. Additionally, Vanco Energy announced plans to drill at least six deepwater exploration wells off Morocco and other West African countries, starting in late 2002.

CMS Energy Corp.'s independent power unit, CMS Generation Co., reported commercial operation of Unit 4 at the $1.5 billion, 1,356-MW Jorf Lasfar power plant in Morocco. The milestone marked completion of construction at the coal-fueled facility, the largest independent power plant in Africa and the Middle East. The 348-MW Unit 4 began supplying commercial power in early 2001 to Office Nationale de l'Ectricite, the Moroccan national utility.

NIGERIA

CAPITAL: ABUJA
MONETARY UNIT: NAIRA
REFINING CAPACITY: 438,750 B/CD
OIL PRODUCTION: 2.1 MILLION B/D
OIL RESERVES: 24 BILLION BBL
GAS RESERVES: 124 TCF

When Nigeria's new cabinet was selected in early 2001, Rilwan Lukman kept his position as Presidential Adviser on Petroleum and Energy, but no one was appointed to head the Petroleum Ministry. In late 2001, the Nigerian federal government agreed with state and local governments to share oil profits estimated at $1.1 billion from the first half of 2001. Hearings were scheduled to determine whether the federal government will be allowed to seize control of the country's offshore oil and gas resources.

Nigeria remained heavily dependent on the oil sector, which accounted for nearly 80% of government revenues and over 90% of foreign exchange earnings. Privatization and other economic reforms moved slowly, and in early 2002 Nigerian National Petroleum Co. (NNPC) E&P subsidiary Nigerian Petroleum Development Co. was removed from the list of companies slated for privatization.

Bonga field development

Shell Nigeria Exploration & Production Co. Ltd. pursued development of deepwater Bonga oil and gas field in Block OPL 212 off Nigeria, which lies in 3,117-3,773 ft (950-1,150 m) of water 75 miles (120 km) off the Niger Delta southwest coastline and has estimated recoverable reserves of 600-750 million bbl. Production was scheduled to begin in 2003 and reach a peak of 350,000 b/d. Shell Nigeria Exploration & Production owned 55% of Bonga and was operating the field on behalf of NNPC. Esso Exploration & Production Co. Nigeria Ltd. had a 20% share, Nigeria Agip Exploration Ltd. 12.2%, and Elf Petroleum Nigeria Ltd. 12.5%.

Shell's Bonga development envisions high-rate horizontal and conventional wells produced through subsea well clusters connected to an FPSO. Initial plans called for producing the unconsolidated sand reservoirs through 20 producing wells (conventional ones frac-packed and horizontal ones openhole gravel-packed). To maintain reservoir pressure, Shell planned to start water injection from the beginning of the project through 16 wells with a 300,000 bw/d capacity.

In early 2001 Shell Nigeria E&P, on behalf of NNPC, awarded a string of major contracts to four European companies for the $2.4 billion Bonga project. The FPSO vessel's 17,000-tonne topsides was to be designed and built by AMEC. Shell required that the facility process 225,000 bo/d, 100,000 bw/d, and 170 MMcfd of associated gas, part of which will go to Nigeria's Bonny LNG terminal. The FPSO would have a storage capacity of 2 million bbl. The topsides were to be installed on the FPSO's $120 million, 300,000-tonne hull when it arrives from South Korea's Samsung Heavy Industries' yard in late 2002.

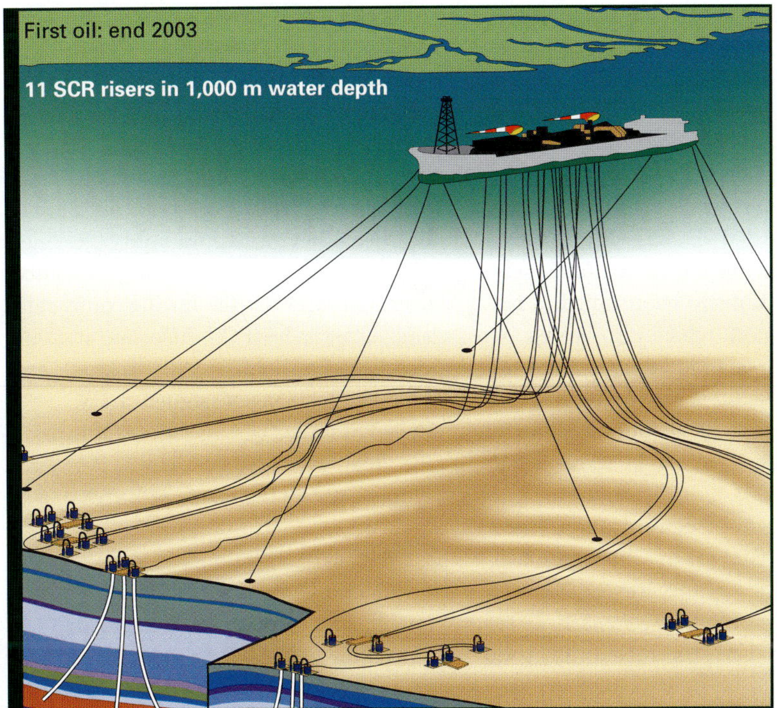

Development of Bonga field off Nigeria was planned to include high-rate horizontal and conventional wells produced through subsea well clusters connected to an FPSO with 2 million bbl storage capacity

Like other FPSOs operating off West Africa, the Bonga system will use lighter spread mooring instead of the heavier turret type. Lighter mooring is sufficient due to the area's milder environment – average wind speed of 82 ft/s (25 m/s), maximum wave height of 23.3 ft (7.1 m), and no hurricanes.

Bonga's gas gathering system would be laid by a consortium led by Stolt Offshore under a contract with Shell Petroleum Development Co., Nigeria announced in mid-2001. Contractor Stolt's partner was DSNL, a subsidiary of the Nigerian Adamac Group. Stolt's share of the contract was $135 million. Shell's offshore gas gathering system included installation of 55 miles (88 km) of 24-inch line from the Forcados-Yokri area to the Bonga riser platform and 11 miles (18 km) of 16-inch line from South Forcados to the same platform.

Also, the companies planned to install a 164-mile (264-km), 32-inch trunkline (including 5 miles, or 8 km, onshore) to connect to the Bonny Island LNG plant. Stolt said this would be the largest trunkline constructed off West Africa so far. Work was to begin in third quarter 2002. Adamac would build the onshore receiving facilities and provide pipe coating, storage, and loading services from its base in Port Harcourt.

In mid-2001 Coflexip Stena Offshore (CSO) International was awarded a portion of the mooring and installation package for the Bonga field development. The contract included installation of three dynamic 19-inch internal diameter, 1.4-mile (2.3-km) flexible oil offloading risers between the FPSO vessel and the single point mooring buoy. CSO said this would be the first use of flexibles of such length as export risers. CSO would also install more than 34 miles (55 km) of gas lift risers and control umbilicals, plus five large manifolds. The offshore work was planned for summer 2003.

In May 2001 Shell announced a second discovery on the block, Bonga South West, also with an estimated 600 million bbl of reserves. Drilling of the second area was to be completed in 2003 with first oil production anticipated for 2005.

Amenam/Kpono field development

After years of delay, Elf Petroleum Nigeria Ltd.'s development of Amenam/Kpono field swung into action during 2001, with production planned to start by mid-2003 at 125,000 b/d. The field, the largest shallow water development in the Niger Delta, straddles Blocks OML 99 and 70 in 131 ft (40 m) of water 19 miles (30 km) from the Nigerian coast. Production from Amenam/Kpono was expected to total 500 million bbl over 25 years, from reserves estimated at more than 1 billion boe. Elf operates the field for NNPC and Mobil Producing Nigeria Unltd., a subsidiary of ExxonMobil Corp.

Elf's development plans called for 31 wells: 18 production, 6 gas injectors, 3 deviated water injectors, and 4 water producer wells to provide injection water from the shallow water-bearing sands. Most of the 614 MMcfd (16.5 million cu m/day) of associated gas would be reinjected. Saibos CML had been selected previously to supply one production platform, two tripods, one flare, and three associated bridges.

All the wells would be drilled with two jack-up rigs. In late 2001, Elf committed to two-year contracts with Global Marine Inc. for the Glomar Adriatic I and Glomar Baltic I jack-ups. The proposed $121 million in contracts would include rig upgrades to cost a total of $23 million. The jack-ups were set to begin work on Amenam/Kpono field in second quarter 2002. Both contracts would allow extensions.

Amenam/Kpono's reservoirs lie at depths greater than 11,480 ft (3,500 m) below sea level, requiring thermally insulated, weight-coated pipe for the development. Elf was to begin taking delivery of the pipe in September 2001. Coating and insulation were completed by Logstor Oil & Gas, while weight coating was performed at Vibrodens AS, both with operations in Denmark. Installation was to begin in early 2002.

The 25-mile (40-km) subsea pipeline would connect platforms installed at water depths of 131-230 ft (40-70 m). The pipeline consists of two piping systems: a 16-inch line (21.6 miles or 34.8 km) and a 24-inch line (3 miles or 4.8 km) connecting the platforms. Field joint coatings would be performed on the lay barge by injecting foam, with installation by S-lay from Stolt Offshore SA's lay barge Polaris.

NIGERIAN MAJOR GAS UTILIZATION PROJECTS

Projects	Location	Promoter/ operator	Status
I Gathering & conditioning			
1. Alakiri gas plant	Alakiri, Rivers State	Shell	Operational
2. Soku gas plant	Soku, Bayelsa State	Shell	Operational
3. Obigbo Agg	Obigbo, Rivers State	Shell	In progress
4. Belema Agg	Belema, Rivers State	Shell	In progress
5. Odidi Agg	Odidi, Delta State	Shell	In progress
6. Forcados/Yokri Agg	Forcados, Delta State	Elf	In progress
7. Offshore gas gathering project	Offshore Delta/Bayelsa State	Elf	Proposed
8. Obite gas plant	Obite, Rivers State	Elf	Operational

Source: Chex & Associates

Exploration and development

Nigerian waters were slated to receive at least 10 of the 34 FPSOs being designed or planned worldwide in mid-2001; only Angola would get more. Two FPSOs for Nigeria were in the bid/design stage and another eight being planned or studied. In addition to Bonga, Nigerian fields to be developed with FPSOs included Agbami, Abo, Akpo, Bosi, Chota, Ikija, N'golo, Nnwa, and Ukot.

Shell and its partners expected to spend $1 billion to develop EA field (Block OML 79), which lies 43 miles off Forcados in the Niger Delta and holds an estimated 350 million boe in reserves. When fully developed, the field was anticipated to yield 120,000 b/d and 100 MMcfd of gas. EA field was due on stream in the second half of 2002. In early 2001 Shell contractor Stolt Offshore Services awarded J. Ray McDermott Middle East a subcontract to build three decks for platforms to be installed in EA field. McDermott, which also would be responsible for loadout and transportation of the structures, started fabrication work at its yard in Jebel Ali Free Zone in Dubai. The decks were to be transported in December 2001.

Oil production began from Okono field in 330 ft (100 m) of water on OML 119, announced Agip Energy & Natural Resources (part of Eni SPA) in early 2002. In 2001 Agip had contracted for two FPSO vessels to be used on Okono and Okpoho field. The $356 million contract was awarded to a joint venture of Eni unit Saipem SPA and Single Buoy Moorings Inc.

In late 2001, Eni also signed a PSA with NNPC for offshore Block 244 off Nigeria, operated by Eni affiliate Nigeria Agip Exploration with 90% interest. A seismic survey and drilling of two exploration wells were planned. Block 244 is 124 miles (200 km) south of Nigeria's Brass River crude terminal in 4,900-9,800 ft (1,500-3,000 m) of water. Eni was working on a series of projects to exploit Nigerian associated gas reserves, including gas liquefaction plants and power generation plants.

Other Nigerian fields with production being planned included ExxonMobil's deep offshore Erha field in Block OPL 209 and its Yoho and Awawa discoveries in shallow-water Block OML 104 (both slated for late 2004 production) and Agip's Abo discovery in Block OPL 316 (late 2002).

In late 2001, NNPC signed a PSA with Chevron Nigeria Deepwater Ltd., Shell Nigeria Offshore Prospecting Ltd., and Brasoil Services Co. for 10-year oil exploration lease and a 20-year oil mining license. Chevron would hold 50%, Shell 35%, and Brasoil 15%. The production sharing ratio was 70% for the contractors and 30% for NNPC. Also late in 2001, Petroleo Brasileiro SA (Petrobras) and partner Horizon Oil signed a PSC with the government of Nigeria for OPL 324 in the deepwater Niger Delta. Petrobras would operate the block with 75%.

Natural gas use

Nigeria's gas reserves will last much longer than its oil reserves (120 years vs. 35), so the country was planning in 2001 to exploit its nonassociated gas, which accounts for half of gas reserves, and stop flaring its associated gas production. Several gas gathering and conditioning projects were operating or being constructed or planned.

Interest surged in using natural gas in independent power plants and as an industrial fuel. Several gas distribution schemes were planned to increase Nigeria's domestic gas consumption, which has been negligible. The government also envisioned gas exports from Nigerian fields to Benin, Togo, and Ghana through the West African Gas Pipeline being sponsored by ChevronTexaco, Shell, and the four countries' state companies. As of early 2002, construction of the 373-mile (600-km), 400-MMcfd pipeline was to start in 2003 and operation before the end of 2004.

These gas export plans represented the later phases of Nigeria's Escravos gas project, which would also include using gas as a feedstock for GTL production in a plant scheduled to come on line in 2005.

In late 2001, Chevron Nigeria Ltd. (a joint venture of ChevronTexaco and NNPC) awarded Foster Wheeler Ltd. a $20 million engineering and design contract for a GTL plant to be built in Escravos. The plant would produce 34,000 b/d of synthetic fuel and naphtha for export. The plant would be the first to use the Sasol slurry phase distillate process

Nigeria in 2001 sought to revive a long-dormant project to build a natural gas trunkline to Algeria. The 2,485-mile (4,000-km) line would cost $7.5 billion and would move Nigerian gas from Abuja fields to the northwest Algerian port of Beni Saf for export to Europe. Both countries pledged to support the project with an extension of the trans-Saharan road and construction of a fiber optic line paralleling the pipeline to improve the area's telecommunications. The World Bank planned a feasibility study for 2002 and also would seek additional financing. To prepare for the project, NNPC was to begin extending the

500-MMscfd line from Ajaokuta in Kogi state to Kaduna state.

LNG developments

In early 2001 Nigeria initiated plans to build an LNG plant west of the Niger Delta to rival the existing Bonny complex. A one-year feasibility study was supported by a group of multinationals including leader ExxonMobil Corp., Conoco Inc., and ChevronTexaco.

In April 2001 Nigeria Liquefied Natural Gas Co. (NLNG) contracted for expansion of Bonny's two-train, 5.9 million tpy LNG plant at Bonny Island. Construction of a third train was under way, with start-up planned for late 2002. Partners in the LNG project planned to invest $7.5 billion to add three more production trains. After completion of the third train and associated facilities, NLNG expected it would be able to process 1.5 bscfd of gas, more than half of the gas that was being flared.

In September 2001 Phillips Oil Co. (Nigeria) Ltd. announced development of another LNG plant in Nigeria with coventurers NNPC and Nigerian Agip Oil Co. The plant, expected to be on stream by 2007, would be located offshore in the Niger Delta near the existing Brass River terminal. When completed, the facility would be the world's first offshore LNG plant.

Refining

Ventech Engineers won a contract in early 2001 to design Nigeria's first privately owned, privately financed oil refinery. The 12,000-b/d facility, to be built adjacent to ExxonMobil's Qua Iboe terminal, would process Qua Iboe crude into naphtha, gasoline, kerosene jet fuel, diesel, gas oil, and fuel oil. Nigeria's four refineries during 2001 were processing 330,000 b/d of crude, said NNPC, the highest refining production level in more than a decade. Still, they cannot meet domestic demand for gasoline, due to increased consumption and smuggling.

In early 2002 the Nigerian government said it would seek international bids for the construction of new refineries as part of its liberalization of the downstream sector. The government could license about five refineries with a capital outlay of at least $1.8 billion.

SOUTH AFRICA

CAPITAL: PRETORIA/CAPE TOWN
MONETARY UNIT: RAND
REFINING CAPACITY: 468,547 B/CD
OIL PRODUCTION: 22,000 B/D
OIL RESERVES: 15.7 MILLION BBL
GAS RESERVES: 1 BCF

The year 2001 was a productive one for companies exploring and exploiting petroleum resources in South Africa, with development projects moving forward and licenses being signed. In midyear, Pioneer Natural Resources Co. and its partners in offshore Sable oil field sanctioned its development, and Soekor E&P Pty. Ltd. planned to spud development wells in late 2002 and begin production in 2003. Pioneer owned 35% of Sable, Soekor 60%, and Petroleum Ltd. 5%. (This Sable development should not be confused with the Sable project offshore Nova Scotia, Canada.)

The field, in 320 ft (100 m) of water 59 miles (95 km) off the country's southern coast, would be developed using six subsea wells (four producers, one gas injector, and a water injector) tied back to Bluewater (UK) Ltd.'s FPSO vessel Glas Dowr, which can process 60,000 b/d of oil, reinject 80 MMcfd of gas, and recover gas liquids. Sable field should produce up to 40,000 b/d of oil initially and use associated gas reinjection to improve liquids recovery, but the gas could later be produced.

Bluewater, which would also operate associated shuttle tankers, hired Coflexip Stena Offshore (CSO) International for the subsea portion of the contract, including installation of 9 miles (15 km) of flexible flowlines and production risers and over 5 miles (9 km) of umbilicals. In late 2001, Soekor awarded FMC a contract to provide subsea equipment and technology solutions for the Sable project.

Boomslang appraisal

Pioneer was also planning to drill an appraisal well to its Boomslang discovery along the southern flank of the Bredasdorp basin. (The South African Boomslang project should not be confused with a similarly-named project in the US Gulf of Mexico.) Pioneer's South African subsidiary found oil and gas with its first well on Boomslang prospect on Block 9 off South Africa in 450 ft of water. The Boomslang No. 2 well is 1.2 miles (2 km) southwest of the discovery well. The company also was to acquire 461 sq mi (1,193 sq km) of 3D seismic data covering Boomslang prospective trend. Pioneer was operator with 49%, and Soekor owned 51%.

The E-DQ-1 discovery well cut 108 ft (128 gross ft) of pay in synrift shallow-marine sands consisting of an oil leg under a gas column at 6,600 ft. On test, a portion of the oil leg interval flowed up to 3,120 b/d of oil and 2.6 MMcfd of gas over a 48-hr period on varying choke sizes. The overlying gas column flowed at a constant rate of 24 MMcfd of gas and 300 b/d of condensate for 5 days, Pioneer said.

After completing Boomslang discovery well tests, Pioneer used the same rig to drill the E-BB2 gas well 75 miles (120 km) south of Mossel Bay. The well found more than 325 ft of gas-bearing sand in five primary zones. Of the three zones that were perforated and tested, one flowed at a sustained 10 MMcfd of gas and 800 b/d of liquids. Pioneer was to act as the operator for the drilling and testing of this gas well and would earn a 40% working interest in the E-BB prospect. Soekor E&P has 60% and is the designated operator.

Other exploration

Also during 2001, the Petroleum Agency of South Africa approved the award of an exploration sublease to Ranger Oil (South Africa) Ltd. for Block 11B/12B off the south coast. Ranger was to acquire and process at least 1,158 sq mi (3,000 sq km) of seismic data, the first significant such collection off the deepwater south coast, in water depths of 1,640-6,562 ft (500-2000 m). Ranger was a wholly owned subsidiary of Canadian Natural Resources Ltd.

In early 2002, South Africa granted final approval to a prospecting sublease agreement for Block 3B/4B to Global Offshore Oil Exploration South Africa Pty. Ltd., a wholly owned affiliate of Global Energy Holdings LLC (Littleton,

CO). Previously, Global conducted a technical cooperation agreement for the area. The 7.5-year agreement would involve acquisition of new seismic data during the first period. Block 3B/4B covers 7.1 million acres off western South Africa in 984-3,937 ft (300-1,200 m) of water.

Global indicated that seismic reprocessing had confirmed a very large prospective structure, dubbed Pinotage prospect, which apparently covers 67,400 acres in 1,968 ft (600 m) of water. Petroleum Geo-Services ASA and Petroleum Agency SA were planning to acquire and market 2D seismic data and interpret new and reprocessed data.

Ibhubesi gas development

With completion of the fourth and final well in its drilling program in Ibhubesi field off South Africa, Forest Oil Corp. (Denver) said in mid-2001 that there are sufficient reserves and deliverability for development of a natural gas market in that country. Initial studies identified potential natural gas markets with takes in the range of 30-200 MMcfd. Ibhubesi field is off the mouth of the Orange River near the border with Namibia.

The final A-Y1 well tested three zones at an aggregate rate of 71.4 MMcfd of natural gas and 1,376 b/d of condensate – or 80 MMcfd of gas equivalent – on an 80/64-inch choke. Three of the four wells drilled under that program, including the A-K1 discovery well, successfully tested at a combined total of 177 MMcfd of gas and 1,900 b/d of condensate, or 190 MMcfd of gas equivalent. Forest Oil was operator with 70% working interest; partners included Anschutz Overseas Corp. and Mvelaphanda Holdings.

Forest Oil planned to book proved reserves and negotiate sales contracts, with delivery to begin as early as 2004, through subsea development. If delineation drilling proves expectations of rich reserves (possibly up to 15 tcf), Forest Oil could develop its own GTL facility and supply gas to the Mossgas GTL plant being considered for Mossel Bay. Mossgas (Pty.) Ltd., South Africa's state-owned gas-to-liquids producing company, expected to deplete its current gas supplies by 2007.

Earlier in 2001, Forest Oil funded a feasibility study by Rentech using that company's technology for possible development of the Ibhubesi field. Rentech estimated the capital cost of a GTL plant in South Africa using Rentech's technology at $252 million for a 10,000-b/d plant, including feed preparation, steam reformer plant, CO^2-removal unit, hydrogen membrane system, Fischer-Tropsch reaction system, power systems, and offsites. Adding facilities to upgrade the Fischer-Tropsch products might require $110 million, depending on the desired product slate.

Other gas, GTL projects

In mid-2001, Sasol Ltd. signed a $1.2 billion gas pipeline deal to import natural gas from Mozambique. The project consists of a gas field development in Mozambique, a 26-inch (660-mm) pipeline extending 537 miles (865 km) to Secunda in South Africa, the conversion of Sasol's current gas pipeline network, and the supply of gas to Sasol customers in South Africa. Earlier, Sasol had agreed to to provide access to Temane and Pande gas fields as well as the granting of pipeline rights in Mozambique. If the project is approved, construction would take three years and completion would be in the first half of 2004.

Sasol will enable the project by converting its existing pipeline gas market to natural gas, switching its Sasolburg plants from coal to gas, and taking gas into Secunda to supplement coal-based growth there. Sasol planned to grow its market to 120 million gigajoules/yr. The South African parliament was expected to consider legislation regulating natural gas sales.

In June 2001 Mossgas awarded Foster Wheeler South Africa (Pty.) Ltd. a contract for a $16.8 million low aromatics distillate plant in South Africa. The plant, to be built at Mossel Bay, would produce 70,000 tpy of low aromatic diesel and kerosene for the export market. The plant is based on an Institut Français du Pétrole license. The project also includes new tanks at a Mossgas refinery and tank farm, plus a 1.2-mile (2-km) subsea pipeline from the tank farm to an existing conventional buoy mooring for loading into tank ships. Work on the project began in 2001, with production scheduled for late 2002.

Refining and marketing

In early 2001, economic analysis at South African Petroleum Refineries (Pty) Ltd.'s (SAPREF) refinery revealed savings of $1.5 million in one year directly attributable to the use of an advanced blend-scheduling decision-support tool called MPMP. The refinery is a Royal Dutch/Shell and BP joint venture in Durban, South Africa.

In early 2002, Enersul Technologies (Calgary) commissioned a second sulfur granulating plant at Sasol Pty.'s Secunda 87,547-b/d refinery in South Africa.

In mid-2001, Shell South Africa tentatively agreed with Thebe Investment that subsidiary Tepco Petroleum would acquire a stake of up to 25% in Shell's South African oil products marketing businesses. It was intended that Thebe would also be a shareholder in the Shell unit, which would hold all of the retail and commercial marketing activities of Shell South Africa and Tepco.

SUDAN

CAPITAL: KHARTOUM
MONETARY UNIT: POUND
REFINING CAPACITY: 121,700 B/CD
OIL PRODUCTION: 200,000 B/D
OIL RESERVES: 563 MILLION BBL
GAS RESERVES: 3 TCF

By year-end 2001, Sudan had increased its oil production to 230,000 b/d and was planning to go for almost 300,000 b/d in 2002 following the development of two oil fields in southern Sudan. The Pantio field in the Wahda region was expected to add 30,000 b/d and the Manka field another 30,000.

In May 2001 Sudan awarded an exploration and PSA for Block 5B covering 7,768 sq mi (20,119 sq km). Petronas Carigali Overseas Sdn. Bhd. and Sudapet Ltd. were designated joint operators with 41% and 10% interest, respectively. Lundin Oil AB's subsidiary Lundin Muglad Ltd. had 24.5% and OMV AG's OMV (Sudan) Exploration GMBH 24.5%.

The block adjoins Block 5A, where Lundin discovered oil earlier in 2001.

The Thar Jath discovery was drilled to 5,971 ft (1,820 m) total depth and encountered 207 ft (63 m) of net pay over two sand reservoirs, the Bentiu and the Aradeiba. The well flowed at 4,260 b/d from three drill stem tests in both zones. Lundin was operator of Block 5A with a 40.4% interest in; Petronas had 28.5%, OMV 26.1%, and Sudapet 5%.

Block 5A also adjoins Blocks 1 and 2, which produce more than 200,000 b/d of oil. All of the blocks lie in Sudan's Muglad basin, estimated to contain total reserve potential of 10-15 billion bbl. The Thar Jath structure is on trend with prolific oil fields operated by Greater Nile Petroleum Operating Co., a consortium led by Talisman Energy Inc. (Calgary) and including a unit of Malaysian state company Petronas, China National Corp., and Sudapet. The Thar Jath discovery confirmed that the prolific oil field trend extendes into Block 5A, said Lundin.

In May 2001 Lundin Oil followed up the discovery with an appraisal well drilled 1.9 miles (3.1 km) from Thar Jath-1 which established 320 ft (100 m) of net pay. The appraisal well flowed 2,000 b/d from separate tests in the Bentiu and the Aradeiba sands. In November 2001 Lundin announced plans to drill two more appraisal wells near the Thar Jath oil discovery and at least one wildcat on Block 5A. The company hoped to complete a conceptual development study on Thar Jath in early 2002.

In early 2001 Talisman announced a large oil discovery at the Shelungo wildcat, the first well drilled on the west side of its 12.2 million acre Greater Nile Oil Project concession. The project was producing 211,000 b/d of crude from six fields for export through a 932-mile buried pipeline to a modern offshore loading and storage facility at Port Sudan on the Red Sea. Other discoveries were scheduled for development. Talisman was planning to spend $120 million (Canadian) during 2001 and increase its share of production to 65,000 b/d in 2002 from 2001's 52,750 b/d. The company said it would drill 17 exploration wells and 25 development wells on Sudanese acreage during 2001.

In mid-2001 Talisman started up Bamboo and Kairat fields. Bamboo was expected to be producing 30,000 b/d during 2001, while Kairat initially produced 10,000 b/d. In August 2001 Talisman said there was no damage to the Heglig oil field in central Sudan after an attack by rebels opposed to the Khartoum regime. The company said production was briefly suspended and then resumed after the attack by a commando unit of the Sudan People's Liberation Army (SPLA). Reportedly there were no injuries to company or security personnel and no damage to facilities, and the Heglig facility was operating normally and was secure. Talisman was operator with a 25% interest in the field, which was producing about 225,000 b/d of crude oil.

TUNISIA

CAPITAL: TUNIS
MONETARY UNIT: DINAR
REFINING CAPACITY: 34,000 B/CD
OIL PRODUCTION: 66,000 B/D
OIL RESERVES: 307.6 MILLION BBL
GAS RESERVES: 2.75 TCF

In late 2001 the IMF issued a generally positive report on Tunisia's economy, stressing the importance of continued privatization. The government was in the process of privatizing some 40 companies, begun in 2000. Tunisia agreed to set up a free trade zone with Egypt, Jordan, and Morocco, ahead of the 2010 target for trade barriers to end.

Due to declining oil production and higher domestic demand, Tunisia became a net oil importer in 2000 for the first time in over 20 years. However, the country's oil minister said that startup of Izis field in the southern Gulf of Gabes would boost oil production in 2001. In September 2001 Enterprises Tuniseienne d'Activites Petrolieres (ETAP) opened tenders for 20 blocks including central and southern onshore blocks and northern and eastern offshore blocks.

In May 2001 a unit of Pioneer Natural Resources Co. began building a position in Tunisia. The company planned to earn a 50% interest in three Tunisian permits from Eurogas Corp. (Calgary) by drilling two wells within two years, spending about $5.2 million. The Bazma, Jorf, and El Hamra permits cover 2.7 million acres in the TAGI sand play of the Ghadames basin. Pioneer Natural Resources Tunisia Ltd. had an option to operate the three permits.

Algerian discoveries Hassi Berkine and Qoubba are on trend with El Borma TAGI field in Tunisia, said Pioneer, and suggest the potential for further Tunisian finds. Pioneer had already spudded Douz No. 1 well in Bazma permit; however, it was dry and will be abandoned, said Eurogas Corp. Two other wells were planned for the area. Pioneer called the area's pipeline infrastructure "well developed" and said it has excess capacity.

In July 2001 Pioneer and two partners in Anaguid permit in the Ghadames basin in southern Tunisia agreed to jointly acquire a 45.83% interest in the permit from a subsidiary of Coho Energy Inc., which said the consideration was $200,000 in cash and assumption of $7 million in obligations under the permit. The deal was approved by a US Bankruptcy Court, but was subject to Tunisian approval. Nuevo Anaguid Ltd. filed under Chapter 11 of the US Bankruptcy Code in February 2001. Pioneer Natural Resources Co. agreed to acquire 30%. The holding of Nuevo Energy Co. would increase to 22.5%. The other partner was Anadarko Tunisia Anaguid Co.

Pioneer planned to join the companies in exploring the 1.1 million acre permit. Pioneer would pay its share of at least one exploration well, reimburse its share of 2001 seismic and geologic costs, and assume other Coho obligations relating to the permit. The group was to target the TAGI sand play with its next well.

In November 2001 Kerr-McGee Corp.'s UK unit said it would use Fred. Olsen Energy ASA's Bredford Dolphin semisubmersible to begin a two-well contract off Tunisia with Nimir Petroleum Co. North Africa Ltd.

In early 2002 Centurion Energy International Inc. (Calgary) resumed oil production from Al Manzah field in Northern Tunisia. Al Manzah-3 was converted to a water injection well, and Al Manzah-2 was producing 2,000 b/d of oil. IPE

FORMER SOVIET UNION

ARMENIA

CAPITAL: YEREVAN
MONETARY UNIT: DRAM
REFINING CAPACITY: NONE
OIL PRODUCTION: NONE
OIL RESERVES: NONE
GAS RESERVES: NONE

Armenia's economy continued to recover during 2001, with gross domestic product (GDP) growth of 6% estimated. The country also managed to slash inflation and privatize most of its medium-sized and smaller companies. Its oil consumption continued to dwindle, and Armenia's chance to be on the route for a new oil pipeline was eliminated with the selection of Tbilisi, Georgia, as a transit point for oil exports from the Caspian Sea region to Turkey and beyond.

All of Armenia's natural gas came from Turkmenistan via the private gas trading company Itera. In early 2001, Armenia and Russia reached an agreement on Armenia's debt for gas shipments dating back to 1999, which Itera supplied on behalf of Russia's Gazprom. Itera and Gazprom owed Armenia their contribution to the incorporation capital of Armrosgazprom, which handles Armenia's gas distribution. In mid-year, Itera agreed to write off Armenia's gas debt in exchange for the Armenian government transferring its gas pipeline property to the joint enterprise towards Russia's share. Armrosgazprom was to invest $6 million from its own funds to rehabilitate Armenia's gas industry.

To diversify its gas supplies, Armenia pushed ahead during 2001 with plans to build an 87-mile (140-km) pipeline to import gas from northern Iran, as well as link Iranian and Russian gas transport systems to allow possible Iranian gas exports to Europe through the Russian grid. In early 2002, Armenia and Iran said they will invite Russian and French companies to invest in the project, estimated to cost $120 million. Gaz de France had announced its intention to invest in the pipeline, and an international consortium was being formed for construction. The pipeline would be capable of supplying about 56 MMcfd of gas (1.5 million cu m/day).

AZERBAIJAN

CAPITAL: BAKU
MONETARY UNIT: MANAT
REFINING CAPACITY: 441,808 B/CD
OIL PRODUCTION: 300,000 B/D
OIL RESERVES: 1.178 BILLION BBL
GAS RESERVES: 4.4 TCF

Azerbaijan's hope for future economic growth rested on successful development of its oil and gas resources in the Caspian Sea region. In April 2001, in an effort to boost foreign investment, Azerbaijan established the Fuel and Energy Ministry to oversee the sector. Azerbaijan was expecting to channel more than $52 billion between 2008-15 into a new Azerbaijan State Oil Fund through its production sharing agreements (PSAs) on the giant Caspian Sea Azeri-Chirag-Guneshli oil fields and Shah-Deniz gas condensate field, where 18 PSA contract sites were under consideration for development.

Oil fund Executive Director Samir Sharifov emphasized the importance that Azerbaijan have an oil wealth management strategy that would create conditions for permanent income from nonoil sectors for the country once the oil has been exhausted. The fund's objective was to finance investment projects aimed at construction and reconstruction of infrastructure with nationwide significance for future generations in Azerbaijan.

In early 2002 US President George W. Bush conditionally waived a decade-old sanction that had blocked official aid to oil-rich Azerbaijan. US oil companies

WESTERN FSU
Barents Sea
Kara Sea
NOVAYA ZEMLYA
Kolguyev Island
Kola Peninsula
White Sea
FINLAND
RUSSIA
FORMER SOVIET UNION
KAZAKHSTAN
UZBEKISTAN
TURKMENISTAN
BELARUS
UKRAINE
ROMANIA
BULGARIA
TURKEY
IRAN
ESTONIA
LATVIA
LITHUANIA
MOLDOVA
GEORGIA
ARMENIA
AZERBAIJAN
Black Sea
Caspian Sea
Aral Sea
Aegean Sea
Gulf of Finland
0 100 200 300 Miles
0 200 400 Km
Moscow
St. Petersburg (Leningrad)
Helsinki
Kiev
Minsk
Arkhangel
Murmanskoye
Naryan-Mar
Vorkuta
Salekhard
Nadym
Urengoi
Surgut
Nizhnevartovsk
Usinsk
Pechora
Ukhta
Serov
Tyumen
Omsk
Kurgan
Chelyabinsk
Sverdlovsk
Perm
Kirov
Izhevsk
Ufa
Kazan
Samara (Kuibyshev)
Saratov
Volgograd
Astrakhan
Atyrau
Orenburg
Orsk
Baku
Grozny
Tbilisi
Yerevan
Ankara
Istanbul
Tehran
Ashgabat (Ashkhabad)
Bucharest
Sofia
Odessa
Novorossiysk
Krasnodar
Rostov
Kharkov
Riga
Tallin
Vilnius
Yaroslavl
Kirishi
Novgorod
Pskov
Smolensk
Kursk
Voronezh
Ryazan
Vladimir
Nizhniy Novgorod
Cherepovets
Petrozavosk
Vyborg
Primorsk
Magnitogorsk
Karaganda
Atbasar
Beineu
Kungrad
Aktau (Shevchenko)
Krasnovodsk
Batumi
Tuapse
Samsun
Burgas
Constanta
Kherson
Kremenchug
Lisichansk
Lugansk
Zaporozhye
Brody
Lvov
Dolina
Uzhgorod
Mozyr
Brest
Grodno
Polotsk
Novopolotsk
Unecha
Orel
Yelets
Michurinsk
Syzran
Ul'yanovsk
Nizhnekamsk
Salavat
Ishimbay
Krasnovolsky
Novyy Port
Punga
Nizhnyaya Tura
Tabriz
Bandar Shah
Astara
Nakhichevan
Evlakh
Kulsary
Alexandrou Gai
Ostrogoshsk
Kishinev
Sochi
Simferopol
Torzhok
Valdai
Kaluga
Kaunas
Tartu
Ventspils
Vinnitsa
Stavropol
Tikhoretsk
Neftekamsk
Okhansk
Admiralteyskaya
Shtockmanovskoye
Tulomskaya
Severo-Kildinskoye
Leningradskoye
Rusanovskoye
Bovanenkovskoye
Yamburg
Medvezhye
Samotlor
Tengiz
Karachaganak
Orenburgskoye
Astrakhan
Shebelinka

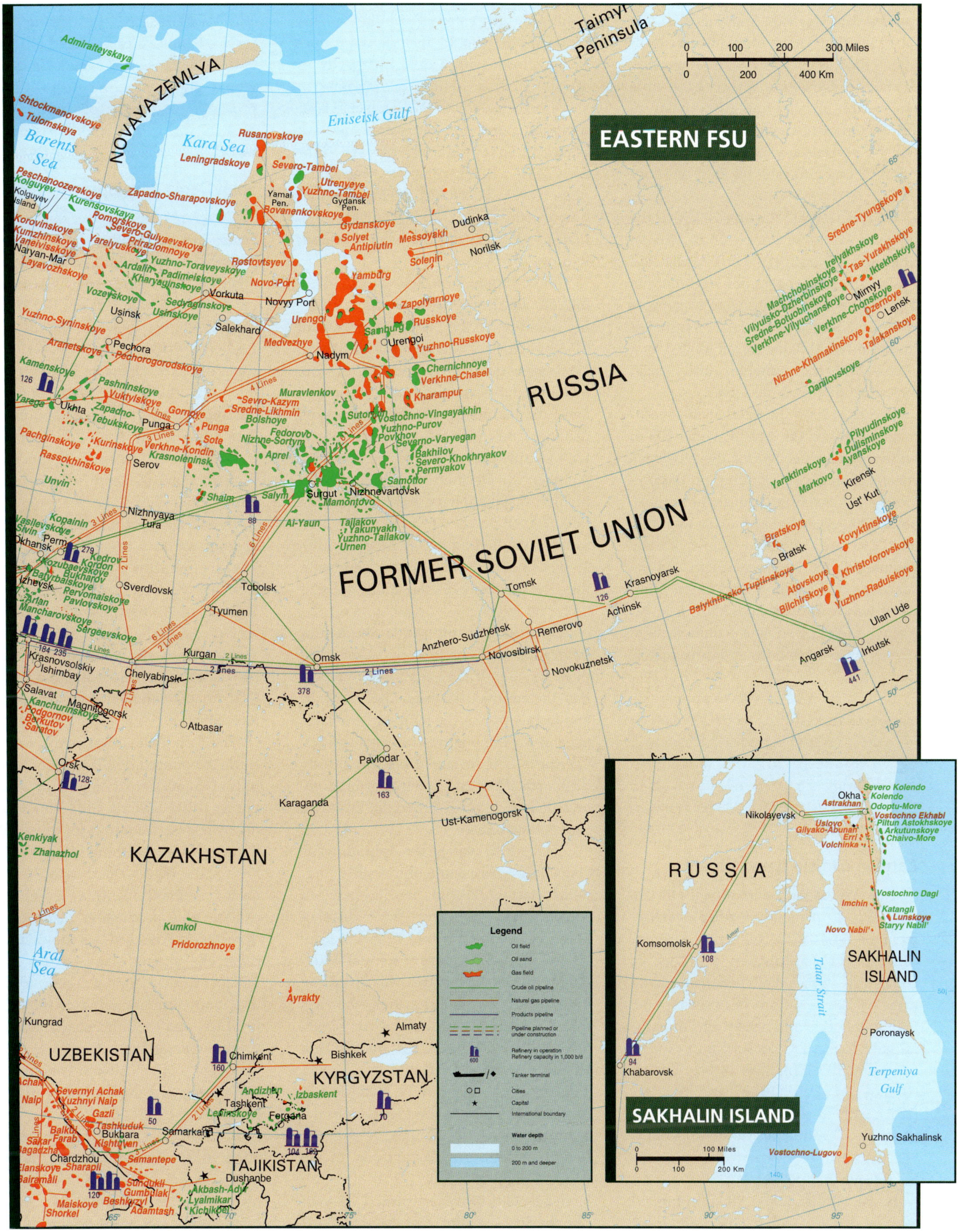
EASTERN FSU
FORMER SOVIET UNION
RUSSIA
KAZAKHSTAN
UZBEKISTAN
KYRGYZSTAN
TAJIKISTAN
NOVAYA ZEMLYA
Taimyr Peninsula
Barents Sea
Kara Sea
Eniseisk Gulf
Aral Sea
0 100 200 300 Miles
0 200 400 Km
Admiralteyskaya
Shtockmanovskoye
Tulomskaya
Rusanovskoye
Leningradskoye
Severo-Tambei
Utrenyeye
Yuzhno-Tambei
Yamal Pen.
Gydansk Pen.
Bovanenkovskoye
Zapadno-Sharapovskoye
Peschanoozerskoye
Kolguyev Island
Kurensovskaya
Pomorskoye
Severo-Gulyaevskoya
Prirazlomnoye
Korovinskoye
Kumzhinskoye
Vaneivisskoye
Yarelyuskoye
Naryan-Mar
Layavozhskoye
Yuzhno-Toraveyskoye
Ardalin
Padimeiskoye
Kharyaginskoye
Vozeyskoye
Vorkuta
Sedyaginskoye
Usinskoye
Usinsk
Yuzhno-Syninskoye
Araneiskoye
Pechora
Pechorogorodskoye
Kamenskoye
Yarega
Ukhta
Pashninskoye
Vuktylskoye
Zapadno-Tebukskoye
Gornoye
Punga
Pachginskoye
Kurinskoye
Rassokhinskoye
Unvin
Serov
Verkhne-Kondin
Krasnoleninsk
Nizhnyaya Tura
Vasilevskoye
Perm
Krasnokamsk
Kedrov
Kordon
Kozubaevskoye
Bukharov
Batyrbaiskoye
Izhevsk
Arlan
Pervomaiskoye
Pavlovskoye
Mancharovskoye
Sergeevskoye
Krasnovsolskiy
Ishimbay
Salavat
Kanchurinskoye
Podgornov
Berkutov
Saratov
Magnitogorsk
Orsk
Sverdlovsk
Chelyabinsk
Kurgan
Tyumen
Tobolsk
Omsk
Atbasar
Gydanskoye
Solyet
Antipiutin
Messoyakh
Solenin
Dudinka
Norilsk
Rostovtsyev
Novo-Port
Yamburg
Novyy Port
Zapolyarnoye
Urengoi
Russkoye
Salekhard
Medvezhye
Samburg
Nadym
Yuzhno-Russkoye
Chernichnoye
Verkhne-Chasel
Kharampur
Muravlenkov
Sevro-Kazym
Sredne-Likhmin
Bolshoye
Sutormin
Vostochno-Vingayakhin
Yuzhno-Purov
Povkhov
Severno-Varyegan
Fedorovo
Nizhne-Sortym
Aprel
Bakhilov
Severo-Khokhryakov
Permyakov
Samotlor
Shaim
Salym
Surgut
Nizhnevartovsk
Mamontovo
Al-Yaun
Tajlakov
Yakunyakh
Yuzhno-Tailakov
Urnen
Tomsk
Krasnoyarsk
Achinsk
Remerovo
Anzhero-Sudzhensk
Novosibirsk
Novokuznetsk
Pavlodar
Karaganda
Ust-Kamenogorsk
Sredne-Tyungskoye
Ireiyakhskoye
Tas-Yurakhskoye
Iktekhskoye
Machchobinskoye
Vilyuisko-Dzherbinskoye
Sredne-Botuobinskoye
Verkhne-Vilyuchanskoye
Verkhne-Chonskoye
Mirnyy
Ozernoye
Lensk
Talakanskoye
Nizhne-Khamakinskoye
Danilovskoye
Pilyudinskoye
Dulisminskoye
Ayanskoye
Yaraktinskoye
Markovo
Kirensk
Ust' Kut
Bratskoye
Bratsk
Kovyktinskoye
Khristoforovskoye
Atovskoye
Bilchirskoye
Yuzhno-Radulskoye
Balykhtinsko-Tuptinskoye
Angarsk
Irkutsk
Ulan Ude
Kenkiyak
Zhanazhol
Kumkol
Pridorozhnoye
Ayrakty
Kungrad
Almaty
Bishkek
Chimkent
Tashkent
Andizhan
Izbaskent
Leninskoye
Fergana
Samarkand
Bukhara
Chardzhou
Dushanbe
Achak
Naip
Severnyi Achak
Yuzhnyi Naip
Gazli
Tashkuduk
Balkui
Farab
Kishtovan
Sakar
Bagadzha
Samantepe
Dianskoye
Sharapli
Bairamali
Sundukli
Gumbulak
Beshkyzyl
Maiskoye
Adamtash
Shorkel
Akbash-Adyr
Lyalmikar
Kichikbel
2 Lines
3 Lines
4 Lines
6 Lines
Legend
Oil field
Oil sand
Gas field
Crude oil pipeline
Natural gas pipeline
Products pipeline
Pipeline planned or under construction
Refinery in operation Refinery capacity in 1,000 b/d
Tanker terminal
Cities
Capital
International boundary
Water depth
0 to 200 m
200 m and deeper
SAKHALIN ISLAND
RUSSIA
SAKHALIN ISLAND
Severo Kolendo
Kolendo
Okha
Astrakhan
Odoptu-More
Vostochno Ekhabi
Piltun Astokhskoye
Arkutunskoye
Chaivo-More
Nikolayevsk
Uslovo
Gilyako-Abunan
Erri
Volchinka
Vostochno Dagi
Imchin
Katangli
Lunskoye
Staryy Nabil'
Novo Nabil'
Komsomolsk
Amur
Tatar Strait
Poronaysk
Terpeniya Gulf
Khabarovsk
Yuzhno Sakhalinsk
Vostochno-Lugovo
0 100 Miles
0 100 200 Km

praised the decision, hoping that the move would create a more positive business climate in Azerbaijan, especially during construction of a 1,083-mile (1,743-km) oil pipeline from Baku to Turkey's port of Ceyhan. The trunkline will be crucial to export from the vast Azeri-Chirag-Guneshli oil field development.

US officials said in early 2002 that the pipeline would be completed in 2005 on time and on budget. US financial aid could expedite construction beginning in 2002. However, the pipeline's sponsors were awaiting a financial commitment from the World Bank, which hinged on a successful environmental and social impact study due in mid-2002.

Earlier in 2001, the Azerbaijan government approved the 1,640-ft wide (500-m wide) corridor needed for the pipeline's construction phase, building on the previous clearance from Turkish authorities. State Oil Co. of Azerbaijan Republic (SOCAR) said that seven months of basic engineering work was complete, and the year-long $150 million field engineering work was to start in mid-2001. The 42-inch, 1 million b/d line would include seven pumping stations and three metering systems.

Azerbaijan also agreed to supply gas exports to Turkey of 3.1 tcf (89.2 billion cu m) over a 15-year period. The gas deal, signed in March 2001, would ramp up exports from 70.6 bcf in 2004 to 106 bcf in 2005, 177 bcf in 2006, and 233 bcf/yr from 2007-2018.

Caspian oil field development

The Azerbaijan International Oil Co. (AIOC) was planning further development of the "ACG" oil fields – Azeri, Chirag, and deepwater portions of Guneshli – which had been producing over 100,000 b/d-from reserves of 4.6-6 billion bbl of oil and nearly 6 tcf of natural gas. The AIOC consortium included operator BP PLC (34.14%), Unocal (10.28%), Lukoil (10%), and SOCAR (10%), with other partners holding smaller shares. Production from ACG fields was expected to be exported through the Baku-Ceyhan pipeline.

In September 2001 AIOC approved the first phase of expanded development costing an estimated $3.5 billion. This initial project would include a production platform with a design capacity of 400,000 b/d, a 30-inch oil pipeline to shore, an extension of the Sangachal land terminal, and a gas and water injection platform. First production was slated for early 2005. When complete, this initial phase would boost ACG fields' production to 500,000 b/d from 120,000 b/d. Plateau production could reach 1 million b/d by 2010.

Construction contracts worth $750 million were awarded in late 2001, bringing spending on the project to more than $1 billion. McDermott Caspian Contractors got the platform topsides job, including fabrication, assembly, hookup, and commissioning of the production, drilling, and quarters integrated deck, with completion scheduled for late 2004. Another McDermott contract would involve installation of a 116-mile (186-km), 30-inch pipeline from the main platform to landfall and a 7½-mile (12-km), 30-inch line for the compressor and water injection platform. Completion was slated for mid-2005.

Bouygues Offshore won a contract for fabrication and loadout of jackets and piles and the drilling template. Saipem had the contract covering the transport and installation of the template and jackets and for the floatover of the topsides. Contracts for other work including drilling facilities and living quarters were awarded to Eiffel Co. and to Emtunga International AS.

Also, AIOC was again upgrading its Dada Gorgud semisubmersible rig, which has worked in the Caspian Sea since it was built in 1980. Caspian Shipyard, a joint venture of Azerbaijan, Russia, and Singapore's Keppel FELS Energy & Infrastructure, was handling the $20 million modernization, the rig's fourth upgrade, with completion slated for March 2002. SOCAR owns Dada Gorgud, which is managed by Caspian Drilling, a joint venture of SOCAR and Santa Fe International Services. Once the semi is ready, AIOC planned to drill the first well in the Azerbaijani part of the ACG oil fields complex.

In late 2001 BP PLC awarded Caspian Geophysical – a joint venture of SOCAR and WesternGeco – a marine seismic services contract for the Azerbaijan sector of the Caspian Sea. Caspian Geophysical would supply marine acquisition and processing services for future seismic projects within any PSA areas operated by BP.

Shah-Deniz development

Development of Azerbaijan's Shah-Deniz gas condensate field was being planned, with drilling and production scheduled to begin in late 2004 and reach 298 bcf/yr (8 billion cu m/yr) by 2007. Shah-Deniz lies in water depths of 160-1,600 ft (50-500 m), 62 miles (100 km) south of Baku. Operator BP held a 25.5% interest in the project; Statoil AS, 25.5%; State Oil Co. of Azerbaijan Republic, 10%; TotalFinaElf SA, 10%; LukAgip, 10%; Iranian firm Oil Industries Engineering & Construction, 10%; and Turkish company Turkiye Petrolleri AO, 9%.

In mid-2001 Technip SA won an engineering contract from BP based on the French contractor's TPG 500 production jack-up concept. BP expected that Technip's self-installing fixed platform would cost in the region of $350 million. A total of 10 wells would be drilled from the platform, which will stand in 330 ft (100 m) of water 50 miles (80 km) off Azerbaijan.

The Technip deal was one of four contracts announced by BP as part of its $2.6 billion Shah-Deniz stage one project. Earlier, KCA Drilling Ltd. won an engineeering design contract for the construction and commissioning of drilling facilities for the field. The development scenario involved using a self-elevating drilling and production platform. BP expected to extend KCA's contract to also cover the second stage of the project, which includes detailed design, construction, and commissioning of the drilling facilities, and the third stage of offshore drilling operations. Contracts for the upstream and midstream concept front-end engineering design had been awarded in 2000.

BP noted that one constraint on operating in Azerbaijan was the dearth of local contractors capable of carrying out high-value service contracts. BP and other companies were seeking to forge alliances that would speed development of the supplier base in the Caspian Sea area. BP was working with the Azerbaijan Enterprise Development Committee to build the local supply chain and was setting up a web-based supplier database with UK SCM specialist First Point Assessment Ltd. to identify local companies with the required skills, resources, and standards. The oil company was also working with the Citizens Democracy Corps development agency on a "train the supplier" program.

Other exploration and development

Azerbaijan's hopes of establishing itself as the hub of a Caspian offshore oil industry – hopes that have been raised and dashed by mixed exploration results in the past – were dealt further blows in 2001. A consortium led by Italy's Agip decided against drilling a third test well in the Kyurdashi sector of the Caspian and pulled out of the project in late 2001. Earlier in the year, ExxonMobil Corp. announced it had drilled a dry hole with is first well on Oguz concession, TotalFinaElf SA walked away from its two-well commitment on the Lenkoran-Talysh-Deniz prospect in the southern Caspian after drilling a dry hole, and Ramco Energy also pulled out of a project.

However, future drilling, always problematic in the landlocked Caspian, will be helped by construction of a deepwater semisubmersible rig, begun in mid-2001 by Maersk Contractors Ltd. under contract to Exxon Azerbaijan Operating Co. LLC and Chevron Overseas Petroleum Azerbaijan Ltd. for a three-year drilling program in the Caspian Sea starting in 2003. The DSS 20-CAS-M semi, capable of drilling in up to 3,300 ft (1,000 m) of water to depths of 30,000 ft, was being built at Keppel FELS' Singapore yard and the Caspian Shipyard Co.'s Baku facility in Azerbaijan.

The four column hull was to be transported by barge from Singapore via the Black Sea, the Don Channel, and Volga River, and then mated with the upper hull at Baku. The majority of the piping, electrical, and mechanical installation, along with the fabrication and outfitting of a 130-person living quarters, would also be handled by the Baku yard of CSC, a joint venture company set up with SOCAR.

The rig design included a variable deckload of 4,000 tonnes; ratings of up to 2 million lb for the derrick, crown-mounted compensator, and traveling block; drawworks rated at 4,000 hp; and riser tensioner rated at 2 million lb. The rig will have three 7,500-psi mud pumps and well control systems incorporating an 18½-inch x 15,000-psi hydraulically operated blowout preventer and a 500-psi diverter system.

To deal with the problems of mooring mobile rigs in the Caspian, caused by undulating soft seabed formations and mud volcanoes, the Maersk semi will have an 8-point wire rope mooring system able to anchor the unit in waters of 250-1,640 ft (75-500 m). Chevron would operate the rig on behalf of partners ExxonMobil, Conoco Inc., TotalFinaElf SA, and SOCAR.

In late 2001 Chadbourne & Parke LLP advised Whitehall International Traders LP and its joint venture with SOCAR, Shirvan Oil, on a 10-year management services agreement with Schlumberger Overseas Services for the onshore Kyurovdag oil field. The Schlumberger Ltd. unit would provide oil field management and contractor services worth $600 million over 10 years but be compensated solely from a percentage of the increased production over an agreed baseline.

Refining

In mid-2001 the US Trade and Development Agency was planning to grant the Azerbaijan Ministry of Fuel and Energy $600,000 for a feasibility study on the upgrading of two refineries and the Port of Dubendi to handle lower grade crude from Kazakhstan, Turkmenistan, and Russia. Completion of the project would free Azerbaijani crude being processed at the refineries for export through the Baku-Ceyhan pipeline. The study would also mull the options for boosting transshipments of crude across the Caspian Sea via the port. The ministry was to choose a US company to conduct the study. Investment in the modernization projects was expected to total $100 million, with potential exports of US goods and services of between $50-80 million, according to the US Embassy in Baku.

GEORGIA

CAPITAL: TBILISI
MONETARY UNIT: LARI
REFINING CAPACITY: 106,436 B/CD
OIL PRODUCTION: 2,000 B/D
OIL RESERVES: 35 MILLION BBL
GAS RESERVES: 300 BCF

Georgia was pinning its hopes for long-term economic recovery on development of an international energy transportation corridor via Tbilisi and the country's key Black Sea ports of Poti and Batumi, as well as foreign investment in Georgia's own nascent oil industry. These hopes took a step forward in 2001 when final agreements were reached to begin constructing the oil pipeline from Baku, Azerbaijan, through Tbilisi to Turkey's port at Ceyhan in 2002.

Georgia reportedly will receive transit fees of 12 cents/bbl in the first five years of operation, increasing to 14 cents/bbl over the next 10 years and then to a minimum of 17 cents. Georgia was already transporting Azerbaijani oil through the Baku-Supsa line, which could be expanded in capacity in the future.

Gas pipeline

Also, a natural gas pipeline came closer to reality during 2001 when Georgia and Azerbaijan reached a tariff agreement in September 2001, paving the way toward construction plans for a pipeline to export natural gas from Azerbaijan's Shah-Deniz field into Turkey. Georgia President Eduard Shevardnadze and Azerbaijan President Heydar Aliyev were joined in the signing by the president of the Georgian International Oil Co., Georgi Chanturia, and Azerbaijan's Minister of Fuel and Energy Majid Karimov.

The intergovernmental agreement covers details of the transit, transportation, and sale of natural gas in the two countries through which the South Caucasus pipeline system will run. The agreement embodies the principles of freedom of transit of gas through the states, the development of related infrastructure, the provision of land rights and the protection of the environment and investments, and the right to purchase gas on a non-discriminatory basis under terms of future purchase and sales agreements.

The Shah-Deniz pipeline would run from Baku via Tbilisi in Georgia to Erzurum in Turkey, bringing gas supplies to regions that are frequently affected by severe energy shortages in winter months. Capacity was planned for about 250 bcf/yr. Under the agreement, Georgia would be paid $0.067/MMcf ($2.50/thousand cu m) in the first five years of operation, rising to $0.134/MMcf ($5/thousand cu m) in the later stage of the 20-year contract. Initially Georgia would take its tariff payment in gas and would also buy up to 18.6 tcf (500 million cu m) at an average price of $1.48/MMcf ($55/thousand cu m).

The gas line's length would total 620 miles (1,000 km), including 174 miles (280 km) within Georgia's borders. Pipe diameter was undecided but was expected to be 26-42 inches. Throughput would start in 2005 and build rapidly to 250 bcf/yr (6.6 billion cu m/yr) of exports to Turkey. Projected development cost was about $1 billion for the pipeline and $1.5 billion for associated offshore structures.

As of late 2001, some important matters remained outstanding, however, most notably the approval of the Georgia, Azerbaijan, and Turkey parliaments and a financial commitment from the shareholders in Shah-Deniz field in the Azerbaijan sector of the Caspian, including downstream marketing agreements with Turkish and world gas markets. BP PLC, operator and major shareholder in the project, said that it will proceed with the pipeline project only if it is commercially sound. BP was hoping for project approval by year-end 2001 and first gas flow late in 2004.

Caspian Region: Progress and Potential

Important progress was made during 2001 on developing and transporting the Caspian area's rich oil and gas resources to energy-hungry markets. On November 27, the 1,083-mile (1,743-km) Caspian oil pipeline was officially dedicated, having come on stream transporting crude from supergiant Tengiz oil field and other western Kazakhstan fields to Russia's new terminal on the Black Sea near Novorossiysk. The pipeline had been in planning stages for nearly a decade, though construction was delayed until 1999. Oil began reaching the Black Sea terminal in mid-2001, and tanker shipments began in October. The $2.65 billion system was expected to reach its full capacity of 600,000 b/d by mid-2002.

The pipeline's largest shipper was ChevronTexaco, which held a 15% interest in the Caspian Pipeline Consortium (CPC) as well as E&P interests in Tengiz field and in supergiant Karachaganak gas, condensate, and oil field in Kazakhstan. Hurricane Hydrocarbons Ltd. (Calgary) also became a player in late 2001 when it acquired a stake in CPC from BP PLC (through BP's Kazakhstan Pipeline Ventures, which held 0.9%). The acquisition would give Hurricane preferential rights to ship 64,000 b/d. Hurricane reportedly was producing nearly 100,000 b/d from several fields in Kazakhstan with proved oil reserves totaling an estimated 178.7 million bbl.

Other CPC partners included Russian Federation (24%), Republic of Kazakhstan (19%), LukArco (12.5%), Sultanate of Oman (7%), Rosneft-Shell Caspian Ventures Ltd. (7.5%), and ExxonMobil Caspian Pipeline Co. (7.5%). BP owned 46% of LukArco, giving it 5.75% of CPC. Pipeline and terminal maintenance contracts worth $40 million were awarded by CPC to Bouygues Offshore subsidiaries in late 2001.

Historically, tensions among the five countries surrounding the Caspian Sea – Kazakhstan, Azerbaijan, Turkmenistan, and the Caspian offshore sectors of Russia and Iran – have inhibited development of oil and gas resources. In early 2002, Turkmenistan said that it will host a long-delayed summit on dividing up the region's riches among these nations. President Niyazov was to set a date when visiting Russian President Putin.

Another piece of the Caspian puzzle began to fall into place during 2001 as plans gelled for construction of another oil pipeline downstream of the Caspian, running from Baku, Azerbaijan, through Tbilisi, Georgia, to the Turkish port of Ceyhan. US officials said in early 2002 that the pipeline would be completed in 2005 on time and on budget.

Rewards and risks

An analysis of Caspian Sea potential by Wood Mackenzie pegged the region's liquids and gas reserves at 39.4 billion bbl and 207 tcf, with the potential to produce 3.8 million b/d and 16.7 bcfd by 2010. The study suggested that natural gas will play an increasingly important role in development of the region, contributing about 44% of production potential in 2014. Kazakhstan and Turkmenistan hold the most gas reserves, while Kazakhstan also contained the most oil. Six key projects dominate the region's liquids reserves: Kashagan, Tengiz, Karachaganak, Azeri-Chirag-Guneshli, Shah-Deniz, and Russia's Severnyi block. Combined, these contain 26.9 billion bbl, or 68% of liquids reserves.

However, Wood Mackenzie pointed out, timely exploitation of these resources is unlikely due to several factors including the complex geopolitics of the region, the distance to international markets, the relatively high levels of risk, and the amount of financing available. With risk and other factors accounted for, more realistic production levels were estimated at 2.8 million b/d of liquids and 12 bcfd of gas by 2014. The singlemost important factor dictating future oil production in the Caspian is the availability of export capacity, said Wood Mackenzie.

Still, the risks associated with a number of key projects were significantly lower in late 2001 than just a year or two earlier: the Kashagan discovery was being appraised, CPC was operational, the commercial framework was in place for the Baku-Ceyhan pipeline, and a gas contract and transit terms with Georgia were approved for Shah-Deniz gas.

E&P activities

CanArgo-Georgia, a Georgian-British oil company, was planning to produce 3,000 b/d of oil in 2001 from the Ninotsminda field through its wholly owned subsidiary Ninotsminda Oil Co. Ltd. In September 2001 CanArgo Energy Corp. (UK) planned to sidetrack its first hole in a Georgian exploration program testing Cretaceous gas prospects. The N100 experienced mechanical problems at 13,254 ft (4,040 m), above the target depth. Because of the sidetrack, the hole was unlikely to be completed before December. The M11, the second of three holes by CanArgo, was drilling at 7,874 ft (2,400 m) toward the Cretaceous at 14,764 ft (4,500 m), but secondary objectives include the shallower Eocene, Oligocene, and Sarmatian. AES Corp. was a partner in the projects.

On a separate license in Georgia, CanArgo and partners planned to spud a wildcat on the Norio block in October 2001. The goal was a large prospect in the Eocene sequence, just north of Samgori oil field. CanArgo, an independent exploration and production company, operates primarily in Georgia and Ukraine.

In March 2001 Anadarko Petroleum Corp. completed its first stage of exploration on the Black Sea shelf, where engineers have discovered roughly 42.5 billion bbl of oil including about 14.7 billion bbl offshore. Anadarko was planning to drill the first test boreholes during 2001.

KAZAKHSTAN

CAPITAL: ASTANA
MONETARY UNIT: TENGE
REFINING CAPACITY: 427,093 B/CD
OIL PRODUCTION: 787,000 B/D
OIL RESERVES: 5.42 BILLION BBL
GAS RESERVES: 65 TCF

By far the largest Central Asian country and its second most populous, Kazakhstan enjoyed record economic growth during 2001, with an increase in real GDP of 13.8% driven by rising oil exports and foreign investment in the booming oil and gas sector. Because of its huge petroleum reserves, Kazakhstan has been called "the new Kuwait." To insulate the economy from volatile energy prices, Kazakhstan set up a National Fund, to which ChevronTexaco gave $660 million in exchange for the state's 5% stake in the Tengiz oil field joint venture.

The government has worked hard to create an attractive business climate through legislation and policy, and the country has avoided the bureaucratic bottlenecks seen in other former Soviet republics. The Kazakhstan government was seeking $140 billion investment in exploration, appraisal, and development over the next 40 years, not counting construction of a transportation infrastructure.

During 2001 the new Caspian pipeline began taking oil from Tengiz and other Kazakh fields to Russia's Black Sea terminal near Novorossiysk. Tengizchevroil agreed to supply German refiner Veba Oel AG with 2 million bbl/month of Tengiz crude starting in mid-2001. ChevronTexaco held 50% interest in Tengizchevroil; other stakeholders were Kazakhoil, the country's national oil firm (20%), ExxonMobil Corp. (25%), and LukArco BV (5%).

Reserves and production potential

Kazakhstan contains 75% of the Caspian region's liquids reserves and 45% of its natural gas reserves, according to Wood Mackenzie in a 2001 analysis. Kashagan field alone holds at least 10 billion bbl of recoverable oil and 25 tcf of gas. Wood Mackenzie estimated that Kazakhstan's liquids production potential could increase to 1.63 million b/d by 2005, 2.246 million b/d by 2010, and 2.484 million b/d by 2015. Much of Kazakhstan's gas is sour and thus costly to develop. Still, the country is capable of producing more than 6 tcf/yr by 2010, or 38% of the Caspian region's potential production of 16-17 tcf that year based on existing discoveries alone.

In early 2002 Kazakhstan reportedly posted a new production record, with crude and natural gas liquids climbing 20,000 b/d to average 870,000 b/d for December 2001. Future oil production growth was expected to come from three massive fields: Tengiz, Karachaganak, and Kashagan. Tengizchevroil was seeking to increase Tengiz field production to about 290,000 b/d in 2001; output could reach 415,000 b/d in 2005 and, given adequate export outlets, a peak of 750,000 b/d by 2010.

Karachaganak Integrated Organization (KIO), a consortium led by BG and Agip, was developing Karachaganak field, with estimated recoverable reserves of 2.25 billion bbl of oil and condensate plus 17.6 tcf of natural gas. In fall 2001 the field's production was suspended over a dispute with Russia on condensate export taxes.

Kashagan field in the Caspian Sea was still in the exploration stage, but it could turn out to be the world's largest oil discovery in the past 40 years. In early 2001 Agip's parent ENI became operator under the new name Agip Kazakhstan North Caspian Operating Co. (Agip KCO). Output at the first stage of Kashagan's development in 2005 was

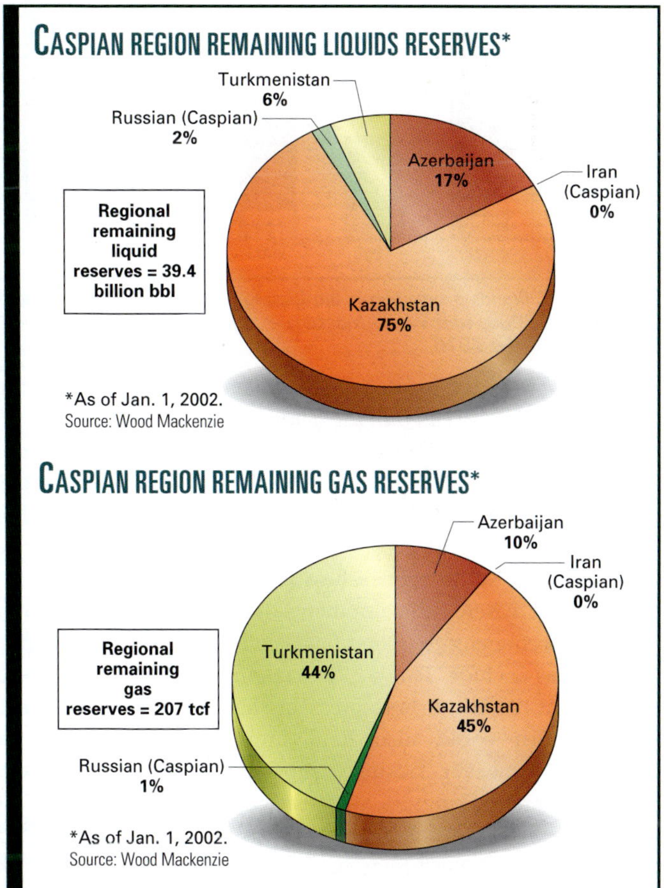

Kazakhstan controls most of the Caspian region's oil reserves and a major portion of its natural gas reserves as well, although much of the Kazakh gas is sour and would be costly to develop, reported Wood Mackenzie.

Caspian Region Liquids Production Potential

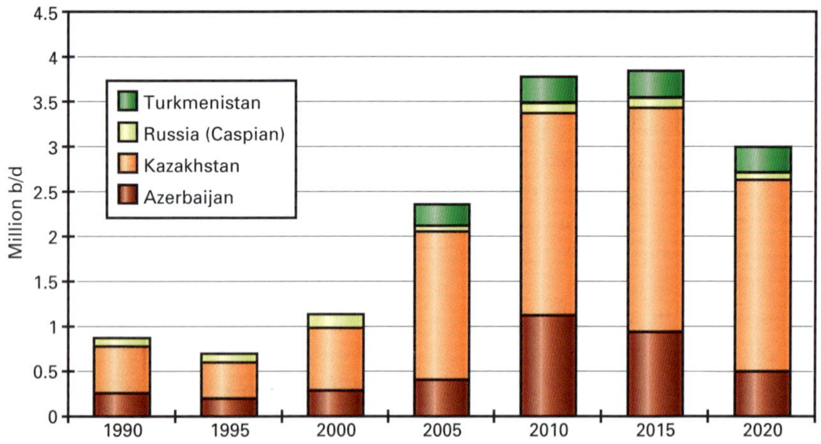

Source: Wood Mackenzie

Caspian Region Gas Production Potential

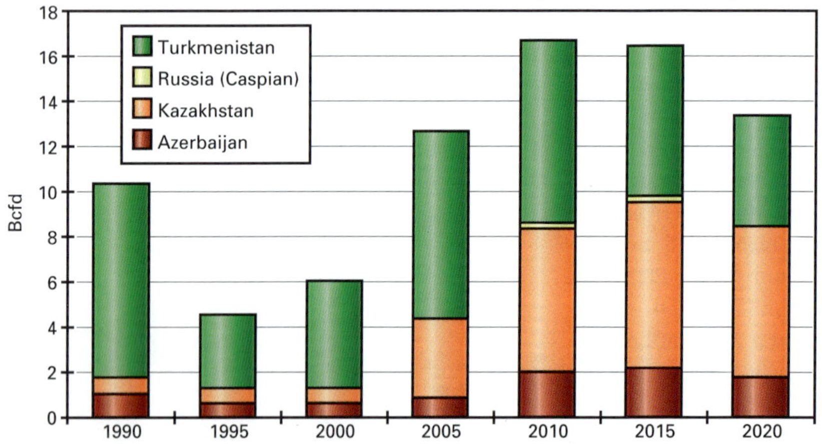

By 2015, liquids production in the Caspian region could more than triple from the level produced in 2000, said Wood Mackenzie, and the area has the potential to produce about 16-17 bcfd of gas by 2010.

anticipated to be 100,000 b/d, and further development could catapult the field into the world's top five oil producers.

Gas imports and production

More than 40% of Kazakhstan's gas reserves lie in Karachaganak field in the northwest, where development has focused on condensate. Other fields containing gas include Tengiz, Zhanazhol, and Uritau, while offshore areas in the Caspian Sea could also hold large amounts. However, lacking a pipeline infrastructure linking its western gas fields to its southern consumers, Kazakhstan must meet domestic demand by importing natural gas, mainly from Uzbekistan, through state company KazTransGaz.

In July 2001, KazTransGaz reached a five-year gas supply agreement with Uzbekistan to purchase 24.5 bcf for the rest of the year and up to 60 bcf in 2002. However, the gas import pipeline passes through neighboring Kyrgyzstan, where illegal tapping caused gas supply interruptions in winter 2001-2002. To alleviate this situation, Kazakhstan planned to tap gas fields in the south, including Amangeldy and Ayrykty in the Zhambyl region, and boost the nation's overall gas production dramatically, to 1.2 tcf in 2005, 1.66 tcf by 2010, and 1.84 tcf in 2015.

Kashagan development

In October 2001 the Kashagan East 2 appraisal well, drilled in the Caspian to 13,589 ft (4,142 m), flowed 7,400 b/d of oil on a constrained test. The second eastern well is 47 miles (75 km) south of Atyrau. After completing Kashagan East 2, the Sunkar drilling rig was to continue appraisal drilling on Kashagan East 1. Previously, Agip KCO (formerly Offshore Kazakhstan International Operation Co., or OKIOC) had discovered oil in Kashagan West 1, a well located 25 miles from the field's first well, Kashagan East 1. The western well brought proved field reserves to 10 billion bbl. The Kashagan structure's surface area totals 300,000 acres, or three times the size of Tengiz field.

Agip KCO consortium partners on the Kashagan production sharing contract were operator ENI (14.28%), TotalFinaElf (18.56%), Royal Dutch/Shell (14.28%), BG PLC (14.28%), ExxonMobil Corp. (14.28%), Phillips Petroleum (7.14%), and Inpex Masela Ltd. (7.14%). The 11 blocks controlled by the group cover a total 21,232 sq mi (55,000 sq km), with a maximum water depth of 33 ft (10 m). Some estimates of Kashagan oil reserves range as high as 50-60 billion bbl. In Kashagan and also Tengiz field, the oil is typically very light (>40º API) with moderate sulfur content, but high in mercaptans and associated with sour gas.

In July 2001 Western GECO, the joint venture company formed by Schlumberger Ltd. and Baker Hughes Inc., began shooting a 3D seismic program over the east flank of Kashagan field, as part of a move to accelerate development. The seismic acquisition, covering 618 sq mi (1,600 km), would be carried out using vessels specially adapted for water depths of 6-33 ft (2-10 m). The first of five appraisal wells was underway, following spudding of exploration wells KW1 and KE1. Drilling was to continue through to 2003 at the field, where a second unit would join the Sunkar rig operating on the 11-block area. ENI hoped to shave several years off the development schedule for Kashagan, bringing the oil/condensate field to first flow in 2005.

Other exploration

In early 2002 Russian firm Yukos Oil Co. announced it would acquire a 77.5% interest in and operatorship of the Federovskoye exploration block in Kazakhstan. First International Oil Corp.

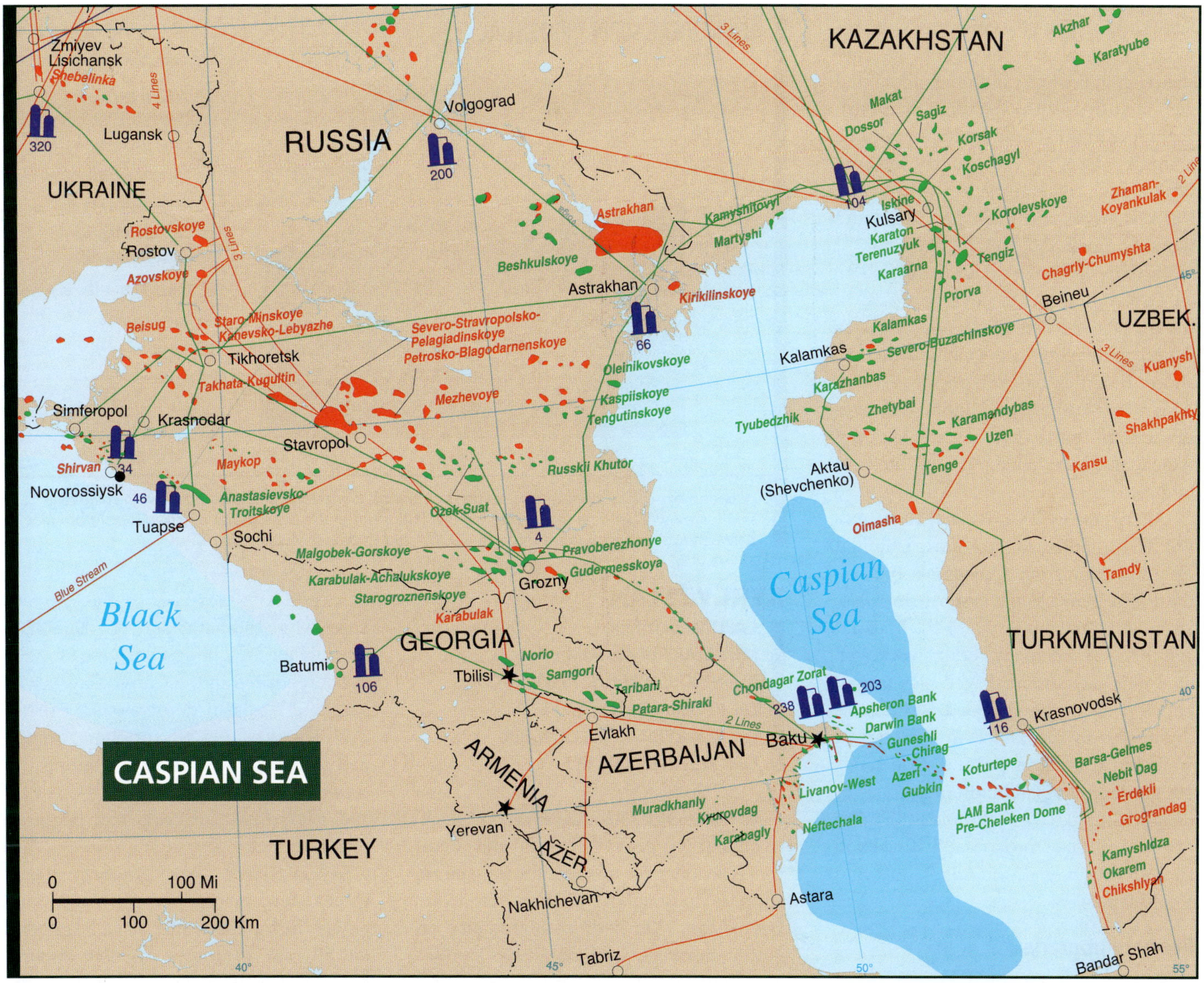

(Houston), which acquired the block in 2000, would participate in exploration activities with Yukos. Previously, Rosneft and Itera Holdings agreed to explore the block with First International, which has identified several attractive structures.

The 1,235 sq mi (3,200 sq km) Federovskoye Block is in north-central Kazakhstan, along the Russian border adjacent to the large Karachaganak oil, gas, and condensate field. Yukos said it has production, transportation, and refining facilities nearby. The minimum work program requires seismic survey and two deep exploration wells over five years. Yukos would carry out a seismic survey in 2002 and drill the first exploration well in 2003.

First International, a private independent, was Kazakhstan's largest acreage holder and has signed agreements relating to exploration and exploitation on several of its own and other blocks in the Precaspian basin. The company was seeking to raise as much as $60 million in equity to fund opportunities in Kazakhstan. Plans included drilling a deep well to Permian clastics on the 1.2 million acre Adaiski block 43 miles (70 km) east of and on trend with supergiant Tengiz oil and gas field. First International agreed with Falcon Energy Overseas Inc. (Delaware) to create First Falcon Oil and Gas Ltd., a 50-50 joint venture. First Falcon would explore for and develop shallow oil fields on First International's 3.2 million acre Sagiski block.

First International was also planning to participate with Potential Oil of Kazakhstan in exploring the 1.1 million acre Begaidar block, which offsets First International's 97.5%-owned Sazankurak oil field. Discoveries could be tied into Sazankurak facilities. As of April 2001, First International was producing 4,000 b/d of oil from two fields with net proved and probable reserves of 70 million bbl of oil and the potential for a further 230 million bbl of oil and oil equivalent. The company held 7.4 million acres in the basin and has invested $100 million there.

In October 2001 France's Technip SA and Kazakhstan's national oil and gas transportation company Transport of Oil & Gas created a joint venture engineering and construction company headquartered in Almaty, Kazakhstan. The companies will hold 50-50 stakes in Technip

Kazakhstan. The new company will provide engineering services for the construction of onshore and offshore oil and gas facilities with Technip's support. Technip Kazakhstan intended to become a major player in the country, helping with the development of Tengiz, Karachaganak, and Kashagan fields and working on oil and gas pipelines.

In early 2002, Kazakhstan offered to let ONGC Videsh, the overseas arm of Indian state-owned Oil & Natural Gas Corp., participate in developing the offshore Karazhanbas oil and gas field. Also, Gas Authority of India Ltd. was invited to participate in developing Amengaldi gas field and indicated an interest.

Also early in 2002, Nelson Resources Ltd. (UK) and Kazakhoil started limited oil production from Alibekmola field on the Precaspian basin's eastern rim. No. 28 and No. 26 wells flowed 1,100 b/d and 500 b/d, respectively, after workover. One new development well was drilled and a second one spudded. The field, discovered by the Soviets in 1987, has 24 wells but was never produced. It holds 40º gravity oil in Carboniferous zones as deep as 10,500 ft. Alibekmola is 25 miles from a rail line. Nelson and Kazakhoil share 50-50 interests in Alibekmola and Kozhasay fields.

Refinery upgrade

Kazakhoil in late 2001 moved ahead with a $235 million modernization of its 104,000 b/d Atyrau refinery, awarding a contract to Marubeni (which had previously done a feasibility study) and JGC. The refinery processes Tengiz crude. The project's goal was to improve the quality of Kazakhoil's gasoline and diesel fuel, as well as making the plant more efficient.

The project would include revamp of the existing crude distillation unit and construction of a naphtha splitter, naphtha hydrotreater, isomerization, diesel hydrotreater, and utility and offsite facilities. Of the total cost, $200 million would be financed by the buyer's credit from the Japan Bank for International Cooperation and the Nippon Export & Investment Insurance.

RUSSIAN GAS SUPPLY POTENTIAL

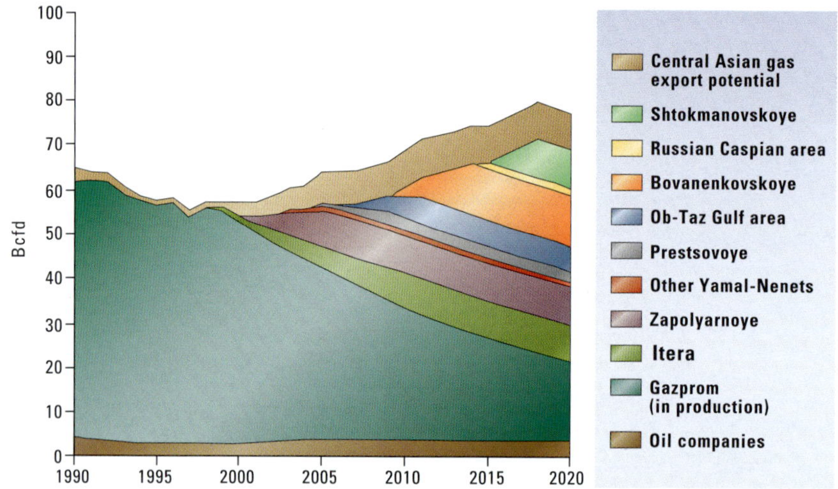

Although exports of natural gas to Russia from the Central Asian states have been relatively small, they will become increasingly more important in the overall gas demand-supply balance between Russia and Europe through 2020, said a Wood Mackenzie report.

KYRGYZSTAN

CAPITAL: BISHKEK
MONETARY UNIT: SOM
REFINING CAPACITY: 10,000 B/CD
OIL PRODUCTION: 1,000 B/D
OIL RESERVES: 40 MILLION BBL
GAS RESERVES: 200 BCF

In the energy sector, Kyrgyzstan's lack of oil and gas reserves has left the country dependent on imports for its energy supplies, particularly natural gas from Uzbekistan. Yet with an abundance of mountain rivers, Kyrgyzstan has significant hydroelectric potential and, with sufficient capital investment, might harness more power to meet the country's energy needs. In July 2000, the World Bank approved a $5-million credit to Kyrgyzstan to reform the fuel and energy sector, strengthen the financial position of Kyrgyzgaz, the state natural gas joint-stock company, and purchase necessary equipment, such as gas meters and spare parts.

Although Kyrgyzstan contains seven developed oil fields and two oil/gas fields, extraction is difficult due to the country's mountainous topography, and water encroachment means that recovery rates are low. In November 2000, a team of specialists which had conducted an oilfield survey for the government forecast that Kyrgyzstan soon could supply itself with enough oil to meet its domestic needs. Oil reserves in the Fergana Valley were estimated at over 700 million bbl, while 1.5-2.2 billion bbl were thought to be deposited in the Chuy, Alay, Issyk-Kul, and At-Bashi depressions. During 2001 Kyrgyzstan's oil suppliers continued to become less reliable, leading the country to plan intensive extraction of oil and gas within 5-6 years and begin to seek funding.

Kyrgyzstan's gas imports from Uzbekistan, which supply heat and power during the winter, have been periodically cut off due to payment arrears. In October 2000, officials from both countries agreed on a payment scheme involving barter of gas, electricity, water, and goods, but in late 2000 and early 2001, gas supplies were cut off twice, leading the Kyrgyz government to seek other suppliers such as Russia's Gazprom and Itera. However, construction of pipelines to import the gas had been stalled due to lack of financing.

The pipeline supplying southern Kazakhstan with natural gas from Uzbekistan crosses Kyrgyzstan, where illegal tapping of the gas has occurred, resulting in supply disruptions to Kazakh cities. According to Kyrgyzgaz, the company took 0.5 bcf of natural gas without permission during late 2001-early 2002 because Uzbekistan's gas monopoly Uztransgaz had failed to fulfill its commitments to supply natural gas to northern Kyrgyzstan.

Refining

Kyrgyzstan crude oil refinery in Dzhalal-abad, about 150 miles south of Bishkek, is run by Kyrgyz Petroleum Co., a joint venture between Kyrgyzneftegaz, the state oil company, and Petrofac Resources International Ltd. (UK), which bought its share from Canadian-based Kyrgoil in June 2000. The 10,000-b/d refinery produces heavy fuel, diesel, and gasoline but has been hamstrung by difficulties in getting reliable supplies of crude oil from neighboring countries.

RUSSIA

CAPITAL: MOSCOW
MONETARY UNIT: RUBLE
REFINING CAPACITY: 5,435,480 B/CD
OIL PRODUCTION: 6.895 MILLION B/D
OIL RESERVES: 48.6 BILLION BBL
GAS RESERVES: 1,680 TCF

Russia's economy in general and its oil and gas industry in particular remained healthy in 2001, benefiting from political stability, economic reforms, and high oil prices. Real GDP was anticipated to grow 4.7% in 2001 after an 8.3% increase in 2000, while oil production was to increase 6% following a 10.5% rise in 2000. Russia was the world's largest exporter of natural gas and second largest exporter of oil, making the economy sensitive to energy prices. Nearly 87% of Russia's oil exports went to countries outside the former Soviet Union (FSU) in 2000.

Although Russia is not a member of the Organization of Oil Exporting Countries (OPEC), it has in the past attempted to coordinate its export strategy with OPEC. During 2001, however, as OPEC cut member production quotas, Russia increased production faster than any other non-OPEC country. As of early 2002, both OPEC and Russia had breached the production cutting agreement of late 2001, according to the International Energy Agency. Russia actually increased its output rather than reducing it by 150,000 b/d, while OPEC's cut was less than half the agreed amount of 1.5 million b/d.

The Russian government took steps toward breaking up Gazprom, the state company that produces nearly 94% of Russia's natural gas, operates the country's 90,000-mile gas pipeline grid and 43 compressor stations, and holds nearly one third of the world's gas reserves. In May 2001 long-time Gazprom chief Vyakhirev was replaced by Aleksei Miller, an ally of President Putin. A restructuring plan being considered would divide Gazprom's upstream operations into separate producing companies, while the government would control transmission pipelines and offer equal access to all gas producers. In addition to gas sector restructuring, stakes in both Rosneft and Slavneft, the last two large oil companies to be privatized, were to be auctioned off in 2001.

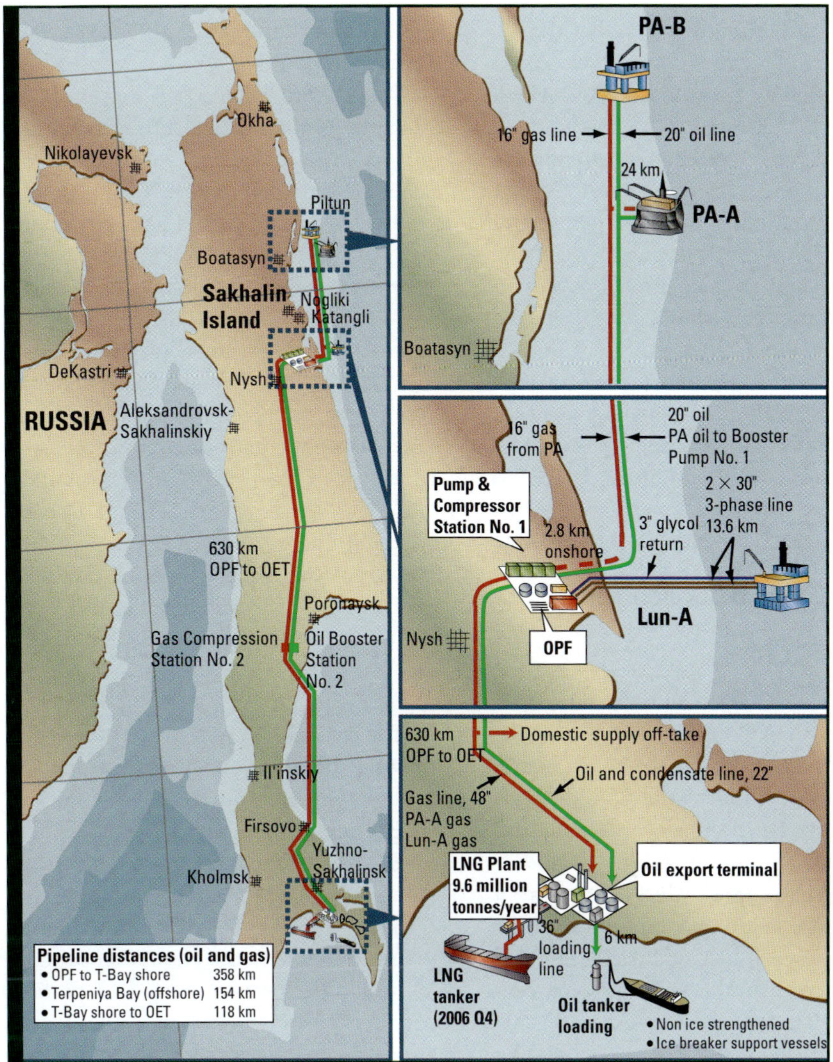

Plans moved ahead during 2001 for Russia's Sakhalin Island phase two, which will incorporate oil and gas production facilities and pipelines, an LNG production plant, and oil and LNG shipping terminals.

Although exports of natural gas to Russia from the Central Asian states have been relatively small, they will become increasingly important in the overall gas demand-supply balance between Russia and Europe through 2020, said a Wood Mackenzie report. Russia's gas supply-demand balance depends on several variables including net production growth, Central Asian gas prices, and the cost and timing of developing Russia's major gas fields. But Asian gas at prices lower than the major developments would be needed to supplement Russia's domestic demand and its export requirements to European markets.

The Russian oil field services market began to develop in line with global trends, said Neft I Kapital. Specialized service companies, most notably PetroAlliance Services Co., have emerged and were making serious progress, especially in providing geological and geophysical services. Still, foreign providers retained their grasp on the market, and in early 2002 Halliburton International Inc.

announced a multi-million dollar deal to provide a full range of oil field services to OAO Sibneft's upstream division, which produces 400,000 b/d.

In mid-2001 TotalFinaElf SA signed a technical and scientific cooperation agreement with the State Polar Academy of Saint Petersburg, through which the French oil and gas company aims to help ensure long term development in the regions of Russia's far north. The deal would cover professional training of specialists; reception and management of students; implementation of environmental, social, and ethnological studies; and the provision of financial and logistical assistance to the State Polar Academy.

Exploration and development

Russian Far East

Russia's OAO Sibneft in 2001 launched an exploration program on the far northeastern Russia shelf. A Sibneft affiliate was drilling a wildcat in the eastern Anadyr basin on a nearshore island in southern Anadyr Gulf. Another Sibneft project would develop gas from an onshore field to replace coal at an Anadyr power plant. Sibneft and OAO Yukos were planning a 50-50 joint venture for late 2001 to reprocess seismic data from the shelves of the Chukotka and East Siberian Seas, then acquire more seismic data and possibly drill offshore wildcats. This area's geological structure was believed to be similar to that of Alaska's North Slope.

Phase one of the Sakhalin-I field development moved forward during 2001 as the ExxonMobil-led consortium purchased a concrete-island drilling system for the offshore Chavyo field east of the island. The system was to be moved in mid-2001 from the Beaufort Sea to the Far East for modifications. Although Sakhalin-I phase one is an oil-only development with a peak production target of 250,000 b/d, natural gas from Sakhalin blocks continued to be marketed to Japan and China. The field's 17 tcf of recoverable reserves could supply 1.5-2 bcfd over a plateau life of decades. ExxonMobil's subsidiary Exxon Neftegas Ltd. as operator held 30% interest in Sakhalin-I; Japan's Sakhalin Oil and Gas Development co. Ltd. had 30%, Russia's Sakhalinmorneftegas-Shelft, 23%, and Rosneft-Sakhalin 17%.

Plans were underway for an LNG plant, the second phase of the Sakhalin-II project on the island, with first LNG scheduled for delivery in 2006. The air-cooled LNG plant would have two trains with a capacity of 4.8 million tonnes per year (tpy) each, making it the world's largest. The plant would employ the Sulfinol-D process for acid gas removal and the Shell Double Mixed Refrigerant process for liquefaction. Plant engineering was being performed by a consortium of OAO NIPIgaspererabotka, Chiyoda Corp., and Fluor Daniel. Sakhalin Energy Investment Co., project manager, was owned by Shell Sakhalin Holdings BV 55%, Mitsui & Co. Ltd. 25%, and Mitsubishi Corp 20% (Diamond Gas Holdings BV).

The two LNG trains would have gas treating, liquefaction, and fractionation facilities with a common gas-inlet station, supporting utilities, and LNG export terminal. Gas from an upstream processing plant, some 1,500 MMscfd, would arrive through a 390-mile (630-km), single-phase trunkline to the plant inlet station.

Halliburton was expected to participate in exploration work in Chukotka, an autonomous far eastern region of Russia, as part of its oil field services deal with Sibneft. Chukotka resources were estimated at 275 million bbl. Sibneft held three onshore blocks, where exploratory drilling started in mid-2001.

Western Siberia

In August 2001 first production began from South Tarasovskoye field on the Urabor Yakhinsky Block in western Siberia. The first well, with a total depth of 9,535 ft, cut a 365-ft gross oil column in multiple productive intervals and produced 1,000 b/d of 40º gravity oil. Two development wells and an appraisal-extension well were being drilled by Geoilbent, a joint venture in which Benton Oil & Gas Co. (Houston) held 34% interest.

Through a long-standing partnership with Gazprom, Wintershall AG was planning to complete an expansion project at Urengoi-Achimov gas fields that would produce 370 bcf/yr of gas (10 billion cu m/yr) and 2.5 million tpy of condensate.

In early 2002 Russia's OAO NK Yukos and Hungary's MOL PLC agreed to jointly further develop Zapadno-Malobalysk oil field in Russia. Plans called for a 50-50 joint venture to invest $300-$350 million in the project under a licensing regime rather than a PSA. Zapadno-Malobalysk is in western Siberia within the Khanty-Mansiysk Autonomous Region, near pipeline and transportation infrastructure. The field, with estimated proven recoverable reserves of at least 20 million tonnes of crude, had been producing 10,000 b/d. Plateau of 55,000 b/d was expected by 2005.

Lukoil took over licenses to develop several fields in Western Siberia's Perm Oblast, including Ozernoye, Tarkhovskoye, Mysinskoye, Borovitskoye, Magovskoye, and Logovskoye. Lukoil Overseas Holding Ltd. acquired 100% of Permtex LLC from SOCO International PLC in two deals during 2001.

In early 2001 Polar Lights Co. announced that Ardalin field had passed the 75 million bbl (10 million ton) production milestone.The field, which began production in the Nenets Autonomous Okrug of the Timan Pechora basin in 1994, is in a harsh arctic tundra environment about 1,000 miles northeast of Moscow. Oil from Ardalin field is moved through a 42-mile line to the Russian Pipeline System. Polar Lights is a joint venture of Conoco Inc. (50%) and Russian partners Arkhangelskgeoldobycha (30%) and Rosneft (20%).

Barents Sea

Wintershall was also planning to develop ice-locked Prirazlomnoye oil field in the Barents Sea, where reserves were estimated to be at least 500 million bbl. Although the field lies in only 100 ft (30 m) of water, it is in one of the globe's most hostile environments. Wintershall believed that the $1 billion project with relatively low profitability would be worthwhile due to the untapped potential of the field's surrounding area for longer term development.

Russian Caspian Sea

Russia's remaining reserves in the Caspian totaled 1.0 billion boe, consisting of 77% liquids and 23% gas, according to a Wood Mackenzie analysis. These amount to only 1% of the region's gas reserves and 2% its liquids reserves, despite Russia's addition of 750 million bbl of oil to reserves in the Yuri Korchagin and Khvalynskoye fields. Production from the Severnyi Block was expected to begin in 2004, initially from Yuri Korchagin and later from the phasing in of Khvalynskoye. Much of the Caspian region's gas reserves are stranded without markets, so Russia continued to dominate Turkmeni and Kazakh gas development.

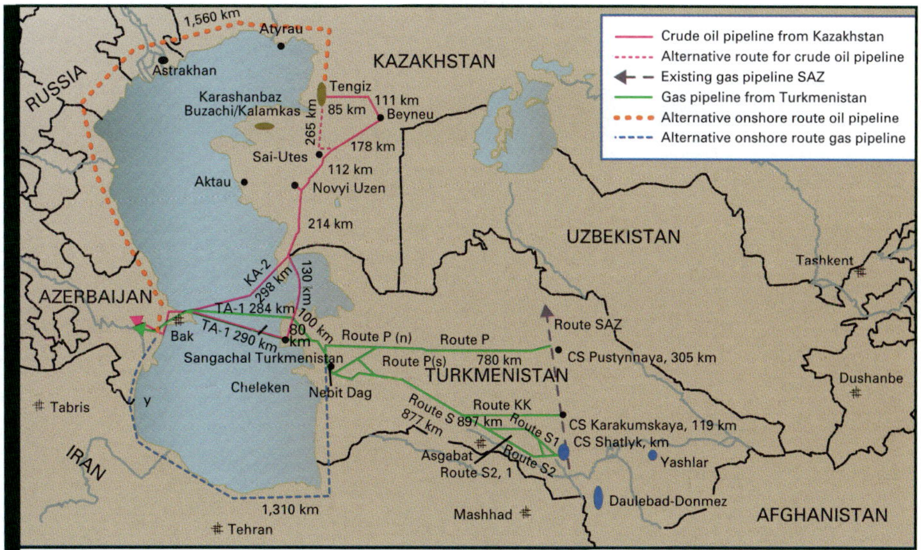

Possible routes for Trans-Caspian pipelines, including a crude oil line from Kazakhstan's Tengiz field and a natural gas line from Turkmenistan's Shatlyk field.

Oil pipelines

The long-awaited Caspian pipeline came on stream in 2001, exporting central Asian oil through Russia's Black Sea port near Novorossiysk. Russia had the largest share in the pipeline project, but its own oil cannot flow into the line until a link is built with Russia's Transneft pipeline system.

In September 2001 Transneft began operating a new 162-mile, 320,000-b/d pipeline from Sukhodolny to Rodionovsky, a route that bypasses Ukraine to avoid high transit fees.

The Baltic Pipeline System, under construction during 2001, would allow Russia to export oil directly from its Baltic Sea port of Primorsk, where a new oil terminal with 240,000 b/d capacity was completed and began tanker loading in late 2001. The 162-mile pipeline to Primorsk from Kirishi was also finished, and the connection from Yaroslavl to Kirishi was being modernized. The project's second stage would build a parallel pipeline and additional storage at Primorsk that would increase the Baltic system's capacity to 600,000 b/d.

In late 2001, Ukraine's parliament ratified a tariff reduction that could enable Russian oil to be exported via a trans-Balkan pipeline network to the Adriatic Sea. Development of the Druzhba-Adria pipeline system would be the first complete pipeline bypass of the Bosporus Straits. The Ukrainian pipeline operator previously had agreed to cut its transit tariff to 64¢ from 72¢/tonne/100 km, bringing it in line with other countries through which the route passes, but Ukraine's parliament had held up approval. The action was regarded as the last major obstacle to this export route, reported the Centre for Global Energy Studies (CGES).

Russian Urals blend would flow 1,987 miles west via the Druzhba and Adria pipeline systems. The 1.2 million b/d Druzhba line has served as Russia's main pipeline export route and has been operating at its highest capacity in years. The Adria pipeline, which runs from Croatia's Adriatic port of Omisalj into Hungary, was being modernized to facilitate integration of the two systems.

The final link in the trans-Balkan pipeline would be the $120 million reversal of a 110.5-mile, 36-inch line from the inland terminal at Sisak, Croatia, to the port of Omisalj. Capacity of the entire line would be 100,000 b/d (5 million tpy), expandable to 200,000 b/d (10 million tpy) after five years and 300,000 b/d (15 million tpy) after 10 years. The Omisalj oil terminal has two berths capable of handling 500,000 dwt tankers. CGES said the terminal has land available to add two berths and to more than double storage capacity.

Larger tankers (VLCCs and ULCCs) would make exports more economic compared to shipping from Novorossiysk through Turkey's Bosporus Straits, which restricts vessel size to 150,000 dwt, said CGES.

Natural gas pipeline

To diversity its natural gas export routes and reach new markets, Russia was planning several new gas pipelines, with the Blue Stream to Turkey as the centerpiece. The 758-mile pipeline, including a 247-mile length under the Black Sea, was scheduled to be operational in 2002 with an initial capacity of 282.5 bcf/yr expected to double to 565 bcf/yr. Gas was expected to cost $3.40/cubic foot or about half again as much as Turkmeni gas. The Saipem 7000 lay barge was laying the subsea section in late 2001, after more than 18 months of retrofitting in preparation for work under more than 6,600 feet (2,000 m) of water.

Russia continued to explore options for a gas pipeline route that would bypass Ukraine, which had been illegally siphoning off natural gas in transit. Ukraine had also accumulated gas debts to Russia, but a repayment deal was struck in August 2001.

Seeking to export gas to China, Russia's Sakhaneftegaz agreed in March 2001 to build a pipeline dedicated to moving gas from Chayandinskoye field with estimated reserves of 43 tcf in eastern Siberia to the Xinjiang region of northern China. Another option would be to build a pipeline linking Irkutsk's Kovykta gas field, with estimated reserves of 49 tcf being developed by the BP-led Russia Petroleum consortium, into northeastern China with a termination in South Korea via a subsea line

your success
VALLOUREC & MANNESMANN TUBES is committed to providing its customers with the best products and services available on the market, made possible by a unique global approach supported by local capabilities.
This commitment is backed by an annual production of 2.5 million tons of steel pipes manufactured in 15 strategically positioned production facilities worldwide, a network of more than 90 VAM® licensee workshops devoted to local support for the VAM® premium connections, and 12,000 skilled professionals working around the clock to provide first-class customer service.
VALLOUREC & MANNESMANN TUBES over 100 years of experience and unrivaled reputation in the oil & gas industry, coupled with a continuous focus on the future, is your guarantee that we are fully dedicated to make your business a success.
VAM® TOP
NEW VAM®
DINO VAM®
VAM® EIS
V&M
VALLOUREC & MANNESMANN TUBES
OCTG Division
Sales & Technical Sales
130, rue de Silly
92100 Boulogne, France
Phone: +33 1 49 09 35 86
Fax: +33 1 49 09 37 13
Internet: www.vamservices.com

across the East China Sea. The feasibility of this route, which might have to bypass Mongolia, was being studied.

Other gas pipelines undergoing feasibility studies during 2001 included the North TransGas to serve Scandinavia and Germany and a pipeline to deliver gas from Russia's Sakhalin field to Japan.

Refining

Oil deliveries to refineries increased during the first half of 2001 by 4%, to 3.58 million b/d, as a result of Russia's linking producers' export pipeline access to meeting their delivery targets. Still, many refineries were operating below capacity, and most did not have financing for improvements. Exceptions were the 220,000-b/d Volgograd refinery, where Lukoil completed a first phase modernization, and the 290,000-b/d Yaroslavl facility, which was undergoing renovation.

TAJIKISTAN

CAPITAL: DUSHANBE
MONETARY UNIT: TAJIK RUBLE
REFINING CAPACITY: 0
OIL PRODUCTION: 0
OIL RESERVES: 12 MILLION BBL
GAS RESERVES: 200 BCF

Like Kyrgyzstan, Tajikistan is a mountainous country with little oil or natural gas but with substantial hydroelectric power, accounting for 98% of generation, and significant undeveloped hydropower potential. It has the lowest per capita GDP in the former Soviet Union, and a long civil war has disrupted the economy and prevented foreign investment. In April 2001, Tajikistan reorganized its state oil and gas production and natural gas distribution enterprises. Aside from small oil production from the northern Leninobod Soghd Region, the country depends on imports from Uzbekistan for 70% of its oil demand.

Tajikistan's natural gas production increased during 2001 following launching of Khoja Sartez gas field in the southern Khatlon Region and greater utilization of the Qizil Tumshuq deposit in southern Khatlon Region's Kolkhozobod District. Still, the country relied heavily on natural gas imports, primarily from Uzbekistan. In mid-2001, after using up more than 65% of the year's projected gas supplies of 12.3 bcf, Tajikistan reached an agreement for Uzbekistan to supply an additional 3.53 bcf (100 million cu m) through September 2001.

TURKMENISTAN

CAPITAL: ASHGABAT
MONETARY UNIT: MANAT
REFINING CAPACITY: 236,970 B/CD
OIL PRODUCTION: 146,500 B/D
OIL RESERVES: 546 MILLION BBL
GAS RESERVES: 101 TCF

Turkemenistan's economy has been stifled by foreign debt and lack of reforms under its former communist president. The country rivals its neighbor Kazakhstan in controlling the most natural gas, holding 44% of the Caspian region's total remaining reserves. An analysis by Wood Mackenzie estimated Turkmeni gas production potential at 8.356 bcfd in 2005 and 8.172 bcfd in 2010, after which levels would fall to 6.6 and 4.8 bcfd in 2015 and 2020.

Turkmenistan's gas reserves, however, were stranded without markets, leaving the country subject to Russia's dominance. Plans for a Trans-Caspian Pipeline to export gas across the sea to Azerbaijan through Georgia and into Turkey fizzled during 2001 as Turkey signed gas purchase agreements with other countries including Azerbaijan. Instead, Turkmenistan focused on shorter term sales to Russia and to Ukraine. Under terms of a May 2001 deal, Turkmenistan will supply Ukraine with 8.83 tcf during 2002-06.

Turkmenistan has managed to increase its oil output, producing close to 150,000 b/d during early 2001. Both oil and gas are sold at fixed prices well below world market levels. Oil swaps have been arranged with Iran, allowing Turkmenistan to export its oil from the Iranian Caspian port of Neka.

In early 2002 Turkmenistan said it will host a long-delayed summit on dividing up the Caspian Sea's oil and gas resources among the five bordering nations.

Exploration and development

In early 2002 development of Adzhiyap Block in southwestern Turkmenistan was being planned through a joint venture of Tracer Petroleum Corp. (Calgary) and Canneft Inc. (Houston). The Adzhiyap Block covers 772 sq mi (2,000 sq km), bordering Iran to the south and the Caspian Sea to the west. Canneft was negotiating a 25-year production-sharing agreement for the block. Canneft and local partner Turkmengeological signed a protocol to form a 75-25 joint operating company for Adzhiyap. Canneft has made an offer to spend up to $25 million over four years to drill 40 wells and tie into the pipeline.

Tracer said the project has estimated reserves of 3-6 tcf of gas. Tests have shown a concentration of 1 bbl of condensate/100 Mcf of gas. It could also contain deeper oil reserves, but the possibility has not yet been evaluated. The 125-mile, 1.18-bcfd Korpedzhe gas line to Iran is only 2½ miles (4 km) from proven gas wells No. 6 and No. 1. Korpedzhe had been carrying only 500 MMcfd.

Also under consideration was an alternate route for a gas pipeline 62 miles (100 km) to the north that would connect to an existing Gazprom line supplying the Ukraine. Tracer and Canneft also intended to study GTL feasibility later in 2002. So far, 12 structures have been identified on the block with 2D seismic. Well No. 1 logged 216 ft of gas pay and No. 6 logged 75 ft.

In May 2001 Turkmengaz started exploration and prospecting on a new field in Darganata, northeastern Turkmenistan.

In early 2001, a western drilling rig was scheduled to begin work for Turkmenneft as part of a 30-well program beginning in October. The rig was manufactured by Bentec GmbH Drilling & Oilfield Systems, a division of Preussag AG (Germany). The 2,000-hp, desert-ready "Turk 1" was to have a 156-ft cantilever mast with a hook load capacity of 1.3 million lb (14 lines) and would be mounted on a 35-ft tall substructure capable of holding 800,000 lb of drill pipe or 1.3 million lb of casing. The drawworks and other systems would be powered by four Caterpillar 3512 engines connected to four 1,900 KVA Leroy-Somer generators.

In early 2001 Dragon Oil PLC (UK) spudded the first of three wells planned

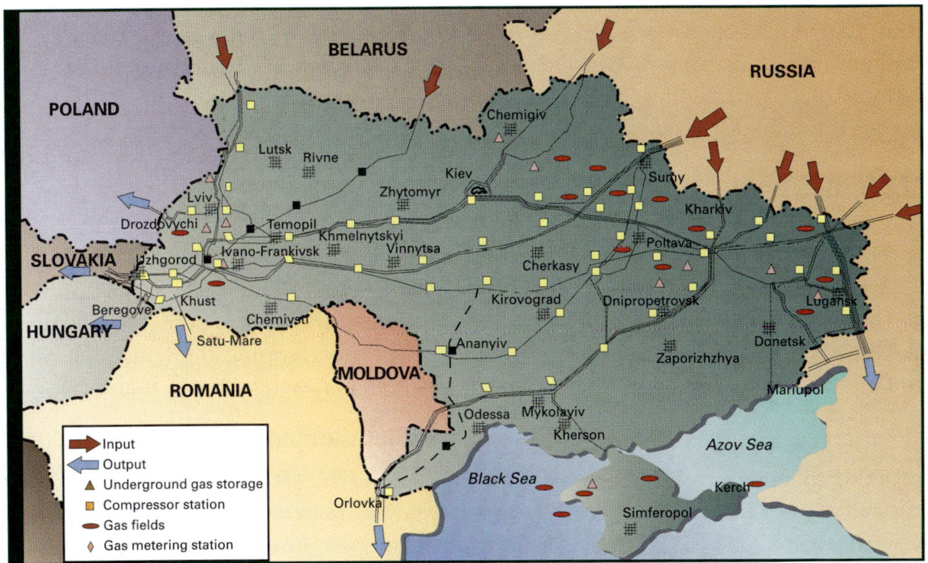

As one of the world's largest gas consumers, Ukraine has a highly developed transmission system, with an inlet capacity of 10.2 tcf/yr and delivery capacity of about 6 tcf/yr.

for Lam field off Turkmenistan in the eastern Caspian. The Cheleken Block could hold 600 million bbl of oil and 82 tcf (2.2 trillion cu m) of gas, Dragon speculated.

Refining

In early 2002 Turkmenneftegas awarded a contract for the design and construction of a diesel hydrotreating plant at its 116,477-b/d refinery at Turkmenbashi to Technip-Coflexip Group for 130 million euros. The work, to be completed in 2004, would include a hydrotreating unit, a sulfur recovery unit based on Technip-Coflexip's proprietary technology, and associated utilities, storage tanks, and control systems.

Previously, Technip-Coflex had participated in the expansion and upgrade of the Turkmenbashi refinery, including design and construction of a 1.8 million tpy catalytic cracker and lube oil plant. The unit was launched in April 2001 with a capacity of 36,150 b/d and was designed to produce 18,478 b/d of high-octane gasoline, 5,120 b/d of diesel, 5,623 b/d of heating oil, and up to 6,929 b/d of LPG.

UKRAINE

CAPITAL: KIEV
MONETARY UNIT: HRYVNIA
REFINING CAPACITY: 1,026,259 B/CD
OIL PRODUCTION: 73,900 B/D
OIL RESERVES: 395 MILLION BBL
GAS RESERVES: 39.6 TCF

Ukraine holds a strategic position as the transit center for Russian oil and natural gas exports to eastern and western Europe. In late 2001, Ukraine's parliament ratified a tariff reduction that could enable Russian oil to be exported via a trans-Balkan pipeline network to the Adriatic Sea. Ukrainian pipeline operator Druzhba previously had agreed to cut its transit tariff to 64¢ from 72¢/tonne/100 km, bringing it in line with other countries through which the route passes, but Ukraine's parliament had held up approval.

In addition to transporting Russian fuels, Ukraine could become a major player in moving oil from Azerbaijan and Kazakhstan to European markets. In 2001 Ukraine opened its Pivdennyi oil terminal on the Black Sea and its 400-mile Odessa-Brody pipeline, which were built to entice Caspian oil exports by avoiding the crowded Bosporus Straits. The terminal and pipeline were seen as the first stage of a project that would allow Ukraine to eventually refine and transport oil from the former Soviet republics of central Asia.

The Odessa-Brody pipeline's initial capacity was 180,000 b/d, expected to rise to 500,000 b/d. In April 2001 Ukraine and Poland agreed to study the feasibility of extending the line by 190 miles to Plotsk, Poland, where it could tie into existing pipelines to allow imports of Caspian crude oil to Poland, Germany, and Baltic states.

Exploration

Europa Oil & Gas Ltd. (UK) drilled its first well in western Ukraine, at Horodok field in early 2001. Horodok 9, drilled to 2,303 ft (702 m) total depth, flowed commercial rates of dry gas from the Upper Dashava at 1,903-1,929 ft (580-588 m). Europa planned to supply gas from four suspended wells to a gathering facility that was under construction. First production of 3 MMcfd was expected in April 2002. Europa operates Horodok and is a partner with Zahidukrgeologia on this and other Ukrainian projects.

In June Europa and Zahidukrgeologia received a license to develop Niklovich gas field in western Ukraine $15^1/_2$ miles (25 km) from the Polish border. The field holds reserves of 100 bcf in two main reservoirs at 3,937-4,921 ft (1,200-1,500 m). On test three suspended wells produced commercial quantities of gas with as much as 131 ft (40 m) of pay in Miocene sands. Commercial production was expected for 2003.

In May 2001 EuroGas Inc. (New York) announced that it would acquire 50% interest in Epic Energy Inc.'s Uzunlarskaya exploration project in Crimea, Ukraine, as part of a loan and investment deal. Apart from its 50% interest in the first well in the Uzunlarskaya prospect, EuroGas would earn 50% participation in the prospect. EuroGas and Epic expected to spud the first well on the Uzunlarskaya prospect, in the southern Kerch Peninsula near the Black Sea coast, in mid-2001. In addition, EuroGas would access the services of Epic's HHO Ltd. subsidiary, which manages Epic's oil and gas operations in Ukraine. HHO would assume management of EuroGas's Ukrainian oil and gas properties.

Carpatsky Petroleum Inc. (Houston) acquired Lateral Vector Resources Inc. (Canada) in May 2001, including Lateral's interest in Bugruvativske oil field in Ukraine, which could cost $25 million to develop. Carpatsky was already developing Rudovsko-Chervonozavodskoye gas and condensate field and Bitkov-Babchensky oil field in the country. Carpatsky had working interests in eight wells with net production of 4 MMscfd of gas equivalent.

Gas transport

Despite fairly large gas reserves, Ukraine consumes much more than it produces, relying on imports from Russia that resulted in substantial debts. Russia has accused Ukraine of illegally siphoning gas during transit, most recently in spring 2001. In August, the two countries finally agreed on the amount of Ukraine's debt ($1.34 billion) and terms of repayment.

Meanwhile, Ukraine turned to other gas suppliers, signing a five-year deal with Turkmenistan in May 2001 to buy 8.83 tcf during 2002-06, beginning with 1.4 tcf in 2002 and increasing to 1.77 tcf in 2003. The agreement was contingent on timely payments to Turkmenistan for the gas and to Russia's Itera for its delivery. Ukraine was also seeking to import gas from Iran through Armenia, where a gas pipeline link between those two countries was under construction.

Ukraine's natural gas system, which moves Russia's exports to eastern and western Europe, was being expanded and upgraded during 2000-2001 to meet greater future demand. Although most of the network's large-diameter pipelines were relatively new, much of the compressor fleet was nearing the end of its service life.

Naftogaz, Ukraine's joint-stock company, reported that gas metering stations had been reconstructed or newly built on Ukrainian borders with Russia, Poland, Slovakia, Romania, and Hungary. An in-line inspection of about 6,200 miles (10,000 km) of pipeline, including almost all transit lines, was scheduled for completion by 2005. Additional underground gas storage was also being planned to accommodate increases in Russian gas volumes through 2010.

Refining

Ukraine's privatization of its six refineries, which had seen dramatic declines in crude supplies, began to reverse the trend in 2000-2001. Russian oil major Tyumen Oil purchase 67% of the Lisichansk refiner, where throughput then increased. Lukoil's purchase of a controlling share of the Odessa refinery was also expected to result in greater supplies of crude through 2004. As of June 2001, Lukoil had invested over $2 million in reconstructing the Odessa refinery and petroleum product sales outlets throughout Ukraine.

UZBEKISTAN

CAPITAL: TASHKENT
MONETARY UNIT: SOUM
REFINING CAPACITY: 222,271 B/CD
OIL PRODUCTION: 143,000 B/D
OIL RESERVES: 594 MILLION BBL
GAS RESERVES: 66.2 TCF

The Uzbekistan government was planning to announce details on the privatization of national oil company Uzbekneftegaz and its subsidiaries in mid-2001. The government was in the process of an economic restructuring aimed at achieving an annual economic growth of 4-5%. Uzbekneftegaz's privatization followed a presidential decree in 2000 regarding new investment legislation in the oil and gas industry.

Plans included selling up to 44% of the stock in Uzneftegazdobycha (oil and gas exploration and production) and Uztransgaz (oil and gas transport). Additionally, up to 39% of Uzneftepererabotka (oil refining) and Uzburneftegaz (drilling) would be put on the block in a tender. Uzbek officials intended to dismantle during 2001 strict currency controls that had hindered foreign investment.

Uzbekistan's gas consumption has kept up with production, holding gas exports stable at about 0.5 tcf. Uzbeki gas goes to Kazakhstan, Kyrgyzstan, Russia, and Tajikistan through the Central Asia-Central Russia pipeline.

Uzbekistan's large gas reserves have allowed it to ramp up production significantly, but much of the gas requires processing due to its high sulfur content. Uzbekneftegaz was planning to commission by year-end 2001 the Shurtan Gas-Chemical Complex, which would include a gas cleaning facility, a booster compressor station, and a petrochemicals plant for producing 125,000 tpy of polyethylene and 140,000 tpy of ethylene. Output from this plant was expected to fulfill the state's domestic demand and be exported to nearby states.

The new petrochemical complex is located near the Shurtan gas fields in southwestern Uzbekistan. It also includes a butene-1 copolymer production plant. The project cost was estimated at more than $1 billion. ABB and three Japanese firms, Mitsui & Co. Ltd., Toyo Engineering Corp., and Nissho Iwai Corp., were responsible for a $600 million turnkey part of the project. Local companies are performing the remaining $400 million of general construction work.

Exploration and Development

In early 2001 the government was gearing up to offer more than 80 fields to prospective investors, in both producing and virgin territory. Eight fields with total remaining reserves of some 1.2 billion boe were open to potential foreign participation, including four in the South West Gissar Basin (Dzharkuduk, Gumbulak, South Kizilbairak, and South Tandircha) and four in the Amuarya region (North Shurtan, Shakarbulak, South Kemachi, and Umid). Uzbekistan was also seeking investment to boost production at existing fields.

In mid-2001 Uzbekneftegaz signed a PSA with Russia's Lukoil and Itera for oil and gas fields in Bukhara-Khivinsky and Gissarsky regions of Uzbekistan and an agreement for exploration in the country. Lukoil would hold 45% of the PSA, Itera 45%, and Uzbekneftegaz 10%. The field development period will be 25 years, which can be extended another 10 years. Maximum gas production under the contract would be 300-370 bcf/yr (8-10 billion cu m/yr). The companies anticipated spending $750 million to bring production from the fields to project capacity.

In July 2001 Uzbekgeofizika was planning to process and image a 108 sq mi (280 sq km) 3D seismic survey acquired over an unspecified oil field in the central Fergana basin. The state firm, which took delivery of new processing and imaging software and hardware from Cie. Generale de Geophysique, said its Tashkent center has the capability to process all Uzbekistan seismic data. IPE

CHINA

CHINA

CAPITAL: BEIJING
MONETARY UNIT: YUAN RENMINBI
REFINING CAPACITY: 4.528 MILLION B/D
OIL PRODUCTION: 3.3 MILLION B/D
OIL RESERVES: 24 BILLION BBL
GAS RESERVES: 48.3 TCF

Throughout 2001, China moved forward at all deliberate speed in oil and gas exploration, discovery, development, and production, spurred by the promise of membership in the World Trade Organization (WTO) and somewhat taken aback that it had not occurred as expected in 2000. In December 2001 China finally became a WTO member and was poised to begin trading. The country's Ministry of Foreign Trade & Economic Cooperation had said earlier it would allow private companies to import 7.2 million tons per year (tpy) of crude and 4 million tons of oil and gas products in the first year after joining the WTO.

China started accepting applications from private companies for crude import licenses the day of its WTO entry. Previously, China had allowed only four state-owned companies to import crude: China National Chemical Import & Export Co., China International United Petroleum & Chemicals Co., China National United Oil Co., and Zhuhai Zhenrong Co. The ministry planned to review the applications and award licenses early in 2002.

China's oil industry plays an important role in China's national economy and social development. It is a giant industry by international standards, ranking fifth in the world in crude oil production (after the US, Saudi Arabia, Russia, and Iran) and third in consumption of petroleum products (after the US and Japan). As China entered 2001, it had a total of 72,255 producing oil wells. Exploration in recent years has shown that the future of China's oil and gas production is in its far northwest provinces and offshore at the bottom of its seas.

China currently engages in no international LNG trade, and natural gas plays only a minor role in its total energy mix, but that situation is changing due to environmental concerns in a country where coal makes up 70% of total energy consumption. More so than coal and oil, however, gas consumption requires infrastructure, and China began improving its gas infrastructure in the late 1990s and is expected to start importing LNG by 2006.

Industry reorganization

Despite its size, China's oil and gas industry was ready for international trading because the country had been preparing itself for WTO entry over the past half-decade. Since the mid-1990s, China had been readying its refineries to handle more sour crudes. In 1998 it reorganized its land-based oil and gas assets into two geographically oriented, vertically integrated companies: China National Petroleum Corp. (CNPC), the parent of PetroChina Co. Ltd., and China Petrochemical Corp. (Sinopec).

A third company, China National Offshore Oil Corp. (CNOOC), has as its operational arm the Hong Kong-based CNOOC Ltd., a 70.6% held subsidiary of CNOOC. CNOOC Ltd. is the sole vehicle through which China carries out oil and gas exploration, development, and production activities offshore China and internationally. In 2001 CNOOC Ltd. was listed on the Hong Kong and New York stock exchanges. Before its IPO in early 2001, to become more efficient and competitive, the company slashed 15,000 jobs, sidelined noncore businesses, and kept only 1,000 workers.

Prior to restructuring, CNPC was the country's exploration and production firm but now controls assets in the north while Sinopec, the former refining and

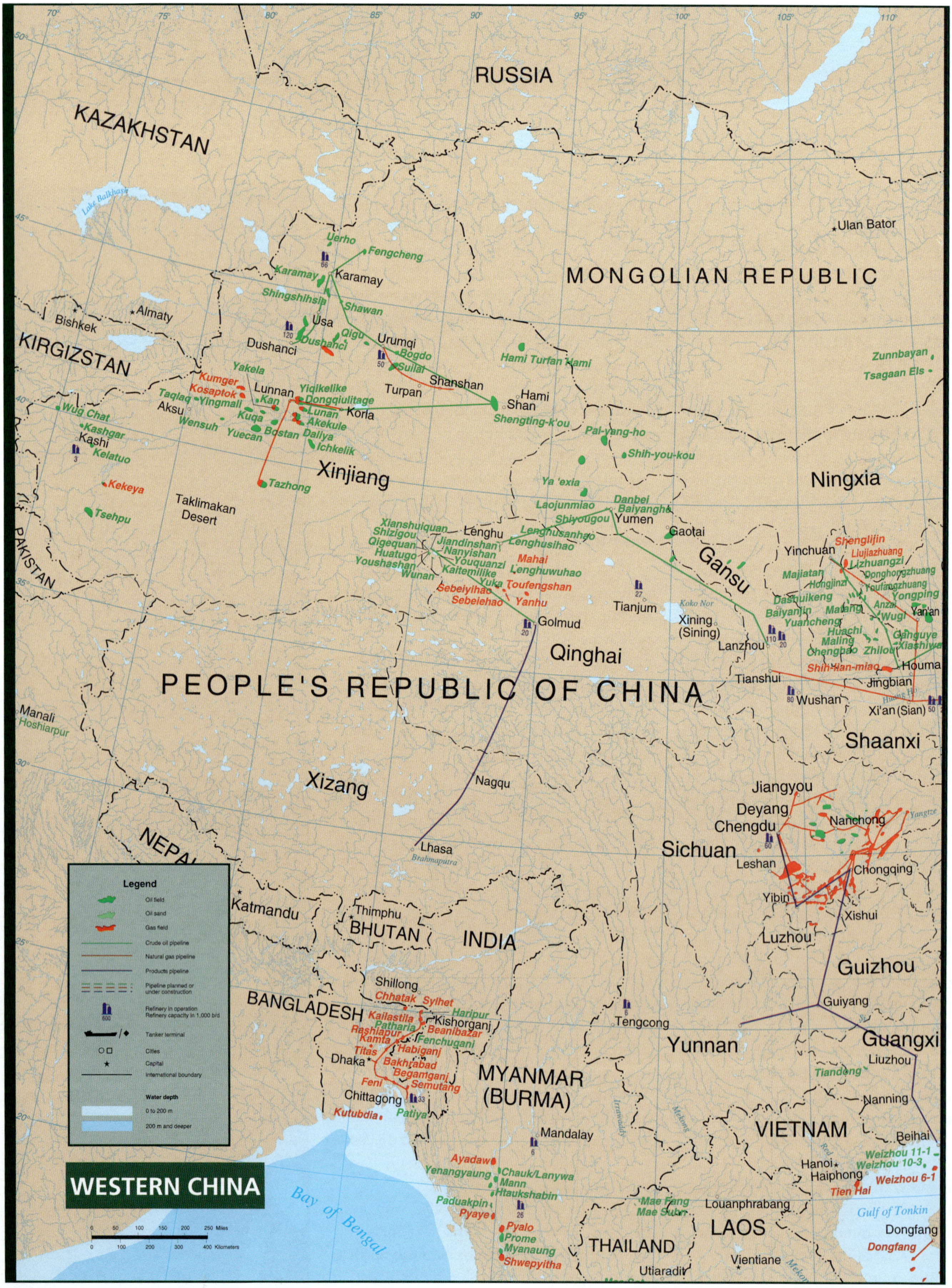
WESTERN CHINA
RUSSIA
KAZAKHSTAN
MONGOLIAN REPUBLIC
KIRGIZSTAN
PAKISTAN
NEPAL
BHUTAN
INDIA
BANGLADESH
MYANMAR (BURMA)
THAILAND
LAOS
VIETNAM
PEOPLE'S REPUBLIC OF CHINA
Xinjiang
Gansu
Ningxia
Qinghai
Xizang
Sichuan
Shaanxi
Guizhou
Yunnan
Guangxi
Taklimakan Desert
Lake Balkhash
Koko Nor
Bay of Bengal
Gulf of Tonkin
Ulan Bator
Almaty
Bishkek
Karamay
Usa
Dushanci
Urumqi
Turpan
Shanshan
Hami
Shan
Lunnan
Korla
Aksu
Kashi
Lenghu
Yumen
Gaotai
Golmud
Tianjum
Xining (Sining)
Lanzhou
Yinchuan
Yan'an
Houma
Jingbian
Tianshui
Wushan
Xi'an (Sian)
Nagqu
Lhasa
Jiangyou
Deyang
Chengdu
Nanchong
Leshan
Chongqing
Yibin
Xishui
Luzhou
Guiyang
Tengcong
Liuzhou
Nanning
Beihai
Hanoi
Haiphong
Louanphrabang
Vientiane
Utiaradit
Dongfang
Mandalay
Manali
Hoshiarpur
Katmandu
Thimphu
Shillong
Kishorganj
Dhaka
Chittagong
Uerho
Fengcheng
Shingshihsia
Shawan
Qigu
Bogdo
Suilai
Hami Turfan Hami
Yakela
Kumger
Kosaptok
Taqiaq
Yingmail
Kan
Kuqa
Wensuh
Yuecan
Bostan
Yiqikelike
Dongqiulitage
Lunan
Akekule
Daliya
Ichkelik
Tazhong
Wug Chat
Kashgar
Kelatuo
Kekeya
Tsehpu
Shengting-k'ou
Pal-yang-ho
Shih-you-kou
Ya 'exia
Laojunmiao
Danbei
Baiyanghe
Shiyougou
Xianshuiquan
Shizigou
Qigequan
Huatugo
Youshashan
Wunan
Jiandinshan
Nanyishan
Youquanzi
Kaitemilike
Lenghusanhao
Lenghusihao
Mahai
Lenghuwuhao
Yuka
Toufengshan
Sebeiyihao
Sebeierhao
Yanhu
Zunnbayan
Tsagaan Els
Shenglijin
Liujiazhuang
Lizhuangzi
Donghongzhuang
Youfangzhuang
Yongping
Majiatan
Hongjinzi
Dasuikeng
Baiyanjin
Mabang
Anzai
Wugi
Yuancheng
Huachi
Maling
Chengpao
Zhilou
Ganguye
Xiashiwa
Shih-tan-miao
Chhatak
Sylhet
Haripur
Kailastila
Patharia
Beanibazar
Rashiapur
Kamta
Fenchuganj
Titas
Habiganj
Bakhrabad
Begamganj
Semutang
Feni
Kutubdia
Patiya
Ayadaw
Yenangyaung
Chauk/Lanywa
Mann
Htaukshabin
Paduakpin
Pyaye
Pyalo
Prome
Myanaung
Shwepyitha
Mae Fang
Mae Suon
Tiandong
Weizhou 11-1
Weizhou 10-3
Weizhou 6-1
Tien Hai
Dongfang
Brahmaputra
Yangtze
Huang He
Irrawaddy
Mekong
Red
Legend
Oil field
Oil sand
Gas field
Crude oil pipeline
Natural gas pipeline
Products pipeline
Pipeline planned or under construction
Refinery in operation Refinery capacity in 1,000 b/d
Tanker terminal
Cities
Capital
International boundary
Water depth
0 to 200 m
200 m and deeper
0 50 100 150 200 250 Miles
0 100 200 300 400 Kilometers

EASTERN CHINA
RUSSIA
MONGOLIAN REPUBLIC
PEOPLE'S REPUBLIC OF CHINA
NORTH KOREA
SOUTH KOREA
JAPAN
TAIWAN
OKINAWA
VIETNAM
LAOS
THAILAND
PHILIPPINES
SEA OF JAPAN
BOHAI SEA
YELLOW SEA
EAST CHINA SEA
SOUTH CHINA SEA
Gulf of Tonkin
HAINAN DAO
Heilongjiang
Jilin
Liaoning
Beijing
Tianjin
Hebei
Nei Mongol
Ningxia
Gansu
Shanxi
Shaanxi
Shandong
Henan
Jiangsu
Shanghai
Anhui
Hubei
Sichuan
Zhejiang
Hunan
Jiangxi
Guizhou
Fujian
Yunnan
Guangxi
Guangdong
Ulan Bator
Komosamolsk
Khabarousk
Qing 'an
Daqing
Harbin
Qianguo
Huadian
Dongfeng
Fusong
Chongjin
Aershan
Saihantala
Zunnbayan
Tsagaan Els
Chifeng
Tieling
Fushun
Shenyang
Liaoyang
Anshan
Jinlin
Yingkou
Dangdong
Chin-hsi
Changli
Jin Xian
Luda
Lushun (Dairen)
Pyongyang
Seoul
Onsan
Ulsan
Pusan
Hiroshima
Yosu
Sasebo
Oita
Kashima
Nagasaki
Qilu
Huanghua
Yanshan
Zhanhua
Kenli
Zhaoyuan
Weifang
Gaoqing
Qingdao
Lianyungang
Xiangshui
Yancheng
Shih-you-kou
Danbei
Yumen
Baiyanghe
Gaotai
Shandan
Tianjum
Koko Nor
Ulan
Xining (Sining)
Lanzhou
Yinchuan
Shengliiin
Liujiazhuang
Yulin
Suide
Yan'an
Pingliang
Shih-lian-miao
Xi'an (Sian)
Luoyang
Anyang
Zhengzhou
Fangcheng
Baoying
Yangzhou
Zhenhai
Ningguo
Hangzhou
Zongyang
Yingshan
Wuhan
Yichang
Mianyang
Duchang
Jiangyou
Chengdu
Nanchong
Suining
Wusheng
Dazu
Leshan
Chongqing
Garze
Batang
Ceshui
Anshun
Kunming
Liuzhou
Bose
Tiandong
Mashan
Nanning
Jiangmen
Lianjiang
Maoming
Guangzhou
Hong Kong
Beihai
Hanoi
Vientiane
Haikou
Dongfang
Yinggehai
Yacheng
Yacheng 13-1
Hemgyang
Fuzhou
Taipei
Taoyuan
Kaohsiung
Pingtung
Donghai
Bo Yuan Ting
Ping Hu
Weizhou 11-1
Weizhou 10-3
Weizhou 6-1
Tien Hai
Wenchang 8-3
Wenchang 9-2
Wenchang 14-3
Wenchang 13-1
Wenchang 19-1
Wenchang 13-2
Jinfeng
Xijiang 24-3
Xijiang 30-2
Lufeng 13-1
Huizhou 26-1
Lufeng 15-1
Huizhou 21-1
Liuhua 4-1
Liuhua 11-1
Mae Suhn
0 100 200 300 Miles
0 200 400 Km

marketing firm, controls assets in the south. In 2000 China closed many small refineries that would have been too costly to renovate to meet its product quality standards.

Additionally, China's reorganization has brought several market-oriented reforms including upgraded management authority to manage the businesses, retail refinery-gate pricing practices, and some moves toward wholesale and retail pricing policies based on international pricing. At the end of 2001, the direction of this progress was still forward, but a slackening pace was attributed to a weakening demand for oil products worldwide.

Legal framework for cooperation

China's government has reaffirmed CNOOC's legal status by allowing it exclusive rights to sign production-sharing contracts (PSCs) with foreign contractors following a loosening of regulations by the State Council. This means the government has also revised rules to remove restrictions on technology transfer and requirements for domestic procurement of components in order to provide a level playing field for all oil field service contractors, domestic and international, under WTO rules. This is expected to benefit CNOOC's exploration and production operations, because it needs foreign capital and technology for sustained offshore oil development.

CNOOC's PSC model calls for foreign companies to cover the cost of exploring the offshore oil and gas blocks licensed to them. If and when they discover economic hydrocarbon reserves, CNOOC has the option to take a 51% stake in field development and production and inject 51% of the capital needed. CNOOC has entered into a total of 148 PSCs and agreements with 70 foreign oil and gas companies from 18 countries and regions. Those contracts were valued at more than $7 billion. Of these, 35 were still valid in late 2001, while the others had expired, and the number of China's new PSCs declined during 2001. Only two were signed, involving a total of $25 million, compared with six in 2000. No new licensing round has been announced since 1999, but CNOOC has begun to offer "reverse PSCs," in which foreign companies join in developing oil and gas reserves discovered by CNOOC to bring them into production more quickly.

Through joint efforts, CNOOC has sunk more than 400 exploratory wells and discovered nearly 100 oil and gas-bearing structures. CNOOC was not planning to offer many new offshore blocks over the next few years because in Bohai Bay and the South China Sea, most of the prospective blocks were already under exploration or development. In the East China Sea, CNOOC and Sinopec expected to tie up with units of Royal Dutch/Shell Group and Unocal Corp. to explore the Xihu trough.

CNOOC's position

CNOOC holds net proven reserves of 1.8 billion boe, comprised of 1.3 billion bbl of crude and condensate and 3.3 tcf of natural gas, and net production of 239,337 boe/d at the start of 2001. It held interests in 45 crude oil and natural gas properties in Bohai Bay, the western and eastern South China Sea, and the East China Sea. CNOOC was producing oil from 25 fields and gas from three fields offshore, yet China's deepwater areas are scarcely touched. The producing field in the deepest water is Liuhua in less than 1,000 feet (300 m) of water in the Pearl River Mouth basin.

If China meets its offshore oil production targets, offshore fields would become the second-largest source of China crude, after Daqing field in northeastern China. Daqing has been producing more than 365 million bbl/year for more than 20 years, but its production was expected to level off and drop by 2010. Obtaining more supplies from offshore is essential for China, where total production has remained largely unchanged for the last five years. Total output was expected to rise by only a small margin if not level off between now and 2005. Demand was forecast at 1.79 billion bbl in 2005, of which 548 million bbl would likely have to be imported even if CNOOC achieves its production targets.

To meet production goals, CNOOC and its PSC partners were developing five fields, including three in Bohai Gulf and two in the South China Sea. The company brought two of the five online during 2001. The three still under development are giant Peng Lai 19-3 in Bohai and Wenchang 13-1/13-2 and Dongfang 1-1 in the South China Sea. Development of these five fields will ultimately bring 726.7 million boe of undeveloped hydrocarbon reserves into production.

CNOOC Ltd. said that 2001 offshore output in the third quarter rose 11% year-on-year, to 260,189 boe/d. The company attributed the higher production figure mainly to a 59% rise in crude oil output in Bohai Bay. China's number of offshore wells nearly doubled, with 9 new wells being added in 2001 to 10 a year earlier. Natural gas output rose from 190 MMcfd in the first half of the year to 202 MMcfd in late 2001. CNOOC's oil and gas revenues were also increasing despite lower average realized oil prices.

CNOOC was planning to drill wells independently at a rate of 15-20/yr through 2006, replacing existing wells with new ones to maintain production growth. The company earmarked $4.7 billion for E&P during 2001-03: $1.046 billion in 2001, $1.681 billion for 2002, and $1.935 billion for 2003. Of its total 2001 expenditures, $600 million went for exploration. The company intended to keep its total production expenses, or lifting costs, below $10/bbl for the next 8-10 years. Its 2000 total production expense was $8.22/bbl.

Daqing field

China's largest oil field, Daqing, has produced above 50 million tpy of crude for the last 41 years, but production is increasingly expensive, with water cuts as high as 85%. In 2001, Daqing oil field produced 51.5 million tons, down 1.5 million tons from 2000. Between 2001 and 2005, the field was expected to produce 45-50 million tpy of crude oil, and output was expected to drop further, to 37-45 million tpy after 2006.

Daqing operator PetroChina planned to explore and produce marginal prospects outlying Daqing's main production blocks during the next 5 years, expecting to identify 300 million tons of probable crude reserves with the goal of recovering 54 million. PetroChina anticipated that gas resources could be 856 bcf (23 billion cu m), of which 596 bcf (16 billion cu m) could be recoverable. Daqing was also China's major source of crude exports and was due to ship 3 million tons to Japan under a government-arranged program.

Xinjiang region

Xinjiang, an autonomous region in the northwestern corner of China, covers one sixth of China's total land area and is the largest provincial region. Xinjiang is a strategically sensitive area, bordering Russia and Mongolia to the north and Kazakhstan, Kyrgyzstan, Tajikistan, Afghanistan, and Pakistan to the west. The Junggar basin, which lies in Xinjiang's northern section, contains one of China's largest producing oil field complexes, Karamay. Surface geology studies have shown that the basin's several structural units are all prospective for oil and gas. Since 1955, when the first commercial oil discovery was made at Karamay, dozens of oil fields have been developed in the vicinity. Combined, the Junggar, Tarim, and Turpan basins provide nearly 20% of the nation's total crude oil output.

Gas reserves found in the Tarim basin of the Xinjiang province were projected to rise 30% to 24.2 tcf (650 billion cu m) by the first half of 2002. PetroChina's oil fields in the province might also hold about 623 million bbl.

Although Xinjiang province's present and projected oil production exceeds its own current and projected oil consumption, national policies require the province to ship 50% of its crude oil output to other provinces within China, resulting in an oil deficit in Xinjiang itself. Regional cooperation is thus important to alleviate potential future energy shortages in the area.

Most basic to realizing the potential oil and gas production in Xinjiang province is transportation to the population centers at Shanghai and throughout the country. Toward this end, the West-East Pipeline, China's second largest infrastructure project, was proposed but not yet underway. China also has ambitious plans to build distribution networks to feed gas into households and industries in the eastern and southern coastal provinces over the next 10 years.

Offshore prospects

Given rising demand, slumping supply in land-based wells, and bleak prospects for higher onshore production, China was looking offshore for much of its future production. The country expected offshore oil production by 2005 to more than double 2000's level, to 392 million boe/yr (40 million tons/year of oil equivalent), including 221 million bbl of crude oil (30.28 million tons) and 406 bcf (10.9 billion cu m) of natural gas.

More than 50% of total oil and gas production will be based on PSCs with foreign companies that operate off China. CNOOC Ltd. expected to produce 176 million boe in 2005, compared with the company's estimated 2001 production of 95 million boe. China was putting emphasis on output from new fields because of rising demand and falling crude oil output at Daqing field.

China planned to step up exploration in frontier basins off its East Coast as part of its drive to establish an integrated gas industry in some of its most populous areas. Although large volumes of gas for consumption in eastern China were expected to originate as far west as Xinjiang's Tarim basin, gas supplies from the East China Sea, LNG imports, and possibly the Taiwan Strait will be needed to supplement onshore gas supplies.

Exploration wells drilled off China Jan.-Sept. 2001

	Well name	Well type	Operator	Area
Bohai Gulf	Caofeidian 11-2-1	Wildcat	Kerr-McGee	Block 04/36
	Caofeidian 12-1-5	Appraisal	Kerr-McGee	Block 04/36
	Bozhong 25-1-2	Wildcat	ChevronTexaco	Block 11/19
South China Sea	Wenchang 6-1E-1	Wildcat	Devon	Block 26/35
	Huizhou 29-1-1	Wildcat	Burlington Resources	Block 16/21
	Songtao 36-1-1	Wildcat	CNOOC	—

China's offshore oil fields in production between 1997-2005

	Name	Year	Field type	Operator
Bohai Gulf				
	Jinzhou 20-2	1997	Gas	CNOOC
	Qikou 17-3; Qikou 18-1	1997	Oil, gas	CNOOC
	Boxi	1998	Gas	CNOOC
	Jinzhou 9-3	1999	Oil	CNOOC
	Qikou 17-2	2000	Oil	CNOOC
	Suizhong 36-1	2000	Oil	CNOOC
	Qinghuangdao 32-6	2001	Oil	CNOOC, BP, ChevronTexaco
	Suizhong 36-1 Phase II	2001	Oil	CNOOC
	Penglai 19-3	2002	Oil	CNOOC, Phillips
	Qinghuangdao 32-6 Phase II	2002	Oil	CNOOC, BP, ChevronTexaco
	Bonan	2003	Oil, gas	CNOOC
	Nanbao 35-2	2004	Oil	CNOOC
	Penglai 19-3 Phase II	2004	Oil	CNOOC, Phillips
South China Sea				
	Xijiang 24-1	1997	Oil	CNOOC, Phillips, Pecton
	Lufeng	1997	Oil	CNOOC, Statoil
	Huizhou 32-5	1999	Oil	CACT
	Weizhou 12-1	1999	Oil	CNOOC
	Weizhou 11-4	1999	Oil	CNOOC
	Huizhou 26-1N	2000	Oil	CACT
	Wenchang 13-1/13-2	2002	Oil	CNOOC, Husky Oil
	Dongfang 1-1	2003	Gas	CNOOC
	Dongfang 1-1 Phase II	2005	Gas	CNOOC
East China Sea				
	Pinghu	1998	Gas, oil	CNOOC, Sinopec

East China Sea

Until recently, CNOOC Ltd. has concentrated on oil development in China's two premier offshore producing areas, the South China Sea and Bohai Bay. Exploration was only beginning in the East China Sea, and Pinghu gas and oil was the only producing property off eastern China. So far, exploration in the area has been disappointing, however. Licenses had been issued for 20 blocks in the East China Sea, and ten foreign companies including Texaco, Shell, and Phillips signed 18 PSCs for the blocks. In the end, they drilled 14 dry holes. Nevertheless, CNOOC said it expects the area to become an important gas-producing base in the future.

Pinghu

Progress on ramping up exploration in the East China Sea was slow during 2001. Discovered in 1992, Pinghu began producing crude oil in 1998 and gas and condensate in 1999. The field covers 93 sq mi (240 sq km) in 287 ft of water. Pinghu has reserves of 402 bcf (10.8 billion cu m) of gas, 13 million bbl of condensate, and 79 million bbl of light crude. The field was under joint ownership of the Shanghai Municipal Government (Shanghai Oil & Gas Corp.) 40%, Sinopec Star Oil & Gas Co. Ltd. 30%, and CNOOC 30% of the 5.1 billion-yuan total investment. Pinghu produced 9.68 bcf (260 million cu m) of gas in 2000, still below its design capacity of 18.1 bcf/yr (486 million cu m/yr).

From the Pinghu platform on the continental shelf, gas is piped about 250 miles (400 km) to Shanghai, where it is consumed in power generation, domestic uses, and the chemical sector. Oil and condensate are piped almost the same distance to Ningbo. Initially, Pinghu production was to continue for 10-15 years. Early in 2001, China made public a plan to double Pinghu's gas production to 71-78 MMcfd by 2003 and double reserves, which would markedly extend field life.

Xihu trough

CNOOC took steps in 2001 to develop fields on the Chunxiao structure, which is intended to supply gas to Zhejiang Province. Chunxiao lies in the Xihu trough 50 miles (80 km) southeast of the Pinghu field in the East China Sea. CNOOC said it had verified gas reserves of 1.9 tcf in the Chunxiao area and planned to start production from Chunxiao by 2004. Production was anticipated to be 71 bcf the first year and 282-353 bcf/yr by 2010. Another Chunxiao area find is Tianwaitian, which was appraised in 2001. Tianwaitian-3 flowed 24 MMcfd of gas and 175 b/d of condensate. It was drilled to 12,336 feet (3,760 m) MD and found more than 10 gas pay zones.

CNOOC was to start talks in late 2001 with units of Royal Dutch/Shell and Unocal Corp. in regard to joint exploration of the Xihu trough, which holds an estimated 11.2 tcf (300 billion cu m) of probable natural gas reserves. CNOOC planned to drill four to six exploration wells in 2002 at a cost of $200 million in a total budget of $1.5 billion. Shell has the option to participate because it signed a strategic alliance in late 2000 covering exploration, development, and production of gas in the Xihu trough. Unocal has conducted a geological survey in the East China Sea with Sinopec Star Petroleum E&P Co.

Vicky discovery

Another undeveloped accumulation in the southwestern East China Sea is Lishui 36-1-1, or Vicky. Primeline Petroleum Corp. made the Vicky discovery in 1997. The discovery well, in 295 ft (90 m) of water 62 miles (100 km) offshore, brought almost 10 MMcfd of gas, 117 b/d of condensate, and 15.7 b/d of water from 108 ft (33 m) of Paleocene Lingfeng sands and shales at about 7,380 ft (2,250 m). The gas at Vicky was analyzed at 55% methane and 34.5% carbon dioxide.

Primeline acquired 3D seismic data for the discovery area and formulated plans for its appraisal for potential gas marketing to the city of Wenzhou. Because of the deposit's relatively small size, substantial price increases would be needed to justify its development. In mid-2000 Primeline agreed to transfer its stake to CNOOC. Primeline retains a 15% interest and estimates Vicky reserves at 705 bcf of gas and 8 million bbl of condensate.

Taiwan Strait

Taiwan has made several attempts to explore portions of the strait to its west during the past decade and in late 2001 revoked a ban on direct trade and investment with mainland China that had been in effect since 1949. CNOOC and Taiwan's Chinese Petroleum Corporation (CPC) also signed a letter of intent to jointly explore for oil and gas in the Taiwan Strait. The historic project involves seismic testing and drilling in the Tainan basin, a 5,946 sq mi (15,400 sq km) area between southern Taiwan and Guangdong and Fujian provinces.

Leading up to the late 2001 letter of intent, CNOOC and CPC had signed a preliminary agreement in 1996, and Taipei ratified the agreement in 1998. The two companies jointly ran geophysical surveys in 1999 and reported structures worthy of evaluation. CPC was to process data acquired jointly with CNOOC since 1996 on the Tainan-Chaoshan block. A CPC spokesman noted that both sides feel the need for exploration considering that initial study showed possible substantial oil and gas reserves.

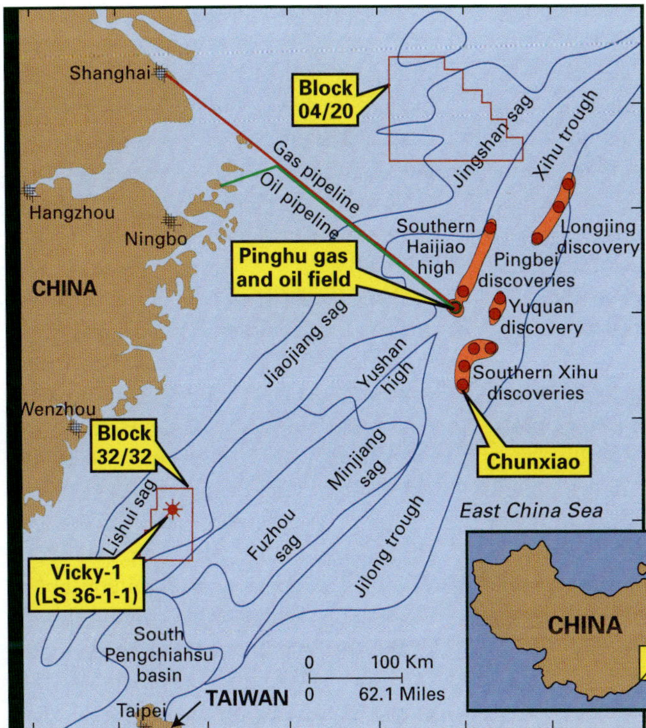

Exploration of East China Sea was still in its infancy, but China planned to step up exploration in these frontier basins off eastern China.

The block to be explored covers almost 6,000 sq mi between Taiwan and Guandong and Fujian provinces. Taiwanese press reports described the area as lying southeast of the Zhu Jiang river estuary, which is in Guangdong Province. The distance between Taiwan and Mainland China is 80 miles at its narrowest point. CNOOC and government-owned CPC have cooperated in conducting geophysical surveys in the area since 1996, and the new agreement is seen as a sign of strengthening economic ties between China and Taiwan.

Offshore drilling

CNOOC originally planned to drill 20 marine exploration wells in 2001, and its foreign partners were to have drilled 27 to 38 wells off China. The company also planned to conduct 6,200 miles (10,000 km) of 2D and 463 sq mi (1,200 sq km) of 3D seismic surveys in 2001. Foreign operators were to conduct 1,118 miles (1,800 km) of 2D and 773 sq mi (2,000 sq km) of 3D seismic surveys in 2001.

CNOOC had completed drilling 9 of the 20 wells as of October 2001, and foreign companies completed 24 of the exploration and appraisal wells planned for the year. CNOOC struck hydrocarbons in four out of the nine, and foreign companies were successful in 17 of the 24 wells drilled. CNOOC and its foreign partners were also drilling five wildcats.

Bohai Gulf

Bohai Gulf continued to be CNOOC's key area for crude oil growth over the near term, following large discoveries made by Phillips. As of mid-2001, CNOOC and its foreign partners had identified 19.7 billion bbl of proven crude oil reserves in a large number of fields in Bohai. Over the next four years, China plans to build 50 new platforms and drill 1,100 wells in Bohai Gulf at a cost of $6 billion, including investments under PSCs, in an effort to raise output to 146 million bbl/year from 26.3 million. Of CNOOC's total $4.7 billion E&P budget for 2001-03, $2 billion was allocated to Bohai.

Peng Lai

Peng Lai 19-3, which Phillips discovered in 1999, remained China's largest offshore oil discovery, with crude reserves of 4.4 billion bbl. CNOOC and Phillips signed an agreement late in 2000 to jointly develop the field. Initial development called for a drilling platform, a floating production, storage, and offloading vessel, off-loading facilities, and two offshore oil pipelines. Peng Lai 19-3 covers 19 sq mi (50 sq km) in 75 ft (23 m) of water near the Longkou coast. Pay zone thickness averages 492 ft (150 m).

Phillips started production at Peng Lai 19-3 in late 2001, with initial output of 40,000 b/d from 24 production wells. Production will eventually be expanded to 120,000-150,000 b/d by 2005 when the second phase is brought on line, which will account for 43% of the total crude production CNOOC expects to produce from Bohai. CNOOC Ltd. owns 51% of Peng Lai and Phillips 49%. Phillips China has invested \$2 billion at Peng Lai field, which analysts said could hold China's second-largest reserves after Daqing.

Qinghuangdao, Suizhong

Bohai's crude output rose in late 2001 with the startup of Qinghuangdao 32-6 and Suizhong 36-1 fields. Qinghuangdao 32-6, 130 km east of Tianjin City, was producing an initial 25,000 b/d of oil. Production was expected to peak at 65,000 b/d when the second phase starts operation in 2002. Discovered in 1995, Qinghuangdao has 103 million bbl of reserves. China entered into a reverse PSC with BP PLC and Texaco in 1998 to develop the field. Again, CNOOC has a 51% interest and BP and ChevronTexaco 24.5% each.

During 2001, CNOOC commissioned three new platforms in the self-financed second phase of Bohai's Suizhong 36-1 field. The new platforms brought Suizhong's oil production to 69,000 b/d. Suizhong 36-1, discovered in 1987 in the southern part of Liaodong Bay, was 100% owned and operated by CNOOC. It is currently China's largest self-financed field, but the amount of the expenditure was not available.

Other Bohai action

Meanwhile, CNOOC has raised its crude reserve portfolio with a new discovery at its Nanbao field in the southern part of Bohai. The Nanbao 35-2-7 appraisal well was expected to increase the reserve size at the Nanbao 35-2 field by 25%. The well flowed 432 b/d of 16° gravity oil when drilling reached a depth of 4,833 ft (1,473 m). It encountered a 210-ft (64-m) pay zone. Nanbao 35-2-7 is less than a mile (1.4 km) northeast of an earlier wildcat well, Nanbao 35-2-1, which led to the discovery of Nanbao 35-2 in 1996. The initial appraisal allowed CNOOC to place the field's reserves at 65 million bbl. CNOOC, which owns the field, was drilling another appraisal well and planned to bring Nanbao on line in 2004.

CNOOC and Shell Exploration China Ltd. planned to start examining development of oil and gas reserves on the Bonan Block in southeastern Bohai. The block contains several oil and gas fields 65 miles north of Longkou in 330 ft of water that hold a combined 1 tcf of gas and 600-700 million bbl of oil in place. Two key structures, Bozhong 26-2 and Bozhong 28-1, and three other fields hold a combined reserve of more than 160 million bbl of liquids and around 300 bcf of gas. The block is close to several metropolitan areas of Shangdong Province, namely Longkou, where a pipeline would make landfall and continue east to Yantai and south to Qingdao on the Yellow Sea.

Late in 2001, Sinopec's Shengli Oil & Gas Co. unit began construction of one drilling and one production platform for the Zhao Dong Block 11/19 in western Bohai just offshore from the Dagang fields. Apache Corp., Houston, ordered the platforms for production of oil in 15 ft (4.5 m) of water on the 49,000-acre block starting in 2002 or 2003.

Noble Affiliates Inc. was to start production in 2002 from Cheng Dao Xi Block A field in 25 ft of water in southwestern Bohai. The \$101 million development plan called for 16 wells, including injectors, to feed facilities with a capacity of 10,000 b/d connected to a 5-mile pipeline to shore infrastructure in the Shengli fields.

South China Sea

The Pearl River Mouth basin in the South China Sea was the mainstay of China's offshore oil production. Four fields – Huizhou, Liuhua, Lufeng, and Xijiang – produce a combined 210,000 b/d of oil. Huizhou was the biggest at 94,000 b/d. Huizhou 32-5, which started up in 1999 on Block 16/08, was the latest field to be brought on line and the first project in the South China Sea to use production technology allowing subsea links to an existing platform, Huihzou 26-1. Three wells at Huihzou 32-5 are tied to Huihzou 26-1, 3.5 km northwest. The crude oil is then processed through the surface facilities on the Huizhou 26-1 platform. Huizhou 32-5 was operated by the CACT Group, a consortium of CNOOC Ltd., Agip China BV, Chevron Overseas Petroleum Ltd., and Texaco China BV.

Husky Oil China Ltd., a unit of Husky Energy Inc., could emerge as a major foreign player in the South China Sea, where it was working several PSCs. Along with CNOOC, Husky was developing the Wenchang 13-1/13-2 blocks in the western South China Sea, 85 miles (136 km) east of Hainan Island. CNOOC licensed the two blocks to Husky in a reverse PSC. The two blocks have 60.6 million bbl of oil reserves, with CNOOC holding a 60% of the working interest and Husky holding the remainder. Wenchang was expected to come on line in the first half of 2002 with a maximum design production capacity of more than 50,000 b/d.

Also with Husky, CNOOC signed in mid-2001 a PSC to explore Block 39/05. The 2,200 sq mi (5,700 sq km) block is 62 miles (100 km) east of Hainan Island. Under PSC terms, Husky committed to drill a minimum of three exploration wells during a seven-year period.

Dongfang, Angsi

In addition to the Wenchang oil fields that CNOOC is developing with Husky, CNOOC in 2001 was developing Dongfang 1-1, a gas-prone field west of Hainan Island. CNOOC was injecting 5 billion yuan to build 59.6 bcf/yr (1.6 billion cu m/yr) of gas production capacity at Dongfang by late 2003, expandable to 89.3 bcf/yr (2.4 billion cu m/yr) by 2005. By then, Dongfang will be China's second largest offshore gas field in the South China Sea after Yacheng 13-1, operated by CNOOC and BP. Dongfang 1-1 traps 3.35 tcf (90 billion cu m) of gas reserves.

For Dongfang gas production, CNOOC has signed a package contract for developing a few downstream projects in Hainan Province, committing about 10 billion yuan to building a chemical plant and gas pipeline grid and retrofitting a Hainan power plant with gas.

In late 2001, ExxonMobil Corp. brought the South China Sea's Angsi field on line, with Petronas Carigali Sdn. Bhd. (a branch of Malaysian national oil company Petronas) beginning first production six months ahead of schedule and 30% below the original approved cost. ExxonMobil Exploration & Production Malaysia Inc. is Petronas Carigali's 50% partner in the development.

Angsi's initial flow was 15,000 b/d of oil and 60 MMscfd of gas. At peak, Angsi should produce 65,000 b/d of oil and 450 MMscfd of gas (10% of Malaysia's current oil production and 17% of its current gas production, respectively). An integrated oil and gas central processing platform is connected to a 52-well drilling platform via a 321-foot (100-m) bridge, as well as a 32-well satellite drilling platform. There are connections to existing gas production facilities at Guntong-D and Seligi-A, a connection to the existing oil production facilities at Tapis, and a new 103-mile (166-km) pipeline from the Angsi complex to the onshore receiving facility.

Discoveries

One of the few oil finds in the South China Sea in 2001 was Liuhua 19-3, about 137 miles (220 km) south of Hong Kong. CNOOC struck oil at the Liuhua 19-3-1 well, TD 4,921 feet (1,500 m). The well brought 1.8 bcfd (49,000 cu m)/day of gas through a 44/64-in. choke. CNOOC planned to drill more wells around the discovery well.

CNOOC also made a discovery during 2001 at the Wenchang structure of Qiongdongnan basin southeast of Hainan island. The CNOOC-financed Wenchang 15-1 well flowed 4,760 b/d of 35° gravity oil and 330 Mcfd of gas on a 16 mm choke from 4,101 feet (1,250 m). The pay zone is 40 ft thick, and the structure covers 7.1 sq mi (18.5 sq km). The company will appraise the find in 2002. It was also drilling the Songtao 36-1-1 wildcat late in the year in the Qiongdongnan basin 37 miles (60 km) southeast of Hainan Island.

Devon Energy Corp., Oklahoma City, made a discovery on Block 16/02 in the eastern South China Sea. The Huizhou 9-2 discovery well flowed 1,602 b/d of oil and 2.1 MMcfd of gas. Also, the CACT Group had an initial flow of 4,700 b/d of oil from the Huizhou 19-2 well.

One of CNOOC's appraisal wells, Weizhou 12-1-6, had a flow of 5,345 b/d of crude oil and 1.97 MMcfd of gas. It is on the Yulin 35 Block in Beibu Gulf in the western part of the South China Sea.

CNOOC signed another PSC in 2001 with Santa Fe Energy Resources Corp., now part of Devon Energy, to explore and develop Block 27/10, which covers 2,529 sq mi (6,546 sq km). The contract was the eighth between Santa Fe and CNOOC, of which six are activated.

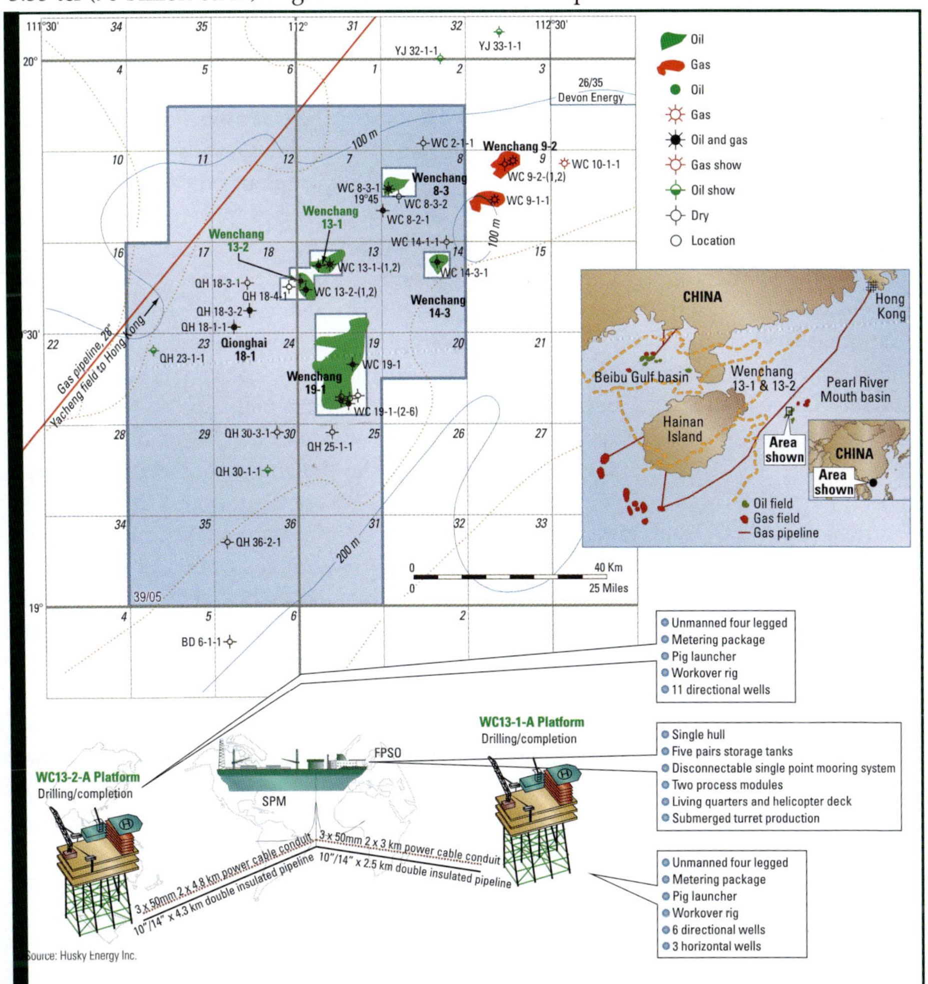

Husky Oil China Ltd. was poised to emerge as a major foreign player in South China Sea, where it is working several production sharing contracts in areas shown here.

Santa Fe's major discoveries in Pearl River Mouth basin are Panyu 4-2 and Panyu 5-1.

CNOOC's gas strategy

CNOOC has an ambitious plan to build distribution networks to feed gas into households and industries in China's eastern and southern coastal provinces over the next 10 years. The company was on track to raise its natural gas production to 406 bcf/yr (10.9 billion cu m/yr) by 2005, up from 171 bcf (4.6 billion cu m) in 2000. An initial plan called for the company to verify 13.8-22.3 tcf (370-600 billion cu m) of gas by 2005. CNOOC's gas production was being transported to users in Hong Kong, Tianjin, Liaoning, and Shanghai.

The company was involved with Shell in conducting a feasibility study regarding construction of a 1,240-mile (2,000-km) gas pipeline along China's southern and eastern coasts. Although the government has discounted CNOOC's ability to have enough gas resources to make the pipeline viable, CNOOC has taken a different approach, that of building the pipeline piecemeal, province by province. It recently struck a deal to form a joint venture with Zhejiang local companies to build a distribution network and has signed agreements with Fujian and Shandong provincial governments to build gas facilities including terminals to import LNG.

To Zhejiang, the company would supply gas to be produced from the Xihu trough in the East China Sea. CNOOC expected to raise gas reserves at Xihu to 5.6-7.4 tcf (150-200 billion cu m) by 2003 from 2 tcf (54 billion cu m). CNOOC would supply part of the gas produced from Bohai to Shandong, and CNOOC and the Shandong government would build an LNG terminal in the province to supplement the Bohai gas. However, the central government was unlikely to approve CNOOC's gas expansion plan until its flagship LNG project in southern China's Guangdong province becomes viable.

LNG project

CNOOC was spearheading a consortium made up of BP and local Guangdong and Hong Kong businesses to conduct a feasibility study to build a 3 million tpy LNG receiving terminal at Shenzhen in Guangdong province. The project had government approval, and CNOOC began an international tender for LNG supply contracts. CNOOC issued bidding documents to Australia, Indonesia, Russia, Qatar, Malaysia, and Yemen. The Guangdong terminal was estimated to cost $616 million. Expected to come on stream in 2005, it would also involve building a 186-mile (300-km) pipeline.

In a bid to secure enough LNG supplies for the terminal, CNOOC signed a memorandum of understanding with Chevron Australia Pty. Ltd. to explore the feasibility of acquiring an equity stake in the Gorgon area off Western Australia. Gorgon field is the largest natural gas field discovered in Australia, with confirmed hydrocarbon reserves in excess of 9.6 tcf.

In a late 2001 agreement between Australia's North West Shelf (NWS) LNG project and CNOOC Ltd., China contracted to invest in the development of the North West Shelf gas project for the sale of LNG to China markets. No financial details of the agreement were released, but it was likely that timing of the agreement would be linked to progress on the Guandong LNG project on mainland China. This marked China's first foray into the international LNG market and signaled its intention to use CNOOC Ltd. as a vehicle for future international investment. Wei Liucheng, CNOOC chairman and CEO, said that CNOOC Ltd. has long expressed the desire to secure upstream assets for LNG import and that it is looking forward to similar agreements with other potential suppliers.

Woodside Energy Ltd. is the operator of Australia's NWS LNG project, and partners include BHP Billiton Pty. Ltd., BP Developments Australia Pty. Ltd., TexacoChevron Australia Pty. Ltd., Japan Australia LNG (MIMI) Pty. Ltd., and Shell Development (Australia) Pty. Ltd. Each partner has a 16.67% stake in the project.

Sebei-Xining-Lanzhou pipeline

The Sebei-Xining-Lanzhou pipeline, which runs 590 miles (950 km) from Qinghai Province to neighboring Gansu Province, is one of the central government's biggest projects to develop the plentiful natural resources of the economically backward western regions. In late 2001, PetroChina said that its $280 million gas pipeline in Northwest China will break even in 2004 and generate profits within four years.

The pipeline started construction in 2000 and began channeling gas in September 2001. It was designed to carry up to 74 bcf (2 billion cu m) of natural gas a year and was expected to reach 18.6 bcf (500 million cu m) in 2002. The pipeline will cover its production cost when capacity reaches 31.7-37.2 bcf (850 million to 1 billion cu m), a level the company anticipated the line could reach by 2005. The pipeline could take 10 years to repay the investment.

The pipeline channels gas from Sebei gas field in Qinghai Province through Xining, the provincial capital, to Lanzhou, capital of Gansu Province. The central government expected the pipeline to assist the two provinces in moving from poverty to economic stability by selling their gas reserves. It also hoped that the gas would replace coal to improve the environment of western industrial cities like Lanzhou, one of the country's most polluted.

According to a PetroChina spokesman, the pipeline also anchors the company's drive to convert its rich gas reserves into cash. PetroChina controls 80% of natural gas reserves in China but has difficulty selling the gas because it lacks extensive transmission networks and because customers have not become accustomed to using natural gas. PetroChina's gas reserves could become a profit center within a 5-10 year timeframe depending on government infrastructure and pricing. The company was expecting its gas production to rise to 1.23 tcf (33 billion cu m) by 2005 from the current 681 bcf (18.3 billion cu m).

West-East Pipeline

PetroChina's ambitious plans to build a 2,500-mile (4,000-km) Xinjiang-

Shanghai gas pipeline, which would link Northwest China's Xinjiang Uygur Autonomous Region to Shanghai on the East Coast, were slowed in 2001 by a dispute over prices and choice of a prime contractor for the project. In September 2001 BP, China's biggest foreign investor, withdrew from negotiations for a stake in the pipeline's construction.

At year's end, two consortia were in negotiations with PetroChina, but there were concerns that a prime contractor had yet to be designated. The question of the price for gas was another main issue. The government caps Chinese gas prices to encourage the fuel's use and reduce reliance on coal. China plans to more than quadruple the use of gas by 2010 in place of coal, which now accounts for 70% of the country's energy. The price ceiling had been discouraging companies from boosting their investments to explore for gas and build pipelines.

In early 2002, Royal Dutch/Shell Group and its partners agreed to take a 45% holding in the $14 billion gas production and pipeline project. However, the second consortium, led by ExxonMobil Corp. with Hong Kong-based CLP Holdings Ltd., was still talking with Chinese authorities and was expected also to emerge as a major shareholder in the project. PetroChina said it would hold the remaining 55% while talks continued. The Shell-led group includes Hong Kong & China Gas Co. and Russia's OAO Gazprom, the world's biggest gas company. For Shell and ExxonMobil, the project represents a chance to increase their presence in China, Asia's fastest-growing energy market, and gain access to China's western oil and gas fields. PetroChina has said it may allow foreign companies to explore for oil and gas in Xinjiang if they agree to invest in the pipeline.

PetroChina previously announced that it aims to complete the pipeline by 2005, as part of an overall project to tap gas reserves in northwestern Xinjiang province, help replace dwindling reserves from older fields, and meet commitments to customers in Shanghai. Delays to the pipeline may benefit rival Chinese producer CNOOC, which planned to develop offshore gas fields east of Shanghai and move the gas by pipeline to cities on the East Coast. The pipeline project is part of a $48 billion government program to develop its 11 western provinces, where a quarter of the country's population generated only a seventh of its wealth.

The pipeline from Xinjiang would also serve a gas field in Inner Mongolia, less than half the distance from Shanghai. Shell had argued that that field should be tapped first to reduce the initial cost until a market for gas has been created in Shanghai. The Chinese government vetoed that proposal and insisted the pipeline should link the Xinjiang fields as early as possible to speed development in the region.

PetroChina's profits from gas sales probably doubled to $24 million in 2001, according to financial analysts in Hong Kong; the profit from that division could jump 40-fold to almost $1 billion in 2005, when the pipeline is due to be completed, if gas production rises as anticipated.

Eastern China pipelines

In late 2001, Sinopec announced plans to build two gas pipelines in eastern China by 2005. The lines would cost a combined $120 million (1 billion yuan) and would be able to move 149 bcf/yr (4 billion cu m/yr) of gas. Sinopec unit Zhongyuan Oil & Gas Co. would manage construction, and Sinopec signed letters of intent with industrial and household users in Jinan, Zibo, Weifang, and Qingdao for the gas transport volumes.

One line was planned to start at Zhongyuan oil and gas field at Puyang, supplying Jinan and Zibo and ending at Qingdao. The second would also start at Puyang, serving Heze and ending at Laiyang. This line would be a branch of the 2,500-mile (4,000-km) Xinjiang-Shanghai gas pipeline that PetroChina plans to build. Sinopec planned to acquire a stake in that line to move gas from its Tarim basin fields.

Sinopec opening retail markets

Demand for retail gasoline and oil was growing at 4.5%/yr within China. Under the terms of China's entry into the WTO, foreign firms will be allowed to own up to 49% of gasoline retailing ventures within 3 years. Sinopec cut its plans for purchase of service stations in 2002 from the 4,000 purchased in 2001 to 1,000 in 2002. China had about 75,000 gasoline stations in 2001, down from 88,000 a year ago. In late 2001, China tightened controls over the construction of gasoline stations. In a circular issued by the State Council, the government said all stations without business licenses or being built without approval from provincial economic and trade commissions would be ordered to stop business or stop construction.

The Council announced it would allow only Sinopec and PetroChina to build future gasoline stations. The government also put the wholesale oil business under the control of Sinopec and PetroChina, saying the two companies should draft a development plan for that segment, a program that would be subject to State Economic and Trade Commission approval.

Later in 2001, however, on the heels of becoming a WTO member, China relaxed its policies against foreign ownership of gasoline stations, and BP PLC gained the first foothold in the lucrative Chinese service station sector. BP planned to set up a chain of 500 service stations during 2002 on mainland China in a joint venture with Sinopec. The joint venture was the first time China has officially allowed a foreign company into the retail market for gasoline and diesel fuel, but it is unlikely to be the last. Under the deal with BP, Sinopec will take an initial 60% slice of the new joint company, but gave no other details.

In 2000, BP spent $378 million to acquire a stake in Sinopec at the time of the company's $1.43 billion flotation on the Shanghai Stock Exchange. BP has also bought a stake in PetroChina Corporation for $578 million.

News of the new joint venture emerged as BP officials were in Beijing for a meeting with Sinopec and the Shanghai Petrochemical Corporation to mark the official formation of the joint venture company that will build a $2.7 billion ethylene cracker and chemical derivatives complex near Shanghai.

BP planned to operate stations in the prosperous east coast province of

Zhejiang. Many would be existing sites operated by Sinopec, and all have recently benefited from substantial infrastructure investment. Royal Dutch/Shell Group and ExxonMobil Corporation were also understood to be involved in negotiations with Sinopec and were expected to announce similar agreements in 2002. Shell was expected to be offered sites in Jiangsu province, and ExxonMobil, in Fujian.

Other downstream activity

Sinopec, which had capital expenditures of $5.3 billion in 2001, cut its expenditures budget 21.6%, to $4.17 billion in the face of weakening demand for 2002, according to a Sinopec official. Previously it had forecast expenditures of $5.2 billion to $5.9 billion per year between 2002 and 2004. In addition, 12 crude oil pipelines exploded in September 2001 at Maoming, China's largest refinery, shutting down production for 10 days before resuming at reduced capacity. Total refining charge capacity (b/cd) at the close of 2001 was crude, 4,528,100; vacuum distillation, 40,000; coking, 306,000; and catalytic cracking, 892,000.

Growth in China's refining output was not sustained in 2001. China continued to close smaller, less productive refineries and to build new ones, but not all refinery figures were reported, and future capacity remained somewhat cloudy.

Petroleum product tariffs were already low and were not driving prices that Chinese consumers pay, according to Chem Systems. China's import duties on crude oil and condensate were 16 yuan/ton. Import fees for naphtha, diesel, and fuel oil were 6%; for gasoline the figure was 9%. Chem Systems did not expect these rates to change through 2005. Naphtha tariffs could ease to 3% by 2010. On the other hand, WTO entry requires that China's average petrochemical and polymer tariffs drop from 15% in 2001 to 10% in 2005. Although the country largely supplied its own needs for gasoline and ethylene, it was a net importer of LPG, middle distillates, fuel oil, and polymers in 2001, and Chem Systems expected that China will import even more of these products by 2005. IPE

ASIA-PACIFIC

AUSTRALIA

CAPITAL: CANBERRA
MONETARY UNIT: DOLLAR
REFINING CAPACITY: 848,250 B/CD
OIL PRODUCTION: 640,000 B/D
OIL RESERVES: 3.5 BILLION BBL
GAS RESERVES: 90 TCF

Australia continued to suffer a mild economic slowdown, with gross domestic product (GDP) growth for 2001 forecast at 1.8%. Compared to other industrial nations, Australia's energy resources remained relatively unexplored. The country offered 42 offshore petroleum exploration areas for open bid in April 2001 to encourage exploration of the continental shelf. The licenses were a mix of relatively mature, underexplored, and frontier areas. Permits would be awarded for an initial term of six years. Bids for 13 of the areas closed in October 2001; the other 29 were to close in April 2002.

Otway gas finds, pipelines

Australia's Otway basin, where gas discoveries in 2001 sparked new interest, straddles onshore and offshore western Victoria and southeastern South Australia, as well as offshore Tasmania. Aided by 3D seismic, companies may have found the basin's "signature," according to producers in the region, after a history of dry holes. Thylacine 1, on Tasmanian Permit T/30P about 43½ miles (70 km) off Victoria in 330 ft (100 m) of water, was regarded as one of the most significant finds since the North West Shelf. Although such a comparison could be premature, 3D seismic data delineate the prospect.

Thylacine, which was ironically named after the extinct Tasmanian tiger, found a 922-ft (281-m) gross gas column in Cretaceous age Waarre sandstone, a prolific reservoir that contains gas in most of the region's smaller fields. Preliminary estimates indicate the basin's production exceeds 600 bcf and could reach 1 tcf. Operator Woodside Petroleum Ltd. held a 50% stake; OriginEnergy Resources Ltd. (Sydney) had 30% and Benaris International Ltd. 20%.

Other companies have had success in the Cretaceous Waarre sand. The Santos Ltd. group, adding to the three fields it had on production, recorded five straight Waarre gas successes onshore Victoria during the first half of 2001: McIntee, Tregony, Croft, Lavers, and Naylor. With 98 ft (30 m) of net pay, Naylor was deemed the most significant with potential production of 10 terajoules (tj)/day. The group comprised Santos 90% and Beach Petroleum NL 10%. Four of the five finds were on Victoria permit PEP154; Tregony was on Santos 100%-owned PEP153. All were near the town of Port Campbell and close to the original North Paaratte-Wallaby Creek-Brumby gas fields discovered previously by Beach.

In June 2001 an Australian wildcat found a gas field in offshore Otway, announced Origin Energy, which was managing Australian exploration permit Vic/P43. Geographe 1 recovered gas in wireline samples at 6,007, 6,220, and 6,611 ft (1,831 m, 1,896 m, and 2,015 m). Preliminary analysis indicated the gas had an inert level of 6-8%. Condensate and LPG content was being determined. Pay zones were not given, but earlier Origin had said gas was indicated in Flaxmans/Waarre sands and shales.

Geographe is 34 miles (55 km) off the Victorian coast and 9 miles (15 km) north of the Thylacine discovery on adjacent permit T/30P. The wildcat was drilled to 7,972 ft (2,430 m) in 280 ft (85 m) of water by the Ocean Bounty semisubmersible. Origin said the structure comprises two related prospects, with Geographe 1 being on the southerly one. The well might have tapped 500 bcf of in-place gas on the southern culmination, which could increase to 1 tcf with further

drilling in the northern culmination, said Origin. Geographe 1 was being cased and suspended for later completion. Partners were Origin with 30%, Woodside Energy Ltd. 55%, and CalEnergy Gas (UK) Ltd. 15%.

Lakes Oil NL also had interests in several permits, including 40% of onshore PEL 57 in South Australia, which contains part of a prospect called McNamara Park. The prospect straddles the boundary with PEL 72, and Lakes would earn 7.5% direct interest in PEL72 by contributing to the McNamara Park 1 being drilled by Origin Energy.

In a more audacious move, Lakes took a 100% interest in offshore South Australian permit EPP24, which contains the Troas prospect drilled in 1992 by BHP Petroleum. Gas had been found there but was thought unlikely to be commercial. Lakes has long disputed this and believes that reserves could be up to 600 bcf, a commercial prospect if additional Otway basin pipelines are built. Lakes planned to redrill Troas early in 2002.

Opportunities for commercializing onshore Otway gas discoveries were improved by development of pipeline infrastructure between Melbourne and the western Victorian town of Warrnambool. Additional pipelines were being proposed to link new fields to markets in South Australia and Adelaide. The South Australian government has wanted a long-term gas supply that offered an alternative to the existing single pipeline from the Cooper basin. Among several projects proposed to the government, Duke Energy Corp.'s plan would build a $250 million (Aus.), 416-mile (670-km) line from BHP Petroleum's undeveloped Minerva field off Victoria, taking in production from other Otway discoveries on the way. Duke said the line would need 28 bcf/yr of demand, but the plan would be to build in a total capacity of 56 bcf/yr. Duke would require commitments from South Australian customers before proceeding.

A rival plan also announced in 2001 involved Australian National Power and Origin Energy, which established the South East Australia Gas Pty. Ltd. joint venture to build a 391-mile (630-km), $200 million (Aus.) line from Port Campbell in Victoria to Adelaide. Australian National Power signed a deal to buy gas from BHP's Minerva over 10 years. Most would go to a planned 300-MW expansion of the company's 500-MW gas-fired power station at Pelican Point, near Adelaide. Another potential market for the gas was a proposed new magnesium smelter to be built near Port Pirie on Spencer Gulf, north of Adelaide.

Only one of these two pipeline projects, both timed for completion around 2004, would be needed in the short term, but one would be enough to sustain the renewed Otway basin exploration effort. Further evidence of the growing geological confidence in the region was the successful float in early 2001 of Essential Petroleum Resources, which held interests in five permits operated by Origin. In June 2001 Victoria awarded Essential Permit Vic/P46 to explore Area V00-1 offshore in the Otway basin.

New Guinea pipeline

A pipeline to carry natural gas from Papua New Guinea to Australia was still uncertain in early 2002 as Santos dropped out of the project. ExxonMobil and other partners signed a new agreement without Santos but were still negotiating how to divide its 12.5% stake. The 1,990-mile (3,200-km) pipeline had been delayed previously by gas revenue arguments and lack of political support. The new agreement replaced one that expired at year-end 2001, but the companies have yet to give final approval to the project.

Legendre producing oil

In May 2001 the Ocean Legend offshore production facility brought the $110 million (Aus.) two-field Legendre development off Australia to first oil, operator Woodside Energy Ltd. reported. Production from Legendre North and Legendre South fields, 22 miles (35 km) southeast of the Wanaea-Cossack developments in production license WA-20-L, would be ratcheted up to a plateau of 50,000 b/d as three more development wells are completed. Legendre's field life was forecast at 3-8 years.

In mid-2001 Legendre field in the Carnarvon basin off Northwest Australia was producing 25,000 b/d of oil from the initial production well, said Woodside. Field partner Apache Energy Ltd. expected the field's production to reach approximately 40,000 b/d later in the year as additional production wells were brought on stream. In June the first cargo of crude oil from the Legendre venture was sold to Shell International Eastern Trading Co. The 630,000-bbl cargo was destined for refineries in eastern Australia. Production from the Ocean Legend offshore facility was 20,000-25,000 b/d.

The Legendre Venture was Woodside's third oil development in Australian waters, following the Laminaria/ Corallina project off northern Australia in the Timor Sea and the North West Shelf venture's Wanaea-Cossack development. Participants and equity interests in License WA-20-L were Woodside (45.94%), Apache (31.5%) and Santos Ltd. (22.56%).

Apache find, other E&P

An oil discovery in the Carnarvon basin off Western Australia was reported in early 2002 by Apache Corp. (Houston), its fourth commercial find there since early 2001. Double Island-1 well found 51 ft of net oil column in the Lower Cretaceous Flag Sand. A wireline test found the oil to be 51º gravity. Apache did not flow test the well because the Flag sandstone's deliverability had been well established in nearby producing fields. The well was owned by the Harriet Joint Venture, which Apache was operating with a 68.5% working interest. Partners were Kufpec Australia Pty. Ltd. with 19.3% and Tap Oil Ltd. with 12.2%.

The Harriet complex was averaging 22,000 b/d of oil, 95% of which was from the Flag sand. Apache was planning to drill four more Flag exploration prospects in the area before mid-2002. Initial oil production from the joint venture's South Plato-Gibson Flag discoveries made during 2001 was expected to be 10,000 b/d by mid-2002. Preliminary plans called for connecting Double Island to the South Plato-Gibson facilities 4 miles away. The oil would then flow to Apache's processing facilities at the Harriet complex on Varanus Island.

The BassGas development received facilitation status from Australia in September 2001, meaning that the federal investment agency would assist project approvals and coordinate with state governments. The $400 million BassGas plan would develop Yolla field in the Bass basin off Victoria and bring the gas onshore for use in the grid and possibly for power generation in South Australia. Production over 15 years was expected to be 13.5 million bbl of condensate and 1 million tonnes of LPG. Commercial gas supply was anticipated for 2004. Australian Worldwide Exploration Ltd. and CalEnergy Gas (UK) Ltd. agreed in early 2001 to sell Yolla gas to partner Origin Energy.

Oceaneering International's Ocean Legend was the mobile offshore production unit operating in Legendre field on the North West Shelf of Western Australia. Photo courtesy of Oceaneering International Inc.

In August 2001 Chevron Australia Pty.

Ltd. agreed to let China explore the feasibility of acquiring an equity stake in the Gorgon area off Western Australia, as detailed in a memorandum of understanding with China National Offshore Oil Corp. (CNOOC). Gorgon is the largest natural gas field discovered in Australia, with confirmed hydrocarbon reserves in excess of 9.6 tcf. China was interested in the possibility of importing Gorgon's gas as LNG, and Chevron was seeking to develop Gorgon as an LNG project.

In mid-2001 a new Carnarvon basin permit was awarded to a joint venture by Australia and Western Australia state. Permit WA-310-P for area W00-56 would be explored by West Oil NL (Perth) and Bounty Oil & Gas NL, which would spend $4 million (Aus.) over six years. The permit was near oil and gas infrastructure and to major gas customers, said the Australian government.

In early 2001 Santos tested Coonaberry-1 well at 7.7 MMcfd (206.7 cu m/day) through a 13-mm choke from the Permian Toolachee at 8,930-8,957 ft (2,722-2,730 m). Coonaberry is in the South West Queensland Unit in the Cooper-Eromanga basins. In addition to operators Santos, partners were Delhi Petroleum Pty. Ltd., Origin Energy, Boral Ltd. unit Oil Co. of Australia, and Niugini Energy Inc.

North West Shelf, LNG

In late 2001 the North West Shelf venture began production from Echo-Yodel fields three months ahead of schedule and within its $205 million (Aus.) budget. The two fields lie northwest of Dampier, Western Australia, on WA-28-P and 14 miles (23 km) southwest of the Goodwyn A platform. They were expected to add 37 million boe to the venture's production over the project's four to five-year life. Two subsea production wells were tied back to the platform by a 12-inch pipeline. Peak production was anticipated to reach 30,000 b/d of condensate and 300 MMscfd of gas. Echo-Yodel production would be commingled with Goodwyn's, helping to offset its decline. Partners on the project estimated Echo-Yodel reserves at 37 million bbl of condensate and 400 bcf of gas.

In early 2002 the North West Shelf venture was planning to install a second gas trunkline, estimated to cost $800 million, linking offshore production facilities to its LNG plant near Karratha, Western Australia, during 2003. Construction was scheduled to be finished in 2004 to coincide with completion of a fourth LNG train, which should begin producing 4.2 million tonnes per year (tpy) of LNG in 2004. The pipeline could also carry enough gas to supply a fifth train. Sales agreements for the fourth train stood at almost 3.9 million tpy.

Woodside Energy was operator of the North West Shelf LNG project with five equal partners: BHP Petroleum (North West Shelf) Pty. Ltd., BP Developments Australia Pty. Ltd., Chevron Australia Pty. Ltd., Japan Australia LNG (MIMI) Pty. Ltd., and Shell Development (Australia) Pty. Ltd., each with a 16.67% stake. Parts of a heat exchange technology developed jointly by Statoil ASA and Linde AG for a Norwegian LNG plant would be used by Shell at the Karratha LNG facility. The technology includes spool-wound heat exchangers that are part of the process of cooling down the gas to its liquefaction temperature of –163ºC.

To install the trunkline, Saipem (Portugal) Comercio Maritimo (an affiliate of Italy's Saipem SPA), under a $100 million (Aus.) contract, planned to use two specialist installation barges, including its own semisubmersible Semac 1. The 42-inch line, which requires a special grade of steel and production techniques not available in Australia, was to be made in Japan and weight-coated with concrete in Indonesia before being transported to Australia.

Saipem was planning to dredge a 12-mile (20-km) section of the offshore pipeline route; collect and deliver the 11,000-ft long (12-m long) weight-coated pipe-sections from Batam Island, Indonesia; weld and lay the 83 miles (134 km) of line with anodes attached; and flood and pressure-test the completed trunkline.

In late 2001, the North West Shelf partners let China invest in developing the project for LNG sales to Chinese markets,

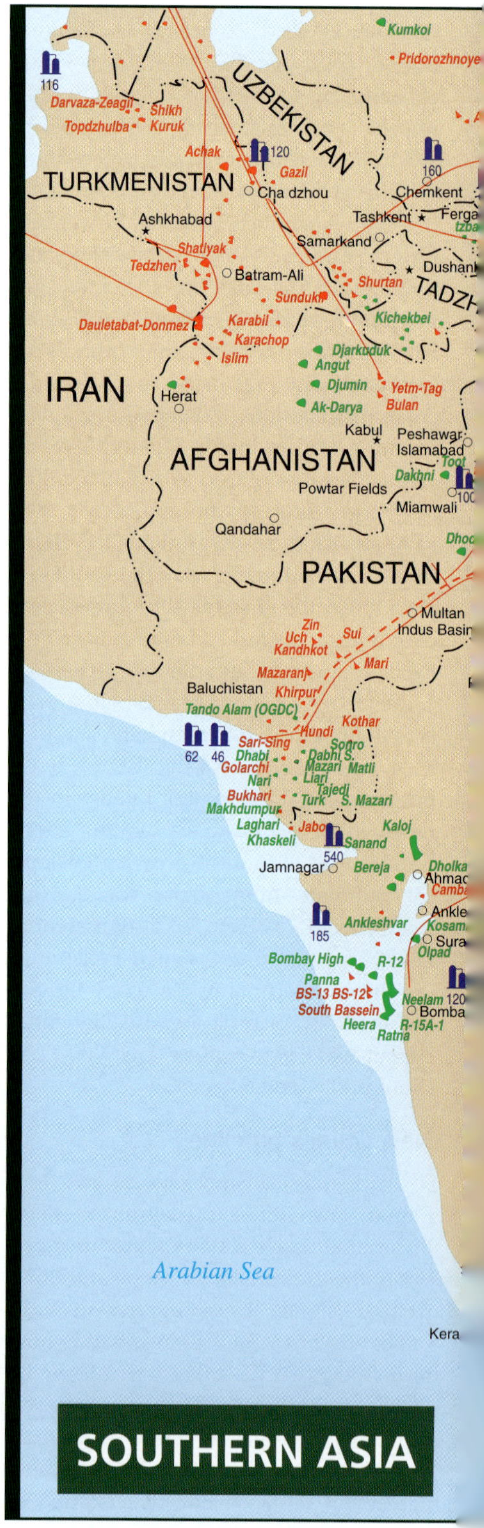

under a contract with CNOOC. The LNG would be supplied to a terminal in the southern Chinese province of Guangdong. The terminal will be taking a total of 3 million tpy of LNG.

In mid-2001 Shell International Gas agreed to key terms of a deal to purchase up to 3.7 million tonnes of LNG from

Australia's North West Shelf project for future delivery to the Americas. The agreement if concluded would run for five years starting in 2004 and would use new LNG carriers, including two ordered by Shell from Mitsubishi Heavy Industries and Daewoo Shipbuilding & Marine Engineering.

Timor Sea

The long-delayed project to build a $1.5 billion pipeline to bring gas from the Bayu-Udan field to Darwin in northern Australia was approved by Phillips Petroleum Co. in late 2001. The project had been held up by a dispute over the tax regime with the East Timor government, which wanted to increase taxes to recoup its refund of exploration and development expenditures. The new tax and fiscal package remained to be ratified by Australia, which would share the East Timor gas project revenues, before contracts could be placed. The Bayu-Undan project coventurers were Phillips as operator with 50.3%, Santos 11.8%, Inpex Ltd. 11.7%, Kerr-McGee Corp. 11.2%, Petroz Ltd. 8.3%, and Agip SPA 6.7%.

The Timor Sea's Bayu-Undan field contains estimated reserves of 400 million bbl of condensate and LPG and 3.4 tcf of gas. It is located in 260 ft (80 m) of water about 300 miles (500 km) northwest of Darwin and 150 miles (250 km) south of Suai, East Timor. Phillips and its partners planned to proceed with a gas reinjection-liquids production recycle project at the field, which had been approved in early 2000 by the Timor Joint Authority created by Australia and Indonesia. The project was to begin full commercial production in 2004.

Coalbed methane, shale oil

Seven onshore exploration licenses were received by Gastar Exploration Ltd. (US) and a partner in mid-2001, with pilot drilling for coalbed methane scheduled to begin during the year. The areas totaled 2 million acres in Victoria's Gippsland basin. Gastar already held another 2 million acres, PEL 238 in New South Wales. Gastar's project area in the Latrobe Valley covers almost all of the onshore portion of the basin.

No organized attempt has been made to produce gas from the Gippsland coal, but oil and gas wells have encountered gas in the seams. The basin's Tertiary-age coals have total thicknesses of as much as 1,000 ft at depths of 3,000 ft or shallower; individual seams are several hundred feet thick.

Up to 260 bcf of coalbed methane could be purchased by Queensland Fertilizer Assets Ltd. from a unit of Tipperary Corp. (Denver) over a 20-year period beginning in 2004. Tipperary's 90%-owned Tipperary Oil & Gas (Australia) Pty. Ltd. subsidiary owned a 65% interest in Australia's Comet Ridge coalbed methane project. The gas would be used at a fertilizer plant to be built in southeastern Queensland.

In May 2001 Southern Pacific Petroleum and Central Pacific Minerals said they were planning to produce 200,000 b/d of shale oil products from their Queensland deposits by 2012. The production would include 85,000 b/d from stage three of the fully developed Stuart shale oil demonstration plant by 2009. Additional production would come from either the Condor or Rundle deposits by 2010.

The $300 (Aus.) million Stuart plant resumed oil production following a two-month maintenance and capital improvement program. Shale processing rates were 160 tonnes/hr, held at 64% of capacity to minimize odor emissions, and shipments began in May 2001. Stage two of Stuart's development was scheduled to begin construction in 2003 with first production in 2005. The two shale oil companies operating the plant and the Queensland deposits agreed in September 2001 to merge, retaining the Southern Pacific Petroleum name. The combined company would have 17.3 billion bbl of oil shale resources in central coastal Queensland and the Stuart plant.

GTL, methanol

The Sweetwater GTL plant in Western Australia in September 2001 cleared one of its last major milestones before the project seeks financing. Syntroleum Corp.'s Syntroleum Sweetwater Operations Ltd. subsidiary signed a $599.5 million contract with German engineering firm Tessag Industrie Anlagen GMBH for engineering, procurement, and construction of the plant on the Burrup Peninsula. The facility would employ the proprietary Syntroleum process to convert natural gas into sulfur-free, synthetic specialty products such as lubricants, industrial fluids, paraffins, and transportation fuels. The plant's design capacity was increased to 11,500 b/d from 10,000 b/d as a result of pilot plant tests demonstrating higher conversion rates than those included in the original design. In April 2001 Syntroleum awarded a $215.4 million operating and maintenance contract to a joint venture of PGS Production Services and Clough Engineering Ltd.

A methanol plant was also being planned for Australia's Burrup Peninsula and secured its gas supply in November 2001. Methanex Corp. (Vancouver) agreed to buy the gas from the North West Shelf participants, and the parties began negotiating a detailed gas sales and purchase agreement. In the project's first phase, Methanex would build what it called the world's largest methanol plant with a capacity of about 2 million tpy to supply customers in the Asia-Pacific region.

In June 2001 the South Australian Cooper Basin Producers joint venture said that the Moomba gas-processing complex was back at 85% of production. An explosion and fire had damaged the liquids pumping station at the Moomba complex earlier in the month, killing one person. The pump station moves up to 40,000 b/d of crude and liquids from Moomba to Port Bonython near Whyalla. Operator Santos said the accident did not affect gas production and supply from Moomba. Investigations into the cause of the accident and needed repairs were continuing.

BANGLADESH

CAPITAL: DHAKA
MONETARY UNIT: TAKA
REFINING CAPACITY: 33,000 B/CD
OIL PRODUCTION: 3,300 B/D
OIL RESERVES: 56.9 MILLION BBL
GAS RESERVES: 10.6 TCF

The High Court of Bangladesh in late 2001 blocked a government attempt to allow major oil companies to export natural gas to neighboring India. The court issued a temporary injunction against exports in effect for at least three months, overriding production-sharing contracts (PSCs) between the government and international oil companies.

The issue of gas exports has been a political controversy for Bangladesh. At its current rate of production, Bangladesh's proven gas reserves would meet demand for 37 years. However, if exports to India were allowed, geologists reported, gas reserves in Bangladesh would last only until 2015. Foreign investors believe potential reserves are significantly higher, although their estimates differ. A survey conducted by a team of Norwegian geologists revealed that the country has 20 tcf of recoverable reserves, with the potential to produce another 40 tcf, according to *The Economist.*

Meanwhile, India's huge gas demand

continued to grow. Although no pipelines connect the two nations, Bangladesh's proximity to India offers an advantage over other suppliers and over India's next-best option, LNG imports. Bangladesh could earn an estimated $300 million to $1.7 billion/yr by exporting gas to India. Several foreign oil firms have asked Bangladesh to allow exports. Unocal Corp. proposed a $500 million, 500-MMcfd export pipeline from Bibiyana field in the northeast to New Delhi.

In October 2001 three Indian government-owned companies agreed to buy gas from South Korean conglomerate Daewoo Corp.'s fields in Myanmar. This would involve building a 342-mile (550-km) pipeline around the continental shelf of Bangladesh, which would not require the permission of that country. Landfall would be at the port town of Haldia in West Bengal. The three Indian companies – Oil & Natural Gas Corp., Indian Oil Corp., and Gas Authority of India Ltd. – are working as a consortium in international oil and gas ventures.

In April 2001 the Bangladesh government awarded exploration rights to a joint venture of ChevronTexaco subsidiaries and British independent Tullow Oil PLC for a 2,656 sq mi (6,880 sq km) tract east of Dhaka. Block 9 surrounds Bakhrabad gas field and is near other major gas-producing areas. Chevron International Bangladesh Ltd., Texaco Exploration Asia Pacific Pathfinding Inc., and Tullow Bangladesh Ltd. each held 30% interest in the block, with Bangladesh Petroleum Exploration Company Ltd. holding 10%.

The award was part of the Bangladeshi government's second licensing round. Tullow was to operate the block, with ChevronTexaco assuming operatorship of any development of a commercial discovery. The venture planned three exploration wells during a three-year period. Several pipelines cross the acreage. Caltex Corp., a ChevronTexaco venture, was developing a plan to supply Bangladesh markets with lubricants during 2001 and had a team studying the feasibility of marketing LPG.

In July 2001 the republic of Bangladesh and its oil and gas company, Petrobangla, signed PSCs for onshore Blocks 10 and 5 with a joint venture made up of Shell Hydrocarbons Holdings BV, Cairn Energy PLC, and Bangladesh Petroleum E&P Company Ltd. Shell and Cairn would be 45% partners in the two blocks, with Bangladesh Petroleum holding the remaining 10% stake.

Block 10, known as greater Noakhali and Bhola, is located in the south of Bangladesh, while Block 5, greater Khulna, is in the southwest. The scope of exploration work for Block 10 would include 1,553 miles (2,500 km) of seismic survey and the drilling of five exploration wells. On Block 5, a 4,040-mile (6,500-km) aerial survey would be conducted, along with 310 miles (500 km) of seismic, and one well drilled.

Shell stressed that no exploration activities would be carried out within the Sundarbans Reserved Forest on Block 5, and that environmental and social studies to determine the possible impact of exploration on both blocks would be completed before any work began.

BRUNEI

CAPITAL: BANDAR SERI BAGAWAN
MONETARY UNIT: DOLLAR
REFINING CAPACITY: 8,600 B/CD
OIL PRODUCTION: 180,000 B/D
OIL RESERVES: 1.35 BILLION BBL
GAS RESERVES: 13.8 TCF

Brunei's small, wealthy economy is based mostly on proceeds from the oil and gas sector, which support a large civil service employing about 75% of the country's workforce. In addition to diversifying into non-hydrocarbon areas, Brunei would like to develop energy-intensive industries such as petrochemicals, refining, and aluminum smelting. The country planned to host the region's major gas industry conference, GASEX 2002. In November 2001 Sultan Bolkiah announced the creation of a national oil company to be wholly owned by the government.

Brunei, already the world's fourth-largest LNG producer, had excellent long-term prospects for further natural gas and LNG development. The Brunei LNG joint venture had a master plan to expand its capacity with an additional 4 million tpy train at its Lamut LNG plant by 2008. The venture consisted of Mitsubishi (25%), Shell (25%), and the Brunei government (50%). The group's plan would depend on finding about 4.2 tcf of gas that remains to be proved. The Lamut plant would also be renovated to extend its operating life to 2033. The additional train and renovation would cost $2.4 billion, with another $2.1 billion to be spent on six new ships to service the new train.

In early 2001 Brunei said 28 companies were interested in PSCs in its deepwater regions. To jump-start the exploration round, the sultanate signed a contract with seismic survey company Petroleum Geo-Services ASA to shoot 3D seismic data. Key prospects were to be delineated by the time companies submitted bids in late 2001. The blocks were in 3,300-9,000 ft (1,000-2,750 m) of water and seem to be related to Malaysian geology where gas discoveries have been made. The first well in the acreage was to be drilled in 2002-03.

Exploration of offshore Block J would be conducted by a group led by TotalFinaElf SA under an early 2002 Brunei contract award, which was dependent on successful negotiation of a production-sharing agreement (PSA). TotalFinaElf would operate the block with 60%, BHP Billiton would hold 25%, and Amerada Hess Corp. 15%. Block J covers 1,900 sq mi (5,000 sq km). It is 62 miles (100 km) from shore in 4,300-5,900 ft (1,300-1,800 m) of water. TotalFinaElf was already operating production of more than 20,000 b/d in Brunei.

Development of Egret gas and oil field was being planned in early 2002 with first gas to flow by mid-2003. Brunei Shell Petroleum Co. Sdn. Bhd. would proceed with first phase development using a 12-slot drilling platform through a 20-inch, 15½-mile (25-km) multiphase pipeline to the Ampa 6 platform. From there the gas would go to Brunei LNG Sdn. Bhd.'s Lamut LNG plant. Egret gas would help satisfy Brunei's domestic and export gas needs for 2003 and beyond.

Brunei Shell awarded Technip-Coflexip Group a a lump-sum $22 million contract related to the $80 million Egret development in early 2002. Main facilities to be engineered and fabricated by Technip would include the platform, with an 860-ton topside and 770-ton jacket, to be installed in 200 ft (60 m) of water; the connecting multiphase pipeline; and a 9-mile (15-km) subsea cable. Completion was expected in late 2003.

In mid-2001 Shell International Trading & Shipping Co. Ltd. said it would manage Brunei Gas Carriers' Abadi LNG vessel, which was to be delivered in mid-2002. It is the eighth vessel Shell will manage for Brunei. Abadi would trade between Brunei and Japan or South Korea.

INDIA

CAPITAL: NEW DELHI
MONETARY UNIT: RUPEE
REFINING CAPACITY: 2,134,625 B/CD
OIL PRODUCTION: 640,000 B/D
OIL RESERVES: 4.84 BILLION BBL
GAS RESERVES: 22.9 TCF

As the world's sixth largest energy consumer, India imports 70% of its crude requirements and in 2001 spent more than $16.5 billion (800 billion rupees) on imported oil. The government was planning major energy infrastructure investments to keep up with increasing demand, particularly for electric power, and possible LNG imports to support power projects. The government was seeking to raise natural gas's share of the energy supply to 20% by 2025 from 8% in 2000.

India's Oil & Natural Gas Corp. (ONGC) announced in late 2001 that it would spend $9.8 billion (469.7 billion rupees) over five years to meet oil and gas production targets. The Tenth Five-Year Plan (2002-07) included drilling 594 exploratory wells and performing additional 3D seismic surveys. Domestic exploration would focus on the deepwater sector, including 35 wells.

India's government said in early 2002 that it will offer 25 oil and gas blocks in the third round of the New Exploration and Licensing Policy before April 2002. Under the first two rounds, 47 blocks had been awarded, including 16 to ONGC and four to Hardy Oil (UK) in partnership with India's Reliance Petroleum. Although the new blocks were unidentified, the ministry indicated several deepwater blocks off the eastern coast would be offered.

In September 2001 US President George W. Bush signed a waiver of all US sanctions against India.

Discoveries off eastern India

Cairn Energy PLC (UK) reported several successful wells during 2001, as well as gas sales for its Lakshmi field development off Gujarat. The company expected to invest $300 million for oil and gas exploration over the next four years in its discovery in the deep waters of the Krishna-Godavari basin, off Andhra Pradesh, eastern India.

In June 2001 a well on Cairn's Block KG-DWN-98/2 off Andhra Pradesh flowed 80 MMscfd on a 1-inch choke from two zones at a combined stabilized rate, constrained by equipment limitations. DWN-R-1, 16.5 miles (26.5 km) southwest of Cairn's earlier N-1 discovery (which reportedly found a cluster of potentially gas-bearing units), was drilled to 7,644 ft (2,330 m) total depth and found several gas pay zones below 6,135 ft (1,870 m). The company estimated that the well discovered 750 MMcf-1 tcf of dry gas. Cairn owned 100% equity interest in the block; ONGC had no back-in right.

In August 2001 Cairn made its third hydrocarbon discovery on the same block with well DWN-P-1. The Prospect P well was drilled 7.8 miles (12.5 km) northwest of the Annapurna (R) gas discovery. DWN-P-1 reached 8,481 ft (2,585 m) total depth and found two hydrocarbon-bearing units. The first was a gas cap with a thin oil leg; the deeper second unit was a thin gas column with an estimated 246-ft (75-m) oil column.

An upper zone, a 56-ft (17-m) gross interval of thinly laminated sandstone and shale, was perforated and flowed 3,648 b/d of oil at a flowing wellhead pressure of 1,020 psi on a 40/64-inch choke with a gas oil ratio of 751 scf/bbl. Cairn then perforated a deeper 23-ft (7-m) interval of sandstone to test both zones together. Because of technical difficulties, testing could not be carried out, but the company said data indicate the interval should test at higher rates than the other zone. The Marine 500 drilling rig was being moved to drill R-2, an appraisal well on the R gas discovery.

Cairn said it terminated attempts to drill a well on Prospect 6 on Block KG-OS off India after encountering a shallow gas reservoir at 1,968 ft (600 m) and water flows from sands just below the seabed. The company planned to drill the prospect again in 2002 with a special top hole casing design and a more specialized drilling rig. Cairn owned a 50% interest in KG-OS/6, with Videocon Petroleum Ltd. holding the balance.

Also in August 2001 Cairn contracted to sell 70 MMcfd of gas over three years to Gujarat Powergen Corp. and 45 MMcfd over five years to BG subsidiary Gujarat Gas Co. The gas would come from Cairn's Lakshmi field development off Gujarat, scheduled to come onstream in mid-2002. Lakshmi is 162 miles (260 km) north of Bombay in the Gulf of Cambay. The Lakshmi project was anticipated to adds 10,000 boe/d to Cairn's previous production level of 20,000 boe/d. Cairn expected to sign other gas sales contracts with buyers in industrialized Gujarat as other discoveries in the block, such as Ambe and Gauri, are sanctioned for development.

In July 2001 Lakshmi field partners awarded an $80 million development contract to a consortium of Australia's Clough Engineering and South Korea's Hyundai Heavy Industries Co. Ltd. Clough would be responsible for all onshore work for Lakshmi, including a new gas treatment plant, while Hyundai would undertake the offshore segment. The gas plant would contain two process trains, each with 75 MMscfd of capacity. Completion was due in September 2002. Partners were Cairn, ONGC, and Tata Petrodyne Ltd.

Clough also had a contract from Cairn to expand Ravva field off eastern India in the Bay of Bengal. The field was partially shut down for satellite gas development operations and because of increased government take. ONGC held 40% interest in Ravva, Videocon 25%, Cairn 12.5% and operatorship, and Marubeni Corp. 12.5%.

Other exploration and development

Essar Oil Ltd. (Mumbai) planned to invest $31 million in an exploration and development program in Rajasthan, northwestern India, the company said in October 2001. Following an oil discovery with Nanuwala-1 in the Bikaner-Nagaur basin, Essar planned a second wildcat near Raisinghnagar. The first wildcat flowed 35º gravity oil and opened production in an unexplored area of Rajasthan. Essar was planning 15-20 wells over two years to develop reserves that could be as large as 150 million bbl. Essar had a two-block PSC with the Indian government; Oil India Ltd. Essar had 75% and Polish Oil and Gas Co. 25% of the contract. Rajasthan has seen little exploration, though the geology is similar to areas with production across the border in Pakistan.

ONGC reported in October 2001 that production was lower than usual during the year, in part because of revamping activities and power shortages in Assam, northeastern India. Crude production at offshore Bombay High field fell short of its target by 14.7%. The redevelopment of Bombay High North, which involves drilling 78 new wells, had begun and was producing reasonably well, according to ONGC. However, Bombay High redevelopment plans were being re-evaluated in late October by India's government, which was questioning the investment since past efforts had failed to produce at predicted levels. Bombay High field, which has been by far India's largest producing field, has 1,724 million tons of oil in place, of which 1,541 million tons are in the L-111 reservoir, for which ONGC had submitted two redevelopment plans.

In late 2001 ONGC reported discovery of an oil and gas bearing structure off the western shore about 4 miles (6 km) east of Vasai gas field. Reserves estimate for the structure was more than 65 million toe. ONGC also found pay with a 7,218-ft (2,200-m) exploratory well at Kadi Taluka, 56 miles (90 km) from Ahmedabad, Gujarat state. A second zone at 6,480 ft (1,975 m) was being tested when well control was lost.

LNG

Lack of progress in India's power generation sector dampened prospects for building LNG import terminals, but several were still being developed. The largest state sector projects were being conducted by Petronet, a joint venture of ONGC, Indian Oil Corp., the Gas Authority of India Ltd. (GAIL), the National Thermal Power Corp., and Gaz de France. Petronet was planning two import terminals in Gujarat state, one at Dahej and one at Cochin. Qatar's RasGas would be the main LNG supplier and was to begin deliveries in 2003. Royal Dutch/Shell Group received approval for an LNG import terminal at Hazira in Gujarat and contracted for LNG supplies from Oman, while Reliance Industries was planning one at Jamnagar near its oil refinery. Neither was under construction.

Another LNG terminal was being planned for the Pipavav port on the Gujarat coast. The terminal would have a capacity of 5.3 million tpy. In September 2001 BG PLC boosted its share of the project to 50% and planned to invest at least $770 million to increase the terminal's capacity to 10 million tpy. Partner National Thermal Power held 26%, with the other 24% to be picked up by the Gujarat government or a financial institution. Completion would be scheduled for 2005 if customers can be signed up after India removes a natural gas price cap and equalizes sales taxes across its states.

BP PLC and partners were involved in reviving plans to build a 2.5 million tpy LNG terminal and power station in the eastern Indian state of Andhra Pradesh, with completion expected by 2007. Partners were Indian Oil, the country's biggest refiner, and Malaysia's state company Petronas; the three participants held equal shares. The LNG terminal was one of nine import projects targeted at meeting India's expected upsurge in gas demand for power generation and domestic use.

Expansion of Enron Corp.'s Dabhol power plant, halted in mid-2001, would have converted the oil-fired station to LNG, and a 5.5 million tpy import terminal was being constructed. The Dabhol LNG terminal was nearly finished and could be completed by another investor once a buyer has been found for Enron's share of the project.

Pipeline from Iran

The feasibility of a proposed $6 billion gas pipeline from Iran to India was to be studied under a late 2001 Iranian government contract with Australia's BHP Kinhill and Italy's Snamprogetti SPA. The 1,660-mile (2,670-km) line, which would begin at a National Iranian Oil Co. gas field in southern Iran, would have to pass through Pakistan to reach India. The proposed 58-inch pipeline would transport 2.46 bcf/yr (66 million cu m/year). Under the proposal, 70% of the line's capacity be contracted to India and the rest to Pakistan. In a bid to reduce security threats, it was suggested that a consortium of multinationals be formed to develop the project. The feasibility report was expected in March 2002.

Meanwhile, the government's GAIL was investing in the pipeline infrastructure needed to support greater use of natural gas. Its plans included doubling the capacity of the main Hazira-Bijaipur-Jagdishpur pipeline to 2.1 bcfd, with work to begin in 2002.

INDONESIA

CAPITAL: JAKARTA
MONETARY UNIT: RUPIAH
REFINING CAPACITY: 992,745 B/CD
OIL PRODUCTION: 1.2 MILLION B/D
OIL RESERVES: 5.0 BILLION BBL
GAS RESERVES: 92.5 TCF

Although Indonesia's oil reserves and production have been declining, its gas reserves are plentiful, and the country is the world's largest exporter of LNG. Indonesia's politicians continued to wrestle with distribution of oil and gas revenues between the central government in Jakarta and the regional governments in producing areas. In July 2001 Megawati Sukarnoputri became president, and in September the IMF released a $395 million loan tranche that had been delayed pending economic reforms. Indonesia's real GDP growth was 4.1% in 2001 and projected at 3.5% in 2002.

In October 2001 Indonesia's legislature passed a law that will remove state oil company Pertamina's monopoly on upstream development, which required it to be included in all PSCs, within two years and its monopoly on petroleum product distribution to be terminated within four years. However, the new law granted a limited authority to regional governments to tax oil companies' profits. The Ministry of Mines and Energy earlier had assumed Pertamina's former role of awarding and supervising PSCs.

To increase oil production, Indonesia signed nine new PSCs in 2001 and was planning to offer 17 new blocks in 2002. The focus of exploration was on frontier regions, particularly eastern Indonesia, since its central and western sections were in production.

During 2001 the separatist movement in Aceh, an energy-rich province in north Sumatra which abuts the strategically important Strait of Malacca, caused security problems that lead ExxonMobil to cease operation of its Arun LNG export terminal and shut in production at the Arun gas fields. The facility reopened in August and resumed supplies to Japan and South Korea. Separatist movements were also active in East Timor, East Kalimantan, Irian Jaya, and Riau.

Discoveries

A gas and condensate discovery was reported late in 2001 by a unit of Repsol-YPF at its Sungai Kenawang well on Jambi Merang Block in Sumatra. The well, drilled to 8,464 ft (2,580 m) total depth, found a 407 ft (124 m) gas and condensate column in the Batujara. On test the well produced 13.4 MMcfd of gas and 370 b/d of 52º gravity condensate. Repsol estimated the well could produce up to 88.3 MMcfd of gas and 3,000 b/d of condensate.

On the same block, the Pulau Gading discovery was commercial, Pertamina said previously, and the new well was considerably larger, according to Repsol. Production from the two discoveries would be processed and marketed together to fulfil a Malaysian contract. YPF Indonesia and Pertamina were joint operators, with 25% and 50% interests, respectively, while Amerada Hess Corp. held 25%.

Two successful appraisal wells, announced in September 2001, were drilled by Unocal Corp. on the Ranggas prospect off East Kalimantan in the southern part of the Rapak PSC area, following a discovery there early in the year. Ranggas-2, drilled to 13,661 ft total depth in 5,192 ft of water about 1 mile southwest of the discovery, had 155 ft of net oil pay and 18 ft of net gas pay. Ranggas-3, drilled in 5,368 ft of water to 13,248 ft vertical depth, encountered 306 ft of net oil pay and 123 ft of net gas pay. The well was 3.4 miles north of the discovery in the central part of the structure.

Unocal estimated gross unrisked resource potential for the Ranggas structure at 190-650 million boe. The company planned to drill another four to eight wells beginning in 2001-02 and to test at least two adjacent prospects. Unocal Rapak was operator of the PSC area with 80%; Lasmo Rapak Ltd. (an ENI SPA subsidiary) held 20%.

In July 2001, 5 tcf of natural gas was discovered in the Luwuk Banggal regency of central Sulawesi, said Pertamina. PT Exspan Tomori Sulawesi made the find following completion of its Senoro No. 2 well, which flowed at a rate of 20 MMcfd of gas and 435 b/d of condensate. The gas discovery could be developed to supply a new power generation plant. Exspan had also discovered oil in the south Sumatera extension block and was testing a discovery in the Sumatera block's Matra field, which could be brought into production by 2003.

Gulf Indonesia Resources Ltd. and partners reported in May 2001 that the Suban-5 delineation well flowed on initial testing at a rate of 31 MMcfd of natural gas. The well, in the Corridor Block PSC area in South Sumatra, followed up on a previous successful delination well, Suban-4, which tested at 80 MMcfd of gas and 420 b/d of condensate – the highest flow rate of any well Gulf had drilled in Indonesia. The Suban-5 test was conducted on pre-Tertiary formations between 7,871 and 9,623 ft through a ?-inch choke at 2,800 psi. Gulf said the results were similar to the Suban-3 well, 2.3 miles to the southeast. Additional test wells and step-outs were planned for 2001.

Gulf Indonesia was operator and had a 54% interest in the Corridor Block; partners were Talisman (Corridor) Ltd. with 36% and Pertamina with 10%. About 1.1 tcf of gas from Suban field was dedicated to Gulf Resources' Caltex 11 gas sales contract. Gulf felt that the Suban field had the potential to be 4 tcf or more in size, which could support additional gas sales of 550 MMcfd from the Corridor Block over and above the current contracted amounts. Conoco Inc. in 2001 acquired a 72% stake in Gulf Indonesia through its acquisition of Gulf Canada Resources Ltd.

In mid-2001 Gulf Indonesia Resources said its third and fourth wells on the Ketapang PSC area off East Java found hydrocarbons. Jenggolo-1 tested at 3,600 b/d of oil from a 7/8-inch choke from perforations between 6,272 and 6,300 ft. A second test from perforations between 5,806 and 5,854 ft flowed 5 MMcfd of gas through a 1-inch choke. A third, between 4,780 and 4,804 ft, flowed 4 MMcfd through a ½-inch choke. Payang-1 well flowed 17 MMcfd of gas through a 1-inch choke from perforations between 4,822 and 4,854 ft. Previously, the first well in the four-well drilling program on Ketapang had gas shows and minor oil shows, but was not tested. The second, Bukit Tua-1, flowed at a combined rate of 7,250 b/d from two tests. Two shallower zones in the Bukit Tua-1 well flowed a combined 6 MMcfd of gas. Gulf Indonesia was operator of Ketapang in a 50-50 partnership with Petronas.

In April 2001 a major oil discovery was reported at the Banyu Urip No. 3 (BU-3) well by ExxonMobil Corp.'s Mobil Cepu Ltd., which said that with estimated reserves of more than 250 million bbl, BU-3 well was one of the most significant oil discoveries in Indonesia in the past decade. The well encountered nearly 1,000 ft of gross oil pay and more than 300 ft of gross gas pay, and oil was tested at a rate of nearly 4,500 b/d from Middle Miocene carbonate and sandstone formations. The BU-3 well is part of the onshore Cepu Technical Assistance Contract operated by Mobil Cepu in Central and East Java provinces. Pertamina took an option to acquire 10% of the project. Planning for a phased, fast-track development was under way. Initial production was expected from the Banyu Urip discovery by 2003 through six to ten wells utilizing an early production system.

In March 2001 Repsol announced the discovery of three oil and gas fields on West Madura Block off Java. Three exploration wells, KE23-1, KE13-1, and KE24-1, were drilled in separate structures in Kujung limestone. In tests, the discoveries produced a total of 6,300 b/d of crude and condensate and 30.1 MMcfd of gas. Preliminary estimates of recoverable reserves for the three fields were 13 million bbl of oil and condensate and more than 153 bcf of gas. KE-23 field was expected to come on stream in late 2001. Repsol affiliate YPF Madura Barat BV was operator with a 25% block interest; partners were Pertamina (50%) and Kodeco Energy Co. Ltd. (25%).

In early 2001 appraisal drilling was successfully completed at Ujung Pangkah field in East Java, confirming reserves on the Pangkah Block structure in excess of 450 bcf of rich gas, or the equivalent of 90 million bbl of oil, said Premier Oil Natuna Sea Ltd. All three wells in the appraisal program (two vertical and a sidetrack by the Trident 17 rig) encountered commercial hydrocarbons. Partners were Premier, the operator with 40%, Amerada Hess (Indonesia-Pangkah) Ltd. 36%, Gulf Resources (Pangkah) Ltd. 12%, and Dana Petroleum PLC 12%.

In early 2001 gas and condensate were discovered in Indonesia's Arafura Sea by the Inpex Masela Ltd. unit of Indonesia Petroleum Ltd. (Tokyo), which held 100% interest. The Abadi-1 wildcat on the Masela PSC Block flowed 25 MMcfd of gas and 260 b/d of condensate. The site is 140 miles northeast of the 1988 Evans Shoal gas discovery and a similar distance southwest of Tanimbar Island. Abadi-1 went to 13,878 ft (4,230 m) in 1,500 ft (457 m) of water east of Kupang, West Timor.

Exploration

The Tanjung Aru Block in the Makassar Straits off Kalimantan was awarded in September 2001 to Amerada Hess and Petronas Carigali Sdn. Bhd., subject to execution of a PSC. Amerada Hess would operate the block with 50%, while Petronas holding the other 50%. Tanjung Aru covers 1,606 sq mi (4,160 sq km) in 490-5,577 ft (150-1,700 m) of water in the Kutei basin, site of Unocal's

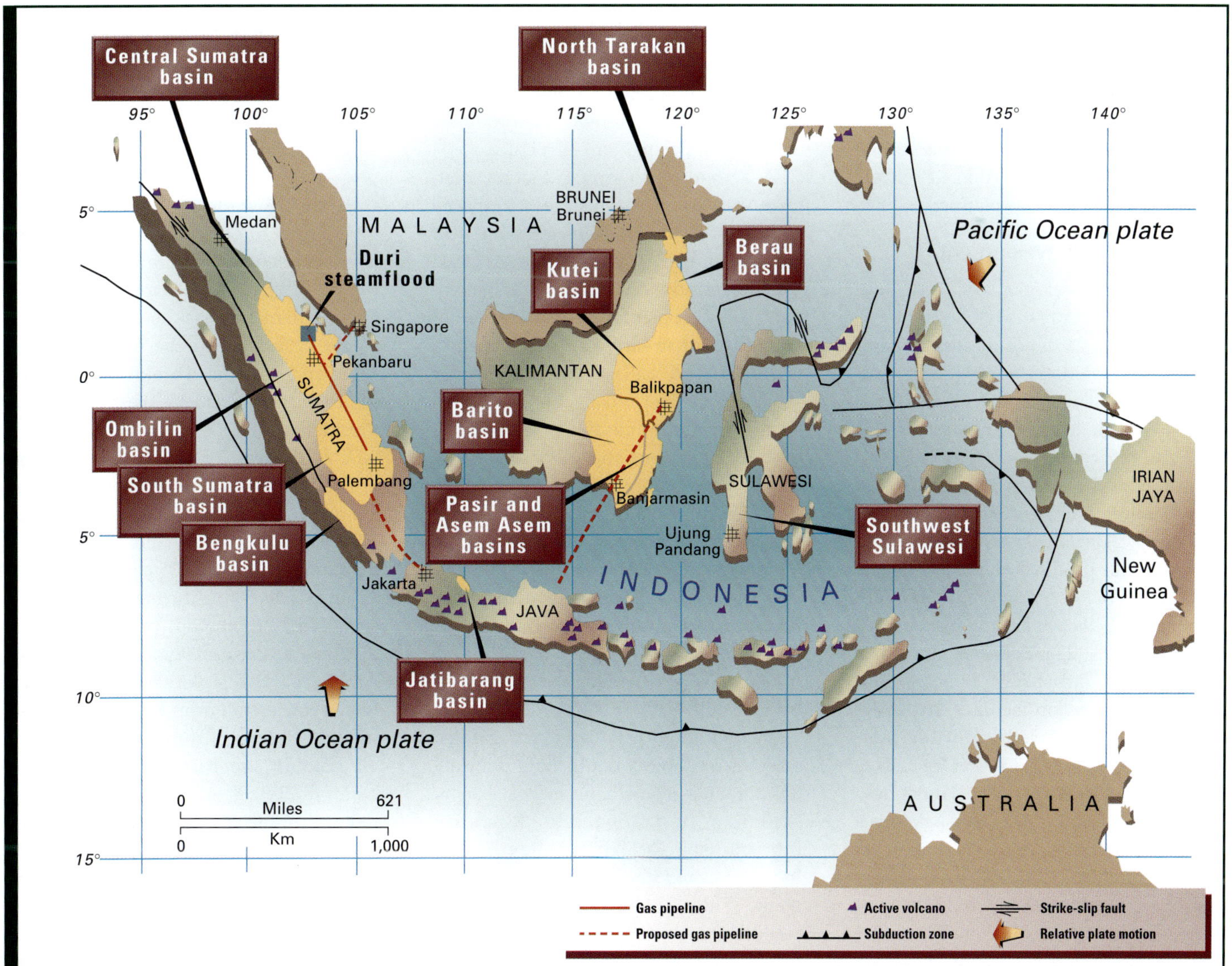

Indonesia's domestic demand for natural gas and its LNG exports might be met by its indigenous coalbed methane resources, which could contain as much as 337 tcf in 11 coal basins.

Ranggas and West Seno discoveries. The partners purchased 570 sq mi (1,477 sq km) of 3D seismic data on the block and planned to drill at least three exploration wells.

The deep offshore Donggala Block north of the Makassar Straits was awarded to a 50-50 consortium of TotalFinaElf SA and Inpex Corp. (Japan) in late 2001. The partners were producing more than 500,000 boe/d in Indonesia, including 60% of the Bontang LNG plant's gas requirements. The 1,475 sq mi (3,820 sq km) block is in 5,900-8,200 ft (1,800-2,500 m) of water 47 miles (75 km) from TotalFinaElf's Tunu field. The block had undergone a 3D seismic study, and operator TotalFinaElf was planning drilling during 2002.

Also in late 2001 a PSC for Nila Block in the West Natuna Sea was awarded to Conoco Nila Ltd. and Inpex Natuna Ltd. Conoco would operate the 2,050 sq mi (5,300 sq km) block, adjacent to its South Natuna Sea Blcok B PSC, with a 65% stake. A three-year program of seismic studies and drilling was planned.

Belanak, Yakin development

Belanak natural gas field in South Natuna Sea Block B was to undergo development by Conoco under a $1.6 billion PSC with Pertamina. Conoco awarded a $587 million contract for an FPSO unit to Brown & Root Indonesia and a $157 million contract for two wellhead platforms and associated pipelines and oil-offloading buoy to J. Ray McDermott Indonesia. The FPSO was to be installed and ready for production in late 2004.

Belanak has expected ultimate production of 530 bcf of natural gas and 100 million bbl of oil, condensate, and LPG. The Belanak structure's FPSO would have LPG extraction facilities, two wellhead platforms, 38 wells, a floating storage and offloading (FSO) unit for LPG, a gas export pipeline, and intra-field pipelines. In addition to Belanak's production, 1.4 tcf of gas and 150 million bbl of oil, condensate, and LPG from at least six other nearby fields would be produced through the Belanak facilities. The FPSO would process maximums of 350 MMcfd of gas, 100,000 b/d of oil and condensate, and 23,000 b/d of LPG. The 1,000-ft vessel would be able to store up to 1 million bbl of oil, which would be off-loaded to tankers.

Belanak development would support two separate gas sales agreements, both supplied by the Conoco-operated South Natuna Sea Block B. The first is a 22-year gas delivery contract between Pertamina and Singapore's SembCorp Gas. The Belanak field is 125 miles from Singapore. The second agreement is a 20-year contract between Pertamina and Malaysia's state oil company Petronas for the delivery of up to 250 MMscfd of gas via pipeline to Malaysia's Duyong field facilities.

At Kerisi field 15 miles north of Belanak, the Kerisi-3 flowed 2,580 b/d of oil and encountered 130 ft of oil-bearing zones in multiple reservoirs, said Conoco in October 2001. The well increases Kerisi field's potential gross recoverable oil and gas to 42 million bbl and 90 bcf, respectively. Kerisi will be produced through the Belanak infrastructure.

A second offshore production platform was installed at Unocal's Yakin oil and gas field 5 miles (8 km) off Balikpapan in East Kalimantan by H&H Utama, an Indonesian construction company, said consultant Sinclair Knight Merz Pty. Ltd. in early 2002. H&H used a float-over and ballast down technique proposed by Merz to install CPP2 platform, which is connected by a 66-ft (20-m) bridge to the existing CPP1 platform. CPP2 will separate 10,000 b/d of wellstream oil and water and 10 MMscfd of associated natural gas. Sinclair Knight Merz designed the platform's double deck topsides and engineered its transportation. Unocal operated the field under a PSC with Pertamina.

Coalbed methane, GTL

In addition to its large natural gas reserves, Indonesia could hold substantial amounts of methane in coal basins, according to Advanced Resources International Inc., PT Caltex Pacific Indonesia, and the Indonesian government. These coalbed methane reserves could represent a lower cost alternative for meeting the country's demand for natural gas and LNG exports. A comprehensive assessment identified about 337 tcf of coalbed methane resource potential in 11 Indonesian coal basins. Although development of the resource faces many challenges, it could compete with Indonesia's stranded conventional gas reserves, some of which are deep, structurally complex, and high in CO_2.

Some of Indonensia's stranded gas could be converted into petroleum products by Pertamina in a GTL plant being considered. Rentech Inc. (Denver) in October 2001 began studying the feasibility of a potential 15,000-b/d project that would use Rentech's proprietary technology to convert the gas into sulfur- and aromatic-free fuels, naphtha, waxes, and other high-value products.

JAPAN

CAPITAL: TOKYO
MONETARY UNIT: YEN
REFINING CAPACITY: 4,786,340 B/CD
OIL PRODUCTION: 12,800 B/D
OIL RESERVES: 58.6 MILLION BBL
GAS RESERVES: 1.4 CF

Japan, the world's second largest energy importer after the US, remained heavily dependent on oil imports from the Middle East in 2001. However, Japan's energy demand has been stagnant in recent years due to the country's slow economic growth. Fereidun Fesharaki, senior fellow at the East-West Center and President of FACTS Inc., said that Japan's GDP could drop 3-5% in the third quarter of 2001. This could affect world oil market growth, since Asia has accounted for 50-100% of global incremental oil demand for the past 20 years.

Japan was trying during 2001 to maintain a strong foothold in overseas oil-producing areas, especially the Middle East. Arabian Oil Co., Japan's largest producer, won a concession from Kuwait to continue operating and drilling Khafji offshore oil field in the Kuwaiti sector of the Neutral Zone. Arabian Oil said that renewal of the contract was crucial since the company had already lost its rights in the Saudi Arabian-controlled side of the same field. Kuwait and Saudi Arabia each had a 10.9% share of the Japanese company, with other shareholders including Tokyo Electric Power Co., Kansai Electric Power Co., and Nippon

Steel Co. The new agreement would allow Arabian Oil to keep exploring the field beyond the January 2003 expiration date.

Projects to import natural gas from the Russian Far East to Japan were active during 2001, including the three Sakhalin Island projects. Factors that could promote greater natural gas consumption in Japan included a deregulated electricity market, the demise of the nuclear industry, the Kyoto greenhouse gas agreement, and an increasing number of suppliers, said the James A. Baker Institute of Public Policy (Houston) in a March 2001 analysis. Whether gas is transported via pipeline or as LNG, Sakhalin Island has emerged as the leading low-cost supplier in the Russian Far East, said the study.

Japanese LNG market

Structural changes in the LNG trade will impact LNG pricing in Japan as well as the nature and duration of Japanese supply contracts, said the Baker Institute in late 2001. Asian markets for LNG are becoming more flexible, with costs becoming as important as security of supply, said the report. Traditional Asian LNG markets could become more similar to Western LNG markets over time, with Japanese customers asking for more flexible terms and US customers seeking longer-term agreements, resulting in a blend of short-term and long-term supply and transportation arrangements that fit customized needs.

The change in emphasis comes as LNG facilities and transport costs have fallen dramatically worldwide and technology has improved, while Atlantic Basin LNG suppliers have taken flexible approaches to pricing, allowing interchangeable supplies and seeking shorter durations to complement long-term contracts. The Pacific Basin is beginning to alter its more staid approach as well, indicated the Baker report. Japanese buyers might not require the current Atlantic Basin model for future LNG commitments, but new, entry-level buyers will be unlikely to provide considerable financial backing for greenfield projects as was customary in the 1970s and 1980s, requiring a change in contract design.

The Trans-Asean gas pipeline network, which connects Malaysia and other Asian countries, could eventually be expanded to tap into the Philippines' Malampaya field and other gas supplies.

The report also predicted that a supply overhang could remain a typical feature of LNG markets in the coming years, so Japan would probably not have to compete with US buyers for limited LNG supplies. Also, Middle Eastern producers could serve in a swing role for a time, delivering LNG to East or West as required. The greater the interaction of these swing suppliers with both markets, the more likely prices would converge over time, said the Baker report.

MALAYSIA

CAPITAL: KUALA LUMPUR
MONETARY UNIT: RINGGIT
REFINING CAPACITY: 514,500 B/CD
OIL PRODUCTION: 720,000 B/D
OIL RESERVES: 3.0 BILLION BBL
GAS RESERVES: 75 TCF

Malaysia has been making an excellent recovery from its financial crisis of the late 1990s, with real GDP growing by 7.5% in 2000 and forecast at 5.6% for 2001, accounting for softened demand for Malaysian exports due to the US economic slowdown. Despite an overall trend of declining oil reserves, Malaysia's oil production has been stable in recent years and increased slightly during 2000-2001.

Development

Petronas Carigali Sdn. Bhd., a branch of Malaysia's state-owned gas and oil company Petronas, began first production from Angsi field in the South China Sea off Malaysia, six months ahead of schedule and 30% below original approved cost. Angsi's initial flow was 15,000 b/d of oil and 60 MMscfd of gas, which should ramp up to 65,000 b/d of oil and 450 MMscfd of gas at peak. ExxonMobil Exploration & Production Malaysia was Petronas Carigali's 50% partner in the development.

ExxonMobil began first production from Seligi H, the first of five satellite platforms being used to develop six fields. Seligi H is 50 miles off Terengganu in the South China Sea. The $240 million satellite platform project would have peak production from all five generic, minimum facility platforms (Seligi H, Raya B, Lawang A, Serudon A, and Irong Barat B) of 40,000 b/d of oil and 50 MMcfd of gas. Total production was estimated at 90 million bbl. Under various PSCs, ExxonMobil had a working interest of 78-80% in the fields, with the remaining interest held by Petronas Carigali.

BP PLC was planning to sell its 25% share of Block A-18 of the Malaysia-

Thailand Joint Development Area (JDA) for about $500 million. The partners wre Trans Thailand Malaysia Co., which was operator with 50%, Petronas Carigali, and PTT Exploration and Production PLC of Thailand. Amerada Hess Corp. held the remaining share.

Murphy Saban, a unit of Murphy Oil Corp., took a farmout of Esso Exploration & Production Sabah Ltd.'s 80% holding in Block H off Sabah. Murphy Sabah will operate the deepwater block, with the other 20% interest held by Petronas Carigali. Block H is contiguous with and shoreward of roughly the eastern half of Murphy-Petronas Block K. At 4.1 million acres, undrilled Block K is the largest block Malaysia has awarded.

Saba Shell Petroleum Co. Ltd. and Petronas reported a commercial deposit of gas pay with oil indications on Block G, shoreward of the western half of Block K. The Kamunsu East-1 is in 2,417 ft of water 10 miles (16 km) north of Kebabangan gas field.

Amerada Hess Malaysia Ltd. discovered oil and gas with its second exploration well, Cendor-1, on Block PM304 off Peninsular Malaysia. The well was spudded in 207 ft (63 m) of water, drilled to 5,873 ft (1,790 m) total depth, and tested at a maximum rate of 2,840 b/d of 42º gravity crude from a single zone. Amerada found hydrocarbons at multiple levels. Amerada was operating the block with a 65.5% interest.

Early in the year, Swedish producer Lundin Oil AB, through its wholly owned subsidiary Lundin Malaysia Ltd., announced an oil discovery on Block PM-3 CAA in the JDA. Seven oil zones were found in the East Bunga Raya No. 1, and the main oil zone flowed at 5,500 b/d in production tests. Lundin held a 41.44% interest. Later in the year, Lundin sold its holdings to Talisman, Calgary.

Carigali-Triton Operating Co. deferred the production start-up of Cakerawala field in the South China Sea by a few months beyond the original mid-2002 schedule because the pipeline has not been built. Cakerawala, on Block A-18 in the JDA, would deliver 390 MMcfd of gas once on stream.

Transport

The pipeline to move gas from the offshore Thai-Malay JDA has been delayed several times, but work must begin this year to meet contract requirements of delivery in mid-2002. For the first five years, Malaysia would take all 390 MMcfd. As planned the 34-inch, 168-

mile line would run offshore from Block A-18. From Songkhla, it would link via a 36-inch, 56-mile line to Kedah in northern Malaysia. Instead, Malaysia has unofficially suggested a line be laid from Block A-18 (operated by a 50-50 joint venture of Triton Energy Ltd. and Petronas Carigali) to Kerteh on the eastern coast of Malay Peninsula. The 34-inch offshore line

would be 193 miles long.

Seven members of the Association of Southeast Asian Nations (ASEAN) were cooperating in the construction of a $6 billion Trans-Asean gas pipeline. The project, planned as a 6,200-mile (10,000-km) gas grid, would link Malaysia, Singapore, Indonesia, the Philippines, Myanmar, Viet Nam, and Thailand. Petronas's 20-year commitment to buy up to 250 MMscfd of gas from Indonesian state company Pertamina would contribute to the project. About 1,578 miles (2,540 km) of ASEAN cross-border pipeline was in place between Malaysia and Singapore, Myanmar and Thailand, and Indonesia and Singapore. The ASEAN Council for Petroleum was overseeing the project development.

Refining, LNG

BASF AG and Petronas opened an integrated chemical complex in Gebeng, Pahang. The 150-hectare Verbund site comprises 12 plants producing acrylic monomers, oxo products, and butanediol. The acrylics complex came on stream in July 2000. A complex for the production of syngas, oxo products, phthalic anhydride and plasticisers became operational in April 2001. A butanediol complex, which was under construction, was scheduled to start up in late 2002. Project investments at the site would total about $900 million. The combined production capacity of the plants would be more than 1 million tpy. More than 80% of the products would be exported to customers in the Asia Pacific region.

Multinational joint venture Malaysia LNG Tiga secured two loans worth $816 million to finance the construction of a third LNG plant and related facilities at Bintulu, Sarawak. The plant, 45% complete, would consist of two 3.4 million tpy trains. Related facilities to be built include an LNG storage facility and an LNG loading jetty. Scheduled to begin operation in 2003, the plant will make the Bintulu LNG complex the world's single largest LNG production facility, with a combined capacity of about 23 million tpy.

Norwegian company Statoil AS sold its remaining 15% equity in Malaysian Refining Co. Sdn. Bhd. (MRC), partly to Petronas and partly to a Conoco Inc. affiliate. MRC owned the second refining train at Petronas's 190,000-b/d refinery complex in Sungai Udang, Melaka state, Malaysia.

BP PLC said it acquired Cairns Ltd., a subsidiary of Genting Berhad of Malaysia, for $107 million and a deferred share of about 3% of net profit from Tangguh field in Indonesia. Genting Berhad is an investment holding and management company.

MONGOLIA

CAPITAL: ULAN BATOR
MONETARY UNIT: TUGHRIK
REFINING CAPACITY: NONE
OIL PRODUCTION: 700 B/D
OIL RESERVES: NONE
GAS RESERVES: NONE

SOCO International PLC (London) concluded its four-well 2001 drilling program in the Tamtsag basin in Mongolia. The 19-13 and 19-14 wells were completed and fracture-stimulated in the Tsagaantsav zone, and the wells tested at initial rates of approximately 350 b/d of oil. These two wells were expected to be put into production in November 2002. The 21-3 well, a rank wildcat, was drilled to 6,890 ft (2,100 m) and abandoned without testing. The 21-4 well, a 1,312-ft (400-m) offset to the SOTAMO 21-2 well drilled in 1997, was drilled to 8,200 ft (2,500 m) and found 246 ft (75 m) of oil shows at 5,469-6,627 ft (1,667-2,020 m).

Huabei Oilfield Services, a subsidiary of PetroChina Co. Ltd., exercised its right to a 10% working interest participation in contract areas 19, 21, and 22. SOCO retained an 85% interest in the areas, with PetroVietnam, the state oil company of Viet Nam, holding 5%.

Mongolia's interest in the $5 billion, 2,485-mile (4,000-km) pipeline linking China's eastern and western regions was dulled when the Chinese government vetoed Shell's proposal that the proposed line first tap a gas field in Inner Mongolia. The Mongolian field is less than half the distance from Shanghai. China insisted the pipeline should link the Xinjiang fields first to speed development in the region. Also, because the concerned parties failed to reach agreement on prices customers are willing to pay for the oil and gas to be transported in a pipeline, construction plans stalled.

Although delayed to date, Mongolia hopes to have its first refinery sometime in 2002.

MYANMAR

CAPITAL: YANGON
MONETARY UNIT: KYAT
REFINING CAPACITY: 32,000 B/CD
OIL PRODUCTION: 8,000 B/D
OIL RESERVES: 50 MILLION BBL
GAS RESERVES: 10 TCF

In 2001, state company Myanmar Oil & Gas Enterprise made two gas discoveries, and Westburne Oil Ltd. made two oil discoveries.

The Yadana and Yetagun fields in Myanmar's Gulf of Martaban averaged 91,899 boe/d compared with 79,820 boe/d for the same quarter last year. First-quarter 2001 sales volume of Yadana gas reached 519 MMcfd; the sales volume of Yetagun field reached 104 MMcfd.

Premier Petroleum Myanmar Ltd. contracted the Singapore office of Norwegian company Kværner ASA to upgrade the Yetagun platform in Myanmar's Gulf of Martaban from 230 MMscfd to 300 MMscfd of capacity. The cost of the upgrade is $30 million.

Early in 2001 Thailand began operating the 96-mile (154-km), 30-inch, Ratchaburi-Wang Noi onshore pipeline that links the 2,000-MW Wang Noi combined-cycle power generator with the Myanmar gas supply. Over the past two years, the Thai state oil and gas company PTT PLC took less Myanmese gas than the contractual rates – 525 MMcfd from Yadana and 200 MMcfd from Yetagun, due largely to the delay in the construction of the Ratchaburi power plant. The $247.9 million line began with deliveries of about 280 MMcfd of Myanese gas. PTT said its take of Myanmese gas will grow to 900 MMcfd in 2001 to feed the Ratchaburi, Wang Noi, and Tri Energy power stations.

One of the first ventures of a newly formed partnership of three Indian companies — Oil & Natural Gas Corp., Indian Oil Corp., and Gas Authority of India Ltd. (GAIL) – was an agreement to buy gas from any Myanmar discoveries made by South Korean conglomerate Daewoo Corp. Daewoo had signed a PSC with state-owned Myanmar Oil & Gas Enterprise for the 960,000-acre Block A-1 off northwestern Myanmar. Block A-1 is near the border with Bangladesh in the Bay of Bengal.

GAIL, the consortium partner with expertise in transport and marketing, would need to build a Myanmar-India pipeline to import gas. The selected route is a 342-mile (550-km) pipeline around the continental shelf of Bangladesh, which comes with an expected cost of around $625 million. The pipeline would not require the permission of Bangladesh.

NEW ZEALAND

CAPITAL: WELLINGTON
MONETARY UNIT: DOLLAR
REFINING CAPACITY: 106,000 B/CD
OIL PRODUCTION: 34,000 B/D
OIL RESERVES: 89.5 MILLION BBL
GAS RESERVES: 2.08 TCF

Following appraisal drilling on its oil and gas discovery, Swift Energy New Zealand Ltd. tested Rimu B-1 well in the Upper Rimu limestone. The well was perforated and acidized prior to flowing at a stabilized test rate of 505 b/d of 46º gravity oil and 2.8 MMcfd of gas. Swift was operating the well with a 90% stake. Antrim Oil & Gas Ltd. and Marabella Enterprises Ltd. each held 5%.

Swift Energy drilled its Kauri-A1 well to evaluate the well's potential in the Tariki sand, Rimu lime, and Cretaceous sands on the large Kauri structure. Plans called for workers to set a 9-5/8 inch casing to a depth of 9,840 ft and drill an 8.5-inch hole to 14,760 ft. Testing found hydrocarbons from the Upper Tariki sand. The Kauri structure is immediately south of the Rimu structure, where previous wells have encountered upper and lower sections of both the Tariki sand and the Rimu lime.

In other drilling activity, Swift Energy awarded Parker Drilling Co. a six-month contract for land rig 246 and a one-well contract for land rig 252. The contract was upgraded later in the year to include up to 16 wells for Rig 252. Marabella Enterprises Ltd. contracted for Parker Drilling's Rig 188 to drill two wells in New Zealand, with an option for one more.

TGS-Nopec ASA (Oslo) began a nonexclusive 4,225-mile (6,800-km) 2D seismic survey covering the deepwater sector of the Taranaki basin off New Zealand. Most discoveries and drilling to date have been onshore.

Reserves estimate

Maui Development Ltd., operator of New Zealand's only major offshore gas field in the Taranaki basin, and the New Zealand government began negotiations to downgrade the reserves estimate for the field and adjust the quantities of gas that can be delivered to New Zealand. The field is near the end of its life, yet it supplies most of New Zealand's gas as well as significant quantities of condensate, LPG, and oil. Maui Development (87.5% owned by Royal Dutch/Shell Group, 12.5% by Todd Petroleum Mining Co. Ltd.) told the government that Taranaki field is expected to end production two years earlier than expected, in mid-2007. The field had estimated remaining reserves of 1.25 tcf of gas and 57 million bbl of condensate.

The redefinition of reserves means that there might be insufficient gas left to meet fulfill take-or-pay contracts between Maui Development and the New Zealand government. The government resells the gas to power companies, industrial users, and wholesalers.

Acquisitions

Swift Energy New Zealand Ltd. agreed to acquire Southern Petroleum (New Zealand) Exploration Ltd. from an affiliate of the Royal Dutch/Shell Group. Swift will pay $55 million (US) in cash for a 96.76% working interest in four onshore producing oil and gas fields (Tariki field, PML 38138; Ahuroa field, PML 38139; Waihapa field, PML 38140; and Ngaere field, PML 38141), hydrocarbon-processing facilities with excess capacity, and pipelines connecting the fields and facilities with terminals and markets, all in the Taranaki basin. The four fields were producing 930 b/d of oil, 23.2 MMcfd of gas, and 450 b/d of gas liquids. Net proved reserves as of late 2001 were 65 bcf, with 74% of the reserves being gas.

Separately, Swift Energy also agreed to buy all of Calgary-based Antrim Energy Ltd.'s New Zealand assets, which included a 5% interest in Rimu-Kauri permit PEP 38719 and a 7.5% interest in Huinga permit PEP 38716.

Swift Energy agreed to sell 38 bcf of natural gas during a 10-year period (3.8 bcf/year) to New Zealand electricity generator Genesis Power Ltd. The gas would come from Swift's Rimu discovery and would be used at Huntly Power Station.

PAKISTAN

CAPITAL: ISLAMABAD
MONETARY UNIT: RUPEE
REFINING CAPACITY: 238,850 B/CD
OIL PRODUCTION: 59,000 B/D
OIL RESERVES: 298.2 MILLION BBL
GAS RESERVES: 25.1 TCF

Pakistan's privatization ministry retained financial adviser JP Morgan Chase & Co. to help it sell 51% and management control of Pakistan State Oil (PSO), the country's company for storage, distribution, and marketing of petroleum products, petrochemicals, LPG, and compressed natural gas. The LPG asset would be sold first. The company had a domestic market share of more than 70% and assets worth $500 million.

The government earlier appointed Merrill Lynch & Co. Inc. the financial advisor for Oil & Gas Development Co. Ltd., the largest state-owned oil and gas exploration and production company. Gaffney, Cline & Associates was hired for the sale of state assets in nine oil and gas fields: Minwal, Pariwali, Turkwal, Badin I, Badin II, Mazarani, Adhi, Ratana, and Dhurnal. Pakistan was netting 9,000 b/d of oil and 60 MMcfd from the nine blocks, which had 50 million boe/d of reserves.

Development

Zamzama gas field, on the Dadu concession between the towns of Dadu and Johi in Sindh Province, was brought on stream less than a year from the signing of a gas sales contract and less than three years from discovery. The joint venture partners were BHP Petroleum Pty Ltd., operator, Premier Oil PLC, Lasmo PLC, and the government of Pakistan's holdings.

Two wells drilled on the structure tested at rates in excess of 90 MMscfd, which was believed to be record flow for gas wells in Pakistan. Zamzama field would produce 70 MMscfd initially with development plans calling for production of more than 400 MMscfd, which would add about 20% to Pakistan's gas supply. When fully developed, the field would supply gas for power generation, backing out oil imports totaling $300 million/yr.

Mari Gas Co. Ltd. discovered more than 1 tcf of gas in southern Sindh province in one of the largest reservoirs in the country. The Pab sandstone formation of Upper Cretaceous age produces or is capable of producing gas in several fields in southern Pakistan, including Mari Deep. Mari Gas was owned by Fauji Foundation, a company 40% owned by retired and serving army personnel. The state-run Oil and Gas Development Corp. and the Pakistani government each had 20% stakes, while the remaining shares were held by the public.

Oil & Gas Development Co. Ltd. made a discovery in Kohat district in the Pakistani North West Frontier province. Chanda Deep No. 1 was drilled to 16,732 ft (5,100 m) and produced 900 b/d of 36º oil and 1,150 b/d of 38º oil, respectively, from two zones. The company was producing about 1,200 b/d and 5.19 MMcfd of natural gas from another well in the same petroleum concession area.

A joint venture led by Petronas Carigali (Pakistan) Ltd. tested a well at a maximum rate of 25.6 MMcfd of gas near Sukkur in Sindh province, Pakistan. Partners with Petronas Carigali (operator, 57% interest), an affiliate of the Malaysian state oil company, were Lasmo Oil Pakistan Ltd. (38%) and Government Holdings (Pvt.) Ltd (5%). The Rehmat-1 well on block 2769-4 (Mubarak Block) reached 11,601 ft (3,536 m) total depth less than three months after being spudded. It is 40 miles (65 km) east of Sukkur.

Pakistan's Ministry of Petroleum and National Resources approved an agreement between the SW-Miano joint venture, led by OMV AG of Austria, and Sui Southern Gas Co. Ltd. for sale of 170 MMcfd of gas from Sawan field in Sindh province. The joint venture would spend $170 million during the next two years to develop the field. Deliveries of the gas via the existing Kadanwari pipeline would begin in late 2002 or early 2003. OMV AG's Pakistani unit operated the venture with 19.7%; Agip SPA owned 23.7%; Moravske Naftove Doly, 7.9%; Pakistan Petroleum Ltd., 26.2%; and the government, 22.5%.

Crescent Petroleum Co. International Ltd., Sharjah, completed feasibility and engineering studies for the proposed $3 billion Qatar-to-Pakistan offshore gas pipeline. Gas for the 1.6 bcfd pipeline could come from TotalFinaElf SA's LNG interests in Qatar.

Premier acquired Shell's 49.9% holding in Premier & Shell Pakistan BV in return for a portion of its holding in the Bhit development project and settlement of liability issues.

Refining

Pak-Arab Refinery Co. Ltd.'s $886 million, 100,000-b/d refinery – at Mehmood Kot, near Multan, Punjab province – began production on a trial basis at about 50% of capacity in early 2001. This refinery will help significantly decrease product imports. Pak Arab Refinery Co. Ltd. (Parco) was a joint venture of the government of Pakistan, with 60% interest, and the government of Abu Dhabi (through Abu Dhabi Petroleum Ltd.) with 40%.

Parco awarded China Petroleum Engineering & Construction Corp. a letter of intent to build a 26-inch, 508-mile (817-km) products line from its refinery at Mehmood Kot in Punjab province to Karachi's Port Qasim. The $480 million pipeline, capable of carrying multiple products up to 12 million tpy, was expected to enter service by December 2002. A joint venture, Parco owned 51%, Shell Pakistan Ltd. 26%, Caltex Pakistan 11%, and Pakistan State Oil 12%.

The $50 million Bosicor (Pakistan) Ltd. refitted oil refinery located in the Balochistan province was slated to start production in 2002. The plant would have a capacity of about 30,000 b/d, purchased from a Qatar-based company, and would produce six petroleum products, including diesel fuel and fuel oil, and 50,000-70,000 tpy of naphtha. Pakistan State Oil would buy all the refinery production. The savings expected from import substitution would be around $25-30 million/year.

Pakistan boosted oil reserves to 45 days' supply from 19, as part of a plan to face any emergency. The country also set up an inter-ministerial management committee to cope with crises, including oil and gas-related ones.

PAPUA NEW GUINEA

CAPITAL: PORT MORESBY
MONETARY UNIT: KINA
REFINING CAPACITY: NONE
OIL PRODUCTION: 58,000 B/D
OIL RESERVES: 238.3 MILLION BBL
GAS RESERVES: 12.2 TCF

InterOil Corp. (Cairns, Australia) set in motion plans to construct the first refinery in Papua New Guinea. It would also be the first refining project for InterOil, a wholly-owned Papua New Guinea subsidiary established in 1996 to finance and construct a niche market refinery in Papua New Guinea.

The $180 million project, located in Napa Napa on the western side of Port Moresby, would construct a hydroskimming (HDS) refinery from relocated, refurbished, and new equipment. With a nameplate crude capacity of 32,500 b/d, the refinery would produce commercial propane, commercial butane, various grades of naphtha for petrochemical feedstocks, unleaded gasoline, automotive diesel fuel, jet A-1 aviation fuel (kerosene), and low-sulfur marine diesel fuel. The HDS and reforming units would produce 3,500 b/d of reformate for gasoline blending, and the gasoline would be blended online. By design, a naphtha isomerization unit can be included in the refinery in the future. Finished products would meet or exceed relevant Australian standards.

The refinery would use indigenous crude Kutubu Blend (0.04% sulfur content and 44º API gravity), which was projected to be available in sufficient production quantities to supply the refinery for more than 10 years. When necessary, the refinery could also use various Asia-Pacific light, sweet crudes. The refinery would meet or exceed environmental standards recommended in the World Bank guidelines for oil refineries and would be self-monitored.

InterOil raised equity of $100 million, $65 million of which had been spent on a second-hand refinery from Chevron, a catalytic reformer, and development costs. In addition, InterOil signed with the Overseas Private Investment Corp. (OPIC, an agency of the US government) for an $85 million loan.

In support of the refinery project and the goal of reducing Papua New Guinea's dependence on foreign petroleum products, the government has agreed to allow InterOil's new refinery to purchase crude oil from Papua New Guinea producers at the producers' export price, to sell refined products at import parity price, and to allow InterOil to supply 100% of Papua New Guinea's refined products demand. In addition, the government gave InterOil a five-year tax-free holiday, effective on the first day of the financial year following refinery start-up, and a customs duty exemption for refinery construction equipment and materials.

Completion of the refinery was slated for the second half of 2002. Kvaerner Process (Houston) and Clough Engineering Ltd. (Perth) performed engineering design for the refinery. Clough Niugini Ltd., the operating arm of Clough Engineering, began engineering, procurement, and construction management of the project.

Papua New Guinea was importing refined products from the South Pacific region, including Australia and Singapore. The lower transportation costs and simpler logistics of a domestic refinery would generate savings to provide lower prices to gas consumers.

InterOil estimated that the country's domestic market would require 65% of the refinery's fuel products, leaving 35% for export. Naphtha would represent 70% of the refinery's export volume.

As part of a larger deal between InterOil and Royal Dutch/Shell Group, Shell International Eastern Trading Co. of Singapore agreed to buy and market the excess naphtha for three years from the Papua New Guinea refinery. InterOil agreed to buy Shell's retail and distribution assets in Papua New Guinea for $18 million in cash and stock. Shell would hold 9% of InterOil's outstanding shares and would buy all of its Papua New Guinea product supplies from the refinery. InterOil would lease the retail and distribution assets back to Shell to facilitate Shell's continuing responsibility for daily operations. InterOil estimated the contract is worth $262 million. The deal would become effective when the refinery begins operations in late 2002.

ExxonMobil Corp. took control from project partner ChevronTexaco of the 1,990-mile (3,200-km) Papua New Guinea-to-Queensland natural gas pipeline project to export gas to Australian consumers. ExxonMobil held more than 30% of the project. The $3.5 billion ($6.5 billion Aus.) proposed pipeline was expected to commence operations in 2003 rather than late 2001.

PHILIPPINES

CAPITAL: MANILA
MONETARY UNIT: PESO
REFINING CAPACITY: 419,500 B/CD
OIL PRODUCTION: 1,000 B/D
OIL RESERVES: 178.1 MILLION BBL
GAS RESERVES: 3.7 TCF

The $4.5 billion Malampaya project off the Philippines delivered first gas to the Batangas gas-handling plant. Major activities in 2001 included the signing of an agreement between Shell and the Philippine government to set up the necessary multipartite environmental and surveillance team and its funding; the completion, delivery, positioning, and connections of the production platform topsides; and the completion of five production wells. At 11,500 tonnes, the topside deck was the heaviest structure ever installed using the "float-over" technique.

Shell Philippines Exploration BV (SPEX, 45% stake in the project) was manager of the joint venture with Texaco Philippines (45% stake) and Philippine National Oil Co.-Exploration Corp. (10%). The project would supply gas from the offshore Malampaya field to fuel three power plants with a total generating capacity of 2,700 MW.

Malampaya field is 31 miles (50 km) northwest of Palawan Island and 373 miles (600 km) southwest of Manila. Discovered on block SC38 in 1992, it has 3 tcf of gas reserves and 120 million bbl of recoverable condensate. First gas production arrived via a 311-mile (500-km), 24-inch pipeline at the Tabangao, Batangas Province, gas plant on Luzon Island. After treatment, it was sold to three new power plants: a 1,200-MW plant operated by the National Power Corp. at Ilijan, a 1,000-MW plant at Santa Rita operated by First Gas Power Corp., and a 500-MW plant at San Lorenzo operated by FGP Corp. All locations are in Batangas province, southern Luzon Island. Construction of these downstream plants cost $2.5 billion.

Shell and its partners were in discussion with potential customers to build a 62-mile (100-km) land pipeline to take Malampaya gas from the shore terminal at Batangas to Manila. A proposed 1,000-MW gas-fired power station there would replace an existing smaller oil-fired unit, with Shell's Intergen subsidiary likely to hold a substantial interest. The proposed Batman pipeline would allow Shell to bring Malampaya to its full production rate of 500 MMcfd. The pipeline would also be a strategic fit with plans to develop a Trans-Asean gas grid.

The Malampaya project was estimated to generate $10 billion in revenues for the Philippines government over the next 20 years and cut the country's dependence on fuel imports, mainly high-sulfur coal and fuel oil, by 20-30%. It also would cut $670 million/year from the nation's energy import bill of $2.5 billion/year, as well as bring in $420 million/year until 2021 in royalty payments from the pipeline connecting the platform to the onshore gas terminal.

Royal Dutch/Shell found a thin oil rim beneath the gas zone in the Malampaya gas reservoir. Tests performed on a deep well drilled through the gas reservoir yielded 8,000 b/d of 30º gravity crude, the highest single-well oil production rate in the Philippines since the 1970s. Flow was limited by the capacity of the processing facilities. Shell predicted 25,000 b/d of unconstrained flow from reserves estimated at 50 million bbl.

At ceremonies celebrating the first gas flow from the Malampaya project, Philippine President Gloria Marcapagal-Arroyo unveiled an energy strategy centered on natural gas. In this strategy, all onshore gas projects would be deregulated, the country's power station network would be partially privatized soon with

MALAMPAYA PROJECT

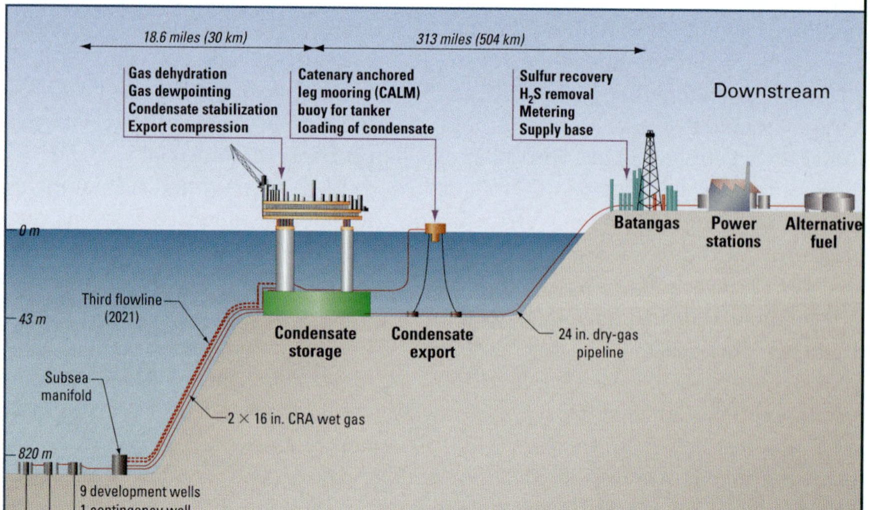

The Malampaya gas development project off the Philippines began production in late 2001, with gas delivered via a 313-mile pipeline to electric power generators on the main island of Luzon.

the condition that new investors convert from burning oil to gas, and the government would start a program to convert all public transportation to compressed natural gas. These policies depended on gas prices being negotiated with existing and new producers.

GNPower, a private power generator based in Manila, signed a letter of intent to buy LNG from the Tangguh joint LNG project held by Pertamina and BP PLC unit BP Indonesia. The Tangguh LNG project is in Berau Bay, Irian Jaya (Papua). It was on schedule to be on stream in 2006. Tangguh consisted of three offshore PSCs and the planned onshore LNG terminal.

The Spanish and Philippine governments signed a $48 million agreement with BP to build a solar energy project to power 150 villages. BP would bring solar power to 35 Agrarian Reform Communities (ARCs) in the Mindanao region of the Philippines in the first phase of the project. The second phase would serve an additional 44 ARCs. Solar power is attractive because of the prohibitive cost of extending power lines and the difficulty of transporting generator fuel to remote, developing areas.

SINGAPORE

CAPITAL: SINGAPORE
MONETARY UNIT: DOLLAR
REFINING CAPACITY: 1,258,500 B/CD
OIL PRODUCTION: NONE
OIL RESERVES: NONE
GAS RESERVES: NONE

The refinery business in Singapore comprised over 12% of its manufacturing sector in 2000, according to the Singaporean Statistical Information Services. As of September 2001, however, refineries were operating at only 63% of their total capacity.

ExxonMobil brought on stream its $2 billion Singapore petrochemical complex on Jurong Island, the largest ExxonMobil investment in the Asia-Pacific region. Work on the complex began in 1997. The complex, which is integrated with ExxonMobil's Singapore refinery, has four plants: an 800,000-tpy steam cracker that produces ethylene, propylene, and other products; a 480,000-tpy polyethylene plant, the largest single reactor plant of its type in the world; a 315,000-tpy polypropylene plant; and a 150,000-tpy oxo alcohol plant. A 155-MW cogeneration unit provides power to the chemical complex and the refinery.

Oil and gas imports

Singapore imports 100% of the natural gas used for power generation and all the oil and gas used as feedstocks for petrochemical production. Early in 2001, Royal Dutch/Shell Group shipped the first cargo of crude from Soroush field off Iran to its refining complex in Singapore.

Natural gas deliveries from Indonesia's West Natuna field to Singapore via the 400-mile (640-km) West Natuna Transportation System (WNTS) began in 2001. This started a 22-year contract between Pertamina, Indonesia's state-owned oil company, and Singapore's SembCorp Gas.

Just weeks after the WNTS started up, PowerGas Ltd. of Singapore signed a 20-year, $9 billion contract with Indonesia's Pertamina to buy 2.27 tcf of South Sumatra natural gas. Singapore would import up to 350 MMcfd through a planned 300-mile (483-km) pipeline, about half of which would be offshore, from south Sumatra to the islands of Batam and Singapore. Deliveries would begin at 150 MMcfd in 2003 and ramp up to 350 MMcfd by 2008.

SOUTH KOREA

CAPITAL: SEOUL
MONETARY UNIT: WON
REFINING CAPACITY: 2,560,100 B/CD
OIL PRODUCTION: NONE
OIL RESERVES: NONE
GAS RESERVES: NONE

On April 19, South Korea was approved for membership in the International Energy Agency (IEA) as the agency's 26th member. South Korea began talks on membership in 1993 and has accumulated emergency oil stocks above 90 days of oil imports, thereby fulfilling a key requirement for all IEA members. The strategic petroleum reserve is managed by the state-owned Korean National Oil Corporation (KNOC).

For the long term, KNOC has entered 19 overseas exploration and production projects in 12 countries and exploration of domestic blocks offshore South Korea. An oil find in August 2001 at the Vung Tau site offshore Viet Nam was estimated to hold 420 million bbl of recoverable reserves that could be developed into production by 2003. The South Korean government planned for KNOC to provide 10% of the country oil needs by 2010.

Korea Gas Corp. (Kogas), the country's sole gas importer, will be separated into three corporations. The two that deal with imports and wholesale marketing would be auctioned off in the second half of 2002 as part of the government's privatization plan. The third corporation would continue as a facilities company operating import terminals and transmission lines. Private sector firms would also be allowed to import and market gas.

South Korea relies on imported LNG for all its natural gas needs, with heavy reliance on supplies from Malaysia and Indonesia and smaller supplies from Brunei, Qatar, and Oman.

Kogas planned to increase the capacity at its Pyontack and Inchon LNG terminals. In addition, Pohang Iron & Steel, the second-largest steelmaker in South Korea, planned to build a $290 million LNG terminal to cut costs and ensure stable fuel supplies for its power plants. Construction of the 1.7 million tpy terminal at Kwangyang, south of Seoul, was expected to begin in June 2002 with completion slated for March 2005.

Longer term, North and South Korea were jointly studying the possibility of importing gas from Russia via pipeline from Eastern Siberia.

Late in 2001, Mirant Corp. reported plans to purchase Hyundai Energy Co. Ltd. (Henco), South Korea, and build a 520-MW combined-cycle power plant in southern Korea. The independent power producer had permits and the land necessary for the development in place and regulatory approvals completed.

The plant would be in the 2,278-acre Yulchon industrial complex in Cholla province, a site with expansion potential for another 1,000 MW. Construction start in the latter half of 2002 would bring operations on line in the third quarter of 2004. The plant's output would be delivered to the Korea Electric Power Co.

TAIWAN

CAPITAL: TAIPEI
MONETARY UNIT: NEW TAIWAN DOLLAR
REFINING CAPACITY: 920,000 B/CD
OIL PRODUCTION: 700 B/D
OIL RESERVES: 4.0 MILLION BBL
GAS RESERVES: 2.7 TCF

The global economic downturn of 2001 hit Taiwan hard and pushed its economy into a recession. Taiwan has very limited domestic energy resources and relies on oil imports, mostly from the Arabian Gulf and also West Africa. Taiwan produced a modest amount of gas, with state-owned Chinese Petroleum Corp. (CPC) reporting 69 onshore wells in production averaging a combined 70 MMcfd of gas at year-end 2000.

Following the entry of Formosa Petrochemical Corp. (FPC), a privately owned subsidiary of Formosa Plastics Corp., into Taiwan's downstream industry in late 2000, Taiwan was on its way to becoming a net refined-product exporter after 2001.

Deregulation of Taiwan's energy sector was fostering competition during 2001, and the country's entry into the World Trade Organization (WTO) with China in November 2001 promised to open the oil industry further. Unlike China, Taiwan was admitted to the WTO as a developed country, with more stringent requirements for reducing barriers to foreign competition. In late 2001 Taiwan revoked a ban on direct trade and investment with mainland China that had been in effect since 1949, and its legislature passed a Petroleum Administration Act in September 2001. The act would permit the eventual sale of a majority stake in CPC, to take place in 2004. The act should also further liberalize petroleum product trade and marketing.

FPC has been building crude distillation units and expected to increase its petroleum product output steadily during 2001-03. FPC was not only Taiwan's first private refiner, but also the first private producer of ethylene. The overall refinery output in Taiwan was expected to reach more than 1.1 million b/d in 2005. CPC and FPC were the only ethylene producers in Taiwan.

FPC's share of the domestic products market has quickly increased to about 25%. Taiwan had about 2,000 retail gas stations, of which CPC owned about 600. The market shares of CPC and FPC are determined through business alliances with retail stations. FPC actually owned only three stations (with a plan to increase the number to seven) but was supplying 400 stations through business alliances and planned to supply 750 by year-end 2001. FPC believed that its share could eventually be more than 50% when all of its crude distillation and other downstream units are online. CPC, which had been supplying products to 1,600 stations, was struggling to maintain its share at a minimum level of 65%.

Foreign investment in Taiwan's retail market remained limited. Only ExxonMobil Corp. (with one station under the Mobil brand) was operating in Taiwan, through joint ventures with local Taiwanese companies. After WTO membership takes effect at the start of 2002, ExxonMobil plans to increase its number of stations to about a dozen. Other oil majors could also enter the retail market after Taiwan relaxes its control of the 35% maximum share of foreign capital in joint ventures.

Taiwan's LNG consumption was expected to increase to about 7.0 million tpy by 2005 and to 10.8 million tpy in 2010. An existing government-owned receiving terminal in southern Taiwan was being expanded. Liberalization of Taiwan's energy sector has made possible the construction of competing LNG import terminals. Several private Taiwanese and Japanese firms have formed a consortium, Tung Ting Gas, to pursue an LNG regasification project in Tao-Yuan county. Construction of the northern terminal was started in mid-2001 by Japan's Chiyoda and Taiwan's CTCI and was scheduled for completion in 2004. CPC's relinquishment of its monopoly on LNG purchasing allowed Tung Ting Gas to sign a preliminary agreement with Qatar's RasGas for LNG supplies.

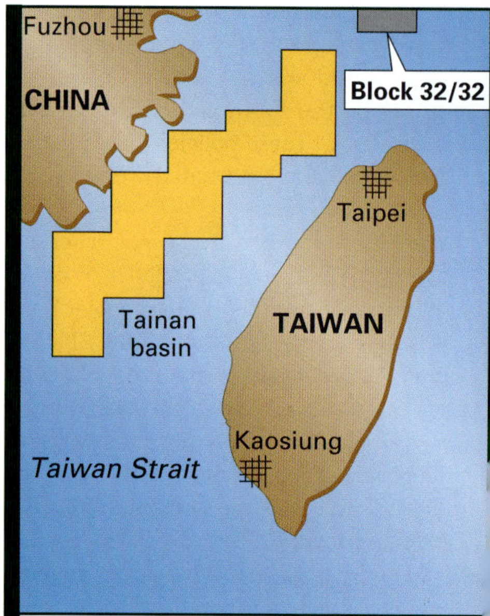

The block China intended to explore jointly with Taiwan covers almost 6,000 square miles between Guandong and Fujian provinces and Taiwan. The distance between the countries is only 80 miles at its narrowest point.

Since Taiwan began allowing direct trade and investment with mainland China, Taiwan's CPC signed a letter of intent in late 2001 with China National Offshore Oil Corp. (CNOOC) to jointly explore for oil and gas in the Taiwan Strait. The historic project would involve seismic testing and drilling in the Tainan basin, a 5,946 sq mi (15,400 sq km) area between southern Taiwan and Guangdong and Fujian provinces.

Previously, despite the lack of formal ties between Taipei and Beijing, the two companies cooperated on geophysical surveys and reported structures worthy of evaluation. CPC was to process data acquired jointly with CNOOC since 1996 on the Tainan-Chaoshan block. CPC noted that both sides feel the need for exploration considering that initial study showed possible substantial oil and gas reserves.

THAILAND

CAPITAL: BANGKOK
MONETARY UNIT: BAHT
REFINING CAPACITY: 681,750 B/CD
OIL PRODUCTION: 114,000 B/D
OIL RESERVES: 515.7 MILLION BBL
GAS RESERVES: 12.7 TCF

Toward the goal of privatization, the Thai state-owned oil and gas firm PTT PLC (formerly Petroleum Authority of Thailand) sold 920 million shares. The Thai Finance Ministry would continue to own 75% of the company, which has a monopoly on gas distribution.

PTT Exploration and Production PLC (PTTEP) of Thailand drilled 10 wells as part of its second exploration campaign at Arthit field in the Gulf of Thailand. Four of the wells were successful. Separately, PTTEP was planning to build a supply base in Songkhla to support natural gas development in the Gulf of Thailand.

PTT approved a $1.55 billion offshore and onshore pipeline to move some 1 bfcd over 544 miles (875 km). The pipeline, to be completed by 2006, would be 36-inch diameter except for one 42-inch section for 275 miles (475 km).

ChevronTexaco and separately Unocal Corp. planned to shift emphasis from gas to oil production. ChevronTexaco would double oil production to 70,000 b/d by 2004. The company has spent more than $500 million for oil and gas development on B8/32 and was investing some $180 million in six minimum facility platforms that can each handle 50 MMcfd of gas, 8,000 b/d of oil, and 5,000 b/d of water.

Unocal began its first oil production in the Gulf of Thailand. It installed a new oil processing platform next to existing gas processing facilities in the Platong field and a new storage and offloading vessel, the SIBEA.

Other oil and natural gas liquids activities included PTTEP developing wells in the Bongkot field that yield more condensate than gas. While the fields' gas production potential was falling, with more than 30% of the estimated 3.7 tcf produced, phase 3B development would involve 47 wells and 2 additional platforms. In related activity, PTTEP would use a tanker to store condensate as an alternative to a more costly subsea concrete unit.

Natural gas demand has eased in Thailand because of a slowdown in the economy. Natural gas fuels 60% of Thailand's power generation. PTT has been pressuring gas companies to ease prices and reconsider take-or-pay contract obligations. In October, Unocal Thailand Ltd. and partners gave PTT a $15 million incentive over a 15 month period to buy 18.3 bcf above the contract minimum. The incentive would effectively lower gas prices one-third to $1.66/Mcf for the particular Thailand fields effected. Other contract pricing would continue unchanged.

Similarly, Shell Exploration & Production Co. reached a new agreement with Electricity Generating Authority of Thailand to cut prices for natural gas by about a third to about $1.30-$1.45/MMbtu with fuel oil as a reference. Shell spent about $73 million toward developing oil reserves.

While natural gas is relatively low cost, to diversify energy sources for power generation there was interest in coal fired power plants using imported coal. One such project was the $700 million Prachuab Khiri Khan power plant.

Separately, BP PLC was planning to sell its major share of a gas producing block off of Malaysia, Block A-18 of the Malaysia-Thailand joint development area (JDA).

Thai Olefins Co. was planning a 300,000-tpy ethylene glycol plant at Rayong using ethane and LPG feedstock.

VIET NAM

CAPITAL: HANOI
MONETARY UNIT: DONG
REFINING CAPACITY: NONE
OIL PRODUCTION: 305,000 B/D
OIL RESERVES: 600 MILLION BBL
GAS RESERVES: 6.8 TCF

In 2001, Conoco Inc. and its partners declared commercial the Sutu Den (Black Lion) oil discovery in the Cuu Long basin on the 1.15 million acre Block 15-1, 120 miles south of Ho Chi Minh City. Discovery well 15-1-SD-1X on the Sutu Den prospect flowed 12,600 bb/d, appraisal well 15-1-SD-2X flowed 11,032 b/d, and a sidetrack well flowed 13,223 b/d of 36º gravity oil. First production was expected in less than three years. Conoco estimated Sutu Den has future gross production of at least 200 million bbl and potential to exceed 400 million bbl.

Four miles southwest of Sutu Den, Conoco drilled the Sutu Vang (Golden Lion) 1X discovery well on Block 15-1. The well tested a stabilized rate of 11,388 b/d of 37º gravity oil. A third prospect in the block, Sutu Trang (White Lion), would be drilled in 2002. The block was being operated by the Cuu Long Joint Operating Co. Conoco owned 23.5%, PetroVietnam 50%, Korean National Oil 14.25%, SK 9%, and Geopetrol 3.5%.

Nippon Mitsubishi Oil Corp.'s subsidiary Nippon Oil Exploration Ltd. found oil at the RD-10X stepout on Block 15-2 in southern Rang Dong field. The well flowed 8,000 b/d and was thought to have a potential of more than 10,000 b/d.

Natural gas production from the Cuu Long basin is largely associated with oil production. An existing 62-mile (100-km) pipeline from the Bach Ho field was operating at capacity; thus, associated gas from the nearby Ruby and Rang Dong fields was being flared.

PetroVietnam Trading Co. awarded Keppel Fels Ltd. and Technics Offshore Engineering Pte. Ltd., both of Singapore, a $46 million contract for a central processing platform for the Bach Ho field. The platform, dubbed CPP-3, would receive well fluids from nearby blocks, process oil and transfer it to floating storage and offloading vessels, treat produced water, and gather and separate associated gas. Its capacity would be 15,000 tons/day for stock tank oil, 19,000 tons/day for liquids, and 12,000 tons/day for produced water.

The state oil companies of Malaysia, Viet Nam, and Indonesia formed a joint venture to explore Blocks 10 and 11.1 in the Nam Con Son basin. This first alliance within the Tripartite Cooperation Arrangement signed in late 2000 would explore 3,956 sq mi (7,915 sq km) on the southern continental shelf of Viet Nam, 130 miles (210 km) southeast of Vung Tau. Con Son Joint Operating Co. was made up of PetroVietnam Investment & Development Co. with 40%, Petronas Carigali Overseas Sdn. Bhd. 30%, and Pertamina 30%.

Noble Affiliates Inc., through subsidiary Samedan Viet Nam Ltd., tested gas at its Swan prospect on Block 12W. The well, in 260 ft of water 230 miles southeast of Ho Chi Minh City in the Nam Con Son basin, was drilled to 14,626 ft total depth and flowed 20 MMcfd of gas with 150 b/d of condensate.

Operator BP was proceeding with a $1.3 billion Nam Con Son integrated gas-to-power development project off the southern province of Ba Ria-Vung Tau. The Nam Con Son project involves the development of the Lan Tay and Lan Do gas fields in block 6/1, construction of a pipeline to bring gas ashore, and a gas-fired power station. It is the country's largest foreign investment so far and its first dedicated gas-to-power project. BP estimated that Nam Con Son contains 2 tcf of gas reserves.

Hyundai Heavy Industries Co. Ltd. began construction of the project's production facilities under a $76 million contract. BP awarded ABB Singapore, a unit of ABB Group, a $1.7 million contract for the integrated control system of the Lan Tay-Lan Do platform.

The gas would be brought ashore at Vung Tau via a new 230-mile (370-km) pipeline and used by three power plants to generate electricity for the Ho Chi Minh City area. From the Dinh Co Gas Terminal, Vung Tau, an extension would move some of the gas inland about 20 miles to the Phu My power plant. The pipeline would cost $565 million and have an ultimate capacity of 247 bcf/yr.

The electricity to be generated from the gas would represent 40% of the country's current demand. BP owned 26.67% of the project; ONGC Videsh (India) 45%, Conoco 13.33% (purchased from Statoil ASA), and state company PetroVietnam 15%. PetroVietnam would own 51% of the pipeline.

Refining

Phu My Plastics & Chemicals Co. Ltd. – a joint venture between Malaysian national oil company Petronas, PetroVietnam, and Tramatsuco – signed with Maybank Hanoi and Bank for Investment and Development of Vietnam (Hanoi) for a $35 million credit facility. The loan would partly finance a 100,000-tpy polyvinyl chloride plant in the petrochemical development zone in Phu My Industrial Area in Baria-Vung Tau province. Petronas held 50% of the joint venture while PetroVietnam had 43%. Tramatsuco, the trading arm of the People's Committee of Baria-Vung Tau province, had the remaining 7%.

Initial construction began at the $1.3 billion VietRoss refinery project in Dung Quat, Quang Ngai province in central Viet Nam. With a capacity of about 140,000 b/d, Viet Nam's first refinery would process crude into gasoline, jet kerosene, diesel, fuel oil, propylene, and LPG. The project was a joint venture between PetroVietnam and the Russian firm Zarubezhneft.

AP Petrochemical Pte Ltd. (APP), a subsidiary of AP Oil International Ltd, planned to build a blending plant to manufacture lubricant oil and specialty chemicals for the Asian Pacific market and a bulk storage terminal with an initial total capacity of 12,000 tonnes. APP gained approval from Vietnamese authorities for the joint venture, which would have APP as 80% owner and Tram Anh Industrial Manufacturing and Trading Co. Ltd. 20% owner. The venture is named AP Petrochemical (Vietnam) Co. Ltd.

Key Stats

OPEC AT A GLANCE

The 11-member Organization of Petroleum Exporting Countries (OPEC) contains most of the world's oil, concentrated in the "OPEC 6" – Iran, Iraq, Kuwait, Saudi Arabia, UAE, and Venezuela.

Current reserves-production ratio

More than three-fourths of the world's current oil reserves belong to OPEC countries, and future oil discoveries are also expected to be larger in OPEC than elsewhere.

OPEC RESERVES-PRODUCTION RATIO

	Reserves, 1/1/01, billion bbl	Production, 2000, 1,000 b/d	Reserves-production ratio, years
Algeria	9.20	809	31
Indonesia	4.98	1,267	11
Iran	89.70	3,682	67
Iraq	112.50	2,567	120
Libya	29.50	1,414	57
Kuwait	96.50	2,099	126
Nigeria	22.50	2,034	30
Qatar	13.16	688	52
Saudi Arabia	261.70	8,264	87
UAE	97.80	2,233	120
Venezuela	76.86	3,028	70
OPEC 11	**814.40**	**28,085**	**79**
Rest of the world	**214.06**	**38,837**	**15**

Sources: Oil & Gas Journal, Centre for Global Energy Studies

Current ratio of cumulative production to ultimate reserves

The five smaller producers – Algeria, Indonesia, Nigeria, Libya, and Qatar – have already produced at least half of their ultimate reserves.

OPEC RATIO OF CUMULATIVE PRODUCTION[1] TO ULTIMATE RESERVES[2]

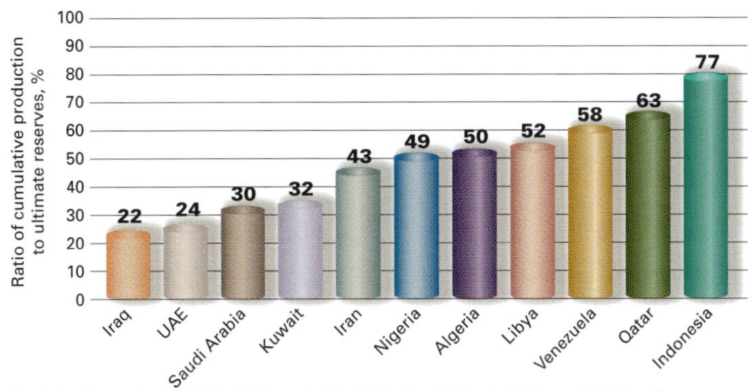

[1]Production data as of yearend 2000. [2]Medium probability (P50), after C.J. Campbell, OGJ, Apr. 7, 1997, p. 52; also, Petroleum Economist, October 1995, p. 22.

Historical and current crude oil production capacity

Saudi Arabia now produces over 30% of OPEC capacity, and the OPEC 6 control about four-fifths.

OPEC'S CRUDE OIL PRODUCTION CAPACITY

	1974[1]	1990[2]	1994[3]	2001[4] (1Q)
	Million b/d, 1Q			
Algeria	1.1	0.85	0.78	0.90
Indonesia	1.5	1.41	1.40	1.35
Iran	6.5	3.30	3.80	3.80
Iraq	2.5	3.66	2.50	3.10
Kuwait[5]	3.8	2.55	2.00	2.20
Libya	3.0	1.50	1.50	1.45
Nigeria	2.4	1.90	1.95	2.30
Qatar	0.7	0.46	0.44	0.75
Saudi Arabia[5]	9.7	9.08	8.50	10.0-10.5
UAE	2.3	2.50	2.30	2.60
Venezuela	3.5	1.54	2.38	3.05
Total OPEC	**37.0**	**28.75**	**27.55**	**31.50-32.00**
Total OPEC 6[6] (**)	28.3	21.23	21.48	24.75-25.25
Share OPEC 6[6], %	76.5	73.8	78.0	78.6-78.9%

[1]M.A. Adelman, The Genie Out of the Bottle, MIT Press (1995), p. 245. [2]M.A. al-Sahlawi, "Expanding OPEC production capacity, OPEC Review, Autumn 1992, p. 242. [3]I.A.H. Ismail, "OPEC Production—1, OGJ, May 2, 1994, p. 98). [4]US Energy Information Administration, www.eia.doe.gov/cabs/opec.html#CAPACITY. [5]Figures for Kuwait and Saudi Arabia include their respective half-shares of the Neutral Zone. [6]OPEC 6 consists of the organization's five founding members (Iran, Iraq, Kuwait, Saudi Arabia and Venezuela) in addition to the UAE.

Conventional crude oil endowment, cumulative production

Future production capacity increases will come mainly from the OPEC 6, which hold close to 90% of the cartel's undiscovered reserves.

OPEC'S CONVENTIONAL CRUDE OIL ENDOWMENT, CUMULATIVE PRODUCTION

	Reserves reported (yearend 2000)[1]	Campbell's median prob. (P50)[2]	—— Ultimate recovery —— Disc.[3]	Undisc.[3]	Total[3]	Cumul. prod. (yearend 2000)[4]
Algeria	9.2	9.4	17.1	5.9	23.0	11.6
Indonesia	5.0	6.5	23.8	1.2	25.0	19.3
Iran	89.7	58.6	92.4	27.6	120.0	51.8
Iraq	112.5	70.9	86.1	28.9	115.0	25.7
Kuwait[5]	96.5	62.8	80.6	20.4	101.0	32.7
Libya	29.5	22.8	40.1	4.9	45.0	21.9
Nigeria	22.5	18.6	32.9	7.1	40.0	20.9
Qatar	13.2	4.2	9.0	1.0	10.0	6.3
Saudi Arabia[5]	261.7	204.5	261.4	44.6	306.0	92.0
UAE	97.7	64.3	73.7	12.3	86.0	20.7
Venezuela	76.9	35.0	84.0	6.0	90.0	52.3
Total OPEC	**814.4**	**557.6**	**801.1**	**159.9**	**961.0**	**355.2**
Total OPEC 6[6]	735.0	496.1	678.2	139.8	818.0	275.2
Share OPEC 6[6], %	90.3	89.0	84.7	87.4	85.1	77.5

[1]OGJ, Dec. 18, 2000, pp. 122-23. [2]OGJ, Apr. 7, 1997, p. 52. [3]Ibid.; also, Petroleum Economist, October 1995, p. 22. [4]Ibid.: for cumulative production (close of 1996); production for 1997-2000 was retrieved from OGJ yearend statistics and added on.
[5]Figures for Kuwait and Saudi Arabia include their respective half shares of the Neutral Zone. [6]The OPEC 6 consists of Iran, Iraq, Kuwait, Saudi Arabia, UAE, and Venezuela.

OPEC AT A GLANCE

As global oil demand slumped during 2001, OPEC wrangled over production cuts with non-OPEC oil exporters.

Compliance with OPEC quotas

The 10 countries subject to OPEC quotas (all but Iraq) averaged well over their quotas in late 2001, and history suggests that OPEC generally realizes about 60-70% of its cuts.

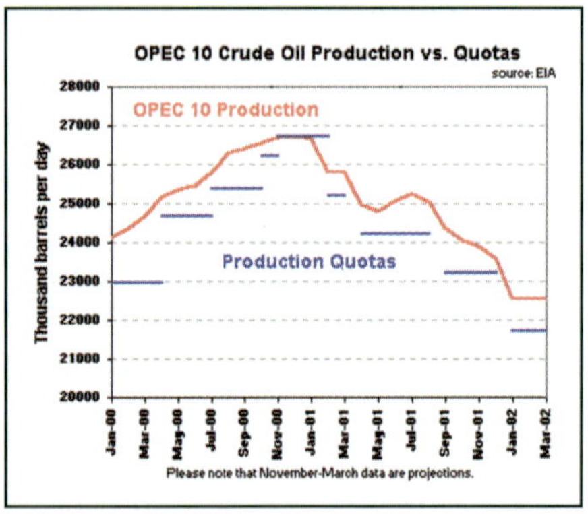

Russia's production and exports

Non-OPEC Russia, the world's second-largest oil exporter, has committed to reduce shipments rather than production.

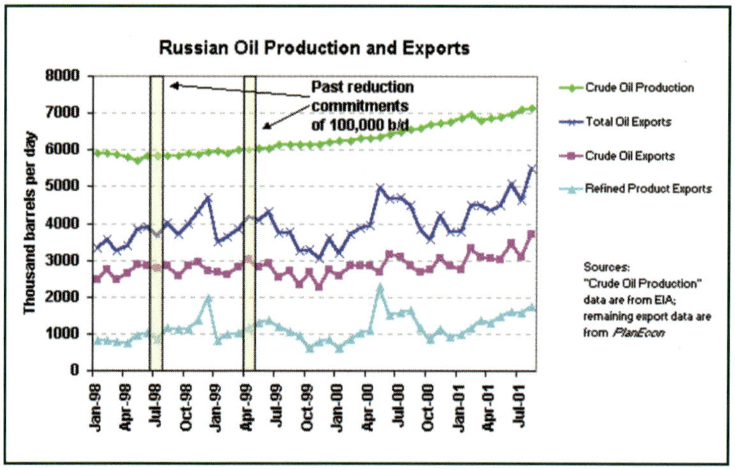

Norway's production

Non-OPEC Norway, the world's third-largest oil exporter, agreed in late 2001 to reduce production by 150,000 b/d for six months.

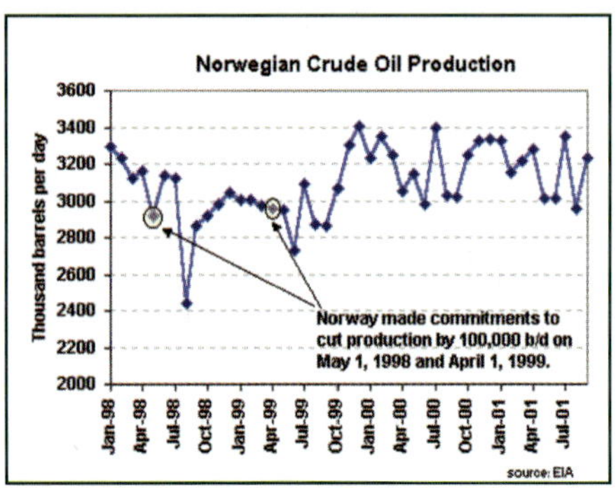

Mexico's production and exports

In late 2001, non-OPEC Mexico promised to restrain exports to 1.66 million b/d, or 100,000 b/d below what had been planned.

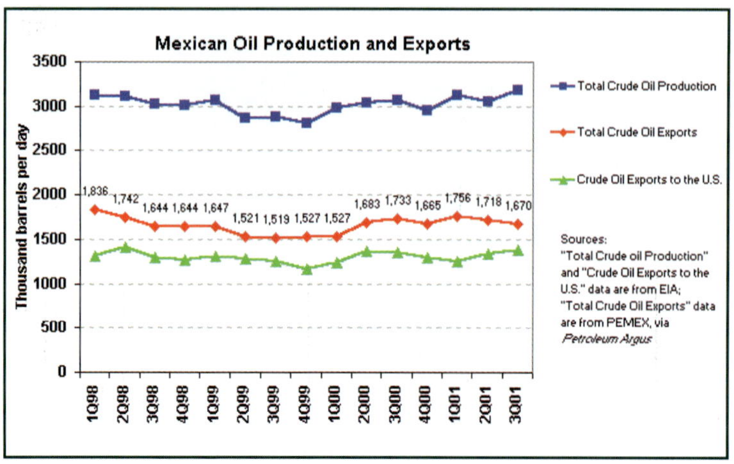

LNG DEVELOPMENTS

The liquefied natural gas (LNG) industry has evolved from a regional, niche player to a practical, global energy business.

Plant evolution over the decades

Four distinct periods mark the progress of LNG technological changes, which are the foundation for next-generation developments.

LNG PLANT EVOLUTION: TURBINE, COMPRESSOR CONFIGURATIONS*

	Decade	Precool	Liquefaction	Per train, million tpy
Brunei	1970s	1 × steam turbine	2 × steam turbine	1.1
Australia	1980s	2 × GE Frame 5	2 × GE Frame 5	2.2
MLNG Dua, NLNG, OLNG	1990s	GT C M 1 × GE Frame 6	1 × GE Frame 7	2.6–3.4
MLNG Tiga, NWS4	Under construction	1 × GE Frame 7	1 × GE Frame 7	3.9–4.2
Shell GameChanger plant		Aero derivative Gas turbines or electric motors	Aero derivative Gas turbines or electric motors	4.5–5.5

*GT = gas turbine; C = compressor; M = electric starter–helper motor

Source: Shell

Cost reduction driven by technology

These decades of technical innovation and evolution have reduced costs by 50% in real terms.

LNG TECHNOLOGY, COST REDUCTION

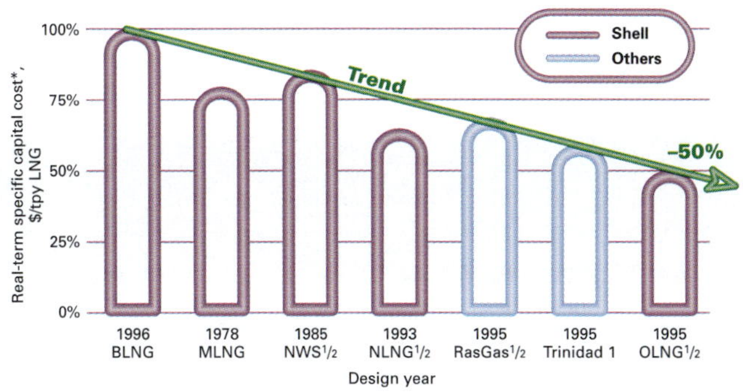

*Normalized to common location.

Source: Shell

Shell's recent liquefaction technology developments

Unit costs are far lower than the best single-train plant and are being driven below the existing lowest-cost two-train benchmark of $200/tonne.

RECENT SHELL LIQUEFACTION TECHNOLOGY DEVELOPMENTS

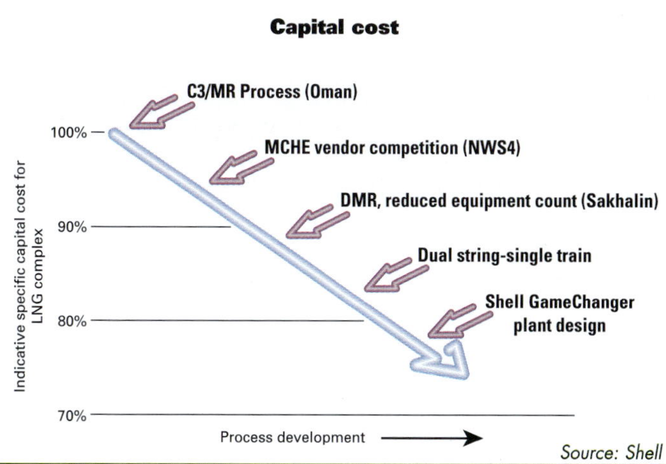

Source: Shell

LNG DEVELOPMENTS

High natural gas prices recently sparked interest in US LNG projects, but the increasing global demand for LNG, especially in Asia, is much less dependent on gas prices.

World demand for LNG

Steady growth has resulted from evolution of the LNG business model from long-term, fixed-price contracts to deals based on oil-indexed prices that limit volatility.

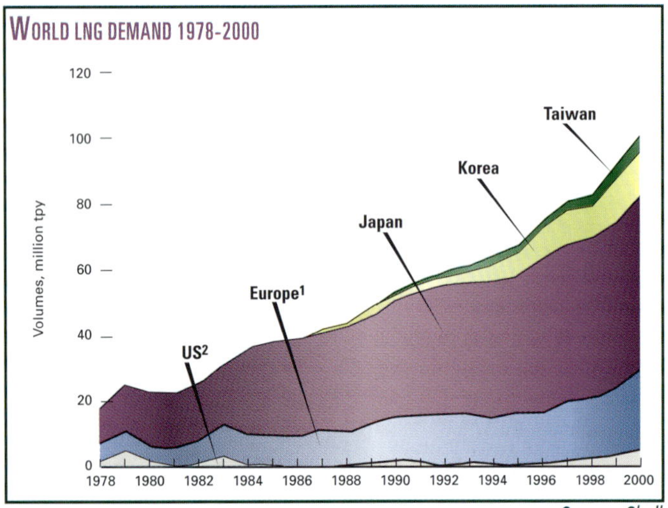

Source: Shell

Growth of short-term LNG trade

Trades of two years' duration or less represent a small portion of globally traded LNG, but this short-term trade has been growing gradually.

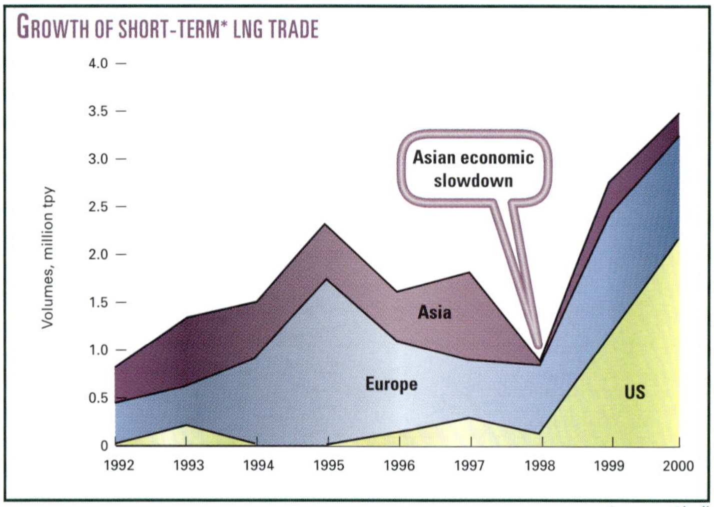

Source: Shell

Asia-Pacific LNG demand growth

The need for flexible contracts is becoming more widely recognized in this region, where the greatest growth is expected in India and South Korea.

ASIA-PACIFIC LNG DEMAND

	1996	1997	1998	1999	2000	2001	2005			2010		
	Million tpy						Low	Base	High	Low	Base	High
Japan	45.9	47.2	49.1	51.3	53.3	54.1	55.0	55.9	60.2	56.5	60.9	66.7
South Korea	9.6	11.6	10.6	13.0	14.6	16.2	16.9	18.2	19.9	18.2	25.7	27.7
Taiwan	2.6	3.1	3.8	4.2	4.3	4.8	6.0	7.0	8.0	9.5	10.8	13.5
India						—	5.0	7.5	10.5	13.0	18.0	25.6
China						—	—	—	3.0	3.0	5.0	8.0
Total	**58.1**	**61.9**	**63.5**	**68.5**	**72.2**	**75.9**	**82.9**	**88.6**	**101.6**	**100.2**	**120.4**	**141.5**

Source: Facts Inc.

GTL PROJECTS

During 2001 several countries with surplus natural gas were planning to build facilities for converting it to synthetic petroleum liquids by developing gas-to-liquids (GTL) projects.

WORLDWIDE GTL CAPACITY

Location; company	Capacity, b/d	Status	Expected Completion	Contractor; contract type	Notes
Argentina-Tierra del Fuego; Shell Gas & Power	–	Planning	–	–	–
Australia-Northern Territory; Shell Gas & Power	–	Planning	–	–	–
Australia-Burrup peninsula; Shell Gas & Power	11,500	Engineering	2004	Tessag Industrie-Anlagen GMBH; eng.	New
Chile-Punta Arenas; ENAP	10,000	Planning	2005	–	New
Egypt-West Damiatta; Shell Int'l Gas/EGPC	75,000	Planning	Dec. 2005	–	New; $1.7 billion
Nigeria-Escravos; Sasol Chevron Ltd.	33,000	Planning	2005	–	New
Qatar-Ras Laffan; Sasol Chevron Ltd.	34,000	Planning	2005	Foster Wheeler Corp.; eng./design	New; $30 million FEED, $800 million total
South Africa-Mossel Bay; Mossgas (Pty.) Ltd.	70,000 tpy	Planning	Oct. 2002	–	–
Trinidad & Tobago-Point Lisas; Reema International Corp.	10,000	Engineering	2004	Parsons Energy & Chemicals Group; eng./constr.	New; $300 million
US-Nikiski, AK; BP	300	Planning	June 2002	–	Test facility
US-Ponca City, OK; Conoco Inc.	400	Planning	June 2002	–	Pilot plant; $75 million
US-Commerce City, CO; Rentech Inc.	800	Planning	2002	–	Methanol plant conversion

GTL PROJECTS

GTL technology has held promise for a long time but has recently become more practical.

GTL COST ANALYSIS

	$/bbl
Feedstock cost	4.50
Operating costs	4.40
Capital repayment	9
Total production cost	18

Source: Foster Wheeler Energy Ltd.

Cost of generic plant

A 30,000 b/sd GTL plant based on Fischer-Tropsch technology could be built at a coastal site for less than $25,000/daily bbl in about 30-33 months.

GTL CAPITAL EXPENDITURES

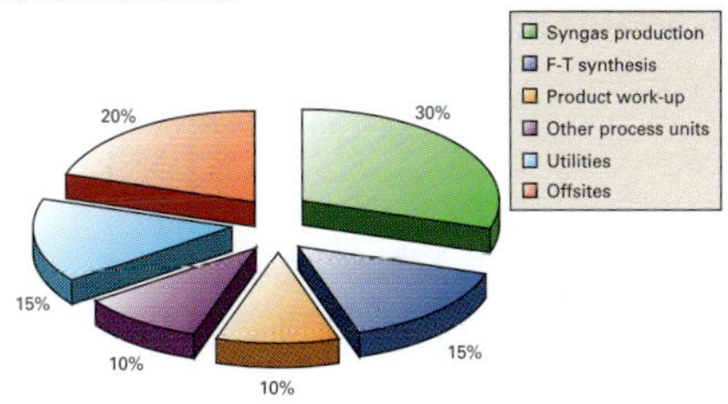

Source: Foster Wheeler Energy Ltd.

Economics of small and large plants

Raytheon Engineers & Constructors Inc. analyzed 15,000 b/d (short-term) and 50,000 b/d (long-term) GTL facilities to use Venezuelan gas, in a study funded by TDA Research Inc.

ECONOMIC ANALYSIS—SHORT-TERM CASE

Description	ExxonMobil	Shell	Sasol[3]	Syntroleum[3,4]	Rentech[5]	Intevep[5]
Total installed cost (TIC)[1], $ million	—	—	395	455	468	373
Gross revenue, $ million	—	—	86.2	77.5	97.8	94.1
Projected internal rate of return (IRR), %	12.9	12.5	14.5	11.2	13.9	16.6
Net present value[6], $ million	85	75	108	33	110	156
Total liquid yield, b/d	—	—	15,300	12,000	16,450	15,300
TIC in $/bbl	29,000[2]	30,000[2]	25,800	37,920	28,450	24,380

[1]Total installed cost includes initial charge of catalyst and license fee. [2]Exxon and Shell declined to participate in the study. The figures provided here were estimated based on an article by Morgan Stanley Dean Witter, Sept. 9, 1997. This article provided the cost for larger-size GTL plants. Based on the $/bbl difference between the long-term and short-term cases as estimated by Raytheon, ExxonMobil's and Shell's $/bbl for the short-term cases were increased to $29,000/bbl and $30,000/bbl, respectively. [3]TIC provided by Sasol is $395 million and that provided by Syntroleum is $351 million. The estimated TIC for the Sasol case was $389 million. However for the calculation of IRR, TIC provided for Sasol has been used. [4]Revenue for Syntroleum includes sale of 150 mw export power. [5]Rentech and Intevep only provided Fischer-Tropsch unit TIC. All other unit investment costs were estimated by Raysponse. [6]NPV was calculated using a discount factor of 10%.

ECONOMIC ANALYSIS—LONG-TERM CASE

Description	ExxonMobil		Shell		Sasol[3]		Syntroleum[3,4]		Rentech[5]		Intevep[5]	
	No lube	Lube	No lube	Lube	No lube	Lube	No lube	Lube	No lube	Lube	No lube	Lube
Total installed cost[1], $ billion	—	—	—	—	1.039	1.095	1.258	1.302	1.268	1.324	0.997	1.05
Gross revenue, $ million	—	—	—	—	264	367	205	294	295	398	282	385
Projected internal rate of return (IRR), %	14.3	18.2	13.2	16.9	16.7	21.3	10.7	15.0	15.4	19.4	18.4	23.0
Net present value[6], $ million	361	755	293	686	442	839	52	400	426	823	547	946
Total liquid yield, b/d	—	—	—	—	50,900	50,900	40,000	40,000	54,900	54,900	50,900	50,900
TIC in $/bbl	24,000[2]	25,000	26,000[2]	27,000	20,410	21,510	31,450	32,550	23,100	24,120	19,590	20,630

[1]Total installed cost includes initial charge of catalyst and license fee. [2]Exxon and Shell declined to participate in the study. The figures provided here were estimated based on an article by Morgan Stanley Dean Witter, Sept. 9, 1997. [3]TIC provided by Sasol is $972 million for 465 MMcfd natural gas feed. The TIC provided by Syntroleum is $1.098 billion. [4]Revenue for Syntroleum includes sale of 150 mw export power. [5]Rentech and Intevep only provided Fischer-Tropsch unit TIC. All other unit investment costs were estimated by Raysponse. [6]NPV was calculated using a discount factor of 10%.

Alaska pipeline extension

The US Department of Energy reported that GTL production could extend the life of the Trans-Alaska Pipeline System by nine years.

NORTH SLOPE OIL PRODUCTION

Fig. 1

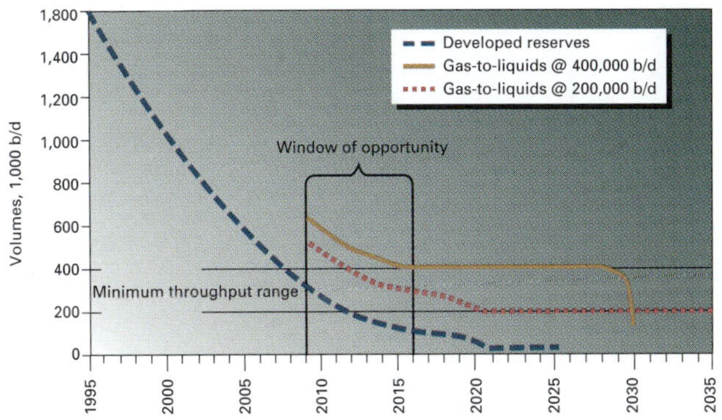

GTL PROJECTS

Syntroleum Sweetwater Operations Ltd. was under contract to build an 11,500 b/d GTL plant on western Australia's Burrup peninsula for Tessag Industrie-Anlagen GMBH, based on this design.

Sweetwater GTL Plant, Western Australia

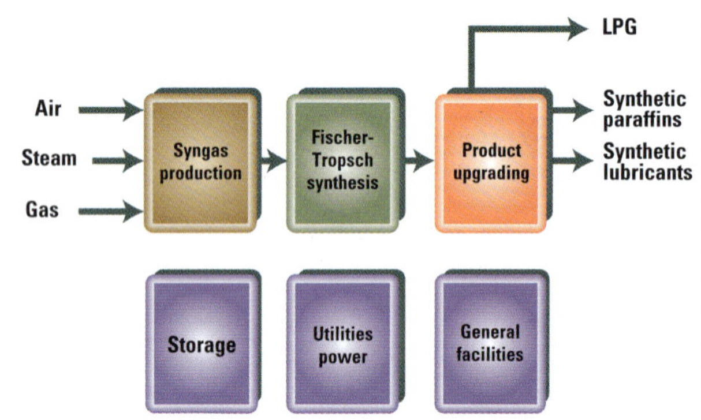

Syngas Production Area

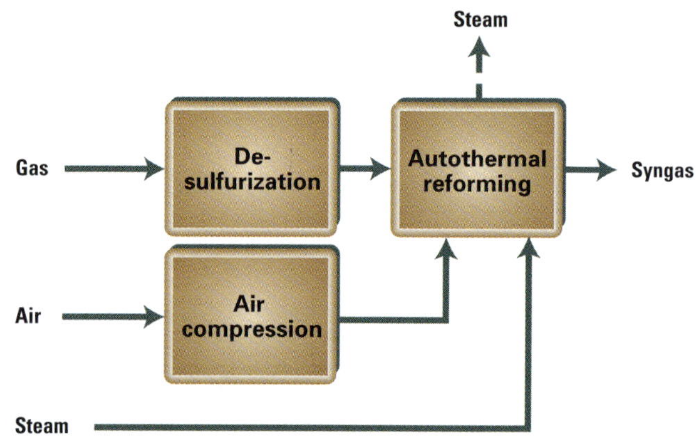

Fischer-Tropsch Synthesis Area

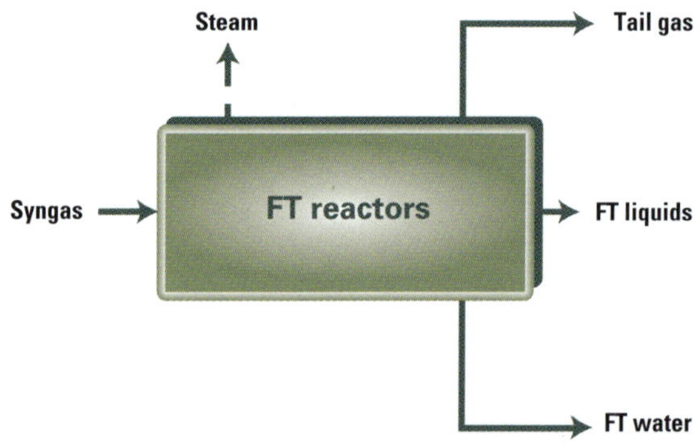

Product Upgrading Area

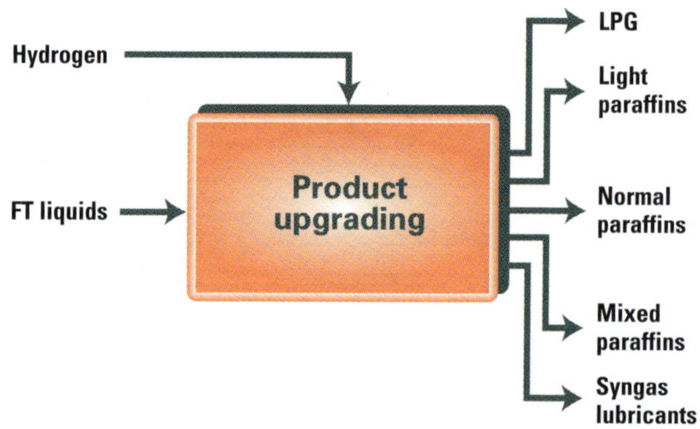

LPG MARKETS

Worldwide demand for liquefied petroleum gas (LPG) has long outpaced total petroleum demand and could continue to grow by slightly more than 3.5%/yr through 2010, Purvin & Gertz Inc. estimated.

World demand shifted east

The East of Suez region in 2001 became a net LPG importer for the first time, due to growing LPG demand for in the Far East and petrochemical demand in the Middle East.

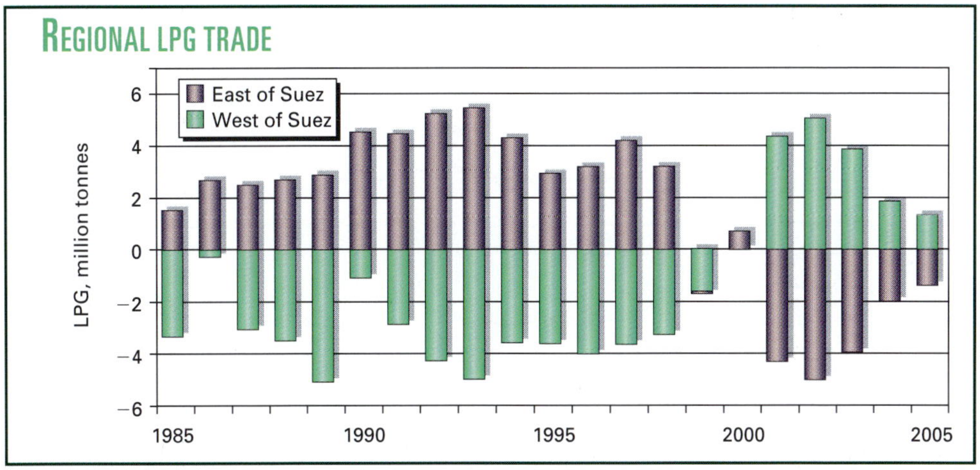

Demand increasing worldwide

Since 1985, global demand for LPG has grown from 116 million tonnes/yr to almost 200 million tonnes in 2000.

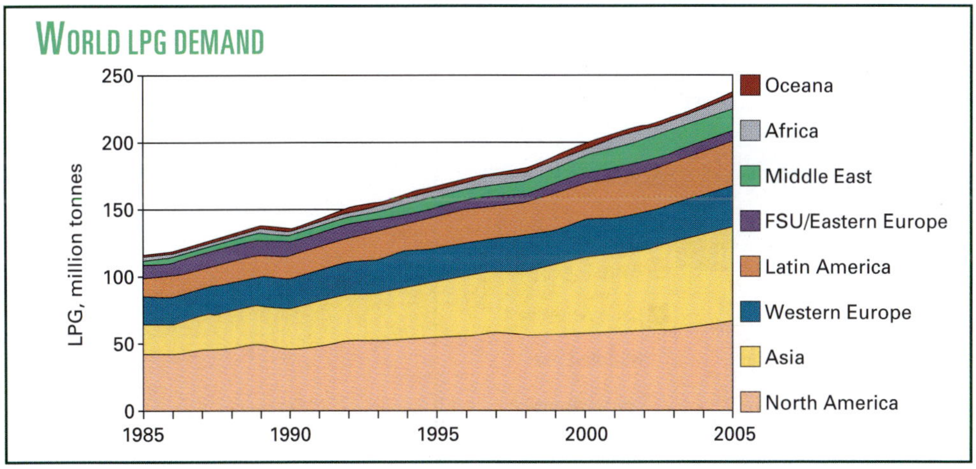

Annual growth rates

During the 1990s, world LPG demand increased by nearly 3.7%/yr, versus about 1.5% for total petroleum demand.

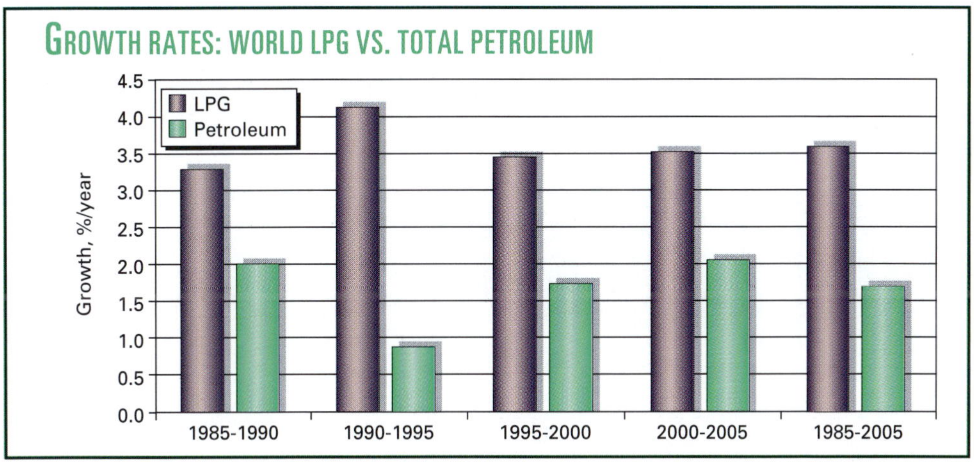

LPG MARKETS

Imports and exports will play an increasingly important role in US LPG supply-demand balances, said Purvin & Gertz.

End-use trends

Residential-commercial and chemical consumption of LPG will increase as auto fuel and industrial applications for LPG shrink.

LPG DEMAND BY END-USE

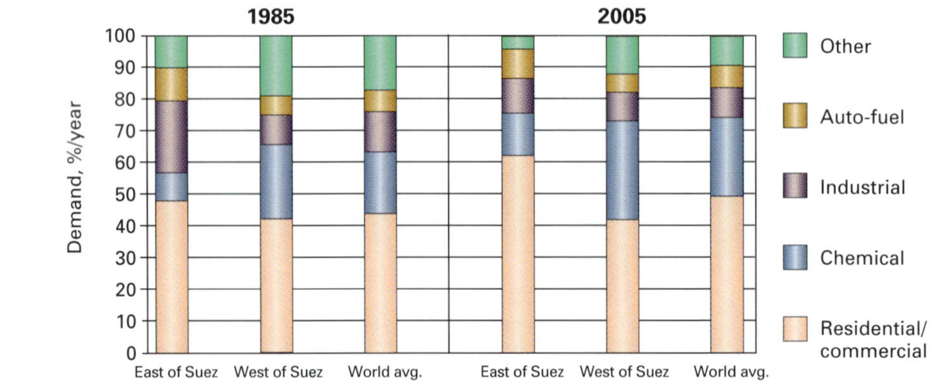

Origins of supply

While global LPG supplies have grown dramatically, an increasing portion – about 37% of production in 2005 – will come from sources East of Suez, compared with 25% in 1985.

WORLD LPG SUPPLY

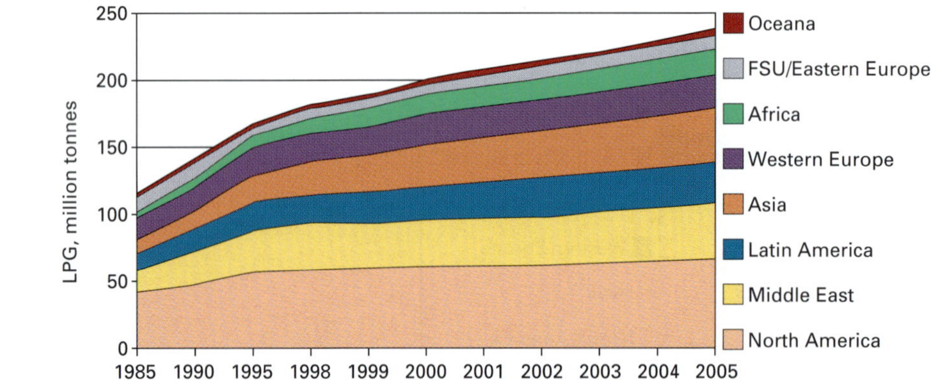

Export trends

Even as the Atlantic Basin's LPG surplus grows through 2005, much of this product will be used to meet the growing deficit East of Suez and will be unavailable for import into the US.

REGIONAL LPG TRADE

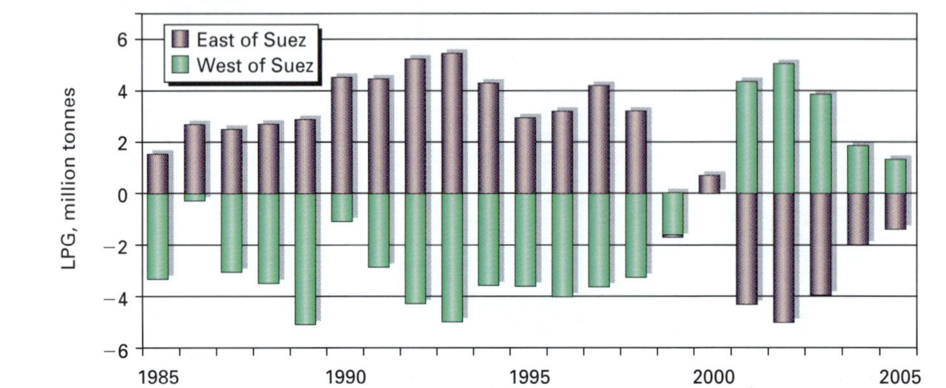

TANKER TRENDS

The profitability outlook for very large crude carriers (VLCCs) and other oil tankers dimmed during 2001, yet prospects for gas carriers, especially those transporting LNG, remained bright, said Clarkson Research Studies.

Oil tanker classification

Tanker deliveries during 2001 reached 49 million deadweight tons (dwt).

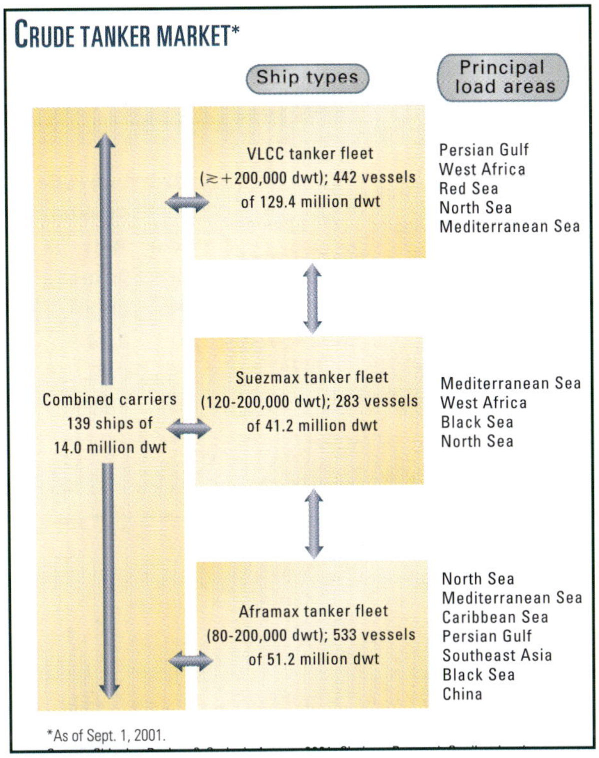

Historical construction and demolition

Tanker demolition activity languished during 2001 as owners sought to extend their ships' useful life.

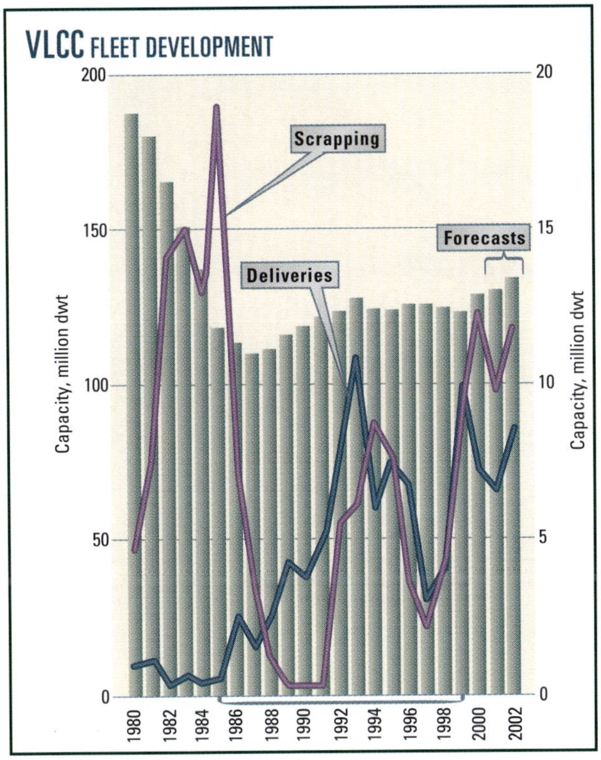

Day rates

VLCC rates dipped to $36,000/day for a modern vessel, while the Aframax sector managed to average almost $30,000/day.

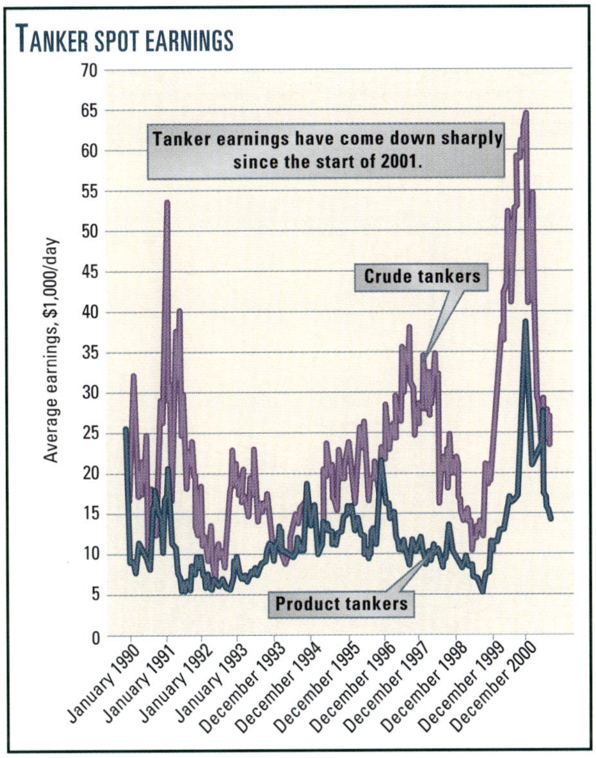

Source (All figures this page): Shipping Review & Outlook Autumn 2001; Clarkson Research Studies, London

TANKER TRENDS

Demand for very large gas carriers (VLGCs) was strengthened by growth in the global market for LNG, said Clarkson Research Studies.

Growing old

Tougher regulations were in the works to phase out tankers at the age of 26, versus the existing maximum of 30 years, and to make other rules more stringent.

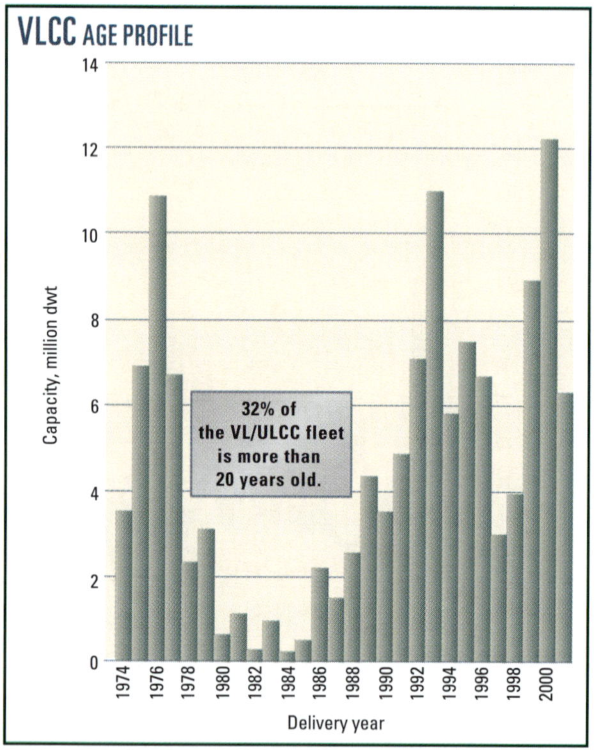

History of LPG carrier capacity

The LPG fleet grew by an estimated 8% during 2001.

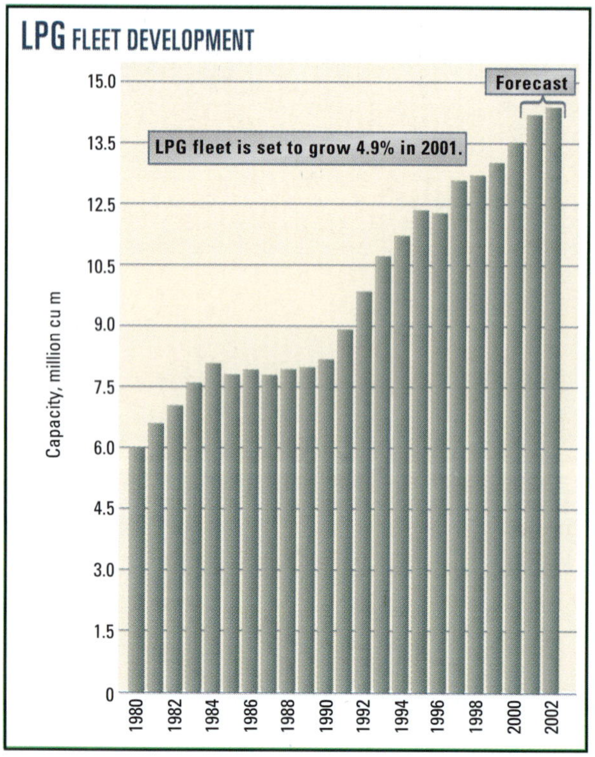

Global LNG trade

Japan remained the world's biggest LNG importer, but the US was the fastest growing market.

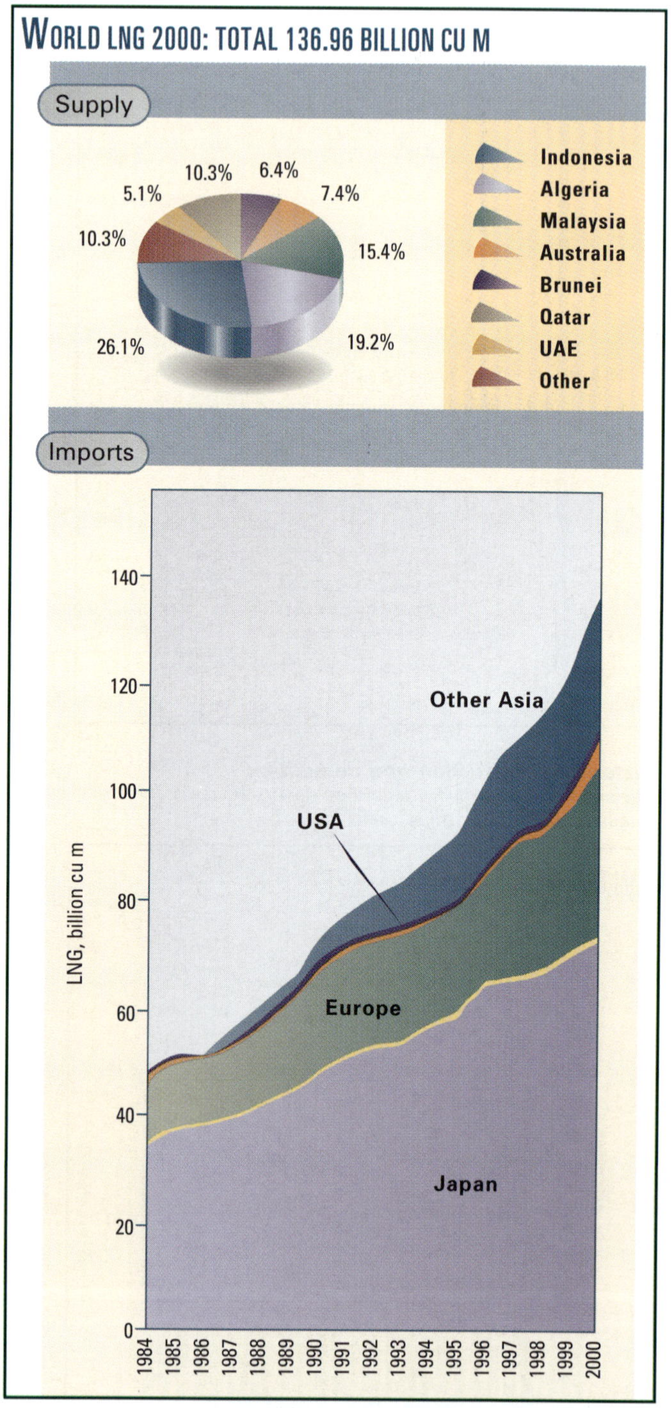

Source (All figures this page): Shipping Review & Outlook Autumn 2001; Clarkson Research Studies, London

GASOLINE ADDITIVES

Regulatory changes affecting oxygenate additives could increase the risk of gasoline supply shortfalls and price spikes, reported Petroleum Industry Research Inc. (PIRC)..

Price trends

The difference between retail gasoline prices and global crude oil prices (WTI) has been growing.

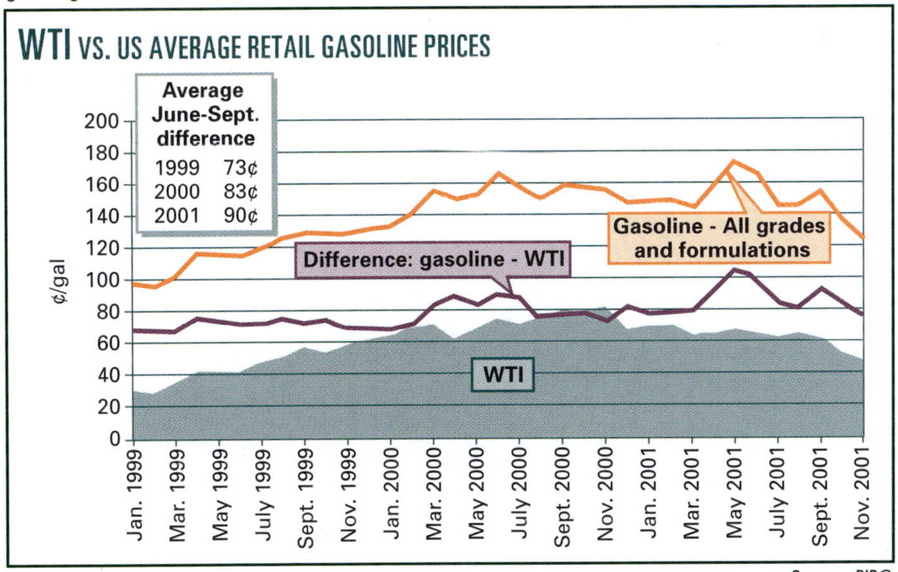

Source: PIRC

Gasoline market shares

US refiners were selling 10 times as much RFG as oxygenated gasoline in 2000.

OXYGENATE, RFG SHARES OF US REFINER GASOLINE SALES

	1994 (%)	2000 (%)
Oxygenated gasoline	9	3
Reformulated gasoline	2	32

Population in nonattainment areas Aug. 2001, million people

CO 30.5
Of which 17.6 in "serious" (13 in Los Angeles Basin)

Ozone 104.9
Of which 29.0 in "serious
42.1 in "severe"
13.0 in "extreme"(Los Angeles Basin)

Source: PIRC

Three RFG markets

The price of reformulated gasoline (RFG) was volatile in California but even more so in Chicago, where RFG is subject to seasonal-use regulations.

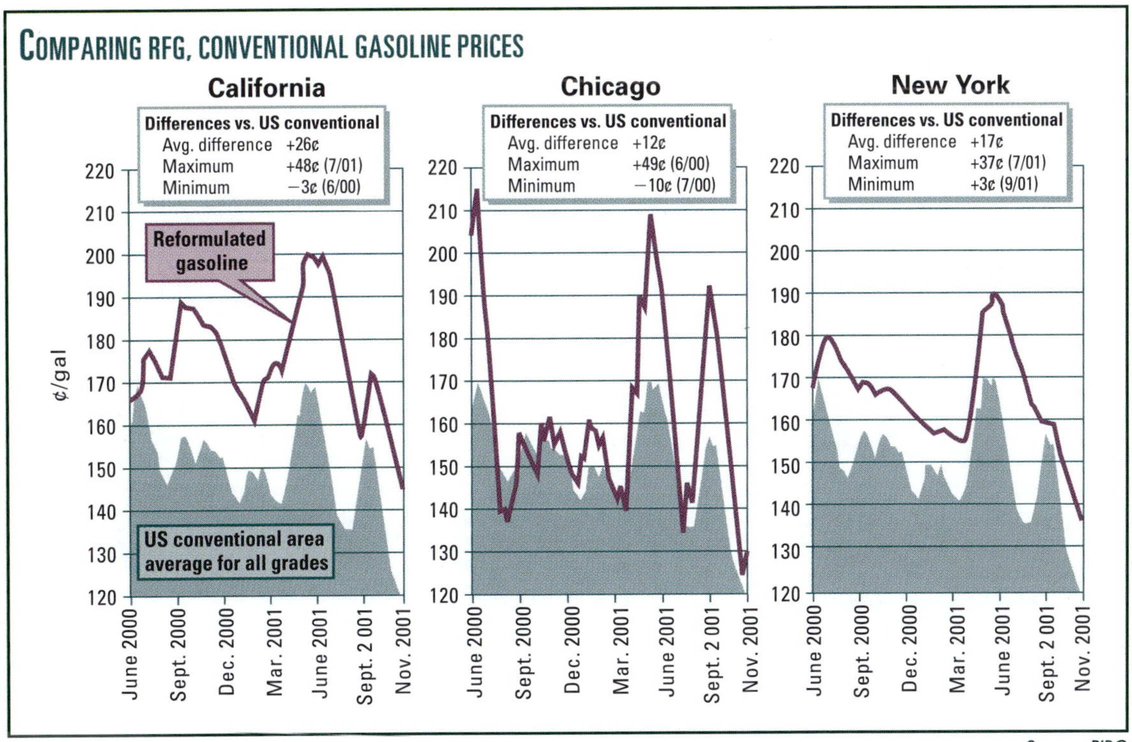

Source: PIRC

GASOLINE ADDITIVES

MTBE and ethanol were by far the most widely used oxygenates during 2001, making up almost 5% of the gasoline barrel, with MTBE volumes roughly triple those of ethanol.

History of additive supply

The favorable regulatory environment – plus, in the case of ethanol, tax incentives – has led to a surge in volume for both oxygenates over the years.

TRENDS IN US ETHANOL AND MTBE SUPPLY

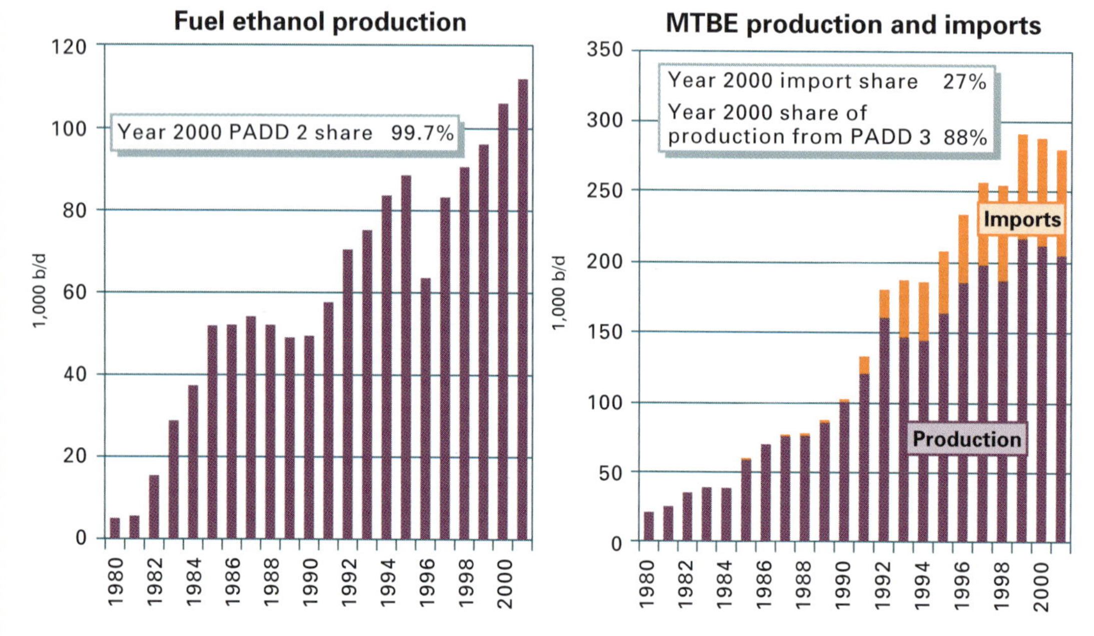

Source: PIRC

Where ethanol is used

Nearly 75% of the ethanol used in gasohol was in PADD 2, where 30% of all gasoline was classified as gasohol. Such geographic differences could complicate attempts to replace MTBE with ethanol in a relatively short time.

DISTRIBUTION OF US ETHANOL USE IN GASOHOL*

	2000 use in gasohol	Control area (1997) oxygenate	Gasohol as % of total 2000 gasoline
	1,000 b/d		
US total	96	45	12
PADD 1	7	2	2
PADD 2	70	32	30
PADD 3	5	2	4
PADD 4	4	2.5	17
Colorado	3.5	2	31
PADD 5	9	7	9
California	4	—	7

Percent change in ethanol use in gasohol from 1999 +12,000 b/d or +14% US 2000 fuel ethanol production 106,000 b/d. PADD 2 share 99.7%. US mid-2001 ethanol nameplate capacity 144,000 b/d. PADD 2 share 97%.

*Prior to the Energy Policy Act of 1992 (EPACT) gasohol was defined as a blend of gasoline and at least 10 vol % fuel alcohol. EPACT created two additional categories, 7.7% gasohol (7.7-10 vol %) and 5.7% gasohol (5.7-7.7 vol %). The 7.7% gasohol is generally used to meet requirements for winter oxygenated gasoline. The 5.7% category meets the federal 2 wt % minimum oxygenate requirement for reformulated gasoline.

Source: PIRC

Oxygenate price trends

In early 2001 when natural gas prices were quite high, MTBE prices approached the pre-tax-credit prices for ethanol.

ETHANOL, MTBE PRICES

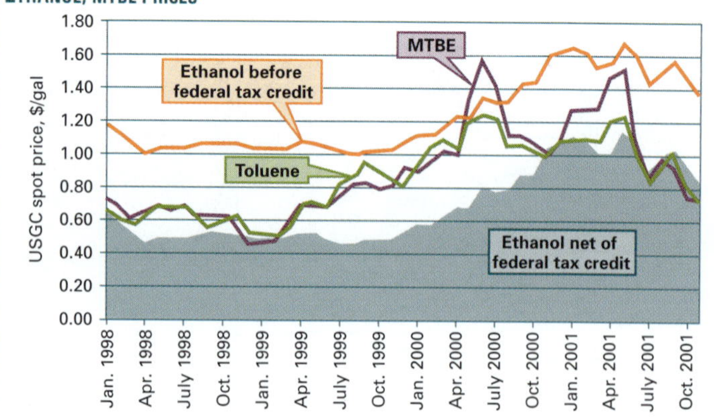

Source: PIRC

WORLDWIDE REFINING

After remaining flat for the past three years, global refining capacity in 2001 declined slightly, by 85,000 b/cd, to 81.2 million b/cd.

Top refining companies

The merger between Chevron and Texaco during 2001 vaulted the combined company to fourth on the worldwide list, primarily because Caltex refineries were included.

HOW THE WORLD'S LARGEST REFINERS RANK

Rank (1-1-01)	Rank (1-1-00)	Company	Crude capacity, b/cd[1]
1	1	ExxonMobil Corp.	5,308,000
2	2	Royal Dutch/Shell	3,889,000
3	3	BP PLC	3,163,000
4	12, 13	ChevronTexaco Corp.[2]	2,950,000
5	4	Sinopec	[3]2,665,000
6	5	Petroleos de Venezuela SA	2,656,000
7	6	TotalFinaElf SA	2,498,000
8	7	Saudi Aramco	1,991,000
9	9	Petroleo Brasileiro SA	1,832,000
10	8	China National Petroleum Corp.	[3]1,763,000
11	43, 14	Phillips Petroleum Co.	1,728,000
12	10	Petroleos Mexicanos	1,692,000
13	11	National Iranian oil Co.	1,484,000
14	16	Nippon Mitsubishi Petroleum Refining Co. Ltd.	1,223,000
15	15	Repsol-YPF	1,150,000
16	17	Agip Petroli SpA	1,043,000
17	18	Pertamina	993,000
18	19	Kuwait National Petroleum Co.	966,000
19	20	Marathon Ashland Petroleum LLC	935,000
20	22	Conoco Inc.	852,000
21	24	Sunoco Inc.	823,000
22	23	SK Corp.	817,000
23	25	Idemitsu Kosan Co. Ltd.	784,000
24	26	Chinese Petroleum Co.	770,000
25	27	Ultramar Diamond Shamrock	753,000

[1]Includes partial interests in refineries not wholly owned by the company. [2]Includes holdings in Caltex. [3]Estimated values.

Top plants

The order of the world's largest refineries remained unchanged from 2000 to 2001.

WORLD'S LARGEST REFINERIES*

	Company	Location	Crude capacity, b/cd
1	Paraguana Refining Center	Cardon/Judibana, Falcon, Venezuela	940,000
2	SK Corp.	Ulsan, South Korea	817,000
3	LG-Caltex	Yosu, South Korea	633,600
4	ExxonMobil Refining & Supply Co.	Jurong/Pulau Ayer Chawan, Singapore	568,500
5	Reliance Petroleum Ltd.	Jamnagar, India	540,000
6	Hovensa LLC	St. Croix, Virgin Islands	525,000
7	S-Oil Corp.	Onsan, South Korea	520,000
8	ExxonMobil Refining & Supply Co.	Baytown, Tex.	516,500
9	ExxonMobil Refining & Supply Co.	Baton Rouge, La.	488,500
10	Russian Investment Co.	Angarsk, Russia	440,700
11	BP PLC	Texas City, Tex.	437,000
12	BP PLC	Whiting, Ind.	410,000
13	Shell Eastern Petroleum (Pte.) Ltd.	Pulau Bukom, Singapore	405,000
14	National Iranian Oil Co.	Abadan, Iran	400,000
15	Saudi Arabian Oil Co. (Saudi Aramco)	Rabigh, Saudi Arabia	400,000

*Plants with 400,000 b/cd or more capacity.

Regional crude capacities

The biggest change in crude capacity occurred in the Middle East, due to the startup of a refinery in the United Arab Emirates.

REGIONAL LOOK AT WORLDWIDE REFINING OPERATIONS

Region	No. of refineries	Crude distillation, b/cd	Vacuum distillation, b/cd	Catalytic cracking, b/cd	Catalytic reforming, b/cd	Catalytic hydrocracking, b/cd	Catalytic hydrorefining, b/cd	Coke, tonnes/day
Africa	46	3,202,262	460,544	195,000	387,273	28,708	835,399	781
Asia	203	20,183,860	3,746,166	2,673,165	2,007,986	715,455	7,906,992	12,513
Eastern Europe	94	10,680,553	3,640,848	847,499	1,434,160	178,729	4,082,450	12,498
Middle East	46	6,072,990	1,952,760	311,500	569,859	582,400	1,746,026	3,000
North America	170	20,033,383	8,849,190	6,464,960	4,073,794	1,751,270	12,555,506	104,687
South America	68	6,490,315	2,781,786	1,232,096	398,234	162,400	1,802,208	16,607
Western Europe	105	14,502,945	5,152,243	2,141,610	2,132,559	882,848	8,090,050	10,731
Total Regions	**732**	**81,166,308**	**26,583,537**	**13,865,830**	**11,003,865**	**4,301,810**	**37,018,631**	**160,817**

WORLDWIDE REFINING

Clean fuels requirements and lower crude prices were expected to drive refining investment in 2002 toward sulfur-removal and octane-upgrading processes and away from heavy oil upgrading processes such as coking and hydrocracking.

Evolving fuel sulfur specifications

Reducing sulfur in diesel fuel is a greater technical challenge than reducing it in gasoline.

GASOLINE AND DIESEL SULFUR STANDARDS

Region	Effective year: Gasoline	Effective year: Diesel	Sulfur standard, ppm — Gasoline: Average	Gasoline: Cap	Onroad diesel: Average	Onroad diesel: Cap
US (excluding California)	2005	2006	30	80	—	15
California[1]	2003	2006	15	40	—	15
Canada[2]	2005	2007	30	80	—	15
European Union[3]	2005	2005	—	50	—	50

[1]The indicated gasoline sulfur standard is part of the California Phase 3 RFG program; the onroad diesel standard is the US national Tier 2 standard. [2]Canada also has an interim standard for gasoline sulfur content. In the July 2002-December 2004 period, a refiner's average sulfur content (over time) must not exceed 150 ppm. [3]The EU recently set a 10 ppm cap on gasoline and diesel fuel sulfur, to be phased in between 2005 and 2011.

Capacity utilization near all-time highs

Because of high utilization rates, any additional loss in US refinery capacity could trigger regional price spikes.

US REFINERY CAPACITY RISING YET UTILIZATION REMAINS HIGH

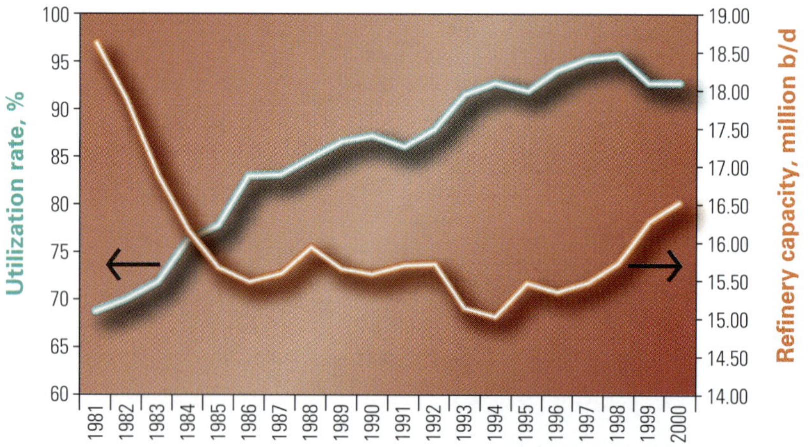

Source: US Department of Energy, Energy Information Administration, Washington, DC

WORLDWIDE ETHYLENE PRODUCTION

Global ethylene production capacity during 2001 rose to 101.5 million tons/yr (tpy), including new units in Argentina and Saudi Arabia.

New capacity

After increasing dramatically in 2000, growth in ethylene capacity during 2001 slowed a bit but was still strong.

GLOBAL ETHYLENE CAPACITY ADDITIONS

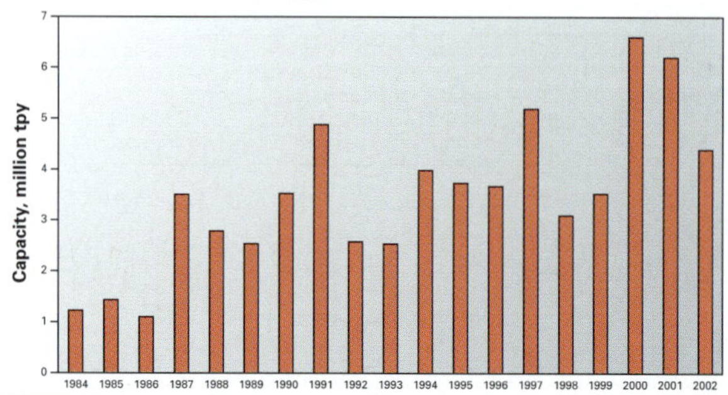

Source: Chem Systems and OGJ.

Biggest plants

Of the world's 10 largest ethylene facilities, six are on the US Gulf Coast, including five in Texas alone.

TOP TEN ETHYLENE COMPLEXES*

	Company	Location	Capacity, tpy
1	Nova Chemicals Corp.	Joffre, Alta.	2,815,000
2	Arabian Petrochemical Co.	Jubail, Saudi Arabia	2,250,000
3	Chevron Phillips Chemical Co.	Sweeny, Tex.	2,086,000
4	ExxonMobil Chemical Co.	Baytown, Tex.	1,890,000
5	Equistar Chemicals LP	Channelview, Tex.	1,750,000
6	Yanbu Petrochemical Co.	Yanbu, Saudi Arabia	1,600,000
7	Shell Chemicals Ltd.	Norco, La.	1,556,000
8	Dow Chemical Co.	Freeport, Tex.	1,540,000
9	BP	Chocolate Bayou, Tex.	1,451,000
10	Formosa Petrochemical Corp.	Mailiao, Taiwan	1,350,000

*As of Apr. 1, 2001.

World capacity by region

Ethylene production capacity grew fastest in the Middle East, led by Saudi Arabia's expansion of its Petrokemya complex and addition of a new unit.

REGIONAL CAPACITY BREAKDOWN

	— Ethylene capacity, tpy —			
	Jan. 1, 2001	Jan. 1, 2000	Change, tpy	Change, %
Asia-Pacific	25,504,000	[1]23,749,000	1,755,000	7.4
Eastern Europe/FSU	7,065,000	7,337,000	–272,000	–3.7
Middle East/Africa	8,582,000	[2]6,612,000	1,970,000	29.8
North America	33,742,000	[3]31,715,000	2,027,000	6.4
South America	3,918,000	3,456,000	462,000	13.4
Western Europe	21,788,000	21,147,000	641,000	3.0
Total capacity	**100,599,000**	**94,016,000**	**6,583,000**	**7.0**

[1]Adjusted to reflect 100,000 tpy expansion by Keiyo Ethylene in 1999, 102,000 tpy expansion by Tosoh in 1998, 300,000 tpy new unit by Gas Authority of India in 1999, and 100,000 tpy Nagothane expansion by Indian Petrochemicals in 1998. [2]Adjusted to reflect new information from Nigeria. [3]Adjusted to reflect 410,000 Taft, Tex., expansion by Union Carbide in 1999.

WORLDWIDE ETHYLENE PRODUCTION

Petrochemicals mergers have better positioned companies for a downturn, according to Chem Systems.

World's largest producers

During 2001 Union Carbide Corp. became a wholly owned subsidiary of Dow Chemical Co., but even without the merger, Dow would have ranked first.

TOP TEN ETHYLENE PRODUCERS*

			— Capacity, tpy —	
	Company	No. of sites	Of entire complexes	With only company partial interests
1	Dow Chemical Co.	16	12,467,000	10,076,500
2	ExxonMobil Chemical Co.	14	10,609,000	7,071,200
3	Equistar Chemicals LP	7	5,265,000	5,265,000
4	Shell Chemicals Ltd.	6	6,188,000	4,539,000
5	Chevron Phillips Chemical Co. LP	3	3,674,000	3,674,000
6	Saudi Basic Industries Corp.	4	5,650,000	3,950,000
7	BP	5	4,151,000	3,036,000
8	Nova Chemicals Corp.	2	3,540,000	2,968,500
9	Atofina	7	3,725,000	2,378,000
10	Enichem SPA	7	3,005,000	2,196,500

*As of Apr. 1, 2001.

Market trend

A handful of companies has captured larger shares of the North American ethylene market.

TOP SHAREHOLDERS OF NORTH AMERICAN CAPACITY

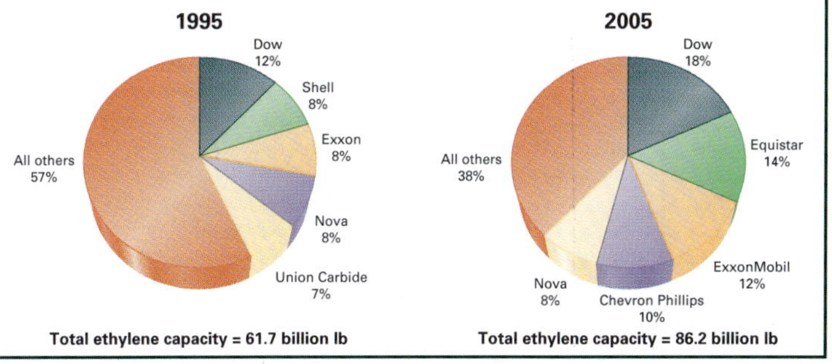

Source: CMAI

US mergers

The number of companies producing ethylene, low linear density polyethylene, and benzene has decreased, resulting in cost savings.

MERGERS HAVE LED TO SAVINGS

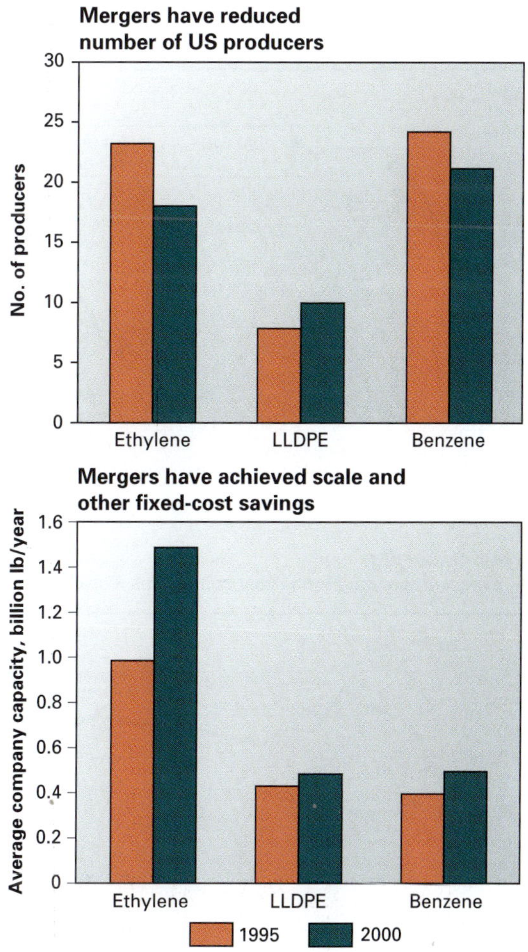

Source: Chem Systems

WORLDWIDE ETHYLENE PRODUCTION

Global supply of ethylene has exceeded global demand, with polyethylene being the most common derivative.

Consumption growing

Worldwide demand for ethylene was expected to reach 100 million tonnes in 2002, according to CMAI (Chemical Market Association Inc.), with Asia accounting for a significant portion of new demand.

GLOBAL ETHYLENE DEMAND GROWTH

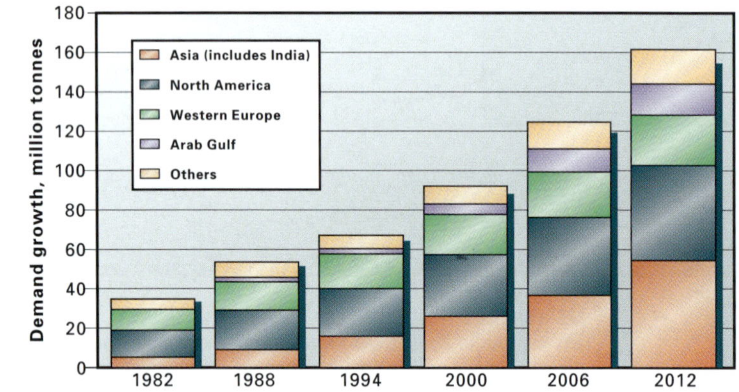

Source: CMAI

Extra capacity

Weak market conditions occurred in early 2001, when NGLs were diverted from ethylene feedstocks due to high natural gas prices.

NORTH AMERICAN SURPLUS ETHYLENE CAPACITY

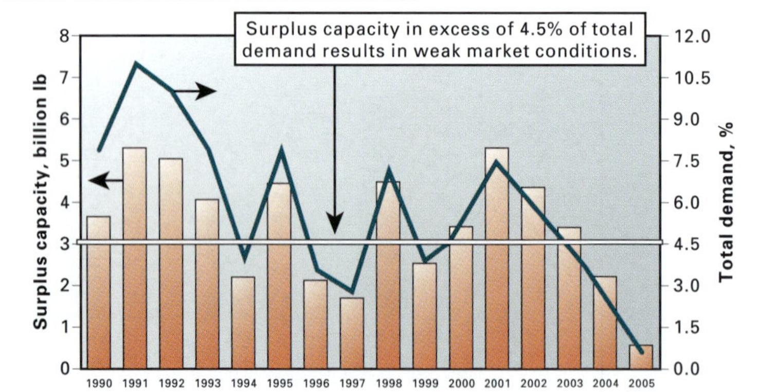

Source: CMAI

Narrowing surplus

CMAI expected that the North American surplus would begin to dwindle in 2004.

US ETHYLENE ECONOMICS

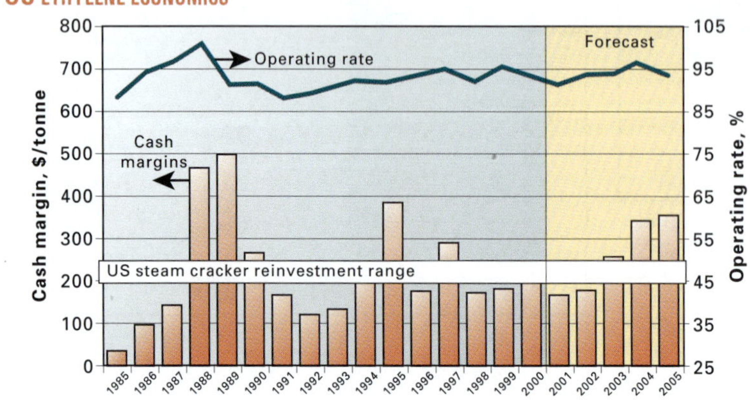

Source: CMAI

WORLDWIDE ETHYLENE PRODUCTION

A total of 6.2 million tpy of new capacity was scheduled to come on stream during 2001, including Formosa Plastic Point Corp.'s Point Comfort project in Texas.

ETHYLENE EXPANSIONS, 2001-2005

Location	Company	Ethylene capacity, tpy 2001	2002	2003	2004	2005
Skikda, Algeria	Sonatrach		100,000			
Bahia Blanca, Argentina	Dow Chemical Co.	420,000*				
Novopolotsk, Belarus	Production Association Polymir	60,000				
Antwerp, Belgium	Atofina Petrochemicals Inc.	30,000				
Antwerp, Belgium	BASF AG	170,000				
Camacari, Brazil	Copene Petroquim Nordeste		80,000	130,000		
Duque de Caxias, Brazil	Petroleo Brasileiro			520,000		
Duque de Caxias, Brazil	Rio Polimeros			500,000		
Huizhou, Guangdong, China	Shell Petrochemicals Co. Ltd.					800,000
Fujian Province, China	Fujian Petrochemical Co. Ltd.					600,000
Jilin, China	Jilin Chemical Industry Co.		300,000			
Guangdong, China	Maoming Petrochemical			850,000		
Jinshan, China	Shanghai Petrochemical Co.		300,000			
Shanghai, China	Shanghai Petrochemical Co.					900,000
Nanjing, China	Yangzi Petrochemical Corp.					600,000
Litvinov, Czech Republic	Chemopetrol AS		57,000		25,000	
Porvoo, Finland	Borealis AS		28,500			
Novo Urengoy, Russia, FSU	Gazkomplektempex			340,000		
Shurtan, Uzbekistan, FSU	Uzbekneftegaz	140,000				
Carling, France	Atofina Petrochemicals Inc.	88,000				
Cologne, Germany	Basell Polyfine GmbH		450,000			
Wesseling, Germany	Basell Polyfine GmbH	38,000				
Dormagen, Germany	EC Erdoelchemie GmbH	60,000				
Dormagen, Germany	EC Erdoelchemie GmbH		180,000			
Gelsenkirchen, Germany	Veba Oel		120,000			
Gelsenkirchen, Germany	Veba Oel				1,000,000	
Tiszauujvaros, Hungary	TVK			300,000		
Tamil Nadu, India	Bharat Petroleum Corp. Ltd.		500,000			
Auraiya, India	Gas Authority of India Ltd.		200,000			
Thane, Maharashtra, India	Indian Petrochemicals Corp. Ltd.	150,000				
Thane, Maharashtra, India	National Organic Chemicals Industries Ltd.			700,000		
Hazira, India	Reliance		150,000			
Lepetkata, Dibrugarh, India	Reliance Assam Petrochemicals					300,000
Bandar Imam, Iran	Amir Kabir Petrochemical Co.		520,000			
Bandar Asaluyeh, Iran	Pars Petrochemical Co.			1,000,000		
Bandar Imam, Iran	Marun Petrochemical Co.				1,100,000	
Bandar Asaluyeh, Iran	Jam Petrochemical Co.				1,400,000	
Porto Torres, Italy	Enichem SpA	60,000				
Priolo, Italy	Enichem SpA				80,000	
Kertih, Malaysia	Optimal Olefins	600,000				
Terneuzen, the Netherlands	Dow Chemical Co.	600,000				
Bataan, Philippines	Chinese Petroleum Corp. & Philippine National Oil Co.				350,000	
Mesaieed, Qatar	Q-Chem		500,000			
Yanbu, Saudi Arabia	Yanbu Petrochemical Co.	800,000*				
Pulau Ayer Merbau, Singapore	Petrochemical Corp. of Singapore Pte. Ltd.	80,000				
Pulau Ayer Chawan, Singapore	ExxonMobil Chemical Co.	800,000				
Tarragona, Spain	Dow Chemical Co.		100,000			
Map Ta Phut, Thailand	Rayong Olefins		200,000			
Map Ta Phut, Thailand	Thai Olefins Co. Ltd.			315,000		
Point Fortin, Trinidad & Tobago	National Gas Co.			750,000		
Aliaga, Turkey	Petkim Petrokimya Holding AS	120,000				
Ruwais, UAE	Abu Dhabi Polymers Ltd.		600,000			
Grangemouth, UK	BP	270,000				
Port Arthur, Tex., US	BASF Corp.	860,000				
Point Comfort, Tex., US	Formosa Plastics Co.	850,000				
Deer Park, Tex., US	Shell Chemical Co.			544,000		
Jose, Anzoategui, Venezuela	Pequiven			1,000,000		
Total		**6,196,000**	**4,385,500**	**6,949,000**	**3,955,000**	**3,200,000**

*Already completed.

STRATIGRAPHIC CHARTS

Swan Creek Field, Tennessee, US

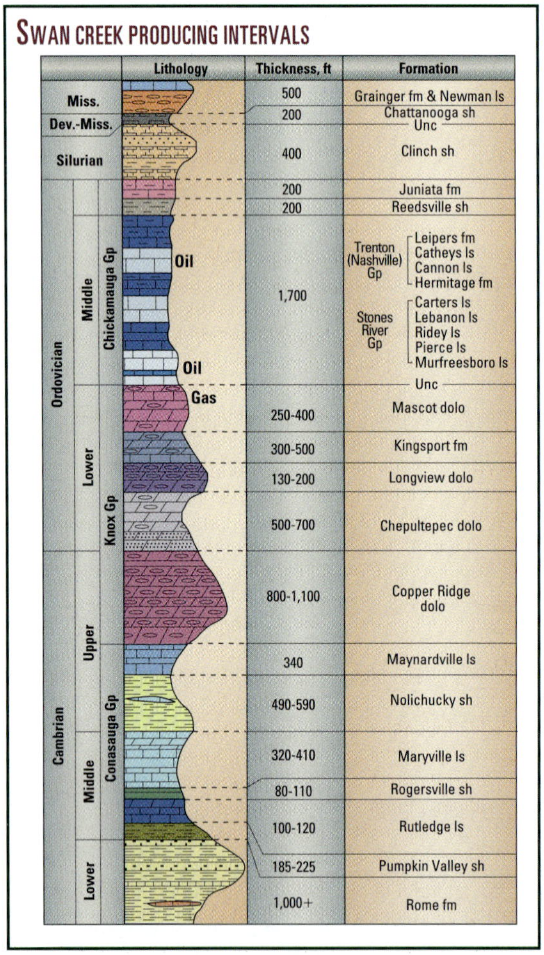

Taranaki Basin, New Zealand

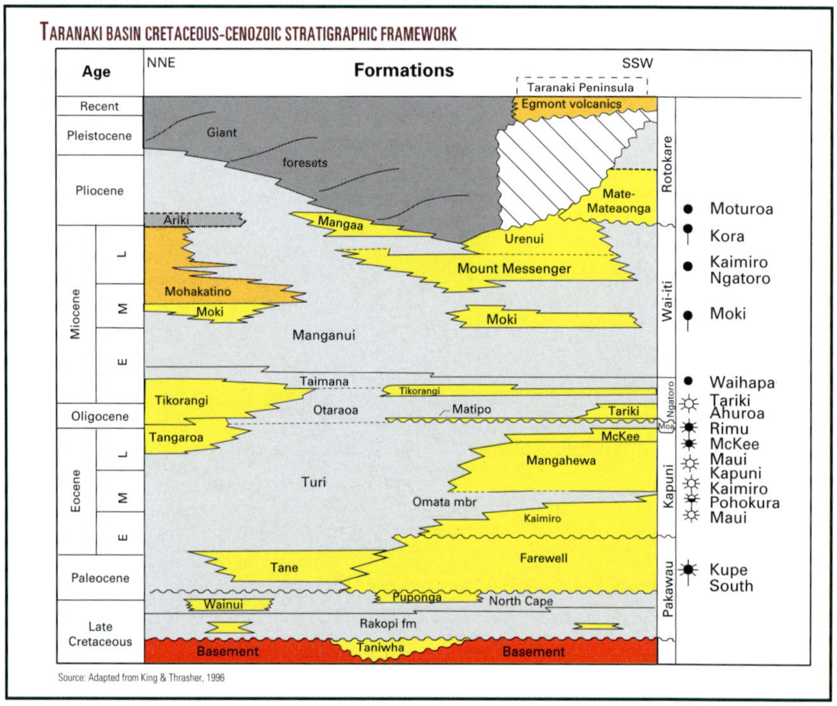

Black Warrior Basin, Mississippi, US

GENERALIZED PALEOZOIC STRATIGRAPHIC SECTION

Mississippi Black Warrior basin

Era	System	Series	Group	Formation
Paleozoic	Permian (absent)			
	Pennsylvanian			Pottsville
	Mississippian	Chesterian		Parkwood-Pennington
				Pride Mountain
	Devonian		Harriman	Devonian
	Silurian		Niagaran	
	Ordovician		Stones River	Stones River ls
				Stones River dolo
			Knox	Knox ls
				Knox dolo
	Cambrian			Copper Ridge dolo
				Conasauga ls
				Rome
				Shady dolo
				Weisner
Precambrian				

STRATIGRAPHIC CHARTS

West Florida Shelf, Gulf of Mexico, US

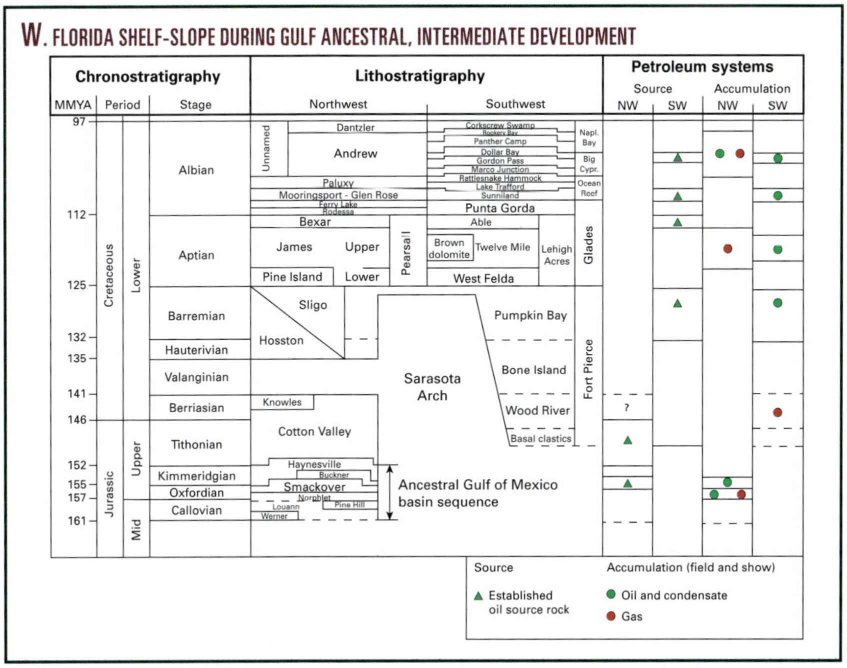

Otway Basin, Australia

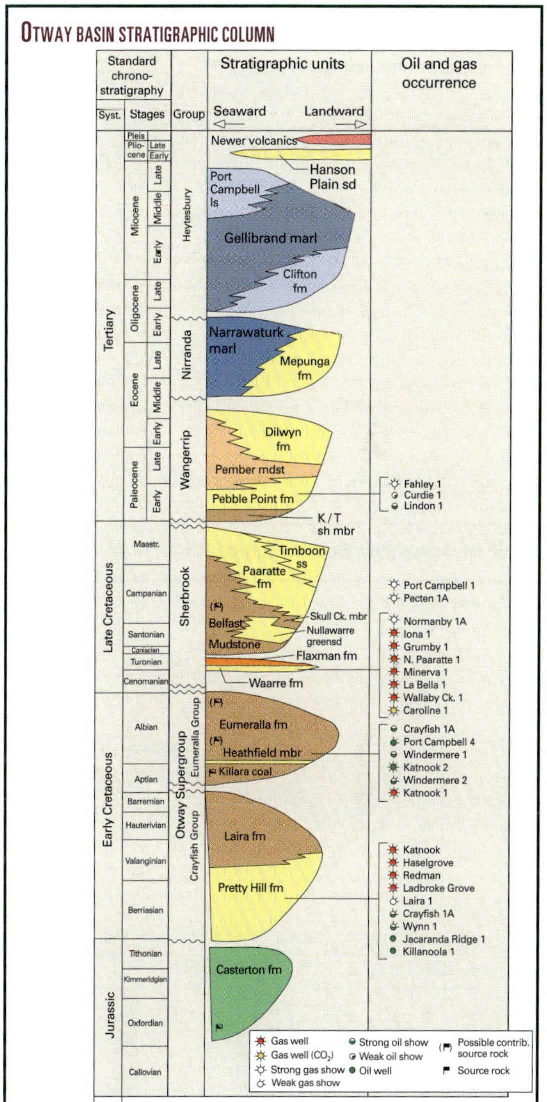

Source: Essential Petroleum Resources NL

Deer Lake Basin, Newfoundland, Canada

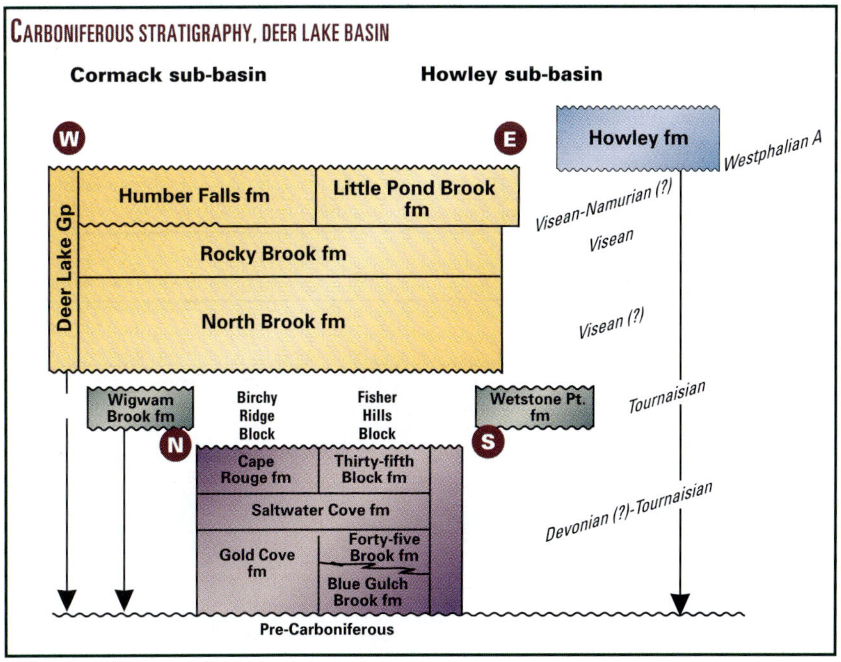

STRATIGRAPHIC CHARTS

Neuquen Basin, Argentina

Subsurface Stratigraphic Scheme

Neuquen basin-Neuquen Province

Period	Groups	Embayment: Formations	Northeast platform	Source rock	Reservoir rock
Tertiary	Malargue	Malargue Gp			
Cretaceous	Neuquen	Neuquen Gp			X
		Rayoso clastic member			X
		Rayoso evaporitic member			
	Rayoso	Huitrin: La Tosca			X
		Upper Troncoso			X
		Lower Troncoso	Rayoso		X
		Upper Chorreado			
		Lower Chorreado			X
	Mendoza	Mulichinco			X
		Agrio	Centenario	X	X-X
		Quintuco	Loma Montosa		X X-X X
		Vaca Muerta	Vaca Muerta	X	
Jurassic		Tordillo/Sierras Blancas	Sierras Blancas		X
	Lotena	Auquilco			
		La Manga/Barda Negra	Lotena	X ?	X-X
		Lotena			X
	Cuyo	Challaco	Punta Rosada		X-X
		Lajas		X	X
		Los Molles	Los Molles		X
Pre-Jurassic	Pre-Cuyo	Puesto Kaufman-Remoredo		X	X
Permo-Triassic	Choiyoi				

Guarico Basin, Venezuela

Guarico Basin General Stratigraphic Column

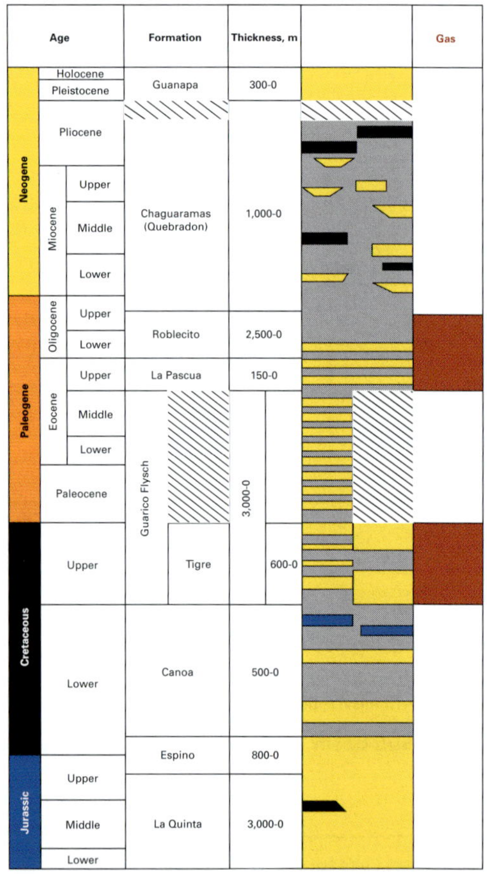

Source: Venezuela Ministry of Energy and Mines

Gulf of Suez Rift Basin, Egypt

Gulf of Suez Stratigraphy

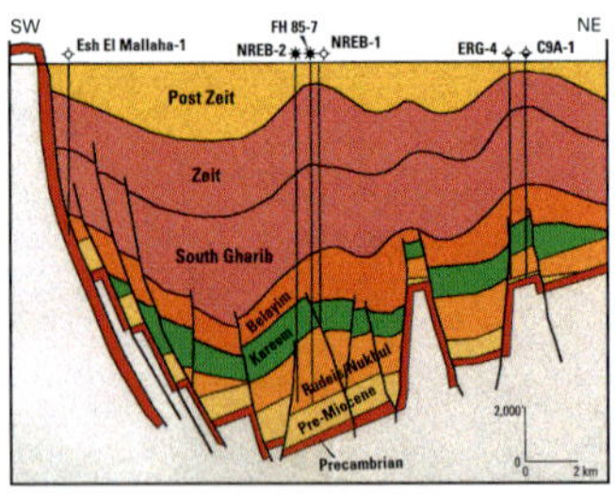

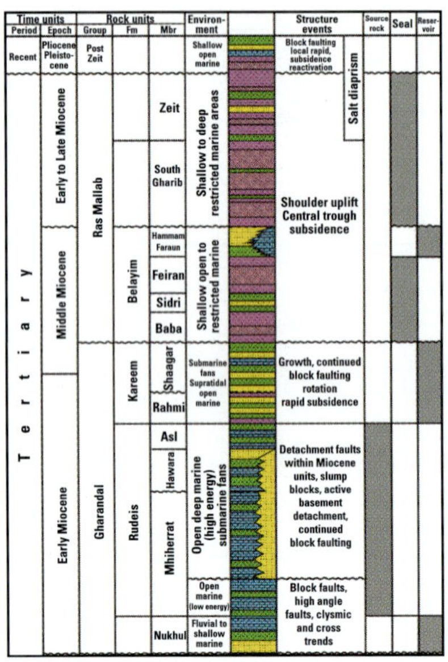

Mozambique Basin, Mozambique

Mozambique Basin Stratigraphy

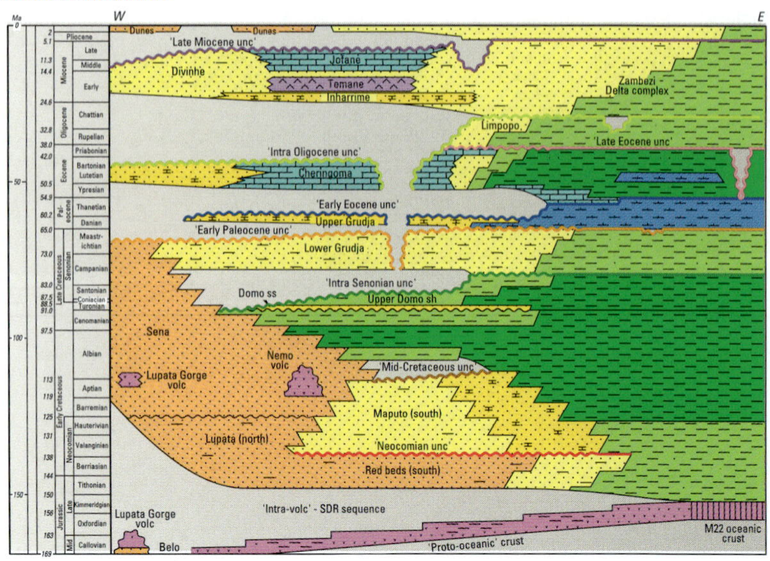

STRATIGRAPHIC CHARTS

North Chukchi Basin, Russia

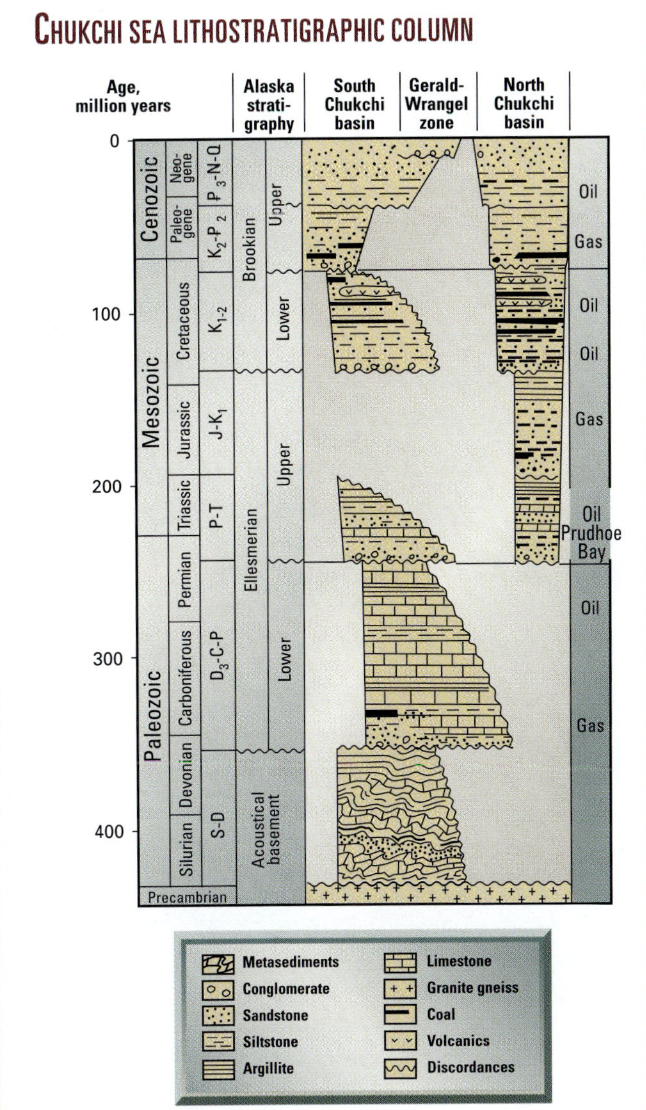

Source: After Orudzheva et al., 1999.

Waropen Basin, Indonesia

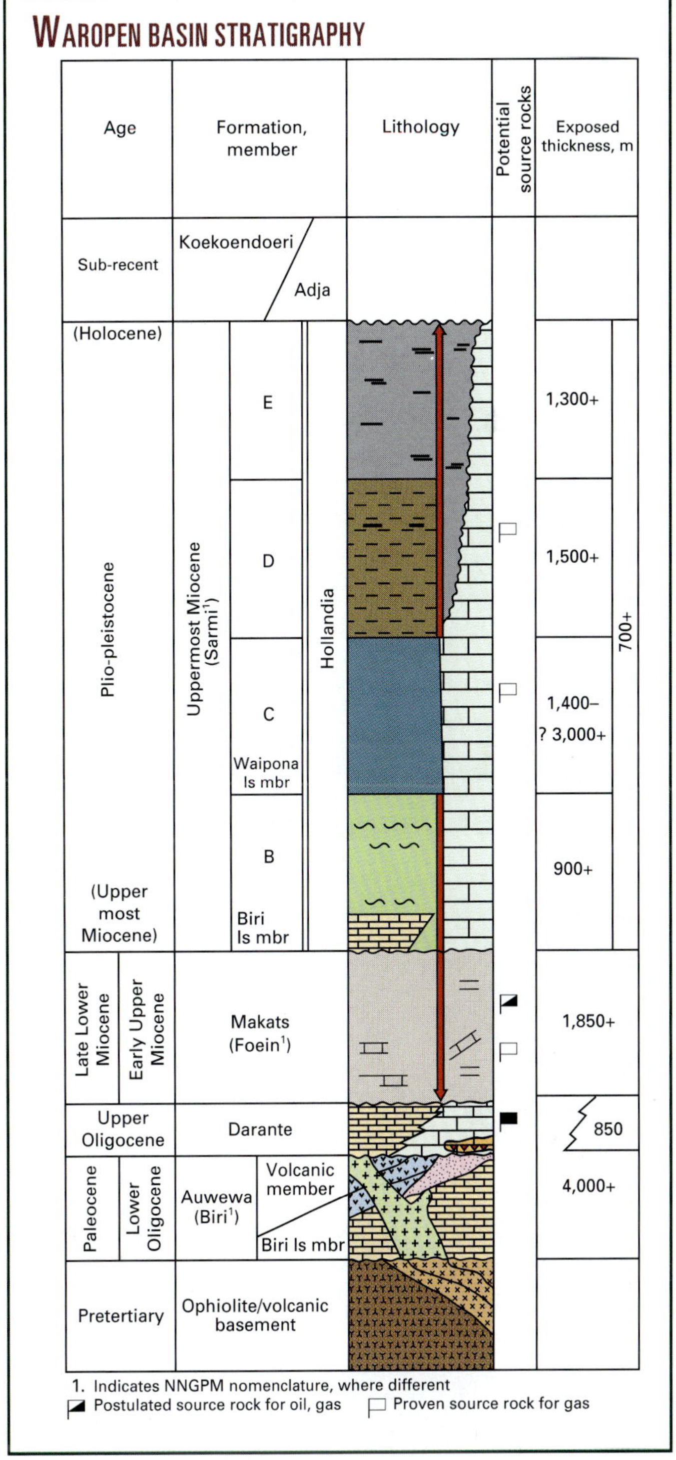

WORLD OIL FLOW MAP

Oil and gas reserves, oil production, refining capacity, and global trade movements for 2000-2001.

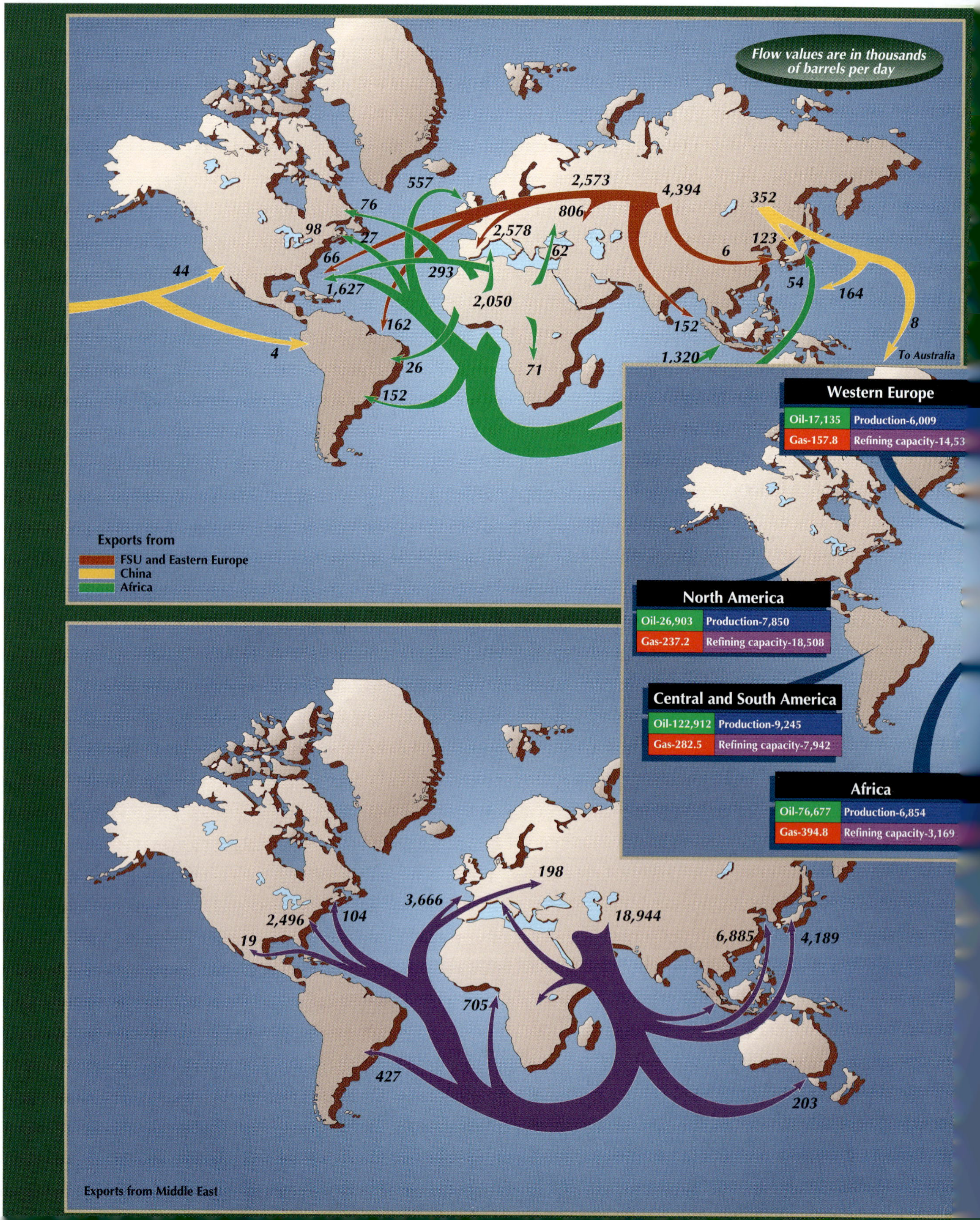

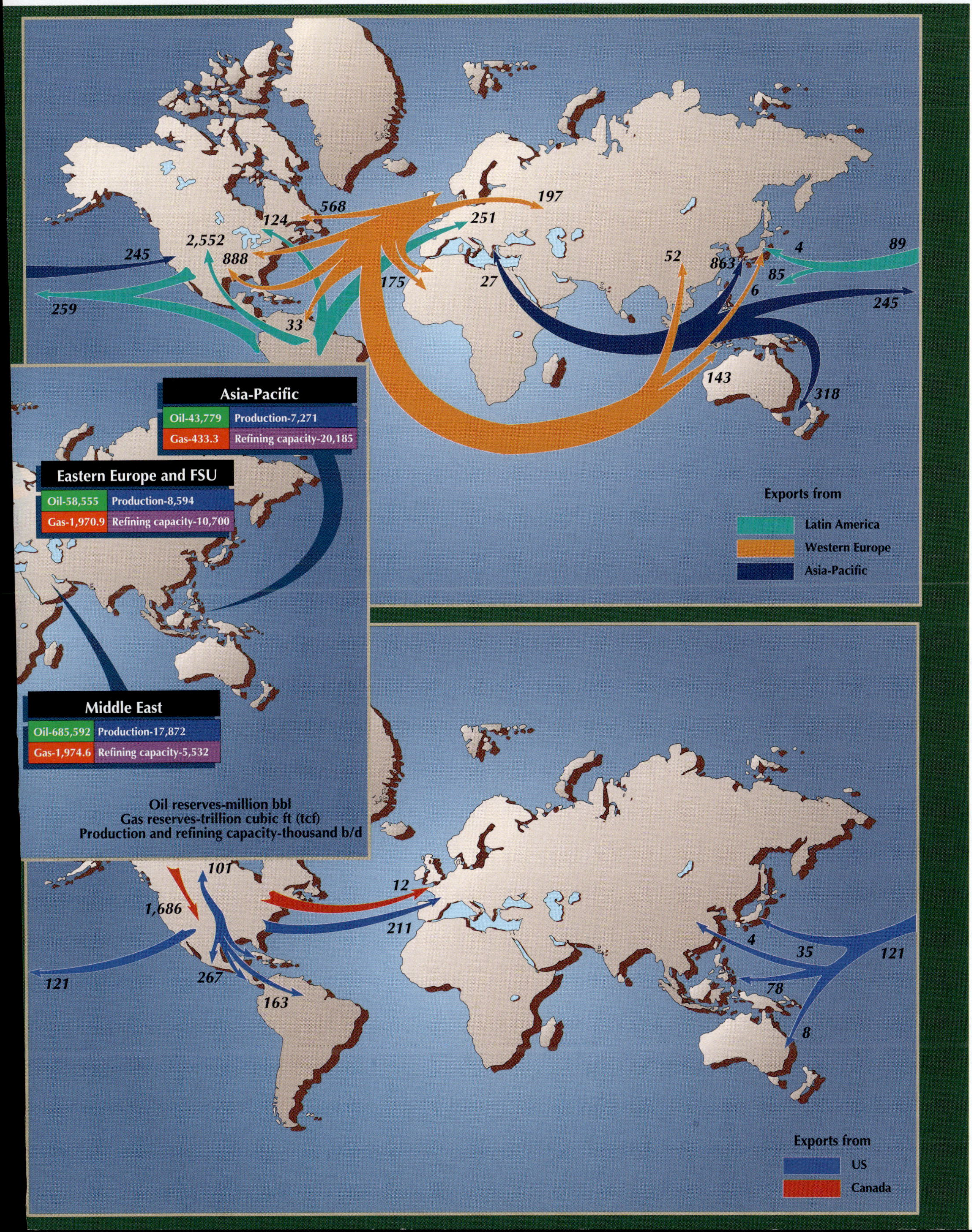
568
124
2,552
888
245
259
33
175
197
251
27
52
863
4
89
85
6
245
143
318
Asia-Pacific
Oil-43,779
Production-7,271
Gas-433.3
Refining capacity-20,185
Eastern Europe and FSU
Oil-58,555
Production-8,594
Gas-1,970.9
Refining capacity-10,700
Middle East
Oil-685,592
Production-17,872
Gas-1,974.6
Refining capacity-5,532
Oil reserves-million bbl
Gas reserves-trillion cubic ft (tcf)
Production and refining capacity-thousand b/d
Exports from
Latin America
Western Europe
Asia-Pacific
101
1,686
12
211
121
267
163
4
35
121
78
8
Exports from
US
Canada

Statistical Tables

WORLDWIDE OIL CONSUMPTION

Upper figures = 1,000 b/d
Lower figures = million t/y

	1998	1999	2000
US	18,030	18,635	18,745
	863.8	888.9	897.4
Canada	1,820	1,795	1,775
	83.5	82.9	82.9
Mexico	1,770	1,765	1,840
	81.4	80.8	84.3
Total North America	**21,620**	**22,195**	**22,360**
	1,028.7	**1,052.6**	**1,064.6**
Argentina	485	440	430
	23.0	20.8	20.1
Brazil	1,800	1,805	1,825
	83.2	83.2	84.4
Chile	245	255	255
	11.4	11.7	11.8
Colombia	265	240	230
	12.0	10.6	10.5
Equador		130	140
		6.0	6.5
Peru		155	150
		7.4	7.0
Venezuela	475	480	495
	21.6	21.9	22.6
Others	1,405	1,130	1,140
	67.9	55.1	55.8
Total S. & Cent. America	**4,675**	**4,635**	**4,665**
	219.1	**216.7**	**218.7**
Austria	255	250	240
	12.3	12.1	11.5
Belgium & Luxembourg	655	670	685
	31.6	32.4	33.1
Bulgaria	90	100	105
	4.4	4.9	5.0
Czech Republic & Slovakia	245	175	170
	11.7	8.2	7.9
Denmark	225	220	215
	10.7	10.6	10.4
Finland	220	225	220
	10.5	10.7	10.5
France	2,015	2,045	2,010
	95.0	96.4	95.1
Germany	2,915	2,825	2,760
	136.6	132.4	129.5
Greece	375	385	390
	18.2	18.7	19.1
Hungary	155	150	145
	7.4	7.1	6.9
Iceland		20	20
		0.9	0.9
Ireland		170	170
		8.3	8.3
Italy	1,975	1,980	1,945
	94.7	94.4	93.0
Netherlands	855	880	900
	39.4	40.6	41.8
Norway	215	215	205
	10.0	10.1	9.4
Poland	425	430	445
	19.9	19.9	20.6
Portugal	315	320	320
	15.1	15.4	15.3
Romania	240	195	185
	12.0	9.5	8.9
Slov		65	60
		3.0	2.8
Spain	1,380	1,425	1,455
	66.4	68.4	70.1
Sweden	340	335	320
	16.2	16.1	15.2
Switzerland	280	270	265
	13.0	12.6	12.2
Turkey	640	640	690
	29.6	29.5	31.5
United Kingdom	1,745	1,720	1,675
	80.9	79.4	77.6
Others	520	340	330
	25.1	16.3	16.0
Total Europe	**16,080**	**16,050**	**15,925**
	760.7	**757.9**	**752.6**

[1]Organisation for Economic Cooperation and Development.
[2]Emerging Market Economies excluding Central Europe and FSU.

Source: British Petroleum

WORLDWIDE OIL CONSUMPTION-cont'd

Upper figures = 1,000 b/d
Lower figures = million t/y

Azerbaijan	120	125	150
	5.9	6.3	7.4
Belarus	200	125	110
	9.9	6.1	5.4
	1998	**1999**	**2000**
Kazakhstan	170	120	125
	8.5	6.0	6.2
Lith		65	60
		3.1	2.9
Russian Federation	2,485	2,535	2,475
	123.7	126.2	123.5
Turkey		90	95
		4.5	4.8
Ukraine	285	265	210
	14.3	13.3	10.4
Uzbekistan	140	145	135
	7.0	7.1	6.8
Others	255	95	115
	12.8	4.7	5.7
Total Former Soviet Union	**3,655**	**3,565**	**3,475**
	182.1	**177.3**	**173.1**
Iran	1,200	1,210	1,170
	58.0	58.4	56.9
Kuwait	165	170	160
	8.2	8.6	8.0
Qat		25	25
		1.1	1.1
Saudi Arabia	1,265	1,305	1,335
	58.8	60.9	62.4
United Arab Emirates	345	260	280
	17.9	13.0	14.3
Others	1,335	1,355	1,375
	64.1	65.0	66.3
Total Middle East	**4,310**	**4,325**	**4,345**
	207.0	**207.0**	**209.0**
Algeria	195	185	195
	8.4	8.1	8.5
Egypt	560	575	545
	27.3	27.8	26.2
South Africa	450	460	475
	21.3	21.8	22.5
Others	1,175	1,230	1,255
	55.7	58.2	59.5
Total Africa	**2,380**	**2,450**	**2,470**
	112.7	**115.9**	**116.7**
Australia	825	840	870
	37.0	38.0	38.7
Bangladesh		65	65
		3.2	3.3
China	4,160	4,415	4,840
	190.3	207.2	226.9
Chech		195	200
		9.3	9.6
India	1,835	2,015	2,070
	86.8	95.2	97.6
Indonesia	915	980	1,065
	43.5	46.8	51.1
Japan	5,550	5,620	5,525
	254.9	257.3	253.5
Malaysia	405	440	445
	19.0	20.3	20.5
New Zealand	130	135	135
	6.1	6.3	6.3
Pakistan	350	365	385
	17.6	18.2	19.6
Philippines	390	375	350
	19.1	18.0	16.8
Singapore	560	545	560
	29.2	28.3	29.1
South Korea	2,010	2,160	2,200
	92.9	99.7	101.8
Taiwan	765	820	815
	37.2	39.9	39.8
Thailand	735	735	715
	35.4	35.4	34.1
Others	570	390	425
	27.3	18.6	20.2
Total Asia & Australasia	**19,200**	**20,095**	**20,665**
	896.3	**941.7**	**968.9**
Total World	**71,920**	**73,315**	**73,905**
	3,406.6	**3,469.1**	**3,503.6**
Of which OECD1	45,465	46,300	46,335
	2,143.7	2,178.1	2,184.8
Of which European Union 15	13,420	13,450	13,305
	635.0	635.9	630.5
Of which Other EMEs[2]	22,050	22,750	23,415
	1,044.2	1,079.8	1,112.7

[1]Organisation for Economic Cooperation and Development.
[2]Emerging Market Economies excluding Central Europe and FSU.

Source: British Petroleum

WORLDWIDE PETROLEUM PRODUCT CONSUMPTION

Area	1990	1991	1992	1993	1994	1995	1996	1997	1998	1999	2000
						1,000 b/d					
USA											
Gasolines	7,644.4	7,521.4	7,683.8	7,783.6	7,871.8	8,015.6	8,180.3	8,314.9	8,570.1	8,707.0	8.827.6
Middle distillates	4,840.5	4,731.3	4,806.9	4,928.7	5,172.3	5,220.6	5,447.4	5,596.5	5,640.6	5,796.0	5,966.1
Fuel oil	1,236.5	1,159.2	1,093.0	1,072.7	1,013.8	842.7	840.9	785.7	877.7	822.5	903.4
Others	3,227.3	3,196.0	3,396.4	3,356.7	3,530.0	3,502.8	3,732.4	3,765.8	3,717.8	4,031.6	3,939.3
Total USA	16,948.7	16,607.9	16,980.1	17,141.7	17,587.9	17,581.7	18,201.0	18,462.9	18,806.2	19,357.1	19,536.4
NORTH AMERICA											
Gasolines	8,781.2	8,667.5	8,744.1	8,887.9	9,010.9	9,222.0	9,400.6	9,546.8	9,839.1	9,983.0	10,124.5
Middle distillates	5,602.8	5,474.7	5,598.6	5,739.3	6,039.6	6,035.4	6,314.7	6,505.8	6,560.4	6,738.9	6,946.8
Fuel oil	1,829.0	1,716.7	1,646.8	1,593.4	1,593.4	1,346.9	1,361.6	1,359.8	1,516.2	1,424.2	1,532.7
Others	3,811.8	3,824.3	4,076.9	4,060.2	4,252.2	4,120.7	4,365.0	4,429.7	4,377.5	4,681.6	4,474.4
Total North America	20,024.7	19,683.2	20,066.4	20,280.8	20,896.2	20,725.0	21,441.9	21,842.0	22,293.2	22,827.7	23,078.7
SOUTH & CENTRAL AMERICA											
Gasolines	951.2	979.0	986.0	1,027.8	1,106.6	1,157.7	1,199.4	1,262.1	1,278.3	1,278.3	1,276.0
Middle distillates	1,190.7	1,236.9	1,304.1	1,344.0	1,430.1	1,507.8	1,577.1	1,659.0	1,734.6	1,719.9	1,749.3
Fuel oil	734.2	708.4	677.1	684.5	699.2	713.9	745.2	793.0	807.8	794.9	793.0
Others	601.2	603.3	720.2	695.1	686.8	705.6	739.0	757.8	782.8	762.9	777.5
Total S. & Cent. America	3,477.3	3,527.6	3,687.4	3,751.4	3,922.7	4,085.0	4,260.7	4,471.9	4,603.5	4,556.0	4,595.8

WORLDWIDE PETROLEUM PRODUCT CONSUMPTION-cont'd

EUROPEAN UNION 15											
Gasolines	3,549.6	3,561.2	3,616.9	3,547.3	3,498.6	3,535.7	3,554.2	3,577.4	3,593.7	3,605.3	3,521.8
Middle distillates	4,790.1	5,035.8	5,075.7	5,115.6	5,134.5	5,262.6	5,544.0	5,600.7	5,768.7	5,848.5	5,896.8
Fuel oil	1,889.7	1,860.2	1,897.0	1,845.5	1,801.4	1,795.8	1,764.6	1,707.5	1,737.0	1,661.5	1,556.6
Others	2,008.2	2,064.5	2,164.7	2,131.3	2,206.5	2,206.5	2,210.7	2,296.3	2,317.1	2,332.4	2,368.0
Total European Union 15	12,237.6	12,521.8	12,754.4	12,639.7	12,640.9	12,800.6	13,073.5	13,181.9	13,416.5	13,447.7	13,343.2
TOTAL EUROPE											
Gasolines	4,203.8	4,162.1	4,208.5	4,143.5	4,113.4	4,185.3	4,220.1	4,259.5	4,294.3	4,324.5	4,208.5
Middle distillates	5,726.7	5,871.6	5,917.8	5,968.2	5,978.7	6,140.4	6,474.3	6,522.6	6,730.5	6,789.3	6,831.3
Fuel oil	2,548.4	2,456.4	2,390.2	2,360.7	2,287.1	2,289.0	2,270.6	2,213.5	2,208.0	2,095.8	1,992.7
Others	2,457.0	2,463.3	2,521.7	2,517.5	2,603.1	2,634.4	2,665.7	2,797.3	2,818.1	2,804.8	2,880.0
Total Europe	14,935.9	14,953.3	15,038.1	14,990.0	14,982.3	15,249.1	15,630.7	15,792.9	16,050.9	16,014.4	15,912.5
MIDDLE EAST											
Gasolines	552.2	547.5	619.4	635.7	705.3	763.3	786.5	828.2	832.9	846.8	863.0
Middle distillates	1,190.7	1,192.8	1,184.4	1,234.8	1,333.5	1,442.7	1,495.2	1,503.6	1,509.9	1,554.0	1,545.6
Fuel oil	1,076.4	1,212.6	1,146.3	1,122.4	1,139.0	1,164.7	1,236.5	1,260.4	1,277.0	1,214.4	1,256.7
Others	534.4	482.2	561.5	611.6	584.5	553.2	580.3	597.0	622.1	637.5	622.8
Total Middle East		3,353.7	3,435.1	3,511.7	3,604.5	3,762.2	3,923.9	4,098.5	4,189.3	4,252.7	4,288.1
AFRICA											
Gasolines	487.2	503.4	517.4	524.3	538.2	554.5	552.2	559.1	566.1	575.4	584.6
Middle distillates	785.4	791.7	810.6	829.5	848.4	875.7	903.0	926.1	951.3	991.2	1,020.6
Fuel oil	426.9	428.7	434.2	423.2	426.9	449.0	460.0	478.4	498.6	509.7	483.9
Others	263.0	267.2	269.3	292.3	302.7	308.9	315.2	317.3	327.7	338.6	349.0
Total Africa	1,962.5	1,991.1	2,031.5	2,069.3	2,116.2	2,188.1	2,230.4	2,280.9	2,343.8	2,414.9	2,438.1
ASIA PACIFIC											
Gasolines	3,039.2	3,185.4	3,463.8	3,640.1	3,779.3	4,060.0	4,294.3	4,681.8	4,876.6	5,185.2	5,407.9
Middle distillates	5,006.4	5,283.6	5,697.3	6,071.1	6,316.8	6,783.0	7,240.8	7,488.6	7,240.8	7,690.2	7,956.9
Fuel oil	3,385.6	3,473.9	3,608.2	3,577.0	4,013.0	3,979.9	3,963.4	3,911.8	3,472.1	3,553.0	3,499.7
Others	2,077.1	2,152.2	2,317.1	2,404.8	2,621.9	2,818.1	2,976.8	3,114.6	3,185.5	3,316.8	3,477.8
Total Asia Pacific	13,508.3	14,095.1	15,086.4	15,692.9	16,731.0	17,641.0	18,475.3	19,196.8	18,775.0	19,745.2	20,342.3
TOTAL WORLD (excluding FSU)											
Gasolines	18,014.8	18,045.0	18,539.1	18,859.3	19,253.7	19,942.7	20,453.1	21,137.5	21,687.4	22,193.1	22,464.6
Middle distillates	19,502.7	19,851.3	20,512.8	21,186.9	21,947.1	22,785.0	24,005.1	24,605.7	24,727.5	25,483.5	26,050.5
Fuel Oil	10,000.4	9,996.7	9,902.9	9,761.2	10,158.6	9,943.4	10,037.2	10,017.0	9,779.6	9,591.9	9,558.8
Others	14,979.9	15,053.0	16,027.8	16,069.6	16,787.7	16,850.3	17,585.1	18,075.7	18,148.7	12,542.1	12,581.8
TOTAL WORLD	**62,497.8**	**62,945.9**	**64,982.6**	**65,877.0**	**68,147.1**	**69,521.4**	**72,080.5**	**73,835.8**	**74,343.2**	**69,810.6**	**70,655.7**

Notes:
Total North America includes USA
Total Europe includes European Union 15

Gasolines--aviation and motor gasolines and light distillate feedstock. Middle distillates-jet and heating kerosines, and gas and diesel oils (including marine bunkers). Fuel oil-heavy fuel oil and marine bunkers. Others-refinery gas, LPG's, solvents, petroleum coke, lubricants, bitumen, wax and refinery fuel and loss.

Source: BP Statistical Review of World Energy

Conversion from metric tons to barrels per day by *IPE* editors.

WORLDWIDE OIL PRODUCTION

NORTH AMERICA

Upper figures = 1,000 b/d
Lower figures = million t/y

	1940	1950	1960	1970	1980	1990	1993	1994	1995	1996	1997	1998	1999	2000
Canada.	23.5	79.6	525.6	1,263.6	1,412.0	1,508.0	1,676.6	1,742.0	1,806.0	1,834.0	1,910.3	2,017.0	1,900.8	1,392.5
	1.2	4.0	26.2	62.9	70.3	75.1	83.5	86.8	89.9	91.3	95.1	100.4	94.7	69.3
US	3,707.4	5,407.1	7,054.6	9,630.0	8,569.0	7,220.0	6,846.6	6,661.6	6,559.6	6,464.5	6,452.0	6,252.0	5,881.5	5,821.6
	184.6	269.3	351.3	479.6	426.7	359.6	341.0	331.7	326.7	321.9	321.3	311.3	292.9	289.9
Total.	**3,730.9**	**5,486.7**	**7,580.2**	**10,893.6**	**9,981.0**	**8,728.0**	**8,523.2**	**8,403.6**	**8,365.6**	**8,298.5**	**8,362.3**	**8,269.0**	**7,782.3**	**7,214.1**
	185.8	**273.3**	**377.5**	**542.5**	**497.0**	**434.7**	**424.5**	**418.5**	**416.6**	**413.2**	**416.4**	**4111.7**	**387.6**	**359.2**

MIDDLE EAST

Upper figures = 1,000 b/d
Lower figures = million t/y

	1940	1950	1960	1970	1980	1990	1993	1994	1995	1996	1997	1998	1999	2000
Abu Dhabi	...	...	...	693.8	1,350.0	1,587.0	1,800.0	1,835.0-	1,832.0	1,954.0	1,873.0	1,912.0	1,700.0	1,900.0
	...	...	...	34.6	67.2	79.0	89.6	91.4	91.2	97.3	93.3	95.2	84.7	94.6
Bahrain	19.3	30.2	45.1	76.6	48.0	42.0	106.0	105.0	105.0	103.8	103.9	102.6	102.2	32.3
	1.0	1.5	2.2	3.8	2.4	2.1	5.3	5.2	5.2	5.2	5.2	5.1	5.1	1.6
Dubai	...	...	...	...	349.0	469.0	350.0	337.0	322.0	268.0	310.5	314.0	300.0	280.0
	...	...	...	...	17.4	23.4	17.4	16.8	16.0	13.3	15.5	15.6	14.9	13.9
Iran	181.2	664.3	1,067.6	3,328.8	1,467.0	3,120.0	3,670.8	3,585.0	3,612.0	3,668.0	3,632.7	3,607.5	3,504.2	3,681.7
	9.0	33.1	53.2	165.8	73.1	155.4	182.8	178.5	179.9	182.7	180.9	179.7	174.5	183.4
Iraq	60.7	136.2	1,004.2	1,548.6	2,638.0	2,083.0	447.5	550.0	600.0	576.0	1,147.0	2,110.0	2,525.0	2,566.7
	3.0	6.8	50.0	77.1	131.4	103.7	22.3	27.4	29.9	28.7	57.1	105.1	125.7	127.8
Kuwait	...	344.4	1,628.2	2,734.5	1,382.0	1,080.0	1,690.0	1,811.0	1,792.0	1,807.0	1,836.3	1,800.0	1,653.3	1,765.0
	...	17.2	81.1	136.2	68.8	53.8	84.2	90.2	89.2	90.0	91.4	89.6	82.3	87.9
Neutral Zone	...	...	136.2	500.6	540.0	315.0	360.0	378.0	430.0	483.0	533.0	550.0	590.0	630.0
	...	...	6.8	24.9	26.9	15.7	17.9	18.8	21.4	24.1	26.5	27.4	29.4	31.4
Oman	...	...	...	332.4	283.0	658.0	779.1	806.4	849.0	885.0	904.5	900.0	832.2	933.3
	...	...	...	16.6	14.1	32.8	38.8	40.2	42.3	44.1	45.0	44.8	41.4	46.5
Qatar	...	33.6	175.1	362.4	472.0	387.0	429.2	407.0	454.0	487.0	622.9	660.8	647.7	688.2
	...	1.7	8.7	18.0	23.5	19.3	21.4	20.3	22.6	24.3	31.0	32.9	32.3	34.3
Ras Al Khaimah	...	...	...	...	...	10.0	1.0	1.0	0.8	0.5	0.5	0.5	0.5	500.0
	...	...	...	...	...	0.5	...	...	...	...	...	...	...	24.9
Saudi Arabia	13.9	546.7	1,247.1	3,548.9	9,630.0	6,215.0	7,907.1	7,811.0	7,859.0	7,838.0	8,082.9	8,021.7	7,521.3	7,995.0
	0.7	27.2	62.1	176.7	479.6	309.5	393.8	389.0	391.4	390.3	402.5	399.5	374.6	398.2
Sharjah	...	...	...	...	10.0	35.0	44.0	50.0	50.0	70.0	70.0	70.2	44.0	50.0
	...	...	...	...	0.5	1.7	2.2	2.5	2.5	3.5	3.5	3.5	2.2	2.5
Syria	...	...	...	83.1	165.0	385.0	554.5	574.0	605.0	577.0	570.0	553.3	542.0	522.5
	...	...	...	4.1	8.2	19.2	27.6	28.6	30.1	28.7	28.4	27.6	27.0	26.0
Yemen	...	...	...	...	...	179.0	222.3	338.0	338.0	365.0	370.0	385.0	407.5	354.0
	...	...	...	...	...	8.9	11.1	16.8	16.8	18.2	18.4	19.2	20.3	17.6
Total	**275.1**	**1,755.4**	**5,306.0**	**13,279.9**	**18,335.0**	**16,565.7**	**18,361.7**	**18,588.6**	**18,849.0**	**19,082.5**	**20,057.5**	**20,987.7**	**20,370.0**	**13,438.0**
	13.7	**87.5**	**264.2**	**661.3**	**913.1**	**825.0**	**914.4**	**925.7**	**938.5**	**950.4**	**998.7**	**1,045.2**	**1,014.4**	**669.3**

WESTERN EUROPE

Upper figures = 1,000 b/d
Lower figures = million t/y

	1940	1950	1960	1970	1980	1990	1993	1994	1995	1996	1997	1998	1999	2000
Austria	7.7	27.9	45.4	53.3	29.0	24.0	23.6	23.7	21.1	20.2	20.4	20.7	20.2	19.1
	0.4	1.4	2.3	2.7	1.4	1.2	1.2	1.2	1.1	1.0	1.0	1.0	1.0	1.0
Denmark	...	...	...	...	6.0	118.0	167.6	184.9	186.0	207.2	230.5	238.1	299.3	363.0
	...	...	...	...	0.3	5.9	8.3	9.2	9.3	10.3	11.5	11.9	14.9	18.1
France	1.4	2.5	38.6	45.8	26.0	62.0	55.3	57.9	49.2	42.3	35.8	34.3	30.9	28.5
	0.1	0.1	1.9	2.3	1.3	3.1	2.8	2.9	2.5	2.1	1.8	1.7	1.5	1.4
Germany	20.1	23.2	108.3	148.4	100.0	78.0	60.2	56.7	57.5	56.8	55.6	57.9	54.3	64.6
	1.0	1.2	5.4	7.4	5.0	3.9	3.0	2.8	2.9	2.8	2.8	2.9	2.7	3.2
Greece	...	...	...	...	...	15.0	11.3	10.7	9.0	10.3	8.5	5.6	0.3	4.9
	...	...	...	...	...	0.7	0.6	0.5	0.4	0.5	0.4	0.3	...	0.2
Italy	0.2	0.2	37.5	26.3	39.0	97.0	87.4	88.0	98.7	108.3	110.4	108.0	99.9	90.8
	...	...	1.9	1.3	1.9	4.8	4.4	4.4	4.9	5.4	5.5	5.4	5.0	4.5
Netherlands	...	13.4	36.0	35.8	25.0	71.0	51.9	81.8	68.8	57.0	56.7	55.5	54.7	29.3
	...	0.7	1.8	1.8	1.2	3.5	2.6	4.1	3.4	2.8	2.8	2.8	2.7	1.5
Norway	...	...	...	...	528.0	1,574.0	2,263.6	2,579.5	2,782.0	3,092.0	3,148.8	3,021.0	3,017.6	3,212.7
	...	...	...	...	26.3	78.4	112.7	128.5	138.5	154.0	156.8	150.4	150.3	160.0
Spain	...	...	...	3.1	32.0	14.3	17.3	16.1	15.0	11.4	7.5	13.0	6.1	5.4
	...	...	...	0.2	1.6	0.7	0.9	0.8	0.7	0.6	0.4	0.6	0.3	0.3
Turkey	...	0.3	7.0	67.9	44.0	70.0	75.1	71.0	67.4	66.9	66.5	68.0	56.3	56.0
	...	...	0.3	3.4	2.2	3.5	3.7	3.5	3.4	3.3	3.3	3.4	2.8	2.8
United Kingdom	0.3	0.9	1.7	1.7	1,619.0	1,860.0	1,921.7	2,469.0	2,565.0	2,553.0	2,543.0	2,632.7	2,724.9	2,436.0
	...	...	0.1	0.1	80.6	92.6	95.7	123.0	127.7	127.1	126.6	131.1	135.7	121.3
Total	**29.7**	**68.4**	**274.5**	**382.3**	**2,448.0**	**3,983.3**	**4,735.0**	**5,639.3**	**5,919.7**	**6,225.4**	**6,283.7**	**6,254.8**	**6,364.5**	**6,310.3**
	1.5	**3.4**	**13.7**	**19.2**	**121.8**	**198.3**	**235.9**	**280.9**	**294.8**	**309.9**	**312.9**	**311.5**	**316.9**	**314.3**

AFRICA

Upper figures = 1,000 b/d
Lower figures = million t/y

	1940	1950	1960	1970	1980	1990	1993	1994	1995	1996	1997	1998	1999	2000
Algeria	...	...	182.7	1,029.1	1,016.0	797.0	747.3	750.0	764.0	818.0	849.2	824.2	745.0	808.0
	...	...	9.1	51.2	50.6	39.7	37.2	37.4	38.0	40.7	42.3	41.0	37.1	1.2
Angola (inc. Cabinda)	...	...	1.3	13.7	150.0	480.0	513.0	549.0	621.0	691.0	713.6	734.5	761.8	740.0
	...	...	0.1	0.7	7.5	23.9	25.5	27.3	30.9	34.4	35.5	36.6	37.9	36.8
Cameroon	...	...	...	...	58.0	164.0	117.5	110.0	102.1	110.0	124.3	125.0	100.0	85.0
	...	...	...	...	2.9	8.2	5.9	5.5	5.1	5.5	6.2	6.2	5.0	4.2
Congo (former Zaire)	...	...	...	...	20.0	28.0	22.7	25.0	28.0	29.3	28.3	26.2	23.7	25.0
	...	...	...	...	1.0	1.4	1.1	1.2	1.4	1.5	1.4	1.3	1.2	1.2
Congo (Brazzaville)	...	...	...	...	56.0	161.0	185.5	185.0	175.0	220.0	244.0	238.4	264.3	265.0
	...	...	...	...	2.8	8.0	9.2	9.2	8.7	11.0	12.2	11.9	13.2	13.2
Egypt	18.1	44.9	61.6	327.3	596.0	873.0	884.3	894.0	922.0	855.0	860.0	865.8	815.0	810.0
	0.9	2.2	3.1	16.3	29.7	43.5	44.0	44.5	45.9	42.6	42.8	43.1	40.6	40.3
Equatorial Guinea	...	...	...	...	...	...	4.0	5.0	7.0	22.0	51.7	83.0	94.0	167.5
	...	...	...	...	...	...	0.2	0.2	0.3	1.1	2.6	4.1	4.7	8.3
Gabon	...	...	16.1	108.8	180.0	269.0	312.3	329.0	364.0	366.5	365.2	355.0	340.0	325.0
	...	...	0.8	5.4	9.0	13.4	15.6	16.4	18.1	18.3	18.2	17.7	16.9	16.2
Ivory Coast	...	...	...	...	...	2.1	0.5	0.0	11.0	28.0	15.0	10.0	20.0	12.0
	...	...	...	...	...	0.1	...	...	0.5	1.4	0.7	0.5	1.0	0.6
Libya	...	...	...	3,318.0	1,785.0	1,369.0	1,361.0	1,380.0	1,390.0	1,393.0	1,420.4	1,391.7	1,346.7	1,410.0
	...	...	...	165.2	88.9	68.2	67.8	68.7	69.2	69.4	70.7	69.3	67.1	70.2
Morocco														
	...													
Nigeria	...	...	17.3	1,083.3	2,057.0	1,808.0	1,905.2	1,883.0	1,890.0	2,246.2	2,281.9	2,132.1	1,964.2	2,030.0
	...	...	0.9	53.9	102.4	90.0	94.9	93.8	94.1	111.9	113.6	106.2	97.8	101.0
South Africa	...	...	...	...	...	...	10.0	10.4	10.0	10.0	16.3	17.9	10.0	31.8
	...	...	...	...	...	...	0.5	0.5	0.5	0.5	0.8	0.9	0.5	1.6
Sudan														185.0
	...													9.2
Tunisia	...	...	3.0	87.7	110.0	93.0	96.6	91.0	87.7	87.6	78.4	80.5	81.6	75.9
	...	...	0.1	4.4	5.5	4.6	4.8	4.5	4.4	4.4	3.9	4.0	4.1	3.8
Total	**18.1**	**45.7**	**285.0**	**5,969.4**	**6,032.0**	**6,048.4**	**6,171.1**	**6,222.0**	**6,381.7**	**6,887.2**	**7,066.2**	**6,904.5**	**6,586.5**	**6,970.2**
	0.9	**2.2**	**14.2**	**297.2**	**300.5**	**301.2**	**307.3**	**309.7**	**317.8**	**343.0**	**351.9**	**343.8**	**328.1**	**346.8**

ASIA-PACIFIC

Upper figures = 1,000 b/d
Lower figures = million t/y

	1940	1950	1960	1970	1980	1990	1993	1994	1995	1996	1997	1998	1999	2000
Australia	...	...	...	175.6	379.0	582.0	495.7	535.0	506.2	535.3	566.6	534.9	526.8	700.0
	...	...	...	8.7	18.9	29.0	24.7	26.6	25.2	26.7	28.2	26.6	26.2	34.9
Bangladesh	...	...	...	...	...	0.9	1.0	1.1	1.1	1.6	1.7	2.0	2.8	3.3
	...	...	...	...	...	...	...	0.1	0.1	0.1	0.1	0.1	0.1	0.2
Brunei	19.3	84.8	92.9	148.0	230.0	143.0	157.8	162.0	151.5	155.5	146.9	141.3	165.7	177.9
	1.0	4.2	4.6	7.4	11.5	7.1	7.9	8.1	7.5	7.7	7.3	7.0	8.3	8.9
China	...	4.0	100.0	500.0	2,119.0	2,755.0	2,907.9	2,961.0	3,007.0	3,128.5	3,188.9	3,199.9	3,224.5	3,237.3
	...	0.2	5.0	24.9	105.5	137.2	144.8	147.5	149.7	155.8	158.8	159.4	160.6	161.2
China, Taiwan	...	...	0.4	1.8	5.0	2.5	1.1	1.1	1.0	0.9	0.9	0.9	0.8	0.6
	...	...	...	0.1	0.2	0.1	0.1	0.1	...	...	...	...	...	...
India	3.9	5.1	9.0	138.6	182.0	679.0	533.8	622.0	673.4	656.0	680.0	658.2	651.8	852.6
	0.2	0.3	0.4	6.9	9.1	33.8	26.6	31.0	33.5	32.7	33.9	32.8	32.5	42.5
Indonesia	169.4	132.6	411.2	853.6	1,576.0	1,274.0	1,327.3	1,319.0	1,498.2	1,392.0	1,364.2	1,315.4	1,280.0	1,267.3
	8.4	6.6	2.05	42.5	78.5	63.4	66.1	65.7	74.6	69.3	67.9	65.5	63.7	63.1
Japan	7.2	5.6	10.2	16.3	10.0	10.5	15.7	15.0	14.4	13.9	14.5	13.7	11.6	12.9
	0.4	0.3	0.5	0.8	0.5	0.5	0.8	0.7	0.7	0.7	0.7	0.7	0.6	0.6
Myanmar	21.1	1.5	11.1	16.4	30.0	13.0	14.0	16.0	16.0	11.0	11.0	15.0	15.0	8.0
	1.1	0.1	0.6	0.8	1.5	0.6	0.7	0.8	0.8	0.5	0.5	0.7	0.7	0.4
New Zealand	...	...	...	...	7.0	39.0	40.4	42.0	33.0	43.0	58.5	50.0	45.0	36.0
	...	...	...	...	0.3	1.9	2.0	2.1	1.6	2.1	2.9	2.5	2.2	1.8

Pakistan	2.4	3.5	7.2	9.9	10.0	60.0	61.6	57.0	54.2	58.0	50.9	53.6	53.4	56.6
	0.1	0.2	0.4	0.5	0.5	3.0	3.1	2.8	2.7	2.9	2.5	2.7	2.7	2.8
Papua New Guinea	...	4.8	4.2	...	...	...	127.2	120.6	99.6	115.1	75.9	81.1	88.6	64.4
	...	0.2	0.2	...	...	...	6.3	6.0	5.0	5.7	3.8	4.0	4.4	3.2
Philippines	...	...	...	...	15.0	5.0	9.1	5.0	2.9	1.4	0.8	0.8	0.9	1.1
	...	...	...	...	0.7	0.2	0.5	0.2	0.1	0.1	...	...	...	0.1
Thailand	...	...	...	0.4	...	41.0	52.2	54.4	50.9	61.2	77.0	75.0	81.9	104.6
	...	...	...	...	...	2.0	2.6	2.7	2.5	3.0	3.8	3.7	4.1	5.2
Viet Nam*	...	...	...	...	...	40.0	122.5	140.0	177.0	174.0	180.2	190.0	295.2	304.0
	...	...	...	...	...	2.0	6.1	7.0	8.8	8.7	9.0	9.5	14.7	15.1
Total	**223.3**	**241.9**	**646.2**	**1,860.6**	**4,851.0**	**6,249.9**	**6,512.3**	**6,691.2**	**6,936.4**	**6,993.2**	**7,170.2**	**7,051.8**	**7,167.3**	**6,826.6**
	11.2	**12.1**	**32.2**	**92.6**	**241.5**	**310.9**	**324.4**	**333.3**	**345.2**	**348.2**	**356.9**	**351.1**	**356.8**	**340.0**

*Prior to 1990, included in other Eastern Europe - FSU

LATIN AMERICA

Upper figures = 1,000 b/d
Lower figures = million t/y

	1940	1950	1960	1970	1980	1990	1993	1994	1995	1996	1997	1998	1999	2000
Argentina	56.3	64.0	171.9	382.9	487.0	473.0	588.0	666.0	720.7	783.2	834.6	846.9	801.8	790.8
	2.8	3.2	8.6	19.1	24.3	23.6	29.3	33.2	35.9	39.0	41.6	42.2	39.9	39.4
Bolivia	0.8	1.7	9.8	16.3	30.0	19.2	21.9	25.7	28.4	29.3	31.0	28.0	32.5	27.6
	...	0.1	0.5	0.8	1.5	1.0	1.1	1.3	1.4	1.5	1.5	1.4	1.6	2.7
Brazil	...	0.9	80.9	160.5	182.0	633.0	638.0	665.1	689.6	780.6	841.5	956.1	1,085.9	1,128.0
	...	...	4.0	8.0	9.1	31.5	31.8	33.1	34.3	38.9	41.9	47.6	54.1	56.2
Chile	...	1.7	19.8	34.6	29.0	20.3	14.2	12.3	10.0	9.3	8.8	9.0	9.0	7.0
	...	0.1	1.0	1.7	1.4	1.0	0.7	0.6	0.5	0.5	0.4	0.4	0.4	0.3
Colombia	69.9	93.3	152.7	214.0	125.0	445.0	455.0	456.0	574.8	623.2	652.2	754.3	815.6	714.4
	3.5	4.6	7.6	10.7	6.2	22.2	22.7	22.7	28.6	31.0	32.5	37.6	40.6	35.6
Cuba*	...	...	...	...	...	15.0	19.0	23.0	28.0	33.0	35.6	27.0	27.0	44.0
	...	...	...	...	...	0.7	0.9	1.1	1.4	1.6	1.8	1.3	1.3	2.2
Ecuador	6.4	7.2	7.7	4.1	222.0	287.0	340.1	379.1	386.7	383.7	388.2	377.9	375.7	401.1
	0.3	0.4	0.4	0.2	11.1	14.3	16.9	18.9	19.3	19.1	19.3	18.8	18.7	20.0
Guatemala	...	...	...	...	5.0	4.0	6.9	7.7	9.8	13.0	19.5	25.5	23.2	20.7
	...	...	...	...	0.2	0.2	0.3	0.4	0.5	0.6	1.0	1.3	1.2	1.0
Mexico	120.3	198.5	270.6	430.2	1,936.0	2,633.0	2,673.0	2,685.0	2,722.1	2,961.1	3,022.1	3,071.1	2,906.4	3,012.0
	6.0	9.9	13.5	21.4	96.4	131.1	133.1	133.7	135.6	147.5	150.5	152.9	144.7	150.0
Peru	33.2	41.1	52.6	72.2	191.0	132.0	126.3	128.0	124.0	119.8	105.7	113.5	102.5	99.2
	1.7	2.0	2.6	3.6	9.5	6.6	6.3	6.4	6.2	6.0	5.3	5.7	5.1	4.9
Trinidad & Tobogo	60.9	56.5	115.7	139.8	211.0	151.0	122.3	129.4	129.3	126.9	123.7	122.6	125.0	119.4
	3.0	2.8	5.8	7.0	10.5	7.5	6.1	6.4	6.4	6.3	6.2	6.1	6.2	5.9
Venezuela	507.0	1,498.0	2,846.1	3,708.0	2,167.0	2,118.0	2,335.0	2,463.0	2,609.0	2,938.0	3,181.8	3,121.7	2,786.7	2,897.0
	25.2	74.6	141.7	184.7	107.9	105.5	116.3	122.7	129.9	146.3	158.5	155.5	138.8	144.3
Total	**854.8**	**1,962.9**	**3,727.8**	**5,162.6**	**5,585.0**	**6,935.7**	**7,346.4**	**7,647.9**	**8,040.8**	**8,809.7**	**9,255.1**	**9,465.6**	**9,105.2**	**9,261.2**
	42.5	**97.7**	**185.7**	**257.2**	**278.1**	**345.5**	**365.8**	**380.9**	**400.4**	**438.7**	**461.0**	**471.4**	**453.3**	**462.5**
World Total, excl. E. Europe-FSU	**5,131.9**	**9,561.0**	**17,819.7**	**37,048.4**	**47,232.0**	**48,511.0**	**51,649.7**	**53,192.6**	**54,439.2**	**56,296.5**	**58,195.0**	**58,933.4**	**57,375.8**	**50,020.4**
	255.6	**476.2**	**887.5**	**1,870.0**	**2,352.0**	**2,415.6**	**2,572.3**	**2,649.1**	**2,713.1**	**2,803.6**	**2,897.7**	**2,934.7**	**2,857.1**	**2,492.1**

EASTERN EUROPE

Upper figures = 1,000 b/d
Lower figures = million t/y

	1940	1950	1960	1970	1980	1990	1993	1994	1995	1996	1997	1998	1999	2000
Albania														5.9
														0.3
Bulgaria	...	...	4.0	6.7	6.0	5.0	0.9	1.0	1.0	0.7	0.8	1.0	1.0	1.0
	...	...	0.2	0.3	0.3	0.2	...	...	...	...	...	...	...	...
Croatia	...	...	...	...	...	...	39.5	37.0	31.3	30.0	30.8	26.7	24.6	23.0
	...	...	...	...	...	...	2.0	1.8	1.6	1.5	1.5	1.3	1.2	1.1
Czech Republic &	...	0.8	2.7	4.1	1.9	3.0	3.4	8.0	3.0	9.0	9.0	9.0	8.9	N/A
Slovakia	...	...	0.1	0.2	0.1	0.1	0.2	0.4	0.1	0.4	0.4	0.4	0.4	N/A
Hungary	...	10.1	24.3	38.7	40.6	40.0	58.0	70.0	33.9	30.0	27.6	25.8	24.2	22.7
	...	0.5	1.2	1.9	2.0	2.0	2.9	3.5	1.7	1.5	1.4	1.3	1.2	1.1
Poland	...	3.3	3.9	8.5	6.6	3.0	4.6	13.0	6.0	6.6	5.0	4.8	4.2	14.1
	...	0.2	0.2	0.4	0.3	0.1	0.2	0.6	0.3	0.3	0.2	0.2	0.2	0.7
Romania	...	87.7	231.0	268.6	230.9	160.0	138.0	138.0	138.0	135.0	138.7	130.1	122.9	122.9
	...	4.4	11.5	13.4	11.5	8.0	6.9	6.9	6.9	6.7	6.9	6.5	6.1	6.1
Serbia[2]	0.3	2.1	18.8	57.8	84.5	65.0	22.8	20.0	20.0	21.0	24.0	18.0	18.0	18.0
	...	0.1	0.9	2.9	4.2	3.2	1.1	1.0	1.0	1.0	1.2	0.9	0.9	0.9
Total	**625.2**	**867.3**	**3,268.7**	**7,510.7**	**12,451.5**	**11,806.0**	**8,131.1**	**7,328.5**	**7,362.6**	**6,993.1**	**7,245.9**	**7,230.6**	**7,335.0**	**207.6**
	31.1	**43.2**	**162.7**	**373.9**	**620.0**	**587.8**	**404.9**	**364.9**	**366.7**	**348.0**	**360.7**	**360.0**	**365.1**	**10.2**

[2]Prior to 1993 listed as Yugoslavia.

FORMER SOVIET UNION

Upper figures = 1,000 b/d
Lower figures = million t/y

	1940	1950	1960	1970	1980	1990	1993	1994	1995	1996	1997	1998	1999	2000
Azerbaijan														275.0
														13.7
Belarus														35.0
														1.7
Georgia														2.0
														0.1
Kazakhstan														675.0
														33.6
Kyrgyzstan														1.0
														...
Russia														6,415.5
														319.5
Turkmenistan														140.0
														7.0
Ukraine														73.0
														3.6
Uzbekistan														155.0
														7.7
Total														**7,771.5**
														386.9
Total World														**57,999.5**
														2,889.2

WORLDWIDE OIL RESERVES AND PRODUCTION

COUNTRY	ESTIMATED PROVED RESERVES				OIL PRODUCTION			
	Jan. 1, 2002		Jan. 1, 2001					
	Oil (1,000 bbl)	Gas (bcf)	Oil (1,000 bbl)	Gas (bcf)	Producing oil wells* Dec. 31, 2000	Estimated 2001 (1,000 b/d)	Change from 2000 (%)	Actual 2000 (1,000 b/d)
ASIA-PACIFIC								
Afghanistan	—	3,530	—	3,530	—	—	—	—
Australia	3,500,000	90,000	2,895,000	44,638	1,392	640.0	–8.6	700.0
Bangladesh	56,902	10,615	56,902	10,615	37	3.3	0.6	3.3
Brunei	1,350,000	13,800	1,350,000	13,800	779	180.0	1.2	177.9
China	24,000,000	48,300	24,000,000	48,300	72,255	3,295.0	1.8	3,237.3
China, Taiwan	4,000	2,700	4,000	2,700	73	0.7	9.4	0.6
India	4,840,150	22,865	4,727,850	22,849	3,386	640.0	–0.9	646.0
Indonesia	5,000,000	92,500	4,979,710	72,268	8,535	1,200.0	–5.3	1,267.3
Japan	58,577	1,414	58,577	1,414	172	12.8	0.6	12.7
Malaysia	3,000,000	75,000	3,900,000	81,700	788	720.0	4.3	690.0
Myanmar	50,000	10,000	50,000	10,000	450	8.0	—	8.0
New Zealand	89,533	2,083	127,000	2,450	73	34.0	–5.6	36.0
Pakistan	298,237	25,078	208,000	21,600	225	59.0	10.6	53.3
Papua New Guinea	238,345	12,230	359,800	7,900	39	58.0	–16.8	69.7
Philippines	178,060	3,693	289,000	2,800	8	1.0	–12.3	1.1
Thailand	515,690	12,705	351,635	11,764	706	114.0	2.7	111.0
Viet Nam	600,000	6,800	600,000	6,800	28	305.0	0.3	304.0
Total Asia-Pacific	**43,779494**	**433,312**	**43,957,474**	**365,128**	**88,946**	**7,270.8**	**-0.6**	**7,318.2**
WESTERN EUROPE								
Austria	85,680	915	85,680	915	950	18.5	-3.3	19.1
Denmark	1,113,300	2,719	1,069,268	3,390	193	335.0	-7.7	363.0
France	140,040	403	145,170	506	407	27.5	-3.1	28.4
Germany	364,300	12,088	379,694	11,498	1,138	67.5	8.1	62.5
Greece	9,000	18	10,000	35	12	4.0	-21.8	5.1
Ireland	—	700	—	700	—	—	—	—
Italy	621,763	8,072	621,763	8,072	210	74.0	-18.5	90.8
Netherlands	106,927	62,542	106,927	62,542	196	27.0	-7.8	29.3
Norway	9,447,290	44,037	9,447,290	44,037	742	3,132.5	-2.5	3,212.0
Spain	21,009	18	21,009	18	18	7.0	53.1	4.6
Turkey	295,760	310	295,760	310	846	56.0	—	56.0
United Kingdom	4,930,000	25,956	5,002,795	26,839	1,430	2,260.0	–10.1	2,513.7
Total Western Europe	**17,135,069**	**157,778**	**17,185,356**	**158,863**	**6,142**	**6,009.0**	**-5.9**	**6,384.4**
EASTERN EUROPE and FSU								
Albania	165,000	100	165,000	100	2,275	6.0	1.7	5.9
Azerbaijan	1,178,000	4,400	1,178,000	4,400	2,102	300.0	9.1	275.0
Belarus	198,000	100	198,000	100	—	35.0	—	35.0
Bulgaria	15,000	210	15,000	210	100	1.0	—	1.0
Croatia	92,196	1,237	92,196	1,237	723	21.3	–7.4	23.0
Czech Republic	15,000	140	15,000	140	200	3.5	8.0	3.2
Georgia	35,000	300	35,000	300	—	2.0	0.0	2.0
Hungary	110,919	1,282	109,716	2,871	1,099	23.5	-9.0	25.8
Kazakhstan	5,417,000	65,000	5,417,000	65,000	11,601	787.0	16.6	675.0
Kyrgyzstan	40,000	200	40,000	200	—	1.0	—	1.0
Lithuania	12,000	—	12,000	—	—	0.0	—	0.0
Poland	114,883	5,119	114,883	5,119	1,773	17.0	18.9	14.3
Romania	955,620	3,556	1,426,140	13,200	6,000	120.0	–2.4	122.9
Russia	48,573,000	1,680,000	48,573,000	1,700,000	41,192	6,895.0	9.0	6,325.0
Serbia	77,500	1,700	77,500	1,700	646	17.0	–5.6	18.0
Slovakia	9,000	530	9,000	530	200	1.0	—	1.0
Tajikistan	12,000	200	12,000	200	—	0.0	—	0.0
Turkmenistan	546,000	101,000	546,000	101,000	2,460	146.5	4.6	140.0
Ukraine	395,000	39,600	395,000	39,600	1,353	73.9	1.2	73.0
Uzbekistan	594,000	66,200	594,000	66,200	2,190	143.0	–7.7	155.0
Total Eastern Europe and FSU	**58,555,118**	**1,970,875**	**59,024,435**	**2,002,107**	**73,914**	**8,593.7**	**8.8**	**7,896.2**
MIDDLE EAST								
Abu Dhabi	92,200,000	196,100	92,200,000	196,100	1,200	1,555.0	–18.2	1,900.0
Bahrain	124,560	3,249	148,110	3,875	496	102.5	0.2	102.3
Dubai	4,000,000	4,100	4,000,000	4,100	200	230.0	–17.9	280.0
Iran	89,700,000	812,300	89,700,000	812,300	1,120	3,130.0	–15.0	3,681.7
Iraq	112,500,000	109,800	112,500,000	109,800	1,685	1,960.0	–23.6	2,566.7
Israel	3,840	1,470	3,870	1,470	6	0.1	—	0.1
Jordan	890	230	890	230	4	—	—	—
Kuwait	94,000,000	52,200	94,000,000	52,200	790	1,440.0	–18.4	1,765.0

WORLDWIDE OIL RESERVES AND PRODUCTION-cont'd

COUNTRY	ESTIMATED PROVED RESERVES Jan. 1, 2002 Oil (1,000 bbl)	Jan. 1, 2002 Gas (bcf)	Jan. 1, 2001 Oil (1,000 bbl)	Jan. 1, 2001 Gas (bcf)	OIL PRODUCTION Producing oil wells* Dec. 31, 2000	Estimated 2001 (1,000 b/d)	Change from 2000 (%)	Actual 2000 (1,000 b/d)
Neutral Zone	5,000,000	1,000	5,000,000	1,000	578	540.0	−14.3	630.0
Oman	5,506,000	29,280	5,506,000	29,280	2,298	960.0	2.9	933.3
Qatar	15,207,000	508,540	13,157,000	393,830	417	574.0	−16.6	688.3
Ras al Khaimah	100,000	1,200	100,000	1,200	7	0.5	—	0.5
Saudi Arabia	259,250,000	219,000	259,200,000	213,300	1,560	6,470.0	−19.1	7,995.0
Sharjah	1,500,000	10,700	1,500,000	10,700	49	45.0	−10.0	50.0
Syria	2,500,000	8,500	2,500,000	8,500	132	515.0	−1.4	522.5
Yemen	4,000,000	16,900	4,000,000	16,900	302	350.0	−1.1	354.0
Total Middle East	**685,592,290**	**1,974,569**	**683,515,870**	**1,854,785**	**10,844**	**17,872.1**	**-16.8**	**21,469.4**
AFRICA								
Algeria	9,200,000	159,700	9,200,000	159,700	1,284	820.0	1.5	808.0
Angola	5,412,000	1,620	5,412,000	1,620	561	685.0	−7.4	740.0
Benin	8,210	43	8,210	43	8	1.0	—	1.0
Cameroon	400,000	3,900	400,000	3,900	175	75.0	−11.8	85.0
Congo (former Zaire)	187,000	35	187,000	35	150	24.0	−4.0	25.0
Congo (Brazzaville)	1,505,913	3,200	1,505,913	3,200	400	260.0	−1.9	265.0
Egypt	2,947,560	35,180	2,947,560	35,180	1,329	755.0	−6.8	810.0
Equatorial Guinea	12,000	1,300	12,000	1,300	38	170.0	1.5	167.5
Ethiopia	428	880	428	880	—	—	—	—
Gabon	2,499,000	1,200	2,499,000	1,200	375	300.0	−7.7	325.0
Ghana	16,510	840	16,510	840	3	6.0	—	6.0
Ivory Coast	100,000	1,050	100,000	1,050	7	10.0	−16.7	12.0
Libya	29,500,000	46,400	29,500,000	46,400	1,470	1,360.0	−3.5	1,410.0
Madagascar	—	—	—	70	—	—	—	—
Morocco	1,800	47	1,800	47	8	0.4	33.3	0.3
Mozambique	—	4,500	—	2,000	—	—	—	—
Namibia	—	2,200	—	3,000	—	—	—	—
Nigeria	24,000,000	124,000	22,500,000	124,000	2,374	2,100.0	3.4	2,030.0
Rwanda	—	2,000	—	2,000	—	—	—	—
Somalia	—	200	—	200	—	—	—	—
South Africa	15,680	1	29,362	780	22	22.0	−15.0	25.9
Sudan	563,000	3,000	262,100	3,000	9	200.0	8.1	185.0
Tanzania	—	800	—	980	—	—	—	—
Tunisia	307,560	2,750	307,560	2,750	210	66.0	−13.1	75.9
Total Africa	**76,676,661**	**394,846**	**74,889,443**	**394,175**	**8,423**	**6,854.4**	**−1.7**	**6,971.6**
WESTERN HEMISPHERE								
Argentina	2,973,700	27,460	3,071,195	26,420	14,238	760.0	1.2	750.7
Barbados	2,508	5	2,508	5	117	1.3	−19.3	1.5
Bolivia	440,500	24,000	396,500	18,300	328	30.0	8.7	27.6
Brazil	8,464,744	7,805	8,100,000	8,228	6,888	1,210.0	7.3	1,128.0
Canada	4,858,000	59,733	4,706,104	61,010	53,302	2,040.0	0.2	2,035.0
Chile	150,000	3,460	150,000	3,460	315	7.0	—	7.0
Colombia	1,750,000	4,322	1,972,000	6,937	2,492	625.0	−9.1	687.3
Cuba	750,000	2,500	283,500	636	245	44.0	—	44.0
Ecuador	2,115,000	3,670	2,115,000	3,670	1,041	400.0	−0.3	401.1
Guatemala	526,000	109	526,000	109	20	20.0	−3.6	20.7
Mexico	26,941,000	29,505	28,260,000	30,394	2,991	3,100.0	2.9	3,012.0
Peru	323,393	8,655	310,000	8,700	4,744	93.0	−5.1	98.0
Suriname	74,000	—	74,000	—	317	12.5	2.5	12.2
Trinidad and Tobago	716,000	23,450	686,000	21,351	3,975	112.0	−5.9	119.0
United States	22,045,000	177,427	21,765,000	167,406	539,754	5,810.0	−0.2	5,822.0
Venezuela	77,685,000	147,585	76,862,000	146,800	15,277	2,830.0	−6.5	3,028.0
Total Western Hemisphere	**149,814,845**	**519,685**	**149,279,807**	**503,426**	**646,044**	**17,094.8**	**−0.6**	**17,194.1**
TOTAL WORLD	**1,031,553,477**	**5,451,065**	**1,027,852,385**	**5,278,484**	**834,313**	**63,694.8**	**−5.3**	**67,233.9**
Total OPEC	**818,842,000**	**2,485,125**	**814,398,710**	**2,343,698**	**36,546**	**24,255**	**−13.7**	**28,100**

EDITOR'S NOTE: All reserves figures except those for the FSU (and gas for Canada) are reported as proved reserves recoverable with present technology and prices. FSU figures are "explored reserves" which are understood to be proved plus some probable. Canadian gas figure, under criteria adopted by Canadian Petroleum Association in 1980, includes proved and some probable. *Does not include shut in, injection, or service wells.

HISTORICAL OIL RESERVES

Jan. 1	1970	1980	1990	1991	1992	1993	1994	1995	1996	1997	1998	1999	2000	2001
	billion bbl crude oil													
Asia-Pacific														
Australia	2.0	2.4	1.7	1.6	1.5	1.8	1.6	1.6	1.6	1.8	1.8	2.9	2.9	3.5
Brunei	1.0	1.7	1.4	1.4	1.4	1.4	1.4	1.4	1.4	1.4	1.4	1.4	1.4	1.4
India	—	2.5	7.5	7.9	6.1	6.0	5.9	5.8	5.8	4.3	4.3	3.9	4.8	4.8
Indonesia	10.0	9.5	8.2	11.0	6.6	5.8	5.8	5.8	5.8	4.9	4.9	4.9	4.9	5.0
Malaysia	—	3.0	3.0	2.9	3.0	3.7	4.3	4.3	4.3	4.0	3.9	3.9	3.9	3.0
Europe														
Norway	1.0	5.5	11.5	7.6	7.6	8.8	9.3	9.4	8.4	11.2	10.4	10.9	10.8	9.4
UK	1.0	14.8	4.2	3.8	3.9	4.1	4.6	4.5	4.3	4.5	5.0	5.2	5.2	4.9
Middle East														
Abu Dhabi	11.8	29.0	92.2	92.2	92.2	92.2	92.2	92.2	92.2	92.2	92.2	92.2	92.2	92.2
Dubai	—	1.4	4.0	4.0	4.0	4.0	4.0	4.3	4.3	4.0	4.0	4.0	4.0	4.0
Iran	70.0	57.5	92.8	92.9	92.9	92.9	92.9	89.3	88.2	93.0	93.0	89.7	89.7	89.7
Iraq	32.0	30.0	100.0	100.0	100.0	100.0	100.0	100.0	100.0	112.0	112.5	112.5	112.5	112.5
Kuwait	67.1	64.9	94.5	94.5	94.0	94.0	94.0	94.0	94.0	94.0	94.0	94.0	94.0	94.0
Neutral Zone	25.7	6.0	5.2	5.0	5.0	5.0	5.0	5.0	5.0	5.0	5.0	5.0	5.0	5.0
Oman	—	2.3	4.2	4.3	4.3	4.5	4.7	4.8	5.1	5.1	5.2	5.3	5.3	5.5
Qatar	3.5	3.6	4.5	4.5	3.7	3.7	3.7	3.7	3.7	3.7	3.7	3.7	3.7	15.2
Saudi Arabia	128.5	165.0	254.9	257.5	258.7	257.8	258.7	258.7	258.7	259.0	259.0	259.0	261.0	259.2
Syria	1.0	1.3	1.7	1.7	1.7	1.7	1.7	2.5	2.5	2.5	2.5	2.5	2.5	2.5
Yemen	—	—	3.0	4.0	4.0	4.0	4.0	4.0	4.0	4.0	4.0	4.0	4.0	4.0
Africa														
Algeria	7.0	8.2	9.2	9.2	9.2	9.2	9.2	9.2	9.2	9.2	9.2	9.2	9.2	9.2
Angola-Cabinda	—	1.2	2.0	2.1	1.8	1.5	1.5	5.4	5.4	5.4	5.4	5.4	5.4	5.4
Egypt	4.5	2.9	4.5	4.5	4.5	6.2	6.3	3.3	3.9	3.7	3.8	3.5	2.9	2.9
Gabon	—	—	0.7	0.7	0.7	0.7	0.7	1.3	1.3	1.3	2.5	2.5	2.5	2.5
Libya	29.2	23.0	22.8	22.8	22.8	22.8	22.8	22.8	29.5	29.5	29.5	29.5	29.5	29.5
Nigeria	9.3	16.7	16.0	17.1	17.9	17.9	17.9	17.9	20.8	15.5	16.8	22.5	22.5	24.0
Tunisia	—	1.7	1.8	1.7	1.7	1.7	1.7	0.4	0.4	0.3	0.3	0.3	0.3	0.3
Western Hemisphere														
Argentina	4.5	2.5	2.3	2.3	1.6	1.6	1.6	2.2	2.2	2.4	2.6	2.6	2.8	3.0
Brazil	—	1.3	2.8	2.8	2.8	3.0	3.6	3.8	4.2	4.8	4.8	7.1	7.4	8.5
Canada	10.7	6.4	6.1	5.8	5.6	5.3	5.1	5.0	4.9	4.9	4.8	4.9	4.9	4.9
Colombia	1.7	1.0	2.0	2.0	1.9	1.9	1.9	3.4	3.5	2.8	2.8	2.6	2.6	1.8
Ecuador	—	1.1	1.5	1.4	1.6	1.6	2.0	2.0	2.1	2.1	2.1	2.1	2.1	2.1
Mexico	3.2	44.0	56.4	51.9	51.3	51.3	50.9	50.8	49.8	48.8	40.0	47.8	28.4	26.9
Venezuela	14.0	17.9	58.5	59.0	59.1	62.7	63.3	64.5	64.5	64.9	71.7	72.6	72.6	77.7
US	37.0	26.4	25.9	26.2	26.3	24.7	23.7	22.9	22.5	22.4	22.0	22.5	21.0	22.0
Subtotal[1]	**511.4**	**562.2**	**917.5**	**916.3**	**932.3**	**937.9**	**939.8**	**940.6**	**948.4**	**959.9**	**960.4**	**975.5**	**956.9**	**948.6**
FSU[2]	77.0	63.0	24.0	57.0	57.0	57.0	57.0	57.0	57.0	59.1	59.1	59.1	59.0	59.0
China	—	—	24.0	24.0	24.0	24.0	24.0	24.0	24.0	24.0	24.0	24.0	24.0	24.0
Total world[1]	**611.4**	**648.5**	**1,001.6**	**999.1**	**991.0**	**997.0**	**999.1**	**999.8**	**1,007.5**	**1,018.8**	**1,019.5**	**1,034.3**	**1,016.0**	**1,031.6**

[1]Columns do not add to totals because countries with lesser reserves are omitted. [2]USSR before 1992.

WORLD OIL BALANCE

	1997	1998	1999	1Q00	2Q00	3Q00	4Q00	2000	1Q00	2Q00	3Q00	4Q00	2001
	million b/d												
DEMAND													
OECD*													
North America	22.7	23.1	24.3	23.7	23.8	24.5	24.4	24.1	24.3	24.0	24.6	24.7	24.4
Europe	15.0	15.3	15.6	15.2	14.6	15.2	15.4	15.1	15.1	14.7	15.2	15.7	15.2
Pacific	9.0	8.4	9.1	9.4	8.1	8.3	8.8	8.7	9.4	7.9	8.3	9.0	8.7
TOTAL OECD	**46.7**	**46.8**	**49**	**48.2**	**46.6**	**48.0**	**48.6**	**47.8**	**48.8**	**46.6**	**48.1**	**49.3**	**48.2**
NON-OECD													
Former USSR	3.8	3.7	3.7	3.5	3.4	3.5	3.7	3.5	3.6	3.5	3.4	3.6	3.5
Europe	0.8	0.8	0.7	0.8	0.7	0.7	0.8	0.8	0.8	0.8	0.7	0.8	0.8
China	4.2	4.2	4.6	4.7	4.5	5.1	4.8	4.8	4.7	5.0	5.0	5.3	5.0
Asia	6.7	6.8	7.1	7.1	7.3	7.2	7.1	7.2	7.2	7.3	7.3	7.3	7.3
Latin America	4.7	4.8	4.8	4.7	4.8	4.9	4.8	4.8	4.7	4.9	5.0	4.9	4.9
Middle East	4.0	4.2	4.1	4.3	4.4	4.5	4.3	4.4	4.4	4.6	4.7	4.5	4.5
Africa	2.3	2.3	2.4	2.4	2.3	2.3	2.4	2.3	2.4	2.4	2.4	2.4	2.4
TOTAL NON-OECD	**26.5**	**26.7**	**27.4**	**27.5**	**27.5**	**28.1**	**27.8**	**27.7**	**27.9**	**28.3**	**28.5**	**28.7**	**28.3**
TOTAL DEMAND	**73.1**	**73.5**	**76.4**	**75.6**	**74.1**	**76.1**	**76.5**	**75.6**	**76.7**	**74.8**	**76.6**	**78.1**	**76.5**
SUPPLY													
OECD*													
North America	14.6	14.5	14.1	14.3	14.4	14.3	14.1	14.3	14.2	14.3	14.4	14.6	14.4
Europe	6.7	6.7	7.1	7.1	6.6	6.6	6.9	6.8	6.8	6.4	6.4	6.7	6.6
Pacific	0.7	0.7	0.7	0.9	0.9	0.8	0.8	0.8	0.8	0.8	0.8	0.8	0.8
TOTAL OECD	**22.1**	**21.9**	**22**	**22.3**	**21.8**	**21.7**	**21.8**	**21.9**	**21.8**	**21.4**	**21.5**	**22.1**	**21.7**
NON-OECD													
Former USSR	7.2	7.3	7.6	7.7	7.8	8.0	8.2	7.9	8.3	8.4	8.5	8.5	8.4
Europe	0.2	0.2	0.2	0.2	0.2	0.2	0.2	0.2	0.2	0.2	0.2	0.2	0.2
China	3.2	3.2	3.2	3.3	3.2	3.2	3.2	3.2	3.3	3.3	3.3	3.3	3.3
Asia	2.2	2.2	2.2	2.2	2.3	2.3	2.3	2.3	2.3	2.3	2.3	2.3	2.3
Latin America	3.4	3.6	3.8	3.7	3.7	3.8	3.9	3.8	3.8	3.8	3.8	3.8	3.8
Middle East	1.9	1.9	1.9	1.9	2.0	2.0	2.0	2.0	2.0	2.0	2.0	2.0	2.0
Africa	2.7	2.8	2.9	2.9	2.9	2.8	2.9	2.9	2.9	2.9	2.9	3.0	2.9
Processing Gains	1.6	1.6	1.7	1.7	1.7	1.7	1.7	1.7	1.8	1.7	1.7	1.8	1.8
TOTAL NON-OPEC	**44.5**	**44.7**	**45.5**	**45.9**	**45.5**	**45.8**	**46.3**	**45.9**	**46.3**	**46.1**	**46.4**	**47.1**	**46.5**
OPEC													
Crude	27.2	28	26.1	26.5	27.8	28.4	29.0	27.9	28.4	27.6	26.8	26.5	27.3
NGLs	2.7	2.8	2.8	2.8	2.9	2.9	2.9	2.9	2.9	3.0	3.0	3.1	3.0
TOTAL OPEC	**29.9**	**30.8**	**29**	**29.3**	**30.7**	**31.3**	**31.9**	**30.8**	**31.3**	**30.6**	**29.8**	**29.6**	**30.3**
TOTAL SUPPLY	**74.4**	**75.5**	**74.4**	**75.2**	**76.2**	**77.1**	**78.2**	**76.7**	**77.6**	**76.7**	**76.2**	**76.7**	**76.8**
STOCK CHANGE	**1.2**	**1.9**	**–2.0**	**-0.4**	**2.1**	**1.0**	**1.7**	**1.1**	**0.9**	**1.9**	**-0.4**	**-1.4**	**0.3**

*Starting in 1995 OECD includes Czech Republic, Korea, Mexico and Poland.

Source: International Energy Agency and Oil & Gas Journal

SUMMARY OF OPERATING REFINERIES WORLDWIDE

(capacities as of January 1, 2002)

Country	No. plants	Crude capacity (b/cd)	Catalytic cracking (b/cd)	Catalytic reforming (b/cd)
Albania	2	26,300	....	3,500
Algeria	4	450,000	....	88,900
Angola	1	39,000	....	1,900
Argentina	10	639,075	168,610	58,800
Aruba	1	280,000	....	....
Australia	10	848,250	234,060	198,560
Austria	1	208,600	26,250	32,725
Azerbaijan	2	441,808	57,750	24,466
Bahrain	1	248,900	41,400	15,300
Bangladesh	1	33,000	....	1,800
Belarus	2	493,323	....	92,000
Belgium	5	791,013	113,500	98,300
Bolivia	3	63,000	....	14,560
Brazil	13	1,785,841	424,329	24,058
Brunei	1	8,600	....	5,700
Bulgaria	1	115,240	23,300	4,060
Cameroon	1	42,000	....	7,000
Canada	21	1,943,900	488,130	349,850
Chile	3	204,849	45,757	16,497
China	95	4,528,100	892,000	157,000
China, Taiwan	4	920,000	123,000	115,000
Colombia	5	285,850	90,000	....
Congo (Zaire)	1	15,000	....	3,500
Congo (Brazzaville)	1	21,000	....	2,000
Costa Rica	1	15,000	....	1,200
Croatia	3	292,990	51,005	50,120
Cuba	4	301,400	14,700	20,000
Cyprus	1	27,000	....	5,100
Czech Republizc	4	198,000	26,000	28,140
Denmark	2	176,400	....	21,500
Dominican Republic	2	49,250	....	7,560
Ecuador	3	176,000	18,000	12,800
Egypt	9	726,250	....	40,540
El Salvador	1	22,000	....	3,200
Eritrea	1	14,564	....	1,465
Finland	2	238,800	51,000	44,430
France	13	1,896,493	372,510	270,850
Gabon	1	17,300	....	1,400
Georgia	1	106,436	....	10,276
Germany	17	2,258,800	343,665	385,627
Ghana	1	45,000	....	5,850
Greece	4	406,500	72,450	53,400
Guatemala	1	16,000	....	3,000
Hungary	2	161,000	24,000	29,600
India	17	2,134,625	167,305	43,074
Indonesia	8	992,745	101,450	92,970
Iran	9	1,484,000	30,000	160,570
Iraq	8	417,500	....	43,500
Ireland	1	71,250	....	10,800
Israel	2	220,000	49,500	26,500
Italy	17	2,282,800	305,000	269,250
Ivory Coast	1	65,200	....	12,827
Jamaica	1	34,200	....	3,240
Japan	35	4,786,340	819,350	742,840
Jordan	1	90,400	4,000	10,900
Kazakhstan	3	427,093	38,356	59,452
Kyrgyzstan	1	10,000	....	....
Kenya	1	90,000	....	9,300
Kuwait	3	773,300	41,400	13,500
Lebanon	2	37,500	7,250	7,442
Liberia	1	15,000	....	2,000
Libya	3	343,400	....	20,250
Lithuania	1	263,420	43,692	25,741
Macedonia	1	51,180	....	10,780
Madagascar	1	15,000	....	2,000
Malaysia	6	514,500	....	77,400
Martinique	1	17,000	....	3,019
Mexico	6	1,525,000	368,000	226,000
Morocco	2	154,901	5,040	24,359
Myanmar	2	32,000	....	....
Netherlands	6	1,205,842	101,400	170,923
Netherlands Antilles	1	320,000	50,000	20,000
New Zealand	1	106,000	....	27,672
Nicaragua	1	20,000	....	2,700
Nigeria	4	438,750	82,700	70,070
North Korea	2	71,000	....	7,300
Norway	2	310,000	54,000	37,600
Oman	1	85,000	....	16,000
Pakistan	4	238,850	....	11,500
Panama	1	60,000	....	11,000
Paraguay	1	7,500	....	....
Peru	5	182,250	23,300	....
Philippines	4	419,500	24,500	61,900
Poland	4	382,000	46,000	39,000
Portugal	2	304,172	31,500	50,182
Puerto Rico	1	....	....	....
Qatar	1	57,500	....	11,500
Romania	10	504,037	103,478	63,063
Russia	42	5,435,480	330,817	775,201
Saudi Arabia	8	1,745,000	103,600	193,360
Senegal	1	27,000	....	2,000
Serbia	2	158,246	18,950	20,153
Sierra Leone	1	10,000	....	....
Singapore	3	1,258,500	65,000	146,500
Slovakia	1	115,000	....	22,270
Slovenia	1	13,500	....	....
Somalia	1	10,000	....	....
South Africa	4	468,547	107,260	78,892
South Korea	6	2,560,100	168,000	230,970
Spain	9	1,293,500	179,650	195,110
Sri Lanka	1	50,000	....	5,450
Sudan	3	121,700	....	1,900
Suriname	1	7,000	....	....
Sweden	5	423,500	29,700	69,900
Switzerland	2	132,000	....	28,000
Syria	2	242,140	....	25,812
Tanzania	1	14,900	....	2,500
Thailand	4	681,750	78,500	82,350
Trinidad	1	160,000	26,600	19,100
Tunisia	1	34,000	....	3,300
Turkey	6	719,275	28,935	64,762
Turkmenistan	2	236,970	15,151	52,540
Ukraine	6	1,026,259	69,000	100,311
United Arab Emirates	5	514,750	34,350	25,875
United Kingdom	11	1,784,000	432,050	329,200
United States	143	16,564,483	5,608,830	3,497,944
Uruguay	1	37,000	9,000	3,000
Uzbekistan	3	222,271	....	23,487
Venezuela	5	1,282,100	231,800	49,500
Virgin Islands	1	525,000	130,000	125,000
Yemen	2	130,000	....	14,500
Zambia	1	23,750	....	5,320
Total	**732**	**81,166,308**	**13,865,830**	**11,003,865**

NORTH AMERICA

Upper figures = 1,000 b/d
Lower figures = million t/y

	1950	1960	1970	1980	1990	1994	1995	1996	1997	1998	1999	2000	2001	2002
Canada	329	920	1,400	2,222	1,852	1,880	1,908	1,849	1,852	1,851	1,873	1,912	1,906	1,944
	16.4	45.8	69.7	110.7	92.2	93.6	95.0	92.1	92.2	92.2	93.3	95.2	94.9	96.8
US	6,696	10,250	12,600	17,720	16,244	15,142	15,319	15,354	15,432	15,898	16,423	16,541	16,539	16,564
	333.5	510.5	627.5	882.5	809.0	754.1	762.9	764.6	768.5	791.7	817.9	823.7	823.6	824.9
Total	**7,025**	**11,170**	**14,000**	**19,942**	**18,096**	**17,022**	**17,227**	**17,203**	**17,284**	**17,749**	**18,296**	**18,453**	**18,445**	**18,508**
	349.9	**556.3**	**697.2**	**993.2**	**901.2**	**847.7**	**857.9**	**856.7**	**860.7**	**883.9**	**911.2**	**918.9**	**918.5**	**921.7**

WESTERN EUROPE

Upper figures = 1,000 b/d
Lower figures = million t/y

	1950	1960	1970	1980	1990	1994	1995	1996	1997	1998	1999	2000	2001	2002
Austria	26	47	94	280	204	210	210	210	210	210	209	209	209	209
	1.3	2.3	4.7	13.9	10.2	10.5	10.5	10.5	10.5	10.5	10.4	10.4	10.4	10.4
Belgium	16	171	669	1,064	614	607	614	610	630	630	715	719	768	791
	0.8	8.5	33.3	53.0	30.6	30.2	30.6	30.4	31.4	31.4	35.6	35.8	38.2	39.4
Cyprus	...	...	...	16	18	22	22	22	26	27	27	27	27	27
	...	...	...	0.8	0.9	1.1	1.1	1.1	1.3	1.3	1.3	1.3	1.3	1.3
Denmark	1	1	186	214	187	184	188	189	189	132	135	135	176	176
	...	...	9.3	10.7	9.3	9.2	9.4	9.4	9.4	6.6	6.7	6.7	8.8	8.8
Finland	...	25	174	336	241	200	200	200	200	200	200	200	200	239
	...	1.2	8.7	16.7	12.0	10.0	10.0	10.0	10.0	10.0	10.0	10.0	10.0	11.9
France	306	769	2,317	3,385	1,820	1,861	1,768	1,782	1,786	1,865	1,947	1,902	1,895	1,896
	15.2	38.3	115.4	168.6	90.6	92.7	88.0	88.7	88.9	92.9	97.0	94.7	94.4	94.4
Germany	81	573	2,359	2,986	1,507	2,265	2,317	2,126	2,108	2,184	2,246	2,275	2,259	2,259
	4.0	28.5	117.5	148.7	75.0	112.8	115.4	105.9	105.0	108.8	111.9	113.3	112.5	112.5
Greece	...	30	102	431	385	396	401	401	395	391	387	383	407	407
	...	1.5	5.1	21.5	19.2	19.7	20.0	20.0	19.7	19.5	19.3	19.1	20.2	20.3
Ireland	...	40	55	56	56	53	53	55	65	62	66	71	71	71
	...	2.0	2.7	2.8	2.8	2.6	2.6	2.7	3.2	3.1	3.3	3.5	3.5	3.5
Italy	107	773	2,963	4,131	2,804	2,262	2,260	2,285	2,262	2,453	2,446	2,341	2,359	2,283
	5.3	38.5	147.6	205.7	139.6	112.6	112.5	113.8	112.6	122.2	121.8	116.6	117.5	113.7
Netherlands	77	359	1,362	1,828	1,381	1,184	1,187	1,187	1,187	1,188	1,188	1,188	1,204	1,206
	3.8	17.9	67.8	91.0	68.8	59.0	59.1	59.1	59.1	59.2	59.2	59.2	60.0	60.1
Norway	1	2	123	264	295	287	285	307	307	307	312	358	305	310
	...	0.1	6.1	13.1	14.7	14.3	14.2	15.3	15.3	15.3	15.5	17.8	15.2	15.4
Portugal	8	23	37	378	313	294	304	304	304	304	304	304	304	304
	0.4	1.1	1.8	18.8	15.6	14.6	15.1	15.1	15.1	15.1	15.1	15.1	15.1	15.1
Spain	17	146	685	1,455	1,293	1,283	1,283	1,327	1,296	1,294	1,315	1,316	1,294	1,294
	0.8	7.3	34.1	72.5	64.4	63.9	63.9	66.1	64.5	64.4	65.5	65.5	64.4	64.4
Sweden	26	48	230	458	428	428	428	436	427	427	427	427	423	424
	1.3	2.4	11.5	22.8	21.3	21.3	21.3	21.7	21.3	21.3	21.3	21.3	21.1	21.1
Switzerland	...	...	106	137	132	132	132	132	132	132	132	132	132	132
	...	...	5.3	6.8	6.6	6.6	6.6	6.6	6.6	6.6	6.6	6.6	6.6	6.6
Turkey	1	7	149	356	725	713	713	713	683	708	688	691	694	719
	...	0.3	7.4	17.7	36.1	35.5	35.5	35.5	34.0	35.3	34.3	34.4	34.6	35.8
United Kingdom	197	963	2,301	2,527	1,831	1,867	1,869	1,888	1,941	1,826	1,854	1,785	1,771	1,784
	9.8	48.0	114.6	125.8	91.2	93.0	93.1	94.0	96.7	90.9	92.3	88.9	88.2	88.8
Total	**864**	**3,977**	**13,912**	**20,302**	**14,234**	**14,248**	**14,234**	**14,174**	**14,148**	**14,340**	**14,598**	**14,463**	**14,498**	**14,531**
	42.7	**197.9**	**692.9**	**1,010.9**	**708.9**	**709.6**	**708.9**	**705.9**	**704.6**	**714.4**	**727.1**	**720.2**	**722.0**	**723.5**

WORLDWIDE REFINING CAPACITY-cont'd

LATIN AMERICA

Upper figures = 1,000 b/d
Lower figures = million t/y

	1950	1960	1970	1980	1990	1994	1995	1996	1997	1998	1999	2000	2001	2002
Argentina	152	238	457	676	689	709	665	661	665	666	653	662	639	639
	7.6	11.9	22.8	33.7	34.3	35.3	33.1	32.9	33.1	33.2	32.5	33.0	31.8	31.8
Aruba	...	...	...	...	...	...	...	...	...	...	205	225	280	280
	...	...	...	...	...	...	...	...	...	...	10.2	11.2	13.9	13.9
Bolivia	7	12	12	74	58	45	45	45	48	48	48	48	63	63
	0.3	0.6	0.6	3.7	2.9	2.2	2.2	2.2	2.4	2.4	2.4	2.4	3.1	3.1
Brazil	12	156	502	1,205	1,397	1,253	1,253	1,256	1,256	1,662	1,772	1,783	1,918	1,786
	0.6	7.8	25.0	60.0	69.6	62.4	62.4	62.5	62.5	82.8	88.2	88.8	95.5	88.9
Chile	4	24	91	139	147	165	165	178	192	205	205	205	205	205
	0.2	1.2	4.5	6.9	7.3	8.2	8.2	8.9	9.6	10.2	10.2	10.2	10.2	10.2
Colombia	24	51	137	193	227	249	249	249	249	249	249	286	286	286
	1.2	2.5	6.8	9.6	11.3	12.4	12.4	12.4	12.4	12.4	12.4	14.2	14.2	14.2
Costa Rica	...	...	8	11	16	15	15	15	15	15	15	15	15	15
	...	...	0.4	0.5	0.8	0.7	0.7	0.7	0.7	0.7	0.7	0.7	0.7	0.7
Cuba	7	87	93	160	160	280	280	301	301	301	301	301	301	301
	0.3	4.3	4.6	8.0	8.0	13.9	13.9	15.0	15.0	15.0	15.0	15.0	15.0	15.0
Ecuador	5	7	33	86	145	148	148	148	148	168	176	176	176	176
	0.2	0.3	1.6	4.3	7.2	7.4	7.4	7.4	7.4	8.4	8.8	8.8	8.8	8.8
El Salvador	...	...	14	16	17	18	19	19	20	20	20	20	21	22
	...	...	0.7	0.8	0.8	0.9	0.9	0.9	1.0	1.0	1.0	1.0	1.0	1.1
Guatemala	...	...	21	16	16	20	20	20	20	20	20	20	22	16
	...	...	1.0	0.8	0.8	1.0	1.0	1.0	1.0	1.0	1.0	1.0	1.1	0.8
Jamaica	...	...	32	34	34	36	36	36	36	36	36	34	34	34
	...	...	1.6	1.7	1.7	1.8	1.8	1.8	1.8	1.8	1.8	1.7	1.7	1.7
Mexico	160	357	495	1,394	1,514	1,524	1,524	1,520	1,520	1,520	1,525	1,525	1,525	1,525
	8.0	17.8	24.7	69.4	75.4	75.9	75.9	75.7	75.7	75.7	75.9	75.9	75.9	75.9
Netherlands Antilles	617	680	840	792	320	477	490	490	490	525	250	320	320	320
	30.7	33.9	41.8	39.4	15.9	23.8	24.4	24.4	24.4	26.1	12.5	15.9	15.9	15.9
Nicaragua	...	...	22	14	15	16	17	17	17	17	18	21	20	20
	...	...	1.1	0.7	0.7	0.8	0.8	0.8	0.8	0.8	0.9	1.0	1.0	1.0
Panama	...	...	78	100	100	100	40	60	60	60	60	60	60	60
	...	...	3.9	5.0	5.0	5.0	2.0	3.0	3.0	3.0	3.0	3.0	3.0	3.0
Paraguay	...	...	5	7	8	8	8	8	8	8	8	8	8	8
	...	...	0.2	0.3	0.4	0.4	0.4	0.4	0.4	0.4	0.4	0.4	0.4	0.4
Peru	35	49	92	170	172	184	184	182	182	182	182	182	182	182
	1.7	2.4	4.6	8.5	8.6	9.2	9.2	9.1	9.1	9.1	9.1	9.1	9.1	9.1
Puerto Rico	...	84	155	284	123	127	127	127	134	134	49	49	49	...
	...	4.2	7.7	14.1	6.1	6.3	6.3	6.3	6.7	6.7	2.4	2.4	2.4	...
Trinidad & Tobago	104	182	430	456	300	305	245	245	245	245	160	160	160	160
	5.2	9.1	21.4	22.7	14.9	15.2	12.2	12.2	12.2	12.2	8.0	8.0	8.0	8.0
Uruguay	16	28	40	40	33	35	35	37	40	37	37	37	37	37
	0.8	1.4	2.0	2.0	1.6	1.7	1.7	1.8	2.0	1.8	1.8	1.8	1.8	1.8
Venezuela	254	886	1,313	1,446	1,201	1,167	1,167	1,177	1,177	1,177	1,187	1,239	1,282	1,282
	12.6	44.1	65.4	72.0	59.8	58.1	58.1	58.6	58.6	58.6	59.1	61.7	63.8	63.8
Virgin Islands	...	...	220	728	545	545	545	545	545	545	545	525	525	525
	...	...	11.0	36.3	27.1	27.1	27.1	27.1	27.1	27.1	27.1	26.1	26.1	26.1
Total	**1,397**	**2,841**	**5,114**	**8,620**	**7,315**	**7,507**	**7,358**	**7,420**	**7,452**	**7,916**	**7,795**	**7,973**	**8,201.6**	**7,942**
	69.4	**141.5**	**254.5**	**429.2**	**364.0**	**373.7**	**366.1**	**369.3**	**371.1**	**394.2**	**388.1**	**396.9**	**408.0**	**395.2**

WORLDWIDE REFINING CAPACITY-cont'd

ASIA-PACIFIC

Upper figures = 1,000 b/d
Lower figures = million t/y

	1950	1960	1970	1980	1990	1994	1995	1996	1997	1998	1999	2000	2001	2002
Australia	14	237	594	725	675	705	705	733	771	762	807	812	847	848
	0.7	11.8	29.6	36.1	33.6	35.1	35.1	36.5	38.4	37.9	40.2	40.4	42.2	42.2
Bangladesh	...	...	...	31	31	31	31	31	31	31	33	33	33	33
	...	...	...	1.5	1.5	1.5	1.5	1.5	1.5	1.5	1.6	1.6	1.6	1.6
Brunei-Malaysia-Singapore	10	46	320	1,093	1,049	1,333	1,386	1,498	1,496	1,536	1,654	1,788	1,792.2	1,782
	0.5	2.3	15.9	54.4	52.2	66.4	69.0	74.6	74.5	76.5	82.4	89.0	89.3	88.7
China	8	98	300	1,600	2,200	2,200	2,867	2,867	2,867	2,967	4,347	4,347	4,347	4,528
	0.4	4.9	14.9	79.7	109.6	109.6	142.8	142.8	142.8	147.8	216.5	216.5	216.5	225.5
China, Taiwan	...	28	119	425	570	543	543	543	770	770	770	770	920	920
	...	1.4	5.9	21.2	28.4	27.0	27.0	27.0	38.3	38.3	38.3	38.3	45.8	45.8
India	6	112	446	557	1,080	1,086	1,086	1,086	1,086	1,086	1,141	1,858	2,113	2,135
	0.3	5.6	22.2	27.7	53.8	54.1	54.1	54.1	54.1	54.1	56.8	92.5	105.2	106.3
Indonesia	203	274	268	528	714	860	805	805	805	930	930	993	993	993
	10.1	13.6	13.3	26.3	35.6	42.8	40.1	40.1	40.1	46.3	46.3	49.5	49.4	49.5
Japan	34	640	3,140	5,509	4,198	4,810	4,847	4,867	4,989	4,966	5,059	4,998	4,962	4,786
	1.7	31.9	156.4	274.3	209.1	239.5	241.4	242.4	248.5	247.3	251.9	248.9	247.1	238.3
Korea, North[2]	...	...	...	...	...	42	42	71	71	71	71	71	71	71
	...	...	...	...	...	2.1	2.1	3.5	3.5	3.5	3.5	3.5	3.5	3.5
Korea, South	...	...	180	601	867	1,147	1,170	1,244	2,210	2,540	2,540	2,540	2,560	2,560
	...	...	9.0	29.9	43.2	57.1	58.3	62.0	110.1	126.5	126.5	126.5	127.5	127.5
Myanmar	...	9	26	26	32	32	32	32	32	32	32	32	32	32
	...	0.4	1.3	1.3	1.6	1.6	1.6	1.6	1.6	1.6	1.6	1.6	1.6	1.6
New Zealand	...	...	66	74	95	93	89	91	91	91	98	98	106	106
	...	...	3.3	3.7	4.7	4.6	4.4	4.5	4.5	4.5	4.9	4.9	5.3	5.3
Pakistan	5	6	107	98	121	133	139	137	137	137	139	143	239	239
	0.2	0.3	5.3	4.9	6.0	6.6	6.9	6.8	6.8	6.8	6.9	7.1	11.9	11.9
Philippines	...	22	182	253	284	279	280	323	323	323	389	401	420	420
	...	1.1	9.1	12.6	14.1	13.9	13.9	16.1	16.1	16.1	19.4	20.0	20.9	20.9
Sri Lanka	...	...	44	38	50	50	50	48	48	48	48	48	50	50
	...	...	2.2	1.9	2.5	2.5	2.5	2.4	2.4	2.4	2.4	2.4	2.5	2.5
Thailand	...	1	62	186	215	314	347	426	558	704	713	713	682	682
	...	...	3.1	9.3	10.7	15.6	17.3	21.2	27.8	35.1	35.5	35.5	34.0	34.0
Total	**284**	**1,472**	**5,866**	**11,986**	**12,181**	**13,658**	**14,419**	**14,802**	**16,285**	**16,994**	**18,771**	**19,645**	**20,166**	**20,185**
	14.1	**73.3**	**292.1**	**596.9**	**606.6**	**680.0**	**718.0**	**737.1**	**811.0**	**846.2**	**934.7**	**978.2**	**1,004.3**	**1,005.1**

[1]Prior to 1980, included in others

[2]Prior to 1994, included in other FSU—Eastern Europe

AFRICA

Upper figures = 1,000 b/d
Lower figures = million t/y

	1950	1960	1970	1980	1990	1994	1995	1996	1997	1998	1999	2000	2001	2002
Algeria	...	...	48	122	465	530	465	465	465	483	503	503	503	450
	...	...	2.4	6.1	23.2	26.4	23.2	23.2	23.2	24.1	25.0	25.0	25.0	22.4
Angola	...	2	14	36	32	32	32	32	32	32	39	39	39	39
	...	0.1	0.7	1.8	1.6	1.6	1.6	1.6	1.6	1.6	1.9	1.9	1.9	1.9
Cameroon	...	...	...	...	43	42	42	42	42	45	35	35	42	42
	...	...	...	...	2.1	2.1	2.1	2.1	2.1	2.2	1.7	1.7	2.1	2.1
Congo (Former Zaire)	...	...	21	16	17	17	17	17	17	17	17	17	15	15
	...	...	1.0	0.8	0.8	0.8	0.8	0.8	0.8	0.8	0.8	0.8	0.7	0.7
Congo (Brazzaville)	...	...	14	...	21	21	21	21	21	21	21	21	21	21
	...	...	0.7	...	1.0	1.0	1.0	1.0	1.0	1.0	1.0	1.0	1.0	1.0
Egypt	39	87	175	234	489	532	532	532	546	546	578	578	726	726
	1.9	4.3	8.7	11.7	24.4	26.5	26.5	26.5	27.2	27.2	28.8	28.8	36.2	36.1
Eritrea*	...	...	11	14	18	18	18	18	18	18	18	18	18	15
	...	...	0.5	0.7	0.9	0.9	0.9	0.9	0.9	0.9	0.9	0.9	0.9	0.7
Gabon	...	...	13	20	24	17	17	17	17	17	17	17	17	17
	...	...	0.6	1.0	1.2	0.8	0.8	0.8	0.8	0.8	0.8	0.8	0.9	0.8
Ghana	...	...	29	26	27	27	25	25	27	27	45	45	45	45
	...	...	1.4	1.3	1.3	1.3	1.2	1.2	1.3	1.3	2.2	2.2	2.2	2.2
Ivory Coast	...	...	19	50	69	63	64	64	64	73	69	69	65	65
	...	...	0.9	2.5	3.4	3.1	3.2	3.2	3.2	3.6	3.4	3.4	3.2	3.2
Kenya	...	...	44	95	95	90	90	90	85	86	90	70	90	90
	...	...	2.2	4.7	4.7	4.5	4.5	4.5	4.2	4.3	4.5	3.5	4.5	4.5
Liberia	...	...	10	15	15	15	15	15	15	15	15	15	15	15
	...	...	0.5	0.7	0.7	0.7	0.7	0.7	0.7	0.7	0.7	0.7	0.7	0.7
Libya	...	...	10	138	329	348	348	348	348	348	348	348	343	343
	...	...	0.5	6.9	16.4	17.3	17.3	17.3	17.3	17.3	17.3	17.3	17.1	17.0
Madagascar	...	...	13	11	16	16	16	15	15	15	15	15	15	15
	...	...	0.6	0.5	0.8	0.8	0.8	0.7	0.7	0.7	0.7	0.7	0.7	0.7
Morocco	...	2	35	72	155	155	155	155	156	152	157	157	157	155
	...	0.1	1.7	3.6	7.7	7.7	7.7	7.7	7.8	7.6	7.8	7.8	7.8	7.7
Nigeria	...	...	46	160	433	433	433	433	433	439	439	439	439	439
	...	...	2.3	8.0	21.6	21.6	21.6	21.6	21.6	21.9	21.9	21.9	21.8	21.8
Senegal	...	...	13	20	24	23	17	17	17	17	17	17	27	27
	...	...	0.6	1.0	1.2	1.1	0.8	0.8	0.8	0.8	0.8	0.8	1.3	1.3
Sierra Leone	...	...	10	10	10	10	10	10	10	10	10	10	10	10
	...	...	0.5	0.5	0.5	0.5	0.5	0.5	0.5	0.5	0.5	0.5	0.5	0.5
South Africa	...	25	184	478	434	364	401	401	414	465	468	467	474	469
	...	1.2	9.2	23.8	21.6	18.1	20.0	20.0	20.6	23.2	23.3	23.3	23.6	23.3
Sudan	...	...	21	26	25	22	22	22	22	22	22	82	122	122
	...	...	1.0	1.3	1.2	1.1	1.1	1.1	1.1	1.1	1.1	4.1	6.1	6.1
Tanzania	...	...	14	17	14	17	17	16	16	15	15	15	15	15
	...	...	0.7	0.8	0.7	0.8	0.8	0.8	0.8	0.7	0.7	0.7	0.7	0.7
Tunisia	...	...	23	34	34	34	34	34	34	34	34	34	34	34
	...	...	1.1	1.7	1.7	1.7	1.7	1.7	1.7	1.7	1.7	1.7	1.7	1.7
Total	**39**	**116**	**787**	**1,665**	**2,824**	**2,860**	**2,825**	**2,824**	**2,849**	**2,932**	**3,007**	**3,046**	**3,265**	**3,169**
	1.9	**5.7**	**38.8**	**82.9**	**140.4**	**142.1**	**140.5**	**140.4**	**141.6**	**145.7**	**149.2**	**151.2**	**162.3**	**157.1**

*Prior to 1995 listed as Ethiopia.

WORLDWIDE REFINING CAPACITY-cont'd

MIDDLE EAST

Upper figures = 1,000 b/d Lower figures = million t/y

	1950	1960	1970	1980	1990	1994	1995	1996	1997	1998	1999	2000	2001	2002
Abu Dhabi	...	...	...	14	180	193	193	213	213	213	213	429	444	N/A
	...	...	...	0.7	9.0	9.6	9.6	10.6	10.6	10.6	10.6	21.4	22.1	N/A
Bahrain	150	187	205	250	243	250	250	250	249	249	249	249	249	249
	7.5	9.3	10.2	12.4	12.1	12.4	12.4	12.4	12.4	12.4	12.4	12.4	12.4	12.4
Iran	502	495	645	921	530	1,162	1,184	1,168	1,242	1,358	1,448	1,474	1,484	1,484
	25.0	24.7	32.1	45.9	26.4	57.9	59.0	58.2	61.9	67.6	72.1	73.4	73.9	73.9
Iraq	9	56	104	168	319	348	348	348	348	348	348	348	418	418
	0.4	2.8	5.2	8.4	15.9	17.3	17.3	17.3	17.3	17.3	17.3	17.3	20.8	20.8
Israel	83	87	107	195	180	208	220	220	220	220	220	220	220	220
	4.1	4.3	5.3	9.7	9.0	10.4	11.0	11.0	11.0	11.0	11.0	11.0	11.0	11.0
Jordan	...	...	7	21	100	100	100	100	100	95	90	90	90	90
	...	...	0.3	1.0	5.0	5.0	5.0	5.0	5.0	4.7	4.5	4.5	4.5	4.5
Kuwait*	25	270	569	645	819	624	759	801	824	886	886	865	764	773
	1.2	13.5	28.4	32.1	40.8	31.1	37.8	39.9	41.0	44.1	44.1	43.1	38.0	38.5
Lebanon	6	24	38	53	37	38	38	38	38	38	38	38	38	38
	\0.3	1.2	1.9	2.6	1.8	1.9	1.9	1.9	1.9	1.9	1.9	1.9	1.9	1.9
Oman														85
														4.2
Qatar	...	1	1	11	62	58	58	58	58	58	58	58	58	58
	...	...	...	0.5	3.1	2.9	2.9	2.9	2.9	2.9	2.9	2.9	2.9	2.9
Saudi Arabia*	140	189	377	487	1,484	1,614	1,661	1,656	1,656	1,651	1,685	1,710	1,745	1,745
	7.0	9.4	18.8	24.3	73.9	80.4	82.7	82.5	82.5	82.2	83.9	85.2	86.9	87.0
Syria	...	...	59	223	244	242	242	242	246	242	242	242	242	242
	...	...	2.9	11.1	12.2	12.1	12.1	12.1	12.3	12.1	12.1	12.1	12.1	12.1
Yemen	...	120	178	175	162	120	120	120	120	120	120	120	130	130
	...	6.0	8.9	8.7	8.1	6.0	6.0	6.0	6.0	6.0	6.0	6.0	6.5	6.5
Total	**916**	**1,429**	**2,290**	**3,163**	**4,447**	**5,033**	**5,258**	**5,299**	**5,399**	**5,636**	**5,757**	**5,928**	**5,965**	**5,532**
	45.5	**71.2**	**114.0**	**151.4**	**221.6**	**250.8**	**261.9**	**264.0**	**269.0**	**280.7**	**286.8**	**295.4**	**297.2**	**275.7**

*Includes Neutral Zone.

EASTERN EUROPE

Upper figures = 1,000 b/d   Lower figures = million t/y

	1950	1960	1970	1980	1990	1994	1995	1996	1997	1998	1999	2000	2001	2002
Albania[1]	...	...	...	...	...	40	40	40	40	26	26	26	26	26
	...	...	...	...	...	2.0	2.0	2.0	2.0	1.3	1.3	1.3	1.3	1.3
Bulgaria[1]	...	...	...	...	...	300	300	300	300	325	133	115	115	115
	...	...	...	...	...	14.9	14.9	14.9	14.9	16.2	6.6	5.7	5.7	5.7
Croatia	...	...	...	...	...	293	294	294	294	294	236	236	253	293
	...	...	...	...	...	14.6	14.6	14.6	14.6	14.6	11.8	11.8	12.6	14.6
Czech Republic[2]	16	46	220	455	455	307	307	187	187	187	198	186	198	198
	0.8	2.3	11.0	22.7	22.7	15.3	15.3	9.3	9.3	9.3	9.9	9.3	9.9	9.9
Hungary	20	58	130	290	220	242	242	232	232	232	232	232	232	161
	1.0	2.9	6.5	14.4	11.0	12.1	12.1	11.6	11.6	11.6	11.6	11.6	11.6	8.0
Lithuania	...	...	...	...	...	263	267	263	240	263	263	263	263	263
	...	...	...	...	...	13.1	13.3	13.1	12.0	13.1	13.1	13.1	13.1	13.1
Macedonia	...	...	...	...	...	51	51	51	51	51	51	51	51	51
	...	...	...	...	...	2.5	2.5	2.5	2.5	2.5	2.5	2.5	2.5	2.5
Poland	4	20	156	385	385	333	352	352	352	365	382	382	382	382
	0.2	1.0	7.8	19.2	19.2	16.6	17.5	17.5	17.5	18.2	19.0	19.0	19.0	19.0
Romania	182	260	356	608	617	681	651	655	559	542	522	499	504	504
	9.1	12.9	17.7	30.3	30.7	33.9	32.4	32.6	27.8	27.0	26.0	24.9	25.1	25.1
Slovakia	...	...	...	...	...	193	193	115	115	115	115	115	115	115
	...	...	...	...	...	9.6	9.6	5.7	5.7	5.7	5.7	5.7	5.7	5.7
Slovenia	...	...	...	...	...	15	11	12	12	12	12	14	14	14
	...	...	...	...	...	0.7	0.5	0.6	0.6	0.6	0.6	0.7	0.7	0.7
Total	**1,134**	**3,234**	**7,013**	**13,836**	**15,438**	**12,846**	**12,851**	**12,730**	**12,649**	**12,750**	**12,086**	**12,048**	**10,711**	**2,122**
	56.5	**161.0**	**349.3**	**689.0**	**768.8**	**639.6**	**639.9**	**633.8**	**629.8**	**634.9**	**601.9**	**600**	**533.4**	**105.6**
World Total	**11,658**	**24,239**	**48,982**	**79,514**	**74,535**	**73,174**	**74,172**	**74,452**	**76,066**	**78,317**	**80,310**	**81,556**	**81,252**	
	580.0	**1,206.9**	**2,438.8**	**3,959.5**	**3,711.5**	**3,643.5**	**3,693.2**	**3,707.2**	**3,787.8**	**3,900.0**	**3,999.0**	**4,060.8**	**4,045.7**	

[1]Prior to 1994, included in others. [2]Prior to 1994 listed as Czechoslovakia.

FORMER SOVIET UNION

Upper figures = 1,000 b/d
Lower figures = million t/y

	1940	1950	1960	1970	1980	1990	1993	1994	1995	1996	1997	1998	1999	2000
Azerbaijan														442
														22.0
Belarus														493
														24.6
Georgia														106
														5.3
Kazakhstan														427
														21.3
Kyrgyzstan														10
														0.5
Russia														5,435
														270.7
Turkmenistan														237
														11.8
Ukraine														1,206
														60.1
Uzbekistan														222
														11.1
Total														**8,578**
														427.4

WORLDWIDE REFINERY THROUGHPUT

	1990	1991	1992	1993	1994	1995	1996	1997	1998	1999	2000
	(thousand barrels per calendar day)										
US	13,410	13,300	13,410	13,615	13,865	13,975	14,195	14,660	14,890	14,805	15,080
Canada	1,585	1,490	1,460	1,540	1,580	1,570	1,645	1,695	1,710	1,715	1,770
Mexico	1,490	1,510	1,495	1,540	1,460	1,490	1,490	1,440	1,450	1,390	1,365
South & Central America	4,665	4,625	4,685	4,770	4,855	4,935	5,080	5,305	5,380	5,460	5,525
Europe	13,915	13,805	13,980	14,250	14,370	14,430	14,685	14,840	15,315	14,835	14,880
Former Soviet Union	9,150	8,820	7,155	5,925	5,075	4,890	4,625	4,735	4,480	4,470	4,560
Middle East	4,415	3,945	4,400	4,810	5,225	5,265	5,490	5,550	5,710	5,980	5,800
Africa	2,255	2,245	2,225	2,290	2,220	2,315	2,305	2,440	2,335	2,415	2,445
Australasia	715	745	745	755	785	790	845	870	865	880	880
China	2,155	2,280	2,425	2,570	2,550	2,710	2,850	3,085	3,220	3,685	4,220
Japan	3,435	3,655	3,880	3,980	4,165	4,170	4,170	4,320	4,210	4,150	4,145
Other Asia Pacific	4,885	5,270	5,715	6,050	6,315	6,820	7,545	8,215	8,040	8,340	8,840
Total world	**62,075**	**61,690**	**61,575**	**62,095**	**62,465**	**63,360**	**64,925**	**67,155**	**67,605**	**68,125**	**69,510**
Of which OECD[1]	34,285	34,880	35,820	36,550	37,090	37,415	38,310	39,555	40,020	39,565	39,940
Of which other EMEs[2]	17,535	17,235	18,060	19,005	19,625	20,320	21,305	22,205	22,420	23,490	24,395

[1]Organisation for Economic Cooperation and Development.
[2]Emerging Market Economies excluding Central Europe and Former Soviet Union.
Source: British Petroleum and Parpinelli Tecnon

WORLDWIDE REFINING MARGINS*

	1988	1989	1990	1991	1992	1993	1994	1995	1996	1997	1998	1999	Dec. 2000	Dec. 2001
	$/bbl													
US Gulf Coast														
West Texas sour	3.15	3.05	4.03	3.69	2.47	2.33	2.20	2.19	3.37	3.83	3.31	2.15	5.54	2.67
Arab light	4.63	5.31	5.80	5.91	4.25	4.15	4.40	3.56	3.63	4.57	4.04	3.14	7.27	4.51
Bonny light	3.05	3.66	4.11	4.61	3.11	2.73	2.64	2.37	2.41	2.90	2.83	1.74	6.40	0.65
US PADD II														
Chicago (WTI)	3.10	5.26	6.06	4.50	3.35	3.22	3.70	3.00	3.74	3.78	3.17	2.47	4.60	2.57
US East Coast														
NY Harbor (Arab med)	3.92	4.59	2.98	3.48	1.71	2.53	3.94	3.40	2.89	3.89	3.62	2.50	6.38	4.11
US West Coast														
Los Angeles (ANS)	5.84	6.68	7.56	6.63	6.11	6.98	5.00	4.33	5.13	5.58	5.60	6.55	12.48	3.36
NW Europe														
Rotterdam (Brent)	2.98	3.25	4.68	5.16	3.16	3.33	2.59	2.48	2.78	3.26	3.26	2.19	4.71	0.88
Mediterranean														
Italy (Urals)	1.63	1.86	2.36	3.58	2.52	2.80	1.33	1.67	1.82	2.24	2.73	1.67	3.28	0.07
Far East														
Singapore (Dubai)	2.32	3.14	4.85	5.92	3.43	3.69	2.52	2.52	2.57	2.50	1.61	0.78	3.89	1.25

*Gross margins before costs.

Source: The Pace Consultants, Inc.

WORLD NATURAL GAS PRODUCTION

Country	1993	1994	1995	1996	1997	1998	1999	2000
	Bcf							
Argentina	760.0	786.5	883.2	938.3	966.9	1,048.1	1,186.6	1,330.8
Barbados	0.6	0.6	0.6	0.6	0.6	0.6	0.6	N/A
Bolivia	103.1	104.9	113.4	109.1	111.9	113.4	141.3	115.0
Brazil	145.8	152.2	159.3	178.3	194.9	202.0	215.4	165.6
Canada	4,910.2	5,266.1	5,603.0	5,778.5	5,847.4	5,984.1	6,049.4	7,082.0
Chile	61.8	68.5	67.1	63.9	75.2	70.3	67.1	53.1
Colombia	156.8	162.4	160.7	167.4	210.8	246.1	203.4	211.6
Ecuador	3.5	3.5	4.2	4.2	3.5	3.5	3.5	7.1
Mexico	1,310.7	1,323.0	1,372.1	1,535.8	1,629.8	1,212.4	1,239.2	1,712.7
Peru	33.9	34.6	32.8	34.3	8.5	17.3	16.6	9.9
Trinidad	217.9	249.7	268.0	302.6	328.4	373.6	384.9	344.6
United States	18,982.0	19,710.0	19,506.0	19,751.0	19,895.0	18,906.8	18,925.1	20,069.0
Venezuela	815.1	876.2	889.9	960.6	993.8	990.9	982.1	840.4
Western Hemisphere	27,501.7	28,738.5	29,060.7	29,825.0	30,267.1	29,169.4	29,415.6	31,942.9
Austria	54.4	48.4	52.3	52.6	50.1	54.0	50.9	64.8
Denmark	159.6	172.3	186.8	228.1	273.0	267.7	274.7	264.8
France	124.3	132.4	125.7	108.4	96.8	81.9	77.0	64.6
Germany	659.7	691.8	713.7	752.9	760.0	742.3	792.8	706.8
Ireland	95.0	97.8	102.1	96.8	87.2	65.0	61.1	N/A
Italy	687.6	728.9	719.7	714.1	687.2	676.6	627.9	591.5
Netherlands	2,966.8	2,769.0	2,766.9	3,166.0	2,890.2	2,821.6	2,666.3	2,515.6
Norway	875.9	947.1	981.9	1,320.2	1,505.2	1,631.5	1,741.0	1,782.5
Spain	22.2	6.4	14.8	13.8	5.7	20.5	19.1	1.5
Turkey	7.4	7.1	6.7	8.1	8.1	8.8	8.1	17.,4
United Kingdom	2,313.1	2,466.0	2,665.6	3,176.6	3,031.4	3,165.3	3,475.0	4,090.1
Western Europe	7,968.5	8,069.0	8,337.6	9,638.9	9,396.3	9,536.7	9,795.2	10,199.0
Albania	1.1	1.1	1.1	1.1	0.7	0.7	0.7	0.3
Bulgaria	2.5	2.1	1.8	1.4	1.4	1.1	1.1	N/A
Croatia	0.0	63.6	69.6	62.9	60.7	55.4	55.4	52.6
Czech Republic & Slovakia	12.0	14.1	16.2	14.5	14.1	10.9	10.9	N/A
F.S.U.	26,823.6	25,403.6	24,904.3	25,219.3	23,649.5	24,277.8	24,714.6	24,013.3
Hungary	189.3	172.0	172.3	163.9	154.3	137.7	120.8	103.4
Poland	174.8	163.5	170.2	127.1	116.5	123.6	116.5	N/A
Romania	745.8	691.8	683.3	618.0	554.4	515.6	508.5	732.0
Serbia	0.0	31.4	32.1	25.8	32.8	33.5	32.8	27.6
East Europe and FSU	27,949.1	26,543.2	26,050.9	26,233.9	24,584.7	25,156.4	25,561.5	25,094.7
Algeria	1,902.0	1,806.7	2,051.8	2,194.1	2,433.2	2,604.5	2,838.9	2,310.8
Angola	19.8	18.4	19.8	19.8	20.1	20.5	20.8	20.4
Egypt	398.7	423.4	439.0	473.2	476.7	489.5	519.1	470.6
Gabon	3.5	3.5	3.5	3.5	3.5	3.5	3.5	3.6
Ivory Coast	0.0	0.0	1.4	19.4	21.2	24.4	24.7	N/A
Libya	224.6	225.7	223.9	226.7	232.0	224.6	215.4	215.6
Morocco	0.0	1.1	0.7	0.7	1.8	1.8	1.8	N/A
Nigeria	178.3	160.7	183.3	192.8	206.6	208.4	211.9	123.7
Tunisia	13.1	12.4	11.7	27.9	58.6	67.1	69.9	69.7
Africa	2,804.3	2,722.1	3,006.0	3,222.5	3,517.0	3,700.3	3,964.8	3,284.5
Bahrain	233.8	229.2	229.2	232.0	281.5	292.8	300.2	237.6
Iran	956.0	1,123.0	1,246.6	1,420.0	1,659.8	1,765.7	1,854.0	1,068.8
Iraq	90.1	111.9	111.9	114.4	107.7	104.2	107.7	242,8
Jordan	7.1	9.9	10.2	9.5	10.2	9.9	10.2	N/A
Kuwait	191.4	210.8	210.8	328.4	327.4	327.4	254.6	200.7
Oman	139.8	150.1	147.3	145.5	176.2	247.6	273.3	170.7
Qatar	476.7	476.7	476.7	483.8	614.5	691.5	847.6	727.6
Saudi Arabia	1,267.8	1,331.4	1,343.4	1,459.9	1,601.2	1,653.4	1,631.9	1,126.8
Syria	130.7	134.2	103.8	142.0	161.0	208.4	221.8	136.5
United Arab Emirates	811.9	911.8	1,106.1	1,193.6	1,282.3	1,309.1	1,343.0	824.2
Middle East	4,306.3	4,690.1	4,986.8	5,529.9	6,222.5	6,610.6	6,845.0	4,745.8
Afghanistan	10.6	10.6	7.1	8.1	8.1	8.1	8.1	N/A
Australia	847.3	997.0	1,042.2	1,052.2	1,062.9	1,106.4	1,114.2	1,100.1
Bangladesh	215.8	234.8	260.3	267.0	269.5	282.5	294.5	N/A
Brunei	292.4	297.3	330.2	325.2	324.2	315.0	329.5	374.7
China	584.8	599.3	607.7	695.2	740.4	780.8	861.3	964.6
India	551.6	600.7	691.2	761.4	789.4	760.7	769.2	954.7
Indonesia	1,972.7	2,206.5	2,237.5	2,354.1	2,371.4	2,241.4	2,330.8	2,619.1

WORLD NATURAL GAS PRODUCTION-cont'd

Japan	76.6	78.0	77.7	78.8	80.5	81.6	86.9	83.0
Malaysia	880.4	922.8	1,019.9	1,230.0	1,356.1	1,437.3	1,518.9	853.2
Myanmar (Burma)	38.8	50.5	57.9	56.5	53.3	62.2	66.0	N/A
New Zealand	175.2	170.6	173.4	163.5	162.1	162.8	164.2	168.7
Pakistan	610.9	627.2	645.9	696.4	698.5	711.6	782.6	855.6
Papua New Guinea	2.8	1.8	3.5	4.6	3.9	3.9	3.9	5.0
Taiwan-China	29.3	32.5	33.5	32.1	32.8	32.8	33.5	N/A
Thailand	310.4	341.8	367.6	428.4	537.8	565.0	625.1	674.2
Viet Nam	8.8	8.8	25.1	29.3	8.1	19.4	20.5	40.5
Asia-Pacific	6,608.5	7,180.3	7,581.1	8,183.2	8,499.4	8,571.9	9,009.5	9,129.8
TOTAL WORLD	77,138.4	77,943.2	79,023.1	82,633.4	82,487.0	82,745.4	84,591.6	84,396.7

Source: Oil & Gas Journal, Cedigaz and BPAmoco Statistical Review of World Energy

WORLD NATURAL GAS CONSUMPTION

Country	1990	1991	1992	1993	1994	1995	1996	1997	1998	1999	2000	Change 2000 over 1999 (%)	2000 share of total (%)
	billion cubic meters												
USA	540.3	549.0	563.7	583.2	596.1	620.6	631.7	630.9	613.8	624.1	654.4	4.8	27.2
Canada	61.8	63.0	66.8	68.4	70.8	70.9	74.3	74.8	70.3	72.7	77.8	7.1	3.2
Mexico	27.8	27.7	27.7	28.2	29.4	29.7	31.0	31.7	34.4	33.9	35.5	4.6	1.5
Total North America	**629.9**	**639.7**	**658.2**	**679.8**	**696.3**	**721.2**	**737.0**	**737.4**	**718.5**	**730.7**	**767.7**	**5.1**	**31.9**
Argentina	20.3	22.1	22.3	23.6	24.3	27.0	28.6	28.5	30.5	33.7	33.1	-1.9	1.4
Brazil	3.8	3.9	4.0	4.5	4.5	4.8	5.5	6.0	6.3	7.1	9.4	32.7	0.4
Chile	1.7	1.5	1.7	1.6	1.7	1.6	1.7	2.8	3.3	4.0	5.8	45.0	0.2
Columbia	4.1	4.1	4.0	4.2	4.2	4.4	4.7	5.9	6.2	5.2	5.9	14.4	0.2
Ecuador	0.1	0.1	0.1	0.1	0.1	0.1	0.1	0.1	0.1	0.1	0.1	...	†
Peru	0.4	0.4	0.4	0.4	0.4	0.4	0.4	0.4	0.5	0.5	0.4	-18.4	†
Venezuela	22.0	21.9	21.6	23.3	24.7	27.5	29.7	30.8	32.3	26.8	27.2	1.6	1.1
Other S. & Cent. America	6.0	6.5	6.8	6.8	7.3	7.3	8.2	8.9	9.9	10.4	10.7	3.6	0.5
Total S. & Centr. America	**58.4**	**60.5**	**60.9**	**64.5**	**67.2**	**73.1**	**78.9**	**83.4**	**89.1**	**87.8**	**92.6**	**5.6**	**3.8**
Austria	5.7	6.1	6.0	6.4	6.5	6.8	7.3	7.7	7.8	8.0	7.9	-1.0	0.3
Belguim & Luxembourg	10.6	10.0	10.0	11.0	10.8	11.8	13.1	12.5	13.8	14.7	14.9	0.8	0.6
Bulgaria	5.9	5.0	4.3	4.1	4.1	5.0	5.2	4.1	3.4	3.2	3.1	-1.6	0.1
Czech Republic	5.4	5.9	5.8	5.9	6.3	7.3	8.4	8.5	8.5	8.6	8.6	-0.4	0.4
Denmark	2.0	2.3	2.4	2.8	3.0	3.5	4.1	4.4	4.8	5.0	4.9	-1.9	0.2
Finland	2.5	2.6	2.7	2.8	3.1	3.2	3.3	3.2	3.7	3.7	3.7	2.2	0.2
France	29.3	30.6	31.4	32.3	30.9	32.9	36.1	34.6	37.0	37.7	39.6	5.0	1.6
Germany	59.9	62.9	63.0	66.4	67.9	74.4	83.6	79.2	79.7	80.2	79.2	-1.2	3.3
Greece	0.2	0.1	0.1	0.1	†	†	†	0.2	0.8	1.4	1.7	26.8	0.1
Hungary	9.6	9.6	8.2	9.0	9.4	10.2	11.4	10.8	10.9	10.9	10.7	-2.5	0.4
Iceland	...	...	...	...	...	...	...	...	...	...	...	...	...
Republic of Ireland	2.1	2.1	2.1	2.4	2.4	2.6	3.0	3.1	3.1	3.3	3.8	14.6	0.2
Italy	43.4	46.2	45.7	46.8	45.3	49.9	51.5	53.2	57.2	62.2	63.8	2.6	2.7
Netherlands	34.4	38.1	36.7	37.9	36.9	37.8	41.7	39.1	38.7	37.9	38.3	1.0	1.6
Norway	2.1	2.4	2.6	2.7	2.9	2.9	3.2	3.7	3.8	3.6	3.9	8.3	0.2
Poland	9.9	8.8	8.7	9.0	9.2	9.9	10.6	10.5	10.6	10.3	11.1	7.5	0.5
Portugal	...	...	...	...	...	...	...	0.1	0.8	2.3	6.0	>100	0.2
Romania	30.8	24.7	25.4	25.2	24.2	24.0	24.2	20.0	18.7	17.2	16.2	-5.9	0.7
Slovakia	5.9	5.4	5.5	5.2	5.0	5.7	6.2	6.3	6.4	6.4	5.7	-12.0	0.2
Spain	5.6	6.1	6.5	6.5	7.2	8.3	9.3	12.3	13.1	15.0	16.8	12.0	0.7
Sweden	0.7	0.7	0.7	0.8	0.8	0.8	0.9	0.8	0.9	0.8	0.9	1.3	†
Switzerland	1.8	2.0	2.1	2.2	2.2	2.4	2.6	2.5	2.6	2.7	2.7	-0.9	0.1
Turkey	3.4	4.4	4.5	5.0	6.5	6.8	9.0	9.4	9.9	12.0	14.1	17.5	0.6
United Kingdom	52.4	56.6	56.4	64.2	66.1	70.5	82.1	83.8	86.9	92.0	95.7	4.0	4.0
Other Europe	6.9	6.7	5.6	5.3	4.0	4.1	5.9	6.0	5.9	5.1	5.5	7.6	0.2
Total Europe	**330.5**	**339.3**	**336.4**	**354.0**	**354.7**	**380.8**	**422.7**	**416.0**	**429.0**	**444.2**	**458.8**	**3.3**	**19.1**
Azerbaijan	15.8	15.1	11.8	8.7	8.1	8.0	5.9	5.6	5.2	5.6	5.4	-3.3	0.2
Belarus	13.8	14.5	16.8	15.6	13.6	12.3	13.0	14.8	15.0	15.3	16.2	6.1	0.7
Kazakhstan	12.5	13.2	13.5	13.0	10.3	10.8	9.0	7.1	7.3	7.9	8.0	1.2	0.3
Lithuania	5.6	5.4	3.0	1.7	2.0	2.3	2.5	2.6	2.3	2.4	2.7	13.0	0.1
Russian Federation	420.1	431.1	417.3	416.0	390.9	377.8	379.9	350.4	364.7	362.7	377.2	4.0	15.7
Turkmenistan	9.8	9.6	9.3	9.3	10.2	8.0	10.0	10.1	10.3	11.3	12.6	11.6	0.5
Ukraine	127.8	121.5	103.5	92.9	81.3	76.2	82.5	74.3	68.8	70.6	68.5	-3.0	2.9
Uzbekistan	36.8	37.1	37.3	40.7	41.3	42.4	43.3	45.4	47.0	49.3	49.8	1.1	2.1
Other Former Soviet Union	20.5	18.2	15.7	11.1	9.3	9.1	7.9	8.8	9.2	7.9	7.9	1.0	0.3
Total Former Soviet Union	**662.7**	**665.7**	**628.2**	**609.0**	**567.0**	**546.9**	**554.0**	**519.1**	**529.8**	**533.0**	**548.3**	**2.9**	**22.8**

Country	1990	1991	1992	1993	1994	1995	1996	1997	1998	1999	2000	Change 2000 over 1999 (%)	200 share o total (%
	billion cubic meters												
Iran	22.7	22.7	25.0	26.6	31.8	35.0	40.1	47.1	51.8	55.0	62.9	14.4	2.
Kuwait	6.2	0.5	2.6	5.4	6.0	9.3	9.3	9.3	9.5	8.7	9.6	10.5	0.
Qatar	6.3	7.6	12.6	13.5	13.5	13.5	13.7	14.6	14.8	15.9	14.5	-9.1	0.
Saudi Arabia	33.5	35.2	38.3	40.0	42.8	42.9	44.5	45.8	46.7	46.2	47.0	1.7	2.
United Arab Emirates	16.9	20.4	18.8	19.6	21.7	24.8	27.2	29.0	30.4	31.4	33.4	6.2	1.
Other Middle East	11.9	11.6	13.3	13.9	14.9	16.1	17.3	19.6	20.4	21.2	21.6	1.8	0.
Total Middle East	**97.5**	**98.0**	**110.6**	**119.0**	**130.7**	**141.6**	**152.1**	**165.4**	**173.6**	**178.4**	**189.0**	**5.9**	**7.**
Algeria	16.1	17.0	17.8	18.6	19.6	21.0	21.4	20.2	20.9	22.2	24.4	9.7	1
Egypt	6.8	7.7	8.4	9.7	10.4	11.0	11.3	11.6	12.0	14.3	17.9	24.5	0
South Africa	...	...	...	...	...	...	...	...	...	...	...	...	.
Other Africa	11.0	10.5	11.2	11.6	11.9	12.8	14.3	14.4	14.9	15.6	16.6	6.3	0
Total Africa	**33.9**	**35.2**	**37.4**	**39.9**	**41.9**	**44.8**	**47.0**	**46.2**	**47.8**	**52.1**	**58.9**	**12.8**	**2.**
Australia	18.4	17.0	16.9	17.4	19.4	19.5	19.9	19.6	20.3	19.8	21.3	7.5	0
Bangladesh	4.8	5.3	5.7	6.1	6.6	7.4	7.6	7.6	7.8	8.3	10.3	24.7	0
China	14.7	14.9	15.1	16.2	16.6	17.7	17.7	19.3	19.3	21.4	24.8	16.0	1
China Hong Kong SAR	...	...	...	...	...	†	1.7	2.6	2.5	2.7	2.5	-9.6	0
India	12.5	14.1	15.8	16.3	17.4	19.6	20.6	21.3	23.2	23.8	25.0	5.2	1
Indonesia	20.1	21.7	22.6	23.9	27.3	30.1	31.4	31.9	27.8	27.6	27.8	0.8	1
Japan	51.2	54.7	56.0	56.3	60.3	61.2	66.1	65.1	69.5	74.6	76.2	2.2	3
Malaysia	7.6	9.0	10.6	13.0	13.6	13.7	15.9	16.7	17.4	19.0	21.7	14.0	0
New Zealand	4.3	4.6	4.9	4.7	4.4	4.2	4.7	5.1	4.5	5.2	5.4	4.6	0
Pakistan	11.2	11.1	11.5	12.2	13.3	14.6	15.4	15.6	16.0	17.3	19.0	9.5	0
Philippines	...	...	...	...	†	†	†	†	†	†	†	50.0	
Singapore	...	...	1.1	1.5	1.5	1.5	1.5	1.5	1.5	1.5	1.5	...	0
South Korea	3.4	3.9	5.1	6.4	8.5	10.2	13.5	16.4	15.4	18.7	21.0	12.3	0
Taiwan	1.9	3.0	3.2	3.1	4.0	4.3	4.5	5.1	6.4	6.2	6.9	10.6	0
Thailand	5.5	7.0	7.5	8.4	9.5	10.0	11.8	14.6	15.9	17.4	20.9	20.5	0
Other Asia Pacific	2.7	2.5	2.7	3.0	3.3	3.4	3.8	4.2	4.6	4.9	5.0	1.4	0
Total Asia Pacific	**158.3**	**168.8**	**178.7**	**188.5**	**205.7**	**217.4**	**236.1**	**246.6**	**252.1**	**268.4**	**289.3**	**7.8**	**12**
Total World	**1971.2**	**2007.2**	**2010.4**	**2054.7**	**2063.5**	**2125.8**	**2227.8**	**2214.1**	**2239.9**	**2294.6**	**2404.6**	**4.8**	**100**
of which: OECD	**988.2**	**1017.4**	**1036.7**	**1078.8**	**1106.3**	**1158.3**	**1222.4**	**1223.2**	**1222.8**	**1261.3**	**1319.9**	**4.6**	**54**
European Union 15	**248.8**	**264.4**	**263.7**	**280.4**	**280.9**	**302.5**	**336.0**	**334.2**	**348.3**	**364.2**	**377.2**	**3.6**	**15**
Other EMEs‡	**270.8**	**282.3**	**304.7**	**327.1**	**352.9**	**381.8**	**409.9**	**435.4**	**452.9**	**468.4**	**505.9**	**8.0**	**21**

†Less than 0.05.

‡Excludes Central Europe and Former Soviet Union.

Notes: As far as possible, the data above represents standard cubic meters (measured at 15°C and 1,013 mbar); as they are derived directly from tonnes of oil equivalent using an average c version factor, they do not necessarily equate with gas volumes expressed in specific national terms.

Source: BP Statistical Review of World Energy

WORLD NATURAL GAS IMPORTS AND EXPORTS

Country	1997 Pipelines	1997 LNG by tanker	1997 Total	1998 Pipelines	1998 LNG by tanker	1998 Total	1999 Pipelines	1999 LNG by tanker	1999 Total	2000 Pipelines	2000 LNG by tanker	2000 Total
	Bcf											
IMPORTS												
Austria	212.95	0	212.95	220.72	0	220.72	219.3	0	219.3	220.45	--	220.45
Belgium	344.32	160.68	505	395.52	151.5	547.02	418.13	142.67	560.8	439.41	148.3	587.71
Brazil	--	--	--	--	--	--	--	--	--	74.1	--	74.1
Bulgaria	180.1	0	180.1	134.2	0	134.2	113.01	0	113.01	119.16	--	119.16
Canada	56.15	0	56.15	44.85	0	44.85	31.08	0	31.08	66.28	--	66.28
Chile	23.66	0	23.66	43.79	0	43.79	70.63	0	70.63	151.56	--	151.56
Croatia	37.08	0	37.08	38.85	0	38.85	39.2	0	39.2	41.33	--	41.33
Czech Republic	324.9	0	324.9	330.55	0	330.55	321.36	0	321.36	317.27	--	317.27
Finland	130.66	0	130.66	148.32	0	148.32	148.32	0	148.32	160.12	--	160.12
France	925.25	326.66	1,251.91	934.07	345.38	1,279.45	1,089.81	362.33	1,452.14	1206.88	396.53	1603.41
Germany	2,662.02	0	2,662.02	2,621.41	0	2,621.41	2,586.10	0	2,586.10	2859.88	--	2859.88
Greece	5.65	0	5.65	31.78	0	31.78	52.97	0	52.97	59.58	10.59	70.17
Hungary	282.52	0	282.52	298.41	0	298.41	310.77	0	310.77	342.59	--	342.59
Iran	3.53	0	3.53	61.8	0	61.8	113.01	0	113.01	98.68	--	98.68
Ireland	29.31	0	29.31	52.97	0	52.97	74.16	0	74.16	96.82	--	96.82
Italy	1,315.47	67.1	1,382.57	1,433.78	70.63	1,504.41	1,642.13	100.29	1,742.43	1966.17	168.78	2134.95
Japan	0	2,251.31	2,251.31	0	2,338.54	2,338.54	0	2,446.60	2,446.60	--	2558.56	2558.56
Luxembourg	25.07	0	25.07	26.84	0	26.84	28.25	0	28.25	22.34	--	22.34
Mexico	38.14	0	38.14	54.03	0	54.03	63.92	0	63.92	115.81	--	115.81
Netherlands	208	0	208	187.17	0	187.17	259.21	0	259.21	446.86	--	446.86
Poland	268.39	0	268.39	261.33	0	261.33	261.33	0	261.33	238.01	--	238.01
Portugal	17.66	0	17.66	28.25	0	28.25	67.1	0	67.1	81.92	--	81.92
Romania	158.92	0	158.92	134.2	0	134.2	120.07	0	120.07	126.61	--	126.61
Singapore	52.97	0	52.97	52.97	0	52.97	52.97	0	52.97	55.86	--	55.86
Slovakia	240.14	0	240.14	243.67	0	243.67	261.33	0	261.33	294.18	--	294.18
Slovenia	32.14	0	32.14	33.2	0	33.2	34.96	0	34.96	37.61	--	37.61
South Korea	0	554.79	554.79	0	505.35	505.35	0	618.71	618.71	--	694.9	694.9
Spain	216.48	223.9	440.37	241.55	209.42	450.97	283.93	252.85	536.78	315.41	299.08	614.49
Sweden	30.37	0	30.37	30.37	0	30.37	30.37	0	30.37	40.96	--	40.96
Switzerland	99.59	0	99.59	105.24	0	105.24	104.18	0	104.18	107.99	--	107.99
Taiwan	0	153.97	153.97	0	164.21	164.21	0	188.93	188.93	--	208.33	208.33
Tunisia	32.14	0	32.14	36.73	0	36.73	35.31	0	35.31	37.24	--	37.24
Turkey	238.37	105.94	344.32	238.37	128.19	366.57	310.77	112.3	423.07	383.55	130.65	514.2
United Kingdom	46.62	0	46.62	31.78	0	31.78	48.38	0	48.38	74.48	--	74.48
United States	2,962.20	78.05	3,040.24	3,097.81	85.81	3,183.62	3,399.39	161.39	3,560.78	3791.95	220.33	4012.28
Uruguay	--	--	--	--	--	--	--	--	--	1.49	--	1.49
World Total	**14,694.80**	**3,922.40**	**18,617.20**	**15,056.42**	**3,999.04**	**19,055.46**	**15,854.18**	**4,386.09**	**20,240.27**	**14497.13**	**4836.06**	**19333.19**
EXPORTS												
Algeria	866.98	854.26	1,721.24	972.21	878.63	1,850.84	1,190.11	909.71	2,099.81	1315.25	929.36	2244.61
Argentina	23.66	0	23.66	43.79	0	43.79	70.63	0	70.63	156.4	--	156.4
Australia	0	336.9	336.9	0	354.56	354.56	0	355.62	355.62	--	356.98	356.98
Bolivia	64.63	0	64.63	76.28	0	76.28	76.63	0	76.63	70.75	--	70.75
Brunei	0	293.11	293.11	0	287.11	287.11	0	297	297	--	310.37	310.37
Canada	2,945.25	0	2,945.25	3,082.97	0	3,082.97	3,344.30	0	3,344.30	3785.62	--	3785.62
Denmark	104.88	0	104.88	95.35	0	95.35	93.94	0	93.94	134.06	--	134.06
France	21.19	0	21.19	28.25	0	28.25	30.02	0	30.02	29.79	--	29.79
Germany	107	0	107	108.06	0	108.06	121.84	0	121.84	132.57	--	132.57
Indonesia	0	1,245.90	1,245.90	0	1,273.45	1,273.45	0	1,370.56	1,370.56	--	1260.57	1260.57
Libya	0	39.55	39.55	0	32.14	32.14	0	32.14	32.14	--	28.25	28.25
Malaysia	52.97	713.71	766.68	52.97	684.75	737.72	52.97	725.72	778.69	55.86	742.57	798.43
Mexico	16.95	0	16.95	14.83	0	14.83	55.09	0	55.09	6.33	--	6.33
Netherlands	1,416.47	0	1,416.47	1,284.04	0	1,284.04	1,236.01	0	1,236.01	1363.66	--	1363.66
Nigeria	--	--	--	--	--	--	--	--	--	--	198.09	198.09
Norway	1,494.16	0	1,494.16	1,503.35	0	1,503.35	1,606.47	0	1,606.47	1824.66	--	1824.66
Oman	17.66	0	17.66	15.89	0	15.89	15.89	0	15.89	--	87.22	87.22
Qatar	0	101	101	0	169.16	169.16	0	287.11	287.11	--	495.75	495.75
Russian Fed.	6,708.03	0	6,708.03	6,945.34	0	6,945.34	4,433.41	0	4,433.41	4853.23	--	4853.23
Trinidad & Tobago										--	123.94	123.94
Turkmenistan	264.86	0	264.86	61.8	0	61.8	113.01	0	113.01	98.68	--	98.68
United Arab Emirates	--	--	--	--	--	--	--	--	--	--	244.7	244.7
United Kingdom	64.98	0	64.98	103.83	0	103.83	196	0	196	488.2	--	488.2
United States	94.29	62.15	156.44	98.88	66.04	164.92	95	58.27	153.27	182.09	58.26	240.35
World Total	**14,694.80**	**3,922.40**	**18,617.20**	**15,056.42**	**3,999.04**	**19,055.46**	**13,299.87**	**4,285.79**	**17,585.66**	**14497.13**	**4836.06**	**19333.19**

Source: BP Statistical Review of World Energy

CRUDE OIL PRICES

	$/bbl												
	U.S. refiner acquisition cost			U.S. landed cost of imports							Average export prices		
					Selected countries								
Year	Domestic	Imported	Composite	Total	Canada	Mexico	United Kingdom	Nigeria	Saudi Arabia	Venezuela	Worldwide	OPEC	Non-OPEC
1983	28.87	29.30	28.99	28.93	25.63	25.78	30.87	30.85	29.27	22.94	29.66	29.89	29.13
1984	28.53	28.88	28.63	28.54	26.56	26.85	29.45	30.36	29.20	25.19	28.60	28.61	28.60
1985	26.66	26.99	25.75	26.67	25.71	25.63	28.36	28.96	24.72	24.43	27.60	28.00	26.98
1986	14.82	14.00	14.55	13.49	13.43	12.17	14.63	15.29	12.84	11.52	14.98	15.45	14.21
1987	17.76	18.13	17.90	17.65	17.04	16.69	18.78	19.32	16.81	15.76	17.41	17.31	17.57
1988	14.74	14.56	14.67	14.08	13.50	12.58	15.82	15.88	13.37	13.66	13.79	13.67	14.04
1989	17.87	18.08	17.97	17.68	16.81	16.35	18.74	19.19	17.34	16.78	16.65	16.40	17.11
1990	22.59	21.76	22.22	21.13	20.48	19.64	22.65	23.33	21.82	20.31	21.35	20.98	22.07
1991	19.33	18.70	19.06	18.02	17.16	15.89	21.37	21.39	17.22	15.92	17.82	17.36	18.52
1992	18.63	18.20	18.43	17.75	17.04	15.60	20.63	20.78	17.48	15.13	17.95	17.65	18.51
1993	16.67	16.14	16.41	15.72	15.27	14.11	17.92	18.73	15.40	13.39	15.81	15.45	16.44
1994	15.67	15.51	15.59	15.18	14.83	14.09	16.64	17.21	15.11	13.12	15.28	15.02	15.56
1995	17.33	17.14	17.23	16.78	16.65	16.19	17.91	18.25	16.84	14.81	16.78	16.51	17.11
1996	20.77	20.64	20.71	20.31	19.94	19.64	20.88	21.95	20.49	18.59	20.04	19.51	20.73
1997	19.61	18.53	19.04	18.11	17.63	17.30	20.64	20.64	17.52	16.35	18.38	18.01	18.88
1998	13.18	12.04	12.52	11.84	11.62	11.04	13.55	14.14	11.16	10.16	11.92	11.64	12.30
1999	17.90	17.26	17.51	17.23	17.54	16.12	18.26	17.63	17.48	15.58	17.37	17.14	17.65
2000	29.06	27.69	28.23	27.44	26.69	26.03	29.26	30.04	26.58	26.05	23.60	23.19	24.11

Source: US Department of Energy

PETROLEUM PRODUCT PRICES

	(¢/gallon)											
	Motor gasoline-pump price				Refiner resale (wholesale) product prices							
Year	Unleaded Regular	Unleaded Premium	Average All-Types	Heating Oil Residential	Motor Gasoline	Aviation Gasoline	Kerosene Jet Fuel	Kerosene	No. 2 Fuel Oil	No. 2 Diesel Oil	Propane	Resid
1983	124.1	138.3	122.5	107.8	88.2	117.8	85.4	89.2	81.5	80.8	48.4	60.9
1984	121.2	136.6	119.8	109.1	83.2	116.5	83.0	91.6	82.1	80.3	45.0	65.4
1985	120.2	134.0	119.6	105.3	83.5	113.0	79.4	87.4	77.6	77.2	39.8	57.7
1986	92.7	108.5	93.1	83.6	53.1	91.2	49.5	60.8	48.6	45.2	29.0	30.5
1987	94.8	109.3	95.7	80.3	58.9	85.9	53.8	59.2	52.7	53.4	25.2	38.5
1988	94.6	110.7	96.3	81.3	57.7	85.0	49.5	54.9	47.3	47.3	24.0	30.0
1989	102.1	119.7	106.0	90.0	65.4	95.0	58.3	66.9	56.5	56.7	24.7	36.0
1990	116.4	134.9	121.7	106.3	78.6	106.3	77.3	83.9	69.7	69.4	38.6	41.3
1991	114.0	132.1	119.6	101.9	69.9	100.1	65.0	72.2	62.2	61.5	34.9	31.4
1992	112.7	131.6	119.0	93.4	67.7	99.1	60.5	63.2	57.9	59.0	32.8	30.8
1993	110.8	130.2	117.3	91.1	62.6	96.5	57.7	60.4	54.4	57.0	35.1	29.3
1994	111.2	130.5	117.4	88.4	59.9	93.3	53.4	61.8	50.6	52.9	32.4	31.7
1995	114.7	133.6	120.5	86.7	62.6	97.5	53.9	58.0	51.1	53.8	34.4	36.3
1996	123.1	141.3	128.8	98.9	71.3	105.5	64.6	71.4	63.9	65.9	46.1	42.0
1997	123.4	141.6	129.1	98.4	70.0	106.5	61.3	65.3	59.0	60.6	41.6	38.7
1998	105.9	125.0	111.5	85.2	52.6	91.2	45.0	46.5	42.2	44.4	28.8	28.0
1999	116.5	135.7	122.1	87.6	64.5	100.7	53.3	55.0	49.3	54.6	34.2	35.4
2000	151.0	169.3	156.3	131.1	96.3	133.0	88.0	96.9	88.6	89.8	59.5	56.6

Source: US Department of Energy

COMPARATIVE ENERGY PRICES

Year	Natural gas delivered to consumers ($/MCF) Residential	Commercial	Industrial	Electric Utilities	Consumer Average	Cost of fossil fuels delivered to utilities (¢/Million BTU) Coal	Heavy oil	Natural gas	Electricity retail prices ¢/Kilowatt Hour) Residential	Commercial	Industrial	Total
1983	6.06	5.59	4.18	3.58	4.82	165.6	457.0	347.4	7.20	7.00	5.00	6.30
1984	6.12	5.55	4.22	3.70	4.05	166.4	481.0	360.3	7.15	7.13	4.83	6.25
1985	6.12	5.50	3.95	3.55	4.72	164.8	424.4	344.4	7.39	7.27	4.97	6.44
1986	5.83	5.08	3.23	2.43	4.13	157.9	240.1	235.1	7.42	7.20	4.93	6.44
1987	5.54	4.77	2.94	2.32	4.05	150.6	297.6	224.0	7.45	7.08	4.77	6.37
1988	5.47	4.63	2.95	2.33	4.09	146.6	240.5	226.3	7.48	7.04	4.70	6.35
1989	5.64	4.74	2.96	2.43	4.22	144.5	284.6	235.5	7.65	7.20	4.72	6.45
1990	5.80	4.83	2.93	2.39	4.19	145.5	331.9	232.1	7.83	7.34	4.74	6.57
1991	5.82	4.81	2.69	2.18	4.07	144.7	246.5	215.3	8.04	7.53	4.83	6.75
1992	5.89	4.88	2.84	2.36	4.17	141.2	247.5	232.8	8.21	7.66	4.83	6.82
1993	6.16	5.22	3.07	2.61	4.43	138.5	236.2	256.0	8.32	7.74	4.85	6.93
1994	6.41	5.44	3.05	2.28	4.51	135.5	240.9	223.0	8.38	7.73	4.77	6.91
1995	6.06	5.05	2.71	2.02	4.13	131.8	258.6	198.4	8.40	7.69	4.66	6.89
1996	6.34	5.40	3.42	2.69	4.67	128.9	303.4	264.1	8.36	7.64	4.60	6.86
1997	6.94	5.80	3.59	2.78	4.99	127.3	278.8	276.0	8.43	7.59	4.53	6.85
1998	6.82	5.48	3.14	2.40	4.60	125.2	207.9	238.1	8.26	7.41	4.48	6.74
1999	6.58	5.32	3.13	2.62	4.52	121.6	243.6	257.4	8.16	7.26	4.43	6.66
2000	7.76	6.59	4.48	4.38	5.80	120.8	406.5	310.4	8.22	7.22	4.46	6.68

Source: US Department of Energy

PRICE HISTORY OF OIL, GAS, AND GASOLINE

	Actual current prices						Inflation adjusted real prices				
	Crude oil			Natural gas			Crude oil			Natural gas	
Year	US average wellhead $/bbl	West Texas Intermediate posted $/bbl	Motor* Gasoline pump price $/gal	US average wellhead $/Mcf	Consumer average $/Mcf	Producer Price index 1982-100	US average wellhead $/bbl	West Texas Intermediate posted $/bbl	Motor Gasoline pump price $/gal	US average wellhead $/Mcf	Consumer average $/Mcf
1927	1.30	1.28	0.211	0.088	0.220	16.5	7.879	7.758	1.278	0.533	1.333
1929	1.27	1.45	0.214	0.082	0.216	16.4	7.744	8.841	1.306	0.500	1.317
1931	0.65	0.77	0.170	0.070	0.233	12.6	5.159	6.111	1.348	0.556	1.849
1933	0.67	1.00	0.178	0.062	0.237	11.4	5.877	8.772	1.563	0.544	2.079
1935	0.97	1.00	0.188	0.058	0.224	13.8	7.029	7.246	1.365	0.420	1.623
1937	1.18	1.22	0.200	0.051	0.220	14.9	7.919	8.188	1.342	0.342	1.477
1939	1.02	1.02	0.188	0.049	0.216	13.3	7.669	7.699	1.410	0.368	1.624
1940	1.02	1.02	0.184	0.045	0.217	13.5	7.556	7.556	1.364	0.333	1.607
1941	1.14	1.17	0.192	0.049	0.221	15.1	7.550	7.748	1.274	0.325	1.464
1942	1.19	1.17	0.204	0.051	0.227	17.0	7.000	6.882	1.202	0.300	1.335
1943	1.20	1.17	0.205	0.052	0.223	17.8	6.742	6.573	1.153	0.292	1.253
1944	1.21	1.17	0.206	0.051	0.215	17.9	6.760	6.536	1.150	0.285	1.201
1945	1.22	1.17	0.205	0.049	0.214	18.2	6.703	6.429	1.126	0.269	1.176
1946	1.41	1.36	0.208	0.053	0.220	20.8	6.779	6.524	0.999	0.255	1.058
1947	1.93	1.84	0.231	0.060	0.232	25.6	7.539	7.188	0.903	0.234	0.906
1948	2.60	2.57	0.259	0.065	0.241	27.7	9.386	9.278	0.934	0.235	0.870
1949	2.54	2.57	0.268	0.063	0.254	26.3	9.658	9.772	1.019	0.240	0.966
1950	2.51	2.57	0.268	0.065	0.266	27.3	9.194	9.414	0.980	0.238	0.974
1951	2.53	2.57	0.272	0.073	0.298	30.4	8.322	8.454	0.893	0.240	0.980
1952	2.53	2.57	0.276	0.078	0.332	29.6	8.547	8.682	0.931	0.264	1.122
1953	2.68	2.72	0.287	0.092	0.355	29.2	9.178	9.298	0.983	0.315	1.216
1954	2.78	2.82	0.290	0.101	0.381	29.3	9.488	9.625	0.991	0.345	1.300
1955	2.77	2.82	0.291	0.104	0.400	29.3	9.454	9.625	0.992	0.355	1.365
1956	2.79	2.82	0.299	0.108	0.415	30.3	9.208	9.307	0.988	0.356	1.370
1957	3.09	3.04	0.310	0.113	0.431	31.2	9.904	9.753	0.992	0.362	1.381
1958	3.01	3.06	0.304	0.119	0.462	31.6	9.525	9.677	0.961	0.377	1.462
1959	2.90	2.98	0.305	0.129	0.477	31.7	9.148	9.385	0.962	0.407	1.505
1960	2.88	2.97	0.311	0.140	0.500	31.7	9.085	9.369	0.982	0.442	1.577

PRICE HISTORY OF OIL, GAS, AND GASOLINE–cont'd

1961	2.89	2.97	0.308	0.151	0.510	31.6	9.146	9.399	0.973	0.478	1.614
1962	2.90	2.97	0.306	0.155	0.514	31.7	9.148	9.369	0.967	0.489	1.621
1963	2.89	2.97	0.304	0.158	0.512	31.6	9.146	9.399	0.963	0.500	1.620
1964	2.88	2.95	0.304	0.154	0.519	31.6	9.114	9.320	0.962	0.487	1.642
1965	2.86	2.92	0.312	0.156	0.522	32.3	8.854	9.040	0.966	0.483	1.616
1966	2.88	2.94	0.321	0.157	0.523	33.3	8.649	8.817	0.964	0.471	1.571
1967	2.92	3.03	0.332	0.160	0.520	33.4	8.743	9.060	0.994	0.479	1.557
1968	2.94	3.07	0.337	0.164	0.504	34.2	8.596	8.977	0.985	0.480	1.474
1969	3.09	3.30	0.348	0.167	0.515	35.6	8.680	9.256	0.978	0.469	1.447
1970	3.18	3.35	0.357	0.171	0.550	36.9	8.618	9.079	0.967	0.463	1.491
1971	3.39	3.56	0.364	0.182	0.590	38.0	8.921	9.368	0.958	0.479	1.553
1972	3.39	3.56	0.361	0.186	0.630	39.8	8.518	8.945	0.907	0.467	1.583
1973	3.89	3.87	0.388	0.216	0.680	45.0	8.644	8.604	0.862	0.480	1.511
1974	6.74	10.37	0.532	0.304	0.840	53.5	12.598	19.383	0.994	0.568	1.570
1975	7.56	11.16	0.567	0.445	1.120	58.4	12.945	19.110	0.971	0.762	1.918
1976	8.14	12.65	0.590	0.580	1.370	61.1	13.322	20.704	0.966	0.949	2.242
1977	8.57	14.30	0.622	0.790	1.660	64.9	13.205	22.034	0.958	1.217	2.558
1978	8.96	14.85	0.626	0.905	1.850	69.9	12.818	21.245	0.896	1.295	2.647
1979	12.51	22.40	0.857	1.178	2.210	78.7	15.896	28.463	1.089	1.497	2.808
1980	21.59	37.37	1.191	1.590	2.800	89.7	24.069	41.661	1.328	1.773	3.122
1981	31.77	36.67	1.311	1.980	3.390	98.0	32.418	37.418	1.338	2.020	3.459
1982	28.52	32.75	1.296	2.460	4.150	100.0	28.520	32.750	1.296	2.460	4.150
1983	26.19	30.25	1.241	2.590	4.820	101.2	25.879	29.891	1.226	2.559	4.763
1984	25.88	29.83	1.212	2.660	4.850	103.6	24.981	28.793	1.170	2.568	4.681
1985	24.09	28.08	1.202	2.510	4.720	103.1	23.366	27.236	1.166	2.435	4.578
1986	12.51	16.44	0.927	1.940	4.130	100.1	12.498	16.421	0.926	1.938	4.046
1987	15.40	18.21	0.948	1.670	4.050	102.8	14.981	17.714	0.922	1.625	3.940
1988	12.58	15.52	0.946	1.690	4.090	106.9	11.768	14.518	0.885	1.581	3.826
1989	15.86	18.29	1.021	1.690	4.220	112.2	14.135	16.301	0.910	1.506	3.761
1990	20.03	23.17	1.164	1.710	4.190	116.3	17.223	19.923	1.001	1.470	3.603
1991	16.54	20.42	1.140	1.640	4.071	116.5	14.197	17.528	0.979	1.408	3.494
1992	15.99	19.67	1.127	1.740	4.173	117.2	13.643	16.783	0.962	1.485	3.561
1993	14.25	18.17	1.108	2.040	4.429	118.9	11.985	15.282	0.932	1.716	3.725
1994	13.19	16.46	1.112	1.850	4.509	120.4	10.955	13.671	0.924	1.537	3.745
1995	14.62	17.50	1.147	1.550	4.128	124.7	11.724	14.034	0.920	1.243	3.310
1996	18.46	20.77	1.231	2.170	4.668	127.7	14.456	16.265	0.964	1.699	3.655
1997	17.23	20.33	1.234	2.320	4.988	127.6	13.503	15.933	0.967	1.818	3.909
1998	10.87	14.15	1.059	1.940	4.600	124.4	8.738	11.375	0.851	1.559	3.698
1999	15.56	18.36	1.165	2.080	4.521	125.5	12.398	14.629	0.928	1.657	3.602
2000	26.73	29.16	1.510	3.680	5.800	132.7	20.143	21.974	1.138	2.773	4.371

*Leaded gasoline price prior to 1982, unleaded gasoline price beginning in 1982.
Source: US Department of Energy

INTERNATIONAL ACTIVE RIG COUNT

Region	1994	1995	1996	1997	1998	1999			2000			2001		
						Land	Off.	Total	Land	Off.	Total	Land	Off.	Total
NORTH AMERICA														
Canada	261	231	270	374	259	240	5	245	339	5	344	262	4	266
US[1]	775	723	779	943	827	519	106	625	776	140	916	866	134	1,000
Subtotal	**1,036**	**954**	**1,049**	**1,317**	**1,086**	**759**	**111**	**870**	**1,115**	**145**	**1,260**	**1,128**	**138**	**1,266**
LATIN AMERICA														
Argentina	71	92	68	57	44	34	1	35	57	0	57	62	0	62
Bolivia	10	8	7	5	12	13	0	13	11	0	11	6	0	6
Brazil	23	20	22	21	20	10	9	19	9	14	23	11	16	27
Chile	3	3	2	2	1	0	1	1	0	1	1	1	0	1
Colombia	15	16	14	16	12	12	0	12	14	0	14	16	0	16
Costa Rica	0	0	0	0	0	0	0	0	0	0	0	N/A	N/A	N/A
Ecuador	6	3	3	4	5	3	0	3	6	1	7	12	0	12
Mexico	28	34	46	49	55	33	10	43	38	6	44	54	9	63
Peru	4	6	7	9	5	2	0	2	3	1	4	2	1	3
Trinidad	5	5	4	4	6	0	3	3	1	3	4	0	5	5
Venezuela	69	85	108	110	82	43	14	57	48	15	63	44	16	60
Other	0	0	0	0	0	0	0	0	0	0	0	0	0	0
Subtotal	**234**	**272**	**281**	**277**	**242**	**150**	**38**	**188**	**187**	**41**	**228**	**208**	**47**	**255**
ASIA-PACIFIC														
Australia	12	12	14	14	15	4	6	10	5	5	10	7	3	10
Bangladesh	0	0	0	0	0	0	0	0	0	0	0	N/A	N/A	N/A
Brunei	2	1	2	3	2	0	3	3	0	2	2	1	1	2
China-offshore	12	12	9	9	12	0	9	9	0	9	9	0	8	8
India	82	70	61	60	52	36	10	46	40	9	49	39	11	50
Indonesia	38	35	43	54	50	21	13	34	23	9	32	39	10	49
Japan	7	8	7	5	5	5	0	5	6	0	6	9	1	10
Malaysia	10	8	6	6	8	0	6	6	0	7	7	0	11	11
Myanmar	10	10	11	7	8	7	0	7	9	0	9	6	0	6
New Zealand	2	2	1	1	1	1	0	1	1	0	1	2	0	2
Papua New Guinea	2	2	2	2	2	1	0	1	0	0	0	1	0	1
Philippines	10	7	6	2	1	0	0	0	0	0	0	1	0	1
Taiwan	3	2	2	2	1	0	0	0	0	0	0	0	0	0
Thailand	5	4	7	7	8	1	6	7	1	5	6	1	7	8
Viet Nam	8	6	6	6	7	1	8	9	0	8	8	0	7	7
Other	1	0	1	2	2	1	1	2	0	1	1	0	0	0
Subtotal	**204**	**179**	**178**	**180**	**174**	**78**	**62**	**140**	**85**	**55**	**140**	**106**	**59**	**165**
AFRICA														
Algeria	28	28	28	29	23	13	0	13	15	0	15	20	0	20
Angola	5	7	10	8	6	0	5	5	0	6	6	0	4	4
Congo	2	2	3	5	6	0	3	3	0	3	3	0	1	1
Gabon	2	3	3	5	6	1	1	2	1	1	2	0	1	1
Kenya	0	0	0	0	0	0	0	0	0	0	0	0	0	0
Libya	13	11	13	13	13	8	0	8	7	0	7	4	0	4
Nigeria	7	9	14	13	12	2	6	8	1	7	8	2	7	9
South Africa	1	1	0	1	1	0	1	1	0	1	1	0	1	1
Tunisia	3	3	2	2	2	1	0	1	1	0	1	1	1	2
Other	3	3	4	5	6	1	1	2	0	3	3	0	3	3
Subtotal	**64**	**67**	**77**	**81**	**75**	**26**	**17**	**43**	**25**	**21**	**46**	**27**	**18**	**45**
EUROPE														
Austria	0	1	0	0	0	0	0	0	0	0	0	N/A	N/A	N/A
Croatia[2]	0	0	3	1	0	0	0	0	0	0	0	0	0	0
Denmark	2	2	2	2	3	0	2	2	0	3	3	1	6	7
France	2	1	1	1	2	1	0	1	1	0	1	0	0	0
Germany	8	9	9	4	3	4	0	4	3	0	3	5	0	5
Greece	0	0	0	0	0	0	0	0	0	0	0	N/A	N/A	N/A
Hungary	2	2	2	1	2	1	0	1	1	0	1	1	0	1
Italy	11	11	13	12	9	8	1	9	7	1	8	6	1	7
Netherlands	5	8	11	11	8	1	3	4	1	3	4	2	6	8
Norway	14	13	15	17	17	0	17	17	0	22	22	0	24	24
Poland	18	18	18	16	15	14	0	14	15	0	15	12	1	13
Romania	4	4	3	2	2	2	0	2	0	0	0	2	0	2
Spain	0	0	0	0	0	0	0	0	0	0	0	N/A	N/A	N/A
Turkey	16	15	10	10	7	6	0	6	6	0	6	6	0	6
United Kingdom	23	27	31	34	28	1	17	18	0	17	17	1	22	23
Other	2	2	3	2	4	1	1	2	3	0	3	2	1	3
Subtotal	**107**	**113**	**121**	**113**	**100**	**39**	**41**	**80**	**37**	**46**	**83**	**38**	**61**	**99**

INTERNATIONAL ACTIVE RIG COUNT-cont'd

MIDDLE EAST														
Abu Dhabi	15	14	14	15	15	7	6	13	7	4	11	10	6	16
Dubai	2	3	3	2	3	1	1	2	1	1	2	0	2	2
Egypt	14	17	19	21	22	9	8	17	12	6	18	17	7	24
Iran	21	21	22	22	25	25	2	27	24	3	27	26	7	33
Iraq	0	0	0	0	0	0	0	0	0	0	0	0	0	0
Jordan	0	0	0	0	0	0	0	0	0	0	0	0	0	0
Kuwait	9	7	5	6	8	11	0	11	12	0	12	6	0	6
Oman	16	17	19	23	24	19	0	19	24	0	24	27	0	27
Pakistan	10	11	11	12	11	7	0	7	10	0	10	5	0	5
Qatar	4	4	8	13	12	3	3	6	2	4	6	3	11	14
Saudi Arabia	17	17	17	26	28	18	2	20	21	4	25	27	4	31
Syria	12	13	15	14	14	13	0	13	14	0	14	19	0	19
Yemen	4	3	3	3	4	4	0	4	6	0	6	7	0	7
Other	2	1	1	0	1	1	0	1	1	0	1	4	0	4
Subtotal	**126**	**128**	**137**	**157**	**167**	**118**	**22**	**140**	**134**	**22**	**156**	**151**	**37**	**188**
Total	**1,771**	**1,711**	**1,843**	**2,126**	**1,845**	**1,170**	**291**	**1,461**	**1,583**	**330**	**1,913**			

[1]US land includes inland water rigs: 22 in 1998, 17 in 1999, 18 in 2000. [2]Prior to 1996 listed as Yugoslavia.

Source: Baker Hughes Inc.

IPE Information Sources
Glossary of Abbreviations and Acronyms

A

ACG	Azeri-Chirag-Guneshli (Azerbaijan oil fields)
ADMA-OPCO	Abu Dhabi Marine Operations Co.
ADNOC	Abu Dhabi National Oil Co.
AEP	American Electric Power Co.
A.G.A.	American Gas Association
Agip KCO	Agip Kazakhstan North Caspian Operating Co., formerly OKIOC
AIOC	Azerbaijan International Oil Co.
AMBO	Albanian Macedonian Bulgarian Oil Co.
AMPCO	Atlantic Methanol Production Co. LLC
ANP	National Petroleum Agency (Brazil)
ANWR	Arctic National Wildlife Refuge (Alaska, US)
API	American Petroleum Institute
ASEAN	Association of Southeast Asian Nations
ATHEER	Abu Dhabi Gas Co. (United Arab Emirates)

B

b/d	barrels per day
Banagas	Bahrain National Gas Co.
Banco	Bahrain National Oil Co.
Bapco	Bahrain Petroleum Co.
bbl	barrels
bcf	billion cubic feet
bcfd	billion cubic feet per day
bcfe	billion cubic feet equivalent
bcm	billion cubic meters
bo/d	barrels of oil per day
boe	barrels of oil equivalent
boe/d	barrels of oil equivalent per day
BTC	Baku (Azerbaijan)-Tbilisi (Georgia)-Ceyhan (Turkey) oil pipeline
Btu	British thermal unit
bw/d	barrels of water per day

C

CEO	chief executive officer
CERI	Canadian Energy Research Institute
cfd	cubic feet per day
CGES	Centre for Global Energy Studies
CNOOC	Chinese National Offshore Oil Corp.
CNR	Canadian Natural Resources Ltd.
CO	carbon monoxide
CO_2	carbon dioxide
CPC	Caspian Pipeline Consortium
CPC	Chinese Petroleum Corp. (Taiwan)
CSO	Coflexip Stena Offshore
cu m	cubic meters (metres)

D

DIGP	Dauphin Island Gathering Partners
DOE	Department of Energy
DONG	Dansk Olie ag Naturgas AS (Denmark)
DUC	Dansk Undergrunds Consortium (Denmark)

E

E&D	exploration and development
E&P	exploration and production
EC	European Commission
EDP	Electricidade de Portugal SA
EGAS	Egyptian Natural Holding Gas Co.
EGPC	Egyptian General Petroleum Corp.
EIA	Energy Information Administration (US)
ELNG	Egyptian LNG joint venture
ENAP	Empresa Nacional de Petroleo (Chile)
ENI	Ente Nazionale Idrocarburi (Italy)
EPA	Environmental Protection Agency
ETAP	Enterprises Tuniseienne d'Activites Petrolieres

EU	European Union
EVE	Basque Energy Authority (Spain)

F

FERC	Federal Energy Regulatory Commission (US)
FNA	Frigg Norwegian Association (Norway)
FPC	Formosa Petrochemical Corp.
FPSO	floating production, storage, and off-loading
FSU	Former Soviet Union
ft	foot
FTC	Federal Trade Commission (US)

G

GAIL	Gas Authority of India Ltd.
GASCO	Abu Dhabi Gas Industries Ltd. (United Arab Emirates)
GCC	Gulf Cooperation Council
GdF	Gaz de France SA
GDP	gross domestic product
GEA	Gas Energy Adria
GFU	Gas Negotiating Committee (Norway)
GTI	Gas Technology Institute (US)
GTL	gas-to-liquids
GTLB	GTL Bolivia SA

H

H&P	Helmerich & Payne Inc.
HDPE	high-density polyethylene
HDS	hydroskimming
hp	horsepower
HP	Hellenic Petroleum (Greece)

I

IEA	International Energy Agency
IGCC	integrated gasification combined cycle
IMF	International Monetary Fund
IMO	International Maritime Organization
INA	Industrija Nafte DD Zagreb (Croatia)
INGAA	Interstate Natural Gas Association of America
IPAA	Independent Petroleum Association of America
IPO	initial public offering
IOC	Indian Oil Corp.

J

JDA	Joint Development Area (Malaysia-Thailand)
JV	joint venture

K

KIO	Karachaganak Integrated Organization (consortium)
km	kilometer
KNOC	Korean National Oil Corp.
Kogas	Korea Gas Corp.
KVA	kilo-volt-amperes
KZM	Ku-Zaap-Maloob complex (Mexico)

L

LNG	liquefied natural gas
Logic	leading oil and gas industry competitiveness (UK)
LPG	liquefied petroleum gas

M

m	meter
Mcf	thousand cubic feet
MIDOR	Middle East Oil Refinery Ltd.
MMBtu	million British thermal units
MMcf	million cubic feet
MMcfd	million cubic feet per day
MMS	Minerals Management Service
MOL Rt.	Hungarian Oil and Gas Co.
MTBE	methyl tertiary butyl ether

N

NAM	Nederlandse Aardolie Maatschappij BV (Netherlands)
NETA	New Electricity Trading Arrangements (UK)
NIOC	National Iranian Oil Company
NLNG	Nigeria Liquefied Natural Gas Co.
NNPC	Nigerian National Petroleum Co.
NO_x	nitrogen oxides
NPR-A	National Petroleum Reserve-Alaska
NYMEX	New York Mercantile Exchange
NYSE	New York Stock Exchange

O

OCP	Oleoducto de Crudos Pesados (consortium and pipeline, Ecuador)
OGJ	Oil & Gas Journal
OKIOC	Offshore Kazakhstan International Operating Company, re-named Agip KCO
OLNGC	Oman Liquefied Natural Gas Co.
ONGC	Oil & Natural Gas Corp. (India)
OPEC	Organization of Petroleum Exporting Countries

OPIC — Overseas Private Investment Corp., a US government loan agency

P

Parco — Pak-Arab Refinery Co. Ltd.
PDO — Petroleum Development Oman
PDVSA — Petroleos de Venezuela SA
Pemex — Petroleos Mexicanos
Petrobangla — Bangladesh Oil, Gas & Mineral Corp.
Petrobras — Petroleo Brasileiro SA
Petroecuador — Petroleos del Ecuador
Petrogal — Petroleos de Portugal
Petronas — Petroliam Nasional Berhad (Malaysia)
PetroVietnam — Vietnam Oil & Gas Co.
PG&E — Pacific Gas & Electric Co.
PGC — Potential Gas Committee
PIC — Petrochemical Industries Co. (Kuwait)
PIRC — Petroleum Industry Research Inc.
PKN — Polski Koncern Naftowy (Poland)
POGC — Polskie Gomictwo Naftowe I Gazownictow SA (Polish Oil & Gas Co.)
PSA — production-sharing arrangement
PSC — production-sharing contract
PSO — Pakistan State Oil
PTT PLC — Thai state-owned oil and gas firm (formerly Petroleum Authority of Thailand)
PTTEP — PTT Exploration and Production PLC

R

RasGas — Ras Laffan Liquefied Natural Gas Co. Ltd. (Qatar)
RFG — reformulated gasoline
ROSBOS — Rosetti Marino SPA-Bouygues Offshore joint venture
RR — railroad
RTO — regional transmission organization

S

SAGD — steam-assisted gravity drainage
SAPREF — South African Petroleum Refineries (Pty) Ltd.
scfd — standard cubic feet per day
SDFI — state's direct financial interest (Norway), renamed Petoro
semi — semisubmersible offshore production platform
SO_2 — sulfur dioxide
SO_x — sulfur oxides
SOCAR — State Oil Company of Azerbaijan Republic
SOTE — Trans-Ecuadorian pipeline's Spanish-language acronym
SPLA — Sudan People's Liberation Army
SPP — Slovensky plynarensky priemysel (Slovakia)
SPR — Strategic Petroleum Reserve
SRS — Schmierstoff Raffinerie Salzbergen (German refinery)
Statoil — Den norske stats oljeselskap AS (Norway)

T

tcf — trillion cubic feet
TCP — Trans-Caspian Pipeline
TGS — Transportadora de Gas del Sur (South American pipeline operator)
tj — terajoules
toe — tons or tonnes of oil equivalent
TPAO — Turkiye Petrolleri Anonim Ortakligi (Turkey)
tpy — tons or tonnes per year

U

UAE — United Arab Emirates
UDS — Ultramar Diamond Shamrock Corp.
UES — Unified Energy Systems (Russia)
ULCC — ultra large crude carrier
UN — United Nations
UOG — United Arab Emirates Offsets Group

V

VLCC — very large crude carrier
VP — Vice President

W

WNTS — West Natuna Transportation System (Indonesia to Singapore)
WTO — World Trade Organization

IPE Information Sources
National Oil Companies & Energy Ministries

A

ALBANIA
Albpetrol (formerly DPNG)
c/o Geological Institute for Oil & Gas
Fieri, ALBANIA
Ph: 355 423-4542
Fax: 355 423-4204

National Petroleum Agency
Agjensia Kombetare e Hidrokarbureve
Rruga "Duressit," No. 83
Tirana, ALBANIA
Ph. & Fax: 355 423-1034

ALGERIA
Sonatrach Inc.
Immeuble Mauretania Pl. Perou
80 Ave. Ahmed Ghermoul
Algiers, ALGERIA
Ph: 213 21 54-80-11
Fax: 213 21 54-77-00
E-mail: sonatrach@sonatrach-dz.com
Website: www.sonatrach-dz.com

Ministry of Energy & Mines
80 Rue Ahmed Ghermoul
Algiers, ALGERIA
Ph: 213 (0)21 67 33 00
Fax: 213 (0)21 67 03 66
E-mail: info@mem-algeria.org
Website: www.mem-algeria.org

ANGOLA
Sociedad Nacional de Combustiveis de Angola (SONANGOL)
Box 1316
Luanda, ANGOLA
Ph: 244-2-334448
Fax: 244-2-391782
Website: www.sonangol.co.ao

ARGENTINA
Repsol-YPF
Av. Roque Sáenz Peña, 777
C.P. 1364, Buenos Aires, ARGENTINA
Website: www.repsol-ypf.com

ARMENIA
Republic of Armenia
Ministry of Energy
Petroleum Project Implementation Unit
2 Government House
Republic Square
Yerevan 375010, ARMENIA
Ph: 3742 52-85-34
Fax: 3742 15-16-87

AUSTRIA
Federal Ministry for Economic Affairs
Supreme Mining Authority
Department VII/3
Landstrasser Hauptstrasse 55-57
A-1011, Vienna, AUSTRIA
Ph: 43/1/711 00-0
E-mail: service@bmwa.gv.at
Website: www.bmwa.gv.at

OMV Aktiengesellschaft
Otto Wagner-Platz 5
A-1090 Vienna, AUSTRIA
Ph: 43 222 404-4090
Fax: 43 222 404-4091
E-mail: info@omv.co.at

AZERBAIJAN
State Oil Co. of Azerbaijan (SOCAR)
Neftchilar Ave. 73
Baku 370004, AZERBAIJAN
Ph: 78922 92-35-11
Fax: 788922 92-32-97
E-mail: eakhmedo@aiocaz.com
Website: www.socar-cc.com
Website: www.azer.com

B

BAHRAIN
Bahrain National Gas Company (Banagas)
P.O. Box 29099, BAHRAIN
Ph: (973) 756222
Fax: (973) 756991
Website: www.banagas.com

Bahrain National Oil Co.
PO Box 25504
Awali, BAHRAIN
Ph: 973 75-46-66
Fax: 973 75-32-03

BANGLADESH
Bangladesh Oil, Gas & Mineral Corp. (PETROBANGLA)
Chamber Bldg.
122-124, Motijheel Comm'l Area
Dhaka 1000, BANGLADESH
Ph: 880 2 81-49-72
Fax: 880 2 81-16-13

Minister of Energy & Mineral Resources
Dhaka, BANGLADESH
Website: www.bangladeshgov.org

BOLIVIA
Yacimientos Petrolifero Fiscales Bolivianos (YPFB)
Box 401
La Paz, BOLIVIA
Ph: 591 2 374-468
Fax: 591 2 374-469

Vice-Ministry of Energy and Hydrocarbons
Av. Mariscal Santa Cruz eaq. Oruro
Edificio Palacio de Comunicaciones, Piso 12
La Paz, BOLIVIA
Ph: 591 2 374-050
Fax: 591 2 392-758
E-mail: enerhid@ceibo.entelnet.bo

BRAZIL
Ministry of Mines and Energy
Website: www.mme.gov.br

Petroleo Brasileiro SA (PETROBRAS)
Caixa Postal 809
Avenida Republica do Chile 65
20035-900 Rio de Janeiro, RJ, BRAZIL
Ph: 55 21 534-4477
Fax: 55 21 534-1939
Website: www.petrobras.com.br

Petrobras Internacional SA (BRASPETRO)
Rua Gen'l Canabarro, 500, 10th Floor
20.271-900, Maracana,
Rio de Janeiro, RJ, BRAZIL
Ph: 55 21 566-4477
Fax: 55 21 566-3400/3401
Website: www.petrobras.com.br

BRUNEI
Director of Petroleum Unit
Petroleum Unit
Jabatan Perdana Menteri
Bandar Seri Begawan 2007
Negara BRUNEI Darussalam
Ph: 673 2 28-932
Fax: 673 2 40-270
Website: www.petroleum-unit.gov.bn

BULGARIA
Committee of Geology & Mineral Resources
22 Princess Marie-Louise Blvd.
Sofia 1000, BULGARIA
Ph: 359 2 981-8861, 981-8841
Fax: 359 2 980-5561
E-mail: cgmr-gf@cserv.mgu.bg

C

CAMEROON
Societe Nationale des Hydrocarbures (SNH)
Box 955
Yaounde, CAMEROON
Ph: 237 20-19-10, 20-98-64
Fax: 237 20-46-51

CANADA
Natural Resources Canada
Energy Sector
K1A 0E4 Ottawa, Ontario
CANADA
Ph: (613) 995-0947
Fax: (613) 992-8738
Website: www.NRCan.gc.ca

CHILE
Empresa Nacional del Petróleo
Todos los derechos reservados
Av. Vitacura 2736 piso 10 Las Condes
Santiago, CHILE
Ph: (56-2) 280 3000
Fax: (56-2) 280 3199
Website: www.enap.cl

CHINA
Ministry of Petroleum Industry
PO Box 1411
Beijing
PEOPLE'S REPUBLIC OF CHINA

China National Offshore Oil Corp. (CNOOC)
Jingxin Bldg., Dongsanhuan Bei Rd.
Box 4705
Beijing 100027
PEOPLE'S REPUBLIC OF CHINA
Ph: 86 10 6466-9001, 6466-3696
Fax: 86 10 6466-2994, 6466-9007
Website: www.cnooc.com.cn

China National Petroleum Corp.
PO Box 766
Beijing
PEOPLE'S REPUBLIC OF CHINA
Ph: 86 1 201-5544

China National Petroleum Corp.
One Westchase Center Suite 840
10777 Westheimer Rd.
Houston, Tex. 77042
Ph: 713 784-8598
Fax: 713 784-9197

China National Star Petroleum Corp.
No. 1 Beisihuan Xilu, Haidian Dist.
PO Box 8326
Beijing 100083, CHINA
Ph: 8610 6232-9090
Fax: 8610 6232-5550

COLOMBIA
Empresa Colombiana del Petroles
(ECOPETROL)
Carrera 13, No. 3624
Apartado Aereo 5938
Bogota, COLOMBIA
Ph: 57 1 287-9308
Fax: 57 1 288-0071
E-mail: prozo@infantas.ecp.com
Website: www.ecopetrol.com.co

Ministerio de Minas y Energia de Colombia
Centro Administrative Nacional
Avenida El Dorado
Bogota, COLOMBIA
Ph: 57 1 44-520, 44-525
Website: www.minminas.gov.co

CONGO (BRAZZAVILLE)
General Direction of Hydrocarbons
Ministry of Mines & Energy - S.P. 2120
Brazzaville, CONGO
Ph: 242 83-12-81, 83-58-73, 83-59-74
Fax: 242 83-62-43

Hydro Congo
B.P. 2008
Pointe Noire, Brazzaville, CONGO
Ph: 242 81-35-60, 81-40-23
Fax: 242 83-12-38

CONGO (FORMER ZAIRE)
Petrozaire
B.P. 7617
Kinshasa/Gombe, ZAIRE
Ph: 243 12 20-344, 20-492

CROATIA
INA-NAFTAPLIN
Subiceva 29
10,000 Zagreb, CROATIA
Ph: 385 1 44-38-44, 44-00-11, 44-00-22
Fax: 385 1 459-2360

CUBA
Ministerio de la Industria Basica
Empedarado 113
Havana, CUBA
Website: www.cubagob.cu

Comercial Cupet SA (CUPET)
km 37.5 Via Blanca
Caribe, Sta. Cruz del Norte
Havana 32900, CUBA
Ph: 53 7 62-40-11
Fax: 53 7 33-34-60, 33-80-27, 33-30-72

CZECH REPUBLIC
Ministry of Economic Policy & Development
tr. SNG 65 - 101 60 Prague 10
CZECH REPUBLIC
Ph: 42 2 712-1111
Fax: 42 2 73-13-57
Website: www.czech.cz

D

DENMARK
Danish Energy Agency
Energiministeriet
DK-1216 Copenhagen, DENMARK
Ph: 45 33 92-67-00
Fax: 45 33 11-47-43

Ministry of Energy
Slotsholmsgade 1
DK-1216 Copenhagen, DENMARK
Ph: 45 3 392-3350
Fax: 45 3 392-7508

Ministry of Environment
Højbro Plads 4
1200 København K, DENMARK
Ph: 45 33 92 76 00
Fax: 45 33 32 22 27
E-mail: mem@mem.dk
Website: www.mem.dk

DUBAI
Dubai Petroleum Co. (DUPETCO)
Box 2222
Dubai, UNITED ARAB EMIRATES
Ph: 971 4 44-29-90

E

ECUADOR
Petroleos del Ecuador (Petroecuador)
Alpallana E8-86 y
Av. 6 de Diciembre
P.O. Box 17-11-5007
Quito, ECUADOR
Ph: (593-2) 2563060
(593-2) 2561589
Fax: (593-2) 2503571
Website: www.petroecuador.com.ec

Ministerio de Recursos Naturales y Energeticos
Direccion General de Geologia y Miras
Carrion 1016 y Paez
Casilla 23-A
Quito, ECUADOR
Ph: 593 2 23-12-04

Organizacion Latinoamericana
de Energia (Olade)
Av. Occidental N5863
Sector San Carlos
Quito, ECUADOR
Ph: (593)-(2) 2598122 / 2598280 / 2597995
Fax: 2539684
E-mail: olade@olade.org.ec
Website: www.olade.org.ec

EGYPT
Egyptian General Petroleum Corporation (EGPC)
Palestine Street part 4
New Maadi, Cairo, EGYPT
Tel: 706 59 56
706 59 54
Fax: 702 88 13 703 14 57
E-mail: egpcinf@starnet.com.eg
Website: www.egpc.com.eg

Arab Republic of Egypt
Ministry of Petroleum
1 (A) Ahmed El Zomor str.
8th District Nasr City
Cairo, EGYPT
Ph: 00(202)670-64-35
00(202)670-64-05
Fax: 00(202)670-64-19

EQUATORIAL GUINEA
Ministry of Mines & Energy
Box 778
Malabo, C/12 de Octubre
EQUATORIAL GUINEA
Ph: 240 9 3567, 2086
Fax: 240 9 3353
Ph: (00 240) 7 7502
E-mail: d.shaw@ecqc.com
Website: www.equatorialoil.com
E-mail: mme@intnet.gq

F

FAROE ISLANDS
Faroese Petroleum Administration
Debesartro
100 Torshaun
FAROE ISLANDS
Ph: 298 12-306
Fax: 298 18-438
E-mail: oms@oms.fo

FINLAND
Ministry of Trade & Industry
Aleksanterinkatu 4
P.O. Box 32
FINLAND 00023 GOVERNMENT
Ph: +358 9 1601
Fax: + 358 9 160 3666
Website: www.vn.fi

FRANCE
Service de Conservation des Gisements d'Hydrocarbures
Ministry of Industry
366 Avenue Napoleon Bonaparte
Rueil-Malmaison, FRANCE

G

GABON
Societe Nationale Petroliere Gabonaise (PETROGAB)
Box 564
Libreville, GABON

Ministry of Mines, Energy, Oil and Hydraulic Resources
PO Box 2199
Libreville, GABON
Ph: 241 76-10-71
Fax: 241 72-49-90
E-mail: mcn@internetgabon.com

GREECE
Ministry for Industry, Energy & Technology
80 Michalakopoulos St.
Athens 611, GREECE
Ph: 30 1 770-5111

Public Petroleum Corp. of Greece
199 Kifissias Ave. GR, 151 24
15124 Athens, GREECE
Ph: 30 1 806-9801
Fax: 30 1 806-4212

GUATEMALA
Ministerio de Energia y Minas
General Directorate of Hydrocarbons
Diagonal 17, 29-78 Zona 11
01011 Guatemala City, GUATEMALA
Ph: (502) 477-0382
Fax: (502) 476-8506
E-mail: informatica@mem.gob.gt
Website: www.mem.gob.gt

H

HUNGARY
MOL Hungarian Oil & Gas PLC
H-1502, PO Box 22
Xi, Schonherz Z. U. 18
1117 Budapest
HUNGARY
Ph: 36 1 166-2646
Fax: 36 1 186-8281
E-mail: sczenthe@mol.hu

I

INDIA
Oil & Natural Gas Corp. Ltd. (ONGC)
7th Floor, Bank of Baroda Bldg.
Parliament Street
New Delhi 110 001, INDIA
Ph: 91 11 331-7205, 371-5291
Fax: 91 11 331-6413
E-mail: endongc@del2.vsnl.net.in
Website: www.ongcindia.com

Director General of Hydrocarbons
Ministry of Petroleum & Natural Gas
Hindustan Times House
18-20, Kasturba Gandhi Marg.,
New Delhi 110 001, INDIA
Ph: 91 11 335-2650, 335-2617
Fax: 91 11 331-7081, 335-2649
E-mail: dspdi.png@sb.nic.in
Website: www.petroleum.nic.in

INDONESIA
Perusahaan Pertambangan Minyak Dan Gas Bumi Negara (PERTAMINA)
Jalan Merdeka Timur No. 1
Box 12
Jakarta 10110, INDONESIA
Ph: 62 21 381-5111, 381-6111
Fax: 62 21 384-3882, 384-6865
Website: www.pertamina.co.id

IRAN
National Iranian Oil Co. (NIOC)
Ave. Takhte Jamshid
Box 1863
Tehran, IRAN
Ph: 98 21 6151

Petroleum Ministry of Islamic Republic of Iran
Box 1863
Tehran, IRAN
Ph: 98 21 66-11-20
Fax: 98 21 66-62-93

IRAQ
Iraq National Oil Co. (INOC)
Al-Khullani Square
Box 476
Baghdad, IRAQ
Ph: 964 1 80-066, 80-069

Ministry of Oil
Box 6178
Al-Mansour, Baghdad, IRAQ
Ph: 964 1 541-0031

IRELAND
Department of Public Enterprise
Energy Division
25 Clare St.
Dublin 2, IRELAND
Ph: + 353 1 1890 443311
Website: www.irlgov.ie

Department of Transport, Energy and Communications
Beggars Bush, Haddington Road
Dublin, IRELAND
Ph: 353 1 671-5233
Fax: 353 1 660-4462

ISRAEL
Israel National Oil Co. Ltd.
Beit Gibor, 6 Kaufman St.
Tel-Aviv 68012, ISRAEL
PO Box 50199
Tel-Aviv 61500, ISRAEL
Ph: 972 3 514-2020
Fax: 972 3 514-2061

Ministry of Energy & Infrastructure
234 Yaffo St.
Box 1442/91013
Jerusalem, ISRAEL
Ph: 972 2 31-61-11
Fax: 972 2 38-14-44
Website: www.std.com

ITALY
Ufficio Nazionale Minerario per gli Idrocarburi e la Geotermia
Via Veneto
33 Roma, ITALY
Ph: 06 47887835
E-mail: martino@mica-dgfe.casaccia.enea.it
Website: http://mica-dgfe.casaccia.enea.it

Ente Nazionale Idrocarburi (ENI)
1, Piazzale Enrico Mattei
00144 Rome, ITALY
Ph: 39 6 59002
Fax: 39 6 5900-2141
Website: www.eni.it

IVORY COAST
Director of the Office of the Minister
Ministere des Mines
Box V 50
Abidjan, COTE D'IVOIRE
Ph: 225 21-50-03
Fax: 225 21-53-20

Societe Nationale d'Operations Petrolieres de la Cote d'Ivoire (PETROCI)
Box V 194
Abidjan, COTE D'IVOIRE
Ph: 225 20-25-00, 20-25-10
Fax: 225 21-68-24

J

JAPAN
Japan National Oil Corp. (JNOC)
Fukoku-Seimei Bldg.
2-2 Uchisaiwaicho, 2-chome
Chiyoda-ku, Tokyo 100, JAPAN
Ph: 81 3 3580-5411
Fax: 81 3 3591-0172
Website: www.jnoc.go.jp

K

KAZAKHSTAN
KazakhOil
Republik Ave., 60
473000 Astana
KAZAKHSTAN
Ph: 7 3172 28 0609, 28 0270
Fax: 7 3172 28 0296
Website: www.kazakhoil.kz

Ministry of Fuel & Energy
Bakenbay-Batyj St. 142
480091 Almaty
KAZAKHSTAN
Ph: 7 3272 62-64-10, 62-60-80
Fax: 7 3272 62-66-30

KUWAIT
Kuwait Petroleum Corp. (KPC)
Box 26565
13126 Safat, KUWAIT
Ph: 965 245-5455
Fax: 965 246-7159
Website: www.kpc.com.kw

Ministry of Oil
PO Box 5077
Safat, KUWAIT

L

LIBYA
General Department of Technical Affairs
Ministry of Petroleum
PO Box 256
Tripoli, LIBYA

National Oil Corp. (NOC)
Box 2655
Tripoli, LIBYA
Ph: 218 21 46-181

M

MALAYSIA
Petroliam Nasional Berhad (PETRONAS)
Menara Dayabumi, Komplex Dayabumi
Jalan Sultan Hishamuddin
Box 12444
50778 Kuala Lumpur, MALAYSIA
Ph: 60 3 274-3833, 274-8011, 274-8022
Fax: 60 3 274-0217
Website: www.petronas.com.my

MEXICO
Petroleos Mexicanos (PEMEX)
Marina Nacional
#329 Col. Huasteca
Delegacion Miguel Hidalgo
Mexico C.D. 11311, MEXICO
Website: www.pemex.com

MONGOLIA
Mongol Petroleum Co. (MGT)
Uildverchnii Street 37
Ulaanbaatar, MONGOLIA
Ph: 976 1 38-34-83
Fax: 976 1 33-11-76

Petroleum Authority of Mongolia
Ulaanbaatar-37
Uildverchnii gudamj
MGT, MONGOLIA
Ph: 976 1 38-34-83
Fax: 976 1 33-11-76

MYANMAR
Burma Ministry of Mines
Minister's Office
Yangon, MYANMAR

Myanma Oil & Gas Enterprise
Box 1049
74-80 Min Ye Kyawzwa Rd.
Yangon, MYANMAR
Ph: 95 1 21-394, 21-027
Fax: 95 1 22-964, 22-965

N

NETHERLANDS
Ministry of Economic Affairs
General Energy Policy Division
Postbus 20101
2500 The Hague
NETHERLANDS
Ph: 31 70 79-89-11
Fax: 31 70 79-63-58

NEWFOUNDLAND & LABRADOR
Government of Newfoundland & Labrador
Department of Mines and Energy
PO Box 8700
St. John's, Newfoundland
CANADA A1B 4J6
Ph: 709 729-6877
Fax: 709 729-2508
Website: www.gov.nf.ca

NEW BRUNSWICK
New Brunswick Department of Natural Resources & Energy
Box 6000
Fredericton, N.B., CANADA E3B 5H1
Ph: 506 453-2206
Fax: 506 453-3671
Website: www.gov.nb.ca

NEW SOUTH WALES
Director General
Department of Mineral Resources
PO Box 536
St. Leonards, NSW 2065, AUSTRALIA
Ph: 61 2 9901 8888
Fax: 61 2 9901 8520
Website: www.doe.nsw.gov.au

NEW ZEALAND
New Zealand Crown Minerals
Ministry of Commerce
Box 1473
Wellington, NEW ZEALAND
Ph: 64 4 472-0030
Fax: 64 4 474-2829
E-mail: public@moc.govt.nz
Website: www.govt.nz

NIGERIA
Ministry of Petroleum Resources
44, Eric Moore, Surulere
P.M.B. 12701
Lagos, NIGERIA
Ph: 234 1 60-31-00

Nigerian National Petroleum Corp. (NNPC)
Falomo Office Complex
Private Mail Bag 12701 (Ikoyi)
Lagos, NIGERIA
Ph: 234 1 261-4650, 4228, 4871, 4959

NORTHERN TERRITORY

Secretary, Dept. of Mines & Energy
PO Box 2901
Darwin, N.T. 0800 AUSTRALIA
Ph: 61 089 99 5511
Fax: 089 99 5530
Website: www.dme.nt.gov.au

NORTHWEST TERRITORIES

Northern Oil & Gas Directorate
6th Floor, 10 Wellington St.
Ottawa, Ontario, CANADA K1A 0H4
Ph: 819-997-0878
Fax: 819-953-5828
Website: www.gov.nt.ca

NORWAY

Den norske stats oljeselskap AS
(STATOIL)
Box 300
N-4035 Stavanger, NORWAY
Ph: 47 51 90 00 00
Fax: 47 5199 00 50
Website: www.statoil.com

Royal Ministry of Petroleum & Energy
Ploensgt 8
Box 8148 Dep. 0033
Oslo 1, NORWAY
Ph: 47 2 34-90-90
Fax: 47 2 34-95-65
Website: odin.dep.no

NOVA SCOTIA

Canada-Nova Scotia Offshore Petroleum Board
6th Floor, TD Centre, 1791 Barrington St.
Halifax, NOVA SCOTIA B3J 3K9
Ph: 902 422-5588
Fax: 902 422-1799
E-mail: postmaster@cnsopb.ns.ca
Website: www.gov.ns.ca

O

OMAN

Director General of Petroleum and Gas Affairs
Ministry of Petroleum & Minerals
Sultanate of Oman
Box 551
Muscat, OMAN
Ph: 968 60-33-33

Petroleum Development Oman (PDO)
Box 81
Mina Al-Fahal, OMAN
Ph: 968 67-81-11
Fax: 968 67-71-06
Website: www.pdo.co.om

P

PAKISTAN

Directorate General of Petroleum Concessions
Ministry of Petroleum & Natural Resources
1019-A, Pak Plaza
Fazal-e-Haq Rd., Blue Area
Islamabad, PAKISTAN
Ph: 92 51 82-49-93, 81-83-52
Fax: 92 51 21-32-45

Government Of Pakistan
Ministry of Petroleum and Natural Resources
3rd Floor, A Block, Pak Secretariat
Islamabad, PAKISTAN
Ph: 051-9208233
Fax: 051-9205437
E-mail: info@mpnr.gov.pk
Website: www.mpnr.gov.pk

Oil & Gas Development Corp.
Markaz F-8
Islamabad, PAKISTAN
Ph: 92 51 85-39-74
Fax: 92 51 85-89-39

PAPUA NEW GUINEA

Papua New Guinea Department of Mining & Petroleum (Petroleum Division)
Private Mail Bag
Port Moresby Post Office
PAPUA NEW GUINEA
Ph: 675 22 4200
Fax: 675 22 4222

PARAGUAY

Petroleos Paraguayos (PETROPAR)
Oliva 299 Casi Chile
Edificio Finamericana 3RO y 4TO Piso
Casilla Correo 571
Asuncion, PARAGUAY
Ph: 595 21 44-85-03
Fax: 595 21 49-62-32

PERU

Ministerio de Energia & Minas y Direccion General de Hidrocarburos
Esq. Avenida Javier Prado Este y Avenida Aviacion
Edifico Sol Gas, 5 Piso
Lima 34, PERU
Ph: 51 14 35-29-96, 36-62-36

Perupetro SA
Av. Luis Aldana No. 320
Lima 41, PERU
Ph: 51 1 475-9590
Fax: 51 1 475-9644, 475-7722
Website: www.perupetro.com.pe

PHILIPPINES

Department of Energy
PNPC Complex, Merritt Rd.
Fort Bonifacio
Makati, Metro Manila, PHILIPPINES
Ph: 63 2 844-1021
Fax: 63 2 812-4016
E-mail: oil&gas@doe.gov.ph
Website: www.doe.gov.ph

Ministry of Energy
7901 Makati Ave.
Box 1031 MCC
Makati, Metro Manila, PHILIPPINES
Ph: 63 2 85-88-56/9

Philippine National Oil Co.
PNOC Bldg.
7901 Makati Ave.
Makati, Metro Manila 1200, PHILIPPINES
Ph: 63 2 85-90-61, 85-88-56, 88-75-51
Fax: 63 2 815-3094, 810-6728

POLAND

Bureau of Geological Concessions
Ministry of Environmental Protection, Natural Resources, and Forestry
Ul. Wawelska 52/56
00-922 Warsaw, POLAND
Ph: 48 22 25-15-03
Fax: 48 22 25-39-72
Website: www.mos.gov.pl

Polish Oil & Gas Co. (PGNiG)
6/14 Krucza St.
00-537 Warsaw, POLAND
Ph: 48 22 628-2769
Fax: 48 22 623-5400

Petrobaltic Oil & Gas Co.
Ul. Stary Dwor 9
80-958 Gdansk, POLAND
Ph: 48 58 3130-6169
Fax: 48 58 31-43-11

PORTUGAL

Petroleos de Portugal EP (PETROGAL)
Edificio Europeia
Av. Fontes Pereira de Meio, No. 6-10
1100 Lisboa, PORTUGAL
Ph: 351 1 353-8821
Fax: 351 1 310-2950

Gabinete Para a Pesquisa e Exploracao de Petroleo (GPEP)
Rua Vale de Pereiro, 4
P1200 Lisbon, PORTUGAL
Ph: 351 1 388-3541
Fax: 351 1 388-2775

Q

QATAR

Ministry of Finance & Petroleum
Department of Petroleum Affairs
Box 2233
Doha, QATAR
Ph: 974 41-41-23, 41-35-71

Qatar General Petroleum Corp. (QGPC)
Box 3212
Doha, QATAR
Ph: 974 49-14-91
Fax: 974 83-19-95

QUEENSLAND
Director General
Department of Mines and Energy
GPO Box 194
Brisbane, QLD 4001, AUSTRALIA
Ph: 61 7 3224 2180
Fax: 61 7 3224-8380
Website: www.dme.qld.gov.au

R

ROMANIA
National Agency for Mineral Resources
36-38 Mendeleev St.
70169 Bucharest, ROMANIA
Ph: 40 1 613-2204
Fax: 40 1 210-7440

Petrom RA
109 Calea Victoriei, Dist. 1
Bucharest, ROMANIA
Ph: 40 1 659-5960
Fax: 40 1 650-2629

RUSSIA
Ministry of Fuel & Energy of the Russian Federation (Roskomnedra)
4/6 Bol'shaya Gruzinskaya St.
Moscow 123812, RUSSIA
Ph: 7 95 254-8277
Fax: 7 95 943-0013

S

SAUDI ARABIA
Saudi Arabian Oil Co. (SAUDI ARAMCO)
Administration Bldg. Room 2220
Box 5000
Dhahran 31311, SAUDI ARABIA
Ph: 966 3875-6110
Fax: 966 3873-8190
Website: www.aramco.com

SOUTH AFRICA
Department of Mineral and Energy Affairs
Pretoria, SOUTH AFRICA
Ph: 27 12 317-9121
Fax: 27 12 322-0810
Website: www.dme.gov.za

Petroleum Agency SA
PO Box 1174
Parow 7499
Cape Province
SOUTH AFRICA
Ph: 27 21 938-3500
Fax: 27 21 938-3520
E-mail: plu@petroleumagencysa.com
Website: www.petroleumagencysa.com

SOUTH AUSTRALIA
Director General
Department of Mines and Energy
PO Box 151
Eastwood, S.A. 5063, AUSTRALIA
Ph: 61 8 274-7500
Fax: 61 8 373-3269
Website: www.sa.gov.au

SOUTH KOREA
Korea Petroleum Development Corp.
#1588-14, Kwanyang-dong, Dongan-gu
Anyang, Kyungki-do,
KOREA 431-711
Ph: 82 343-80-2114, 2932, 2933
Fax: 82 343 80-9321, 9322

SPAIN
Jefe del Servicio de Hidrocarburos de la Direccion General de la Energia del Ministerio de Industria y Energia
Madrid 16, SPAIN

Hispanica de Petroleos SA (HISPANOIL)
Box 14355
Madrid 30, SPAIN
Ph: 34 1 274-7200, 274-9700

Repsol-YPF
Paseo de la Castellana, 278 – 280
28046 Madrid, SPAIN
Website: www.repsol-ypf.com

SYRIA
Ministry of Petroleum & Mineral Resources
Government of Syria
Jadet Al Katib
Damascus, SYRIAN ARAB REPUBLIC
Website: www.mopmr-sy.org

Syrian Petroleum Co.
Box 2849
Damascus, SYRIAN ARAB REPUBLIC
Ph: 963 11 221-7944
Fax: 963 11 223-2083

T

TAIWAN
Chinese Petroleum Corp.
83 Chung Hwa Road
Taipei 10031, TAIWAN
Ph: 886 2 361-0221, 371-7121
Fax: 886 2 331-9645
Website: www.cpc.com.tw

THAILAND
Department of Mineral Resources
Ministry of Industry
Rama VI Rd.
Bangkok 10400, THAILAND
Ph: 66 2 511-0215, 511-0749
Fax: 66 2 245-9855

Petroleum Authority of Thailand (PTT)
14 Vighavadi Rangsit Rd.
Bangkok 10900, THAILAND
Ph: 66 2 537-2000
Fax: 66 2 537-3499
Website: www.ptt.or.th

TRINIDAD & TOBAGO
Ministry of Energy & State Enterprises
4th Floor, Salvatori Bldg.
Box 96
Port of Spain, TRINIDAD
Ph: 809 6238-8416
Website: www.energy.gov.tt

Petroleum Co. of Trinidad & Tobago Ltd. (PETROTRIN)
Administration Bldg.
Pointe-a-Pierre, TRINIDAD
Ph: 868 658-4306
Fax: 868 658-1163
E-mail: lramyad@petrotrin.com

TUNISIA
Entreprise Tunisienne d'Activites Petrolieres (ETAP)
27 bis, Ave. Kherreddine Pacha
B.P. 367, 1002 Tunis de Belvedere
TUNISIA
Ph: 216 1 78-22-88, 78-32-33
Fax: 216 1 78-40-92, 78-61-42
E-mail: dexprom@etap.com.tn
Website: www.etap.com.tn

TURKEY
General Directorate of Mineral Research & Exploration (MTA)
Ankara, TURKEY
Ph: 90 41 287-3430
Fax: 90 41 222-2878

Turkiye Petrolleri Anonim Ortakligi (TPAO)
Mustafa Kemal Mahallesi
2. Cadde No. 86
06520 Ankara, TURKEY
Ph: 90 312 286-9100
Fax: 90 312 286-9000
E-mail: tpaocc@petrol.tpao.gov.tr
Website: www.tpao.gov.tr

TURKMENISTAN
Ministry of Oil and Gas
27 Bitarap Turkmenistan St.
Ashgabat, TURKMENISTAN 744000
Ph: 9 9312 39-38-34
Fax: 9 9312 51-04-43

U

UKRAINE
Counsel of Ministries
Vladimirskaya St. 34
252003 Kiev, UKRAINE
Ph: 7 044 226-2007

State Committee on Geology & Utilization of Mineral Resources
34 Volodymyrska St.
252601 Kiev, UKRAINE
Ph: 7 044 228-5671
Fax: 7 044 228-6221
Website: www.brama.com

UNITED ARAB EMIRATES
Abu Dhabi National Oil Co. (ADNOC)
Box 898
Abu Dhabi, UNITED ARAB EMIRATES
Ph: 971 2 602-0000
Fax: 971 2 602-3389
E-mail: adnoc@adnoc.com
Website: www.adnoc.com

Ministry of Petroleum & Mineral Resources
Box 59
Abu Dhabi, UNITED ARAB EMIRATES
Ph: 971 2 62810

UNITED KINGDOM

Department of Trade and Industry
Energy Division
Oil and Gas Directorate
DTI Enquiry Unit
1 Victoria Street
London SW1H 0ET
Website: www.dti.gov/uk/energy

UK Department of Energy
1 Palace St.
London SW1E 5HE
UNITED KINGDOM
Ph: 44 1 238-3000
Fax: 44 1 834-03771

URUGUAY

Administracion Nacional de Combustibles,
Alcohol y Portland (ANCAP)
Casilla 1090
Montevideo, URUGUAY
Ph: 598 2 92-06-08, 98-95-04, 98-93-55
Fax: 598 2 92-11-36
Website: www.ancap.com.uy

Asistencia Reciproca Petrolera
Empresa Latinoamericana (ARPEL)
Javier de Viana 2345
Casilla de Correo 1006
11200 Montevideo, URUGUAY
Ph: 598 2 40-69-93, 40-74-54
Fax: 598 2 23-70-23
Website: www.arpel.org

UZBEKISTAN

Ministry of Foreign Economic Relations
Buyuk Ipak Iuli 75
700077 Tashkent, UZBEKISTAN
Ph: 7 371 2 68-76-31
Fax: 7 371 2 68-72-31, 68-74-77
Website: www.gov.uz

State Committee for Geology & Mineral Resources
T. Shevchenko St., 11
70060 Tashkent, UZBEKISTAN
Ph: 7 371 2 56-19-98
Fax: 7 371 2 56-32-83

Uzneftegazdobycha
Amir Temir Ulitsa 66
700084 Tashkent, UZBEKISTAN
Ph: 371 2 35-83-23
Fax: 371 2 35-83-37

V

VENEZUELA

Ministerio de Energia y Minas
Edificio Parque Central
Caracas, VENEZUELA
Ph: 58 2 483-2033, 483-4033, 483-2317

Petroleos de Venezuela SA (PDVSA)
Avenida Libertador, Torre Este,
La Campina
Apartado 169
Caracas 1010-A, VENEZUELA
Ph: 58 2 708-4111
Fax: 58 2 708-4461, 708-4420
Website: www.pdv.com

VICTORIA

Secretary
Dept. of Natural Resources and Environment
PO Box 2145MDC
Fitzroy, Victoria 3065, AUSTRALIA
Ph: 61 3 9412 7000
Fax: 61 3 9412 7888
Website: www.vic.gov.au

VIET NAM

Ministry of Petroleum & Gas
193/6A Nam Ky Khoi Nghia
Ho Chi Minh City 3, VIET NAM

Vietnam Oil & Gas Co. (PETROVIETNAM)
Oil & Gas General Dept.
80 Nguyen Du Ha Noi
Hanoi, VIET NAM
Ph: 84 4 25-25-26, 25-23-57, 25-29-10
Fax: 84 4 26-59-42

W

WESTERN AUSTRALIA

Director General
Department of Minerals and Energy
Mineral House
100 Plain St.
East Perth, W.A. 6004, AUSTRALIA
Ph: 61 9 222-3333
Fax: 61 9 325-3515
Website: www.wa.gov.au

Y

REPUBLIC OF YEMEN

Ministry of Oil & Mineral Resources
Petroleum Exploration & Production Board
Sanaa, YEMEN
Ph: 967 2 263-246
Fax: 967 2 268-949

Yemen Petroleum Co.
YPC Bldg.
Box 3360
Al Hudaydah
YEMEN

YUKON TERRITORY

Oil and Gas Resources Branch
Yukon Department of Economic Development
Suite 209, 212 Main St.
Whitehorse, Yukon Y1A 2B1
CANADA
Ph: 867 667-3427
Fax: 867 393-6262
E-mail: brian.love@gov.yk.ca
Website: www.gov.yk.ca

IPE Information Sources
Advertiser's Index/Index of Contents

ADVERTISER'S INDEX

INDEX OF CONTENTS

A

B

C

D

E

F

G

H

I

J

K

L

M

N

O

P

Q

R

S

T

U

V

W

X

Y

Z